Handbook of ATPases

Edited by
Masamitsu Futai, Yoh Wada
and Jack H. Kaplan

Further Titles of Interest

R. Benz (ed.)
Bacterial and Eukaryotic Porins
2004
ISBN 3-527-30775-3

K. Nierhaus, D. Wilson (eds.)
Protein Synthesis and Ribosome Structure
2004
ISBN 3-527-30638-2

G. Krauss
Biochemistry of Signal Transduction and Regulation (3rd Ed.)
2003
ISBN 3-527-30591-2

M. Schliwa (ed.)
Molecular Motors
2002
ISBN 3-527-30594-7

G. Winkelmann (ed.)
Microbial Transport Systems
2001
ISBN 3-527-30304-9

Handbook of ATPases

Biochemistry, Cell Biology, Pathophysiology

Edited by
Masamitsu Futai, Yoh Wada and Jack H. Kaplan

WILEY-VCH

WILEY-VCH Verlag GmbH & Co. KGaA

Edited by

Masamitsu Futai
Institute of Scientific and Industrial Research
Osaka University
8-1 Mihogaoka, Ibaraki
Osaka 567-0047
Japan
m-futai@sanken.osaka-u.ac.jp

Yoh Wada
Institute of Scientific and Industrial Research
Osaka University
8-1 Mihogaoka, Ibaraki
Osaka 567-0047
Japan
yohwada@sanken.osaka-u.ac.jp

Jack H. Kaplan
University of Illinois at Chicago
Department of Biochemistry and Molecular Genetics
900 S. Ashland Ave.
Chicago, IL 60607
USA
kaplanj@uic.edu

Library of Congress Card No.: applied for
British Library Cataloguing-in-Publication Data
A catalogue record for this book is available from the British Library.

Bibliographic information published by
Die Deutsche Bibliothek
Die Deutsche Bibliothek lists this publication in the Deutsche Nationalbibliografie; detailed bibliographic data is available in the Internet at http://dnb.ddb.de.

© 2004 WILEY-VCH Verlag GmbH & Co. KGaA, Weinheim

Printed in the Federal Republic of Germany.
Printed on acid-free paper.

Cover Grafik-Design Schulz, Fußgönheim
Typesetting Hagedorn Kommunikation, Viernheim
Printing betz-druck gmbh, Darmstadt
Bookbinding J. Schäffer GmbH & Co. KG, Grünstadt

ISBN 3-527-30689-7

Contents

Handbook of ATPases. Edited by M. Futai, Y. Wada, J. H. Kaplan
Copyright © 2004 WILEY-VCH Verlag GmbH & Co. KGaA, Weinheim
ISBN 3-527-30689-7

Preface

Inorganic ions play vital roles in mammalian physiology, in processes such as neural transduction, energy transformation, nutrient uptake, *etc.* They form unique ionic compartments both inside and outside the cells. The ion pumping ATPases, transporting ions at the expense of ATP, are the major players in ion homeostasis.

The sodium pump, or Na^+/K^+ ATPase, of the plasma membrane was the first pump introduced to the field [1], and was shown to take up two K^+ ions into cells and export three Na^+ ions, coupling these movements with the hydrolysis of one ATP to ADP and phosphate [2]. Similar ion pumps found subsequently include sarcoplasmic or plasma membrane Ca^{2+} ATPases, gastric H^+/K^+ ATPase, Wilsons and Menkes disease Cu^{2+} ATPases, plant or fungal plasma membrane H^+ ATPase, *etc.* The structural similarities and differences between the major groups of P-type ATPases were recently discussed [3]. This group of pumps form acyl phospho-enzyme intermediates from ATP (hence P-type ATPase), prior to hydrolysis and ion translocation, and show two distinct kinetic states, E1 and E2 forms, during their reaction cycle that alternately expose ion binding sites to the intracellular and extracellular compartments, respectively.

Another line of studies introduced ion-pumping ATPases participating in ATP synthesis in mitochondrial, chloroplast, or bacterial membranes. The overall reaction for ATP synthesis is called oxidative phosphorylation or photophosphorylation emphasizing the reaction starting from respiratory or photo-electron transfer chain. Approaches of resolution and reconstitution of mitochondrial membranes showed that oxidative phosphorylation is carried out by a combined reaction of the electron transfer chain and coupling factors [4]. The transmembrane electrochemical proton gradient is formed by electron transfer chain, and drives ATP synthesis by coupling factors, consistent with Peter Mitchells chemiosmotic theory [5]. The F_o and F_1 are the essential elements of these machines, although other associated proteins have been described. The F_o and F_1 constitute an ATP synthase or ATPase, functioning as the proton pathway and catalytic sector, respectively. Thus, the entire FoF$_1$ complex synthesizes ATP from ADP and phosphate coupling the synthesis with the electrochemical proton gradient established by the respiratory or photoelectron transfer chain. Unique FoF$_1$ complexes were found later in bacteria, such as those pumping Na^+ or only hydrolyzing ATP.

Handbook of ATPases. Edited by M. Futai, Y. Wada, J. H. Kaplan
Copyright © 2004 WILEY-VCH Verlag GmbH & Co. KGaA, Weinheim
ISBN 3-527-30689-7

The vacuolar type proton pumping ATPases are the most recent members to join the ion pump family [6]. They were initially identified in plant and fungal vacuoles, and later in organelles and plasma membranes of mammalian cells and plants. The special features of this group are that they are found ubiquitously in lysosomes, endosomes, Golgi apparatus, *etc.* They are also found in specialized organelles such as synaptic vesicles, lamellar bodies, acrosomes, and in the plasma membrane of specific cells such as osteoclasts or renal epithelial cells. The archaebacterial ATPases have some similarities to this group, but they are also significantly different.

The three classes of the ion pump ATPase family were proposed to be called P-ATPase (P-type ATPase), F-ATPase (F-type ATPase) and V-ATPase (V-type ATPase), the nomenclatures being derived from phosphorylated intermediate, factors of oxidative phosphorylation and a pump found initially in vacuoles, respectively [7]. These simple nomenclatures made the ATPase field easier to access, and have appeared in recent biochemistry and cell biology texts. However, this nomenclature provides only a partial picture of this family. It should be noted that such a description of the ATPase family does not include the emerging wide array of ABC (ATP binding cassette) transporters, which transport a variety of substrates including amino acids and multiple drugs, and other xenobiotics, coupled to the hydrolysis of ATP. We believe that an increased awareness of the characteristics of the ATPase family will be of great interest to workers in the ABC field and vice versa. Furthermore, although the nomenclature for the V-type ATPase was derived from its vacuolar location, this group is now found in a wide variety of organelles and plasma membranes. It is probably more appropriate to emphasize that V-type ATPases could signify *"various"* endomembrane organelle ATPases rather than merely those of *vacuoles*.

Entering the period of *Anno Domini,* (AD) from BCE was significant for Western history. Similarly, the progression of molecular biology and biochemistry into era of AD (after DNA) from BCE (before cloning era) greatly affected the ATPase field. We were able to clone ATPase genes, sequence them, introduce mutations, and study the functions of wild-type or genetically modified enzymes. Furthermore, the molecular biological approach led to the discovery of ion pumping ATPases related to human diseases such as Wilsons Disease, Menkes Disease, osteopetrosis, or renal acidosis, as discussed in contributions to this volume. Furthermore, it became possible to create animals (mice) lacking genes for the ATPases or specific subunits, and study their physiological roles in the intact, living organism. The current combination of these strategies will enable us to study physiological roles of ATPases in selected mammalian cells.

The X ray crystal structures of F_1 from bovine F-ATPase [8] and of sarcoplasmic Ca^{2+} ATPase [9] were important breakthroughs for understanding the mechanism of F- and P-ATPases, respectively. The F-ATPase structure is also useful in understanding the V-ATPase because of the considerable homology shared between these two pump classes. We are close to having a picture of how ions are pumped and ATP hydrolyzed at atomic resolution. Earlier models, from biochemical or molecular biological studies can now be subjected to a structural analysis. The structures of the P-type ATPases suggested mechanical movements of their domains during catalytic turnover. In the F-ATPases, the striking observation is the continuous rotation of the γ subunit located at the center of the $\alpha_3\beta_3$ hexamer of F_1 [10], followed by that of the subunit complex $\varepsilon\gamma c_{10}$~$_{14}$ in F-ATPase holoenzyme [11]. This unique mechanical rotation is consistent with Paul Boyer's binding change mechanism of the catalysis led by proton translocation through the interface of the *a* and *c* subunits [12]. Thus, the entire F-ATPase is a biological nanomotor or a so-called power generator, if we use the terminology of the engineers working on nanotechnology. This concept has been extended to V-ATPase, since its subunit complex was also shown to rotate upon ATP hydrolysis followed by H^+ transport.

The advantage of having genome structures and cDNA libraries (ESTs, or expressed sequence tags) prompted us to find striking diversities among V-ATPases. Different subunit isoforms were found in the V-ATPases specific for endomembrane organelles such as lysosomes, or plasma membranes of osteoclast forming resorption lacuna, and those in kidney intercalated cells. The presence of cell- or organelle-specific isoforms leads to the notion that the diverse physiologies of proton translocation are carried out by unique V-ATPases utilizing specific isoforms [13]. The pathophysiological analysis of inside-acidic compartments has been carried out together with studies of their isoform-specific diseases.

This book is dedicated to the current pictures we have of the ion pumping ATPases from different vantage points including biochemistry, cell biology and pathophysiology. Each Chapter focuses on an extensive discussion of the most recent results following a general introduction. To the best of our knowledge, this is the first book entirely dedicated to the three types of ion pumping ATPases. Thus, we hope it will be useful not only for the researchers in biochemistry and physiology, but also for the students of biochemistry, molecular biology and medicine. It may even provide stimulus to the engineers or scientists in the field of nanotechnology and nanoscience in general.

We are grateful to the experts in the ATPase field and in the related cell physiology for contributing to this work. Without their effort and timely submission of the manuscripts, the book project could not have been successful. Finally, but not the least, we would like to thank Dr. Frank Weinreich of Wiley-VCH for his editorial suggestions and support. Without his enthusiasm, this book would be still at the initial stage of planning.

Tokyo, Osaka and Chicago
Spring 2004

Masamitsu Futai
Yoh Wada
Jack H. Kaplan

References*

1. Skou, JC, *Biochim.Biophys. Acta* **1957**, 23, 394–401.
2. Sen, AK, Post, RL, *J. Biol. Chem.* **1964**, 239, 345–352.
3. Lutsenko S, Kaplan, JH, *Biochemistry* **1995**, 34, 15607–15613.
4. Racker E, *A New Look at Mechanism in Bioenergetics*, Academic Press, New York, San Francisco and London **1976**.
5. Mitchell P, *Nature* **1961**, 191, 144–148.
6. Nelson N, Harvey, WR, *Physiol. Rev.* **1999**, 79, 361–385.
7. Pedersen PL, Carafoli E, *Trends Biochem. Sciences* **1987**, 12, 146–150.
8. Abrahams JP, Leslie AGW, Lutter R, Walker J, *Nature* **1994**, 370, 621–628.
9. Toyoshima C, Nakasako M, Nomura H, Ogawa H, *Nature* **2000**, 405, 647–655.
10. Noji M, Yasuda R, Yoshida M, Kinoshita K, *Nature* **1997**, 386, 299–302.
11. Sambongi Y, Iko Y, Tanabe M, Omote H, Iwamoto-Kihara A, Ueda I, Yanagida T, Wada Y, Futai M, *Science* **1999**, 286, 1722–1724.
12. Boyer PD, *Annu. Rev. Biochem.* **1997**, 66, 717–749.
13. Sun-Wada G-H, Wada Y, Futai M, *J. Bioenerg. Biomemb.* **2003**, 35, 347–358.

* Other references can be found elsewhere in this Book

List of Contributors

Marwan K. Al-Shawi
Department of Molecular Physiology
and Biological Physics
University of Virginia Health Sciences
Center
1300 Jefferson Park Avenue
Charlottesville, VA 22908
USA

Katherine Bowers
Cambridge Institute for Medical
Research and
Department of Clinical Biochemistry
University of Cambridge
Addenbrookes Hospital, Hills Road
Cambridge CD2 2XY
United Kingdom

Marisa Brini
Department of Biological Chemistry
University of Padova
Viale G. Colombo, 3
35121 Padova
Italy

Dennis Brown
Program in Membrane Biology and
Renal Unit
Harvard Medical School
Massachusetts General Hospital
Building 149 13th Street
Boston, MA 02129
USA

Ernesto Carafoli
Department of Biological Chemistry
University of Padova
Viale G. Colombo, 3
35121 Padova
Italy

Daniel J. Cipriano
Department of Biochemistry
University of Western Ontario
1151 Richmond Street
London, Ontario N6A 5C1
Canada

Luisa Coletto
Department of Biological Chemistry
University of Padova
Viale G. Colombo, 3
35121 Padova
Italy

Paul A. Del Rizzo
Department of Biochemistry
University of Western Ontario
1151 Richmond Street
London, Ontario N6A 5C1
Canada

Handbook of ATPases. Edited by M. Futai, Y. Wada, J. H. Kaplan
Copyright © 2004 WILEY-VCH Verlag GmbH & Co. KGaA, Weinheim
ISBN 3-527-30689-7

Iva Dostanic
Department of Molecular Genetics,
Biochemistry and Microbiology
University of Cincinnati
231 Albert Sabin Way
Cincinnati, OH 45267-0524
USA

Stanley Dunn
Department of Biochemistry
University of Western Ontario
1151 Richmond Street
London, Ontario N6A 5C1
Canada

Bin Fan
Department of Biochemistry
and Molecular Biology
Wayne State University School
of Medicine
Detroit, MI 48201
USA

Andrew R. Flannery
Cambridge Institute for Medical
Research and
Department of Clinical Biochemistry
University of Cambridge
Addenbrookes Hospital, Hills Road
Cambridge CD2 2XY
United Kingdom

Michael Forgac
Department of Physiology
Tufts University School of Medicine
136 Harrison Avenue
Boston, MA 02111
USA

Masamitsu Futai
Division of Biological Sciences, and
Nanoscience and Nanotechnology
Center
Institute of Scientific and
Industrial Research
Osaka University
8-1 Mihogaoka, Ibaraki
Osaka 567-0047
Japan

Laurie A. Graham
Institute of Molecular Biology
University of Oregon
Eugene, OR 97403-1229
USA

Giuseppe Inesi
Department of Biochemistry
University of Maryland School of
Medicine
108 N. Green Street
Baltimore, MD 21201
USA

Jack H. Kaplan
University Of Illinois at Chicago
Department of Biochemistry and
Molecular Genetics
900 S. Ashland Ave.
Chicago, IL 60607
USA

Evangelia G. Kranias
Department of Pharmacology and Cell
Biophysics
University of Cincinnati College of
Medicine
Cincinnati, OH 45267
USA

Nga Phi Le
Department of Molecular Physiology
and Biological Physics
University of Virginia Health Sciences
Center
1300 Jefferson Park Avenue
Charlottesville, VA 22908
USA

Silvia Lecchi
Department of Genetics and
Department of Cellular and Molecular
Physiology
Yale University School of Medicine
333 Cedar Street
New Haven, CT 06520
USA

Jerry Lingrel
Department of Molecular Genetics,
Biochemistry and Microbiology
University of Cincinnati
231 Albert Sabin Way
Cincinnati, OH 45267-0524
USA

Svetlana Lutsenko
Department of Biochemistry and
Molecular Biology
Oregon Health & Science University
3181 S. W. Sam Jackson Park Rd.
Portland, OR 97239-3098
USA

David H. MacLennan
Banting and Best Department
of Medical Research
University of Toronto
Toronto, Ontario M5G 1L6
Canada

Vladimir Marshansky
Program in Membrane Biology and
Renal Unit
Harvard Medical School
Massachusetts General Hospital
Building 149 13th Street
Boston, MA 02129
USA

Amy E. Moseley
Department of Molecular Genetics,
Biochemistry and Microbiology
University of Cincinnati
231 Albert Sabin Way
Cincinnati, OH 45267-0524
USA

Eiro Muneyuki
Research Laboratory of Resources
Utiliziation
Tokyo Institute of Technology
4259 Nagatsuta, Midori-ku
Yokohama 226-8503
Japan

Keith Munson
Membrane Biology Laboratory
West Los Angeles VA Medical Center
11301 Wilshire Blvd.
Los Angeles, CA 90073
USA

Robert K. Nakamoto
Department of Molecular Physiology
and Biological Physics
University of Virginia Health Sciences
Center
1300 Jefferson Park Avenue
Charlottesville, VA 22908
USA

Jonathan Neumann
Department of Molecular Genetics,
Biochemistry and Microbiology
University of Cincinnati
231 Albert Sabin Way
Cincinnati, OH 45267-0524
USA

Yoshinori Ohsumi
Department of Cell Biology
National Institute for Basic Biology
Nishigonaka 38
Myodaijicho, Okazaki 444-8585
Japan

Tina Purnat
Department of Biochemistry and
Molecular Biology
Oregon Health & Science University
3181 S. W. Sam Jackson Park Rd.
Portland, OR 97239-3098
USA

Barry P. Rosen
Department of Biochemistry and
Molecular Biology
Wayne State University School
of Medicine
Detroit, MI 48201
USA

George Sachs
Membrane Biology Laboratory
West Los Angeles VA Medical Center
11301 Wilshire Blvd.
Los Angeles, CA 90073
USA

Elim Shao
Department of Physiology
Tufts University School of Medicine
136 Harrison Avenue
Boston, MA 02111
USA

Jai Moo Shin
Membrane Biology Laboratory
West Los Angeles VA Medical Center
11301 Wilshire Blvd.
Los Angeles, CA 90073
USA

Carolyn Slayman
Department of Genetics and
Department of Cellular and Molecular
Physiology
Yale University School of Medicine
333 Cedar Street
New Haven, CT 06520
USA

Tom H. Stevens
Institute of Molecular Biology
University of Oregon
Eugene, OR 97403-1229
USA

Ge-Hong Sun-Wada
Division of Biological Sciences, and
Nanoscience and Nanotechnology
Center
Institute of Scientific and
Industrial Research
Osaka University
8-1 Mihogaoka, Ibaraki
Osaka 567-0047
Japan

Chikashi Toyoshima
Institute of Molecular and Cellular
Biosciences
University of Tokyo
Bunkyo-ku
Tokyo
Japan

Ruslan Tsivkovskii
Department of Biochemistry and
Molecular Biology
Oregon Health & Science University
3181 S. W. Sam Jackson Park Rd.
Portland, OR 97239-3098
USA

Olga Vagin
Membrane Biology Laboratory
West Los Angeles VA Medical Center
11301 Wilshire Blvd.
Los Angeles, CA 90073
USA

Yoh Wada
Division of Biological Sciences
Institute of Scientific and
Industrial Research
Osaka University
8-1 Mihogaoka, Ibaraki
Osaka 567-0047
Japan

Marco D. Wong
Department of Biochemistry and
Molecular Biology
Wayne State University School
of Medicine
Detroit, MI 48201
USA

Masasuke Yoshida
Research Laboratory of Resources
Utiliziation
Tokyo Institute of Technology
4259 Nagatsuta, Midori-ku
Yokohama 226-8503
Japan

Part I
P-type ATPases

Handbook of ATPases. Edited by M. Futai, Y. Wada, J. H. Kaplan
Copyright © 2004 WILEY-VCH Verlag GmbH & Co. KGaA, Weinheim
ISBN 3-527-30689-7

1

Yeast Plasma-membrane H$^+$-ATPase: Model System for Studies of Structure, Function, Biogenesis, and Regulation

Silvia Lecchi and *Carolyn W. Slayman*

1.1
Introduction

The single most abundant protein in the fungal plasma membrane is a proton-transporting ATPase, encoded by the *PMA1* gene [1] and hydrolyzing as much as one-quarter of cellular ATP [2]. By forming a large electrochemical gradient across the surface membrane, the H$^+$-ATPase provides energy to an array of proton-coupled cotransporters for sugars, amino acids, and other nutrients; it also contributes to the regulation of intracellular pH (Figure 1.1).

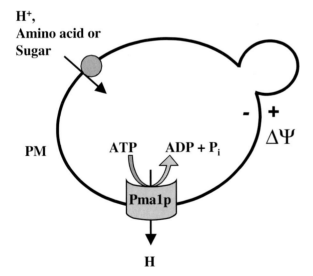

Figure 1.1 Pma1 H$^+$-ATPase plays an essential role in yeast cell physiology. Pumping protons (H$^+$) across the membrane at the expense of ATP hydrolysis, Pma1 H$^+$-ATPase establishes the membrane potential ($\Delta\Psi$) and the pH gradient required for nutrient import and contributes to the regulation of intracellular pH. PM: plasma membrane.

Handbook of ATPases. Edited by M. Futai, Y. Wada, J. H. Kaplan
Copyright © 2004 WILEY-VCH Verlag GmbH & Co. KGaA, Weinheim
ISBN 3-527-30689-7

Direct evidence for the ATPase first came several decades ago from microelectrode studies on the filamentous fungus *Neurospora crassa*, where a large (>200 mV), metabolically sensitive membrane potential was shown to be generated by ATP-dependent proton pumping [3, 4]. By the early 1980s, the H⁺-ATPase had been purified from *Schizosaccharomyces pombe* [5], *Saccharomyces cerevisiae* [6], and *Neurospora crassa* [7, 8] and found to consist of a 100 kDa polypeptide, firmly embedded in the membrane bilayer and requiring detergents for solubilization. Reconstitution into proteoliposomes confirmed that the ATPase did indeed mediate electrogenic proton transport [9–11], with a stoichiometry of 1 H⁺/ATP, as predicted from electrophysiological measurements [12]. In parallel, biochemical studies demonstrated that it split ATP by way of a covalent β-aspartyl intermediate [13–16], the hallmark of a widespread group later named the P-type ATPases. Like other members of that group, the H⁺-ATPase alternated during its reaction cycle between two major conformational states (E_1 and E_2), which could be distinguished by their different patterns of proteolytic fragments following digestion with low concentrations of trypsin [17, 18].

By 1986, cloning of *PMA1* genes from *Saccharomyces cerevisiae* [1] and *Neurospora crassa* [19, 20] had revealed clearcut sequence homology between the fungal H⁺-ATPase, the Kdp K⁺-ATPase from *Escherichia coli* [21], and the plasma-membrane Na⁺,K⁺- and sarcoplasmic reticulum Ca²⁺-ATPases of animal cells [22–24]. More than 200 P-ATPase genes have since been cloned from archaebacteria, eubacteria, fungi, algae, plants, and animals and analyzed to establish their relationship with one another. In 1995, Lutsenko and Kaplan [25] put forward a useful classification scheme that divides P-type ATPases into three groups based on the cation transported, the number and location of hydrophobic transmembrane segments, and the position of three conserved sequences: DKTGT (where the β-aspartyl phosphointermediate forms), TGES (located upstream of the phosphorylation site in a smaller cytoplasmic loop), and GDGXNDXP (close to the ATP-binding site). According to this scheme, heavy metal pumps are classified as P_1-ATPases; H⁺, Na⁺/K⁺, Mg²⁺, and Ca²⁺ pumps, as P_2-ATPases; and the K⁺-ATPase of *E. coli*, as a lone

Table 1.1 Classification of P-Type ATPases.

Lutsenko and Kaplan [25]	Cation specificity	Axelsen and Palmgren [26]	Cation specificity
P1	Heavy metal	Type I	Heavy metal K⁺
P2	Ca²⁺ Na⁺/K⁺ H⁺/K⁺ H⁺ Mg²⁺	Type II	Ca²⁺ Na⁺/K⁺ H⁺/K⁺
		Type III	H⁺ Mg²⁺
P3	K⁺		
		Type IV	Aminophospholipid
		Type V	Unknown

P_3-ATPase (Table 1.1). Several years later, Axelsen and Palmgren [26] developed a more complex classification based on the analysis of eight stretches of amino acid sequence shared by all P-type ATPases (Table 1.1). In this scheme, the heavy metal pumps remain as Type I ATPases, with the K^+-ATPase of *E. coli* distantly related to them. The Na^+,K^+-, H^+,K^+-, and Ca^{2+}-ATPases and H^+-ATPases still fall in the same general sector of the phylogenetic tree, but are separated into Type II and Type III, respectively. Finally, a place is found for two additional groups of P-ATPases that had become known from genome-sequencing projects: Type IV, involved in the transport of aminophospholipids and other hydrophobic compounds, and Type V, whose function is unknown.

Within the P-type group, the P_2- or Type II/III ATPases have been most intensively studied. These ATPases have a characteristic topology, with four membrane-spanning segments (M1-4) towards the N-terminal end of the polypeptide and six such segments (M5-10) towards the C-terminal end (Figure 1.2). The large cytoplasmic loop between M4 and M5 and the smaller cytoplasmic loop between M2 and M3 assemble to form the catalytically active part of the molecule, as discussed below. The N- and C-termini are also exposed at the cytoplasmic surface of the membrane.

This chapter will review how the yeast and Neurospora Pma1 H^+-ATPases have come to serve as simple eukaryotic prototypes of the P_2-ATPases, based on the powerful genetic and cell biological tools that are available to study their reaction mechanism, regulation, and biogenesis.

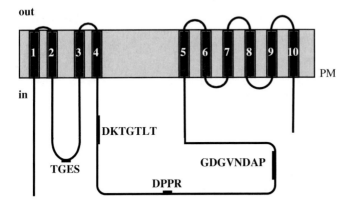

Figure 1.2 Transmembrane topology of Pma1 H^+-ATPase. The ATPase has 10 transmembrane α-helices with both N- and C-termini located in the cytoplasm. Highly conserved amino acid sequences characteristic of Type III ATPases are shown. PM: plasma membrane.

1.2
Structure

1.2.1
Ca^{2+}-ATPase as a Model

A landmark event came in 2000 with the publication by Toyoshima and his colleagues of the first high-resolution structure of a P-type ATPase. They successfully crystallized sarcoplasmic reticulum Ca^{2+}-ATPase in the E$_1$ conformation (with two Ca^{2+} ions bound) and solved the structure at 2.6 Å [27]. As expected from an earlier cryoelectron microscopic map at 8 Å resolution [28], they found the ATPase to be organized into a cytoplasmic headpiece and a transmembrane or *M domain*. Within the latter, four of the ten transmembrane α-helices (M4, M5, M6, and M8) assemble to create a pair of Ca^{2+} binding sites, consistent with the known stoichiometry of the ATPase. Site I is formed by side-chain oxygen atoms in M5 (N768 and E771), M6 (T799 and D800), and M8 (E908), while Site II is formed by main-chain carbonyl oxygens in M4 (V304, A305, I307) and side-chain oxygens in M4 (E309) and M6 (N796 and D800). Local unwinding of transmembrane helices, helped by two Pro residues in M4 and one in M6, plays an essential role in the architecture of both sites.

The headpiece of Ca^{2+}-ATPase is organized into three domains: P (phosphorylation), N (nucleotide-binding), and A (actuator or anchor), quite distinct from one another in the E$_1$Ca conformation. The *P domain* is formed by the two ends of the large M4–M5 cytoplasmic loop and contains the aspartyl residue (D351) that undergoes alternate phosphorylation and dephosphorylation during the reaction cycle. It is a typical Rossman fold, with a seven-stranded parallel β-sheet and eight short α-helices. The large *N domain* consists of the middle portion of the M4–M5 loop, arranged in a seven-stranded antiparallel β-sheet lying between two helix bundles. In crystals soaked with a non-hydrolyzable ATP analog (TMP-AMP), this compound appears within the N domain, surrounded by residues previously implicated in nucleotide binding. Finally, the N-terminal part of the ATPase and the small M2–M3 cytoplasmic loop come together to form the *A domain*, which is connected to the transmembrane part of the ATPase by long loops. The function of the A domain was unclear at the time of the first Toyoshima paper, although there was evidence from proteolytic cleavage studies that its conformation changes substantially during the reaction cycle.

In 2002 the same group published a second high-resolution (3.1 Å) structure of Ca^{2+}-ATPase, this time stabilized in the E$_2$ conformation by the inhibitor thapsigargin [29]. It is strikingly different from the E$_1$Ca structure. On the one hand, six of the ten transmembrane helices (M1–M6) move significantly between E$_1$ and E$_2$, "opening" the M domain, destroying the two Ca^{2+} binding sites, and, presumably, releasing Ca^{2+} into the lumen of the sarcoplasmic reticulum. Simultaneously, the A domain rotates ca. 100° horizontally and the N domain inclines ca. 90° relative to the membrane, shifting the cytoplasmic headpiece into a "closed" conformation. The authors suggest that the return from E$_2$ to E$_1$ requires Ca^{2+} to bind, fixing the membrane helices in a way that prohibits the closed conformation of the cyto-plasmic headpiece. The headpiece opens; the γ-phosphate of bound ATP can now reach and phosphorylate D351; and the cycle continues. Thus, although much remains to be done to understand the reaction mechanism fully, the two high-resolution structures have made it possible to develop a preliminary picture of how the Ca^{2+} pump may work.

1.2.2
Applicability of the Ca^{2+}-ATPase Structure to Other P_2-ATPases, Including the Pma1 H^+-ATPase

Not surprisingly, there has since been a flurry of activity to determine whether the first (and now the second) Toyoshima structure can serve as a legitimate template for other P_2-ATPases. So far, there is reason for optimism. Sweadner and Donnet [30] found that the sequence of Na^+,K^+-ATPase could readily be projected onto the 2.6 Å Ca^{2+}-ATPase structure in a way that superimposes short stretches of homology throughout the cytoplasmic and membrane domains; when this was done, most insertions and deletions appeared (as they presumably should) on the protein surface. The authors concluded that the core regions of both ATPases are likely to fold in the same basic manner. Further support is that conformationally sensitive Fe^{2+} cleavage sites, identified by Karlish and co-workers for the Na^+,K^+-ATPase, can be mapped logically onto the E_1 and E_2 structures of Ca^{2+}-ATPase (reviewed in Ref. 31).

Kühlbrandt and co-workers [32] have recently provided strong evidence that the similarity also extends to the Neurospora and yeast plasma-membrane H^+-ATPases. In this case, the authors made a point-by-point comparison between their earlier 8 Å cryoelectron microscopic map of the Neurospora enzyme [33] and Toyoshima's 2.6 Å crystallographic structure of Ca^{2+}-ATPase [27]. There was an excellent fit between the two in the M and P domains; the A and N domains could also be accommodated by a rigid-body displacement of 10 Å in the first case and a more substantial rotation of 73° in the second case. Within the headpiece, key structural elements involved in ATP binding and formation of the phosphoenzyme reaction intermediate were readily recognizable, and within the membrane, two of the ten cation-ligating residues were conserved in the H^+-ATPase (D730 and E803, corresponding to D800 and E908 of Ca^{2+}-ATPase). Indeed, only the highly divergent N- and C-termini were impossible to overlay on the Ca^{2+}-ATPase template. Thus, it seems reasonable to use the Kühlbrandt H^+-ATPase model as a framework in developing experiments and interpreting results on the Pma1 ATPases of yeast, Neurospora, and other fungi.

1.2.3
H⁺-ATPase Oligomers

By contrast with the steady progress made towards a detailed structure for the 100 kDa catalytic subunit, there has been considerable debate about the oligomeric state of the Pma1 ATPase. For the Neurospora enzyme, Scarborough and co-workers have demonstrated that 100 kDa monomers are fully competent to carry out ATP hydrolysis and ATP-dependent proton transport after reconstitution into proteoliposomes [34]. Under other conditions, however, the same enzyme clearly forms hexamers [35], which allow the ready production of two-dimensional crystals for cryoelectron microscopy [33]. Even larger oligomers of yeast Pma1 H⁺-ATPase (up to dodecamers) are observed when detergent-solubilized cell extracts are analyzed by gradient centrifugation [36], blue native polyacrylamide gel electrophoresis [37], or co-immunoprecipitation of differently tagged 100 kDa polypeptides [38].

The biological significance of the oligomers is not yet clear. Recent studies by the Chang, Schekman, and Simons laboratories [36–38] suggest that they may play a role in biogenesis, since they are formed soon after the 100 kDa ATPase polypeptide is synthesized and appear to be associated with packaging the ATPase into "lipid rafts" for transport to the plasma membrane (see below). Conversely, Kühlbrandt et al. [32] found crystalline patches of rosette-shaped particles in freeze–fracture replicas of starving yeast and Neurospora cells, and consider them to be a down-regulated "storage form" of the enzyme. Further work is required to sort out these possible interpretations, which are not mutually exclusive. It will also be of interest to map the regions of the Pma1 molecule that take part in oligomer formation; based on their structural model of the Neurospora ATPase, Kühlbrandt et al. [32] propose that the C-terminal domain of one monomer (known to be exposed in the cytoplasm [39]) interacts with Q624 and R625 in the P domain of the neighboring monomer.

1.2.4
Associated Proteolipids

Studies of highly purified Pma1 H⁺-ATPase preparations have given no evidence for a tightly bound β subunit of the kind seen in mammalian Na⁺,K⁺- and H⁺,K⁺-ATPases. Conversely, two small proteolipids do associate closely with the Pma1 enzyme and appear to play a physiologically significant role. Information about them began to emerge in 1992, when Navarre and co-workers first observed diffuse 4 and 7.5 kDa bands in Coomassie-stained gels of highly purified yeast H⁺-ATPase [40]. Chloroform–methanol extracts of both bands yielded the same 38-amino acid proteolipid, encoded by a gene that the authors named *PMP1* (for *plasma-membrane proteolipid*). A closely related *PMP2* gene was later cloned by hybridization with a *PMP1* probe [41]. Although the two open reading frames predict slightly different N-termini, post-translational processing yields mature Pmp1 and Pmp2 proteolipids that share all but a single amino acid residue (A21 vs. S21). Their hydropathy profiles are virtually identical, with a hydrophobic stretch of 24 residues followed by a highly basic, 14-amino acid tail. In their overall topology, Pmp1 and Pmp2 call to mind the amphipathic FYXD proteins that co-purify with mammalian Na^+,K^+-ATPases (reviewed in Ref. 42) as well as two regulatory proteins, phospholamban and sarcolipin, that are associated with sarcoplasmic reticulum Ca^{2+}-ATPases [43, 44].

Single deletions of *PMP1* or *PMP2* have only minor effects on the activity of yeast H^+-ATPase, but deletion of both genes leads to a 50% reduction in activity, suggesting some kind of regulatory role for the two proteolipids [41]. Recent NMR studies by Neumann and co-workers have explored the *in vitro* interactions between synthetic fragments of Pmp1 and lipid micelles [45, 46]. Based on their results, the authors propose that Pmp1 extends across the plasma membrane bilayer, with its bulky, hydrophobic N-terminal end located alongside the membrane-embedded domain of the ATPase in the outer, sterol- and sphingolipid-rich leaflet, while its C-terminal part sequesters a subset of negatively charged phosphatidylserine lipids in the inner leaflet to create a proper environment for the ATPase at the cytoplasmic surface of the membrane. The relationship between these structural data and stepwise models for ATPase biogenesis is discussed below.

1.3
Reaction Mechanism

1.3.1
Overview of the Reaction Cycle

This section covers three interrelated aspects of the reaction mechanism of Pma1 H^+-ATPase:

(i) ATP binding and formation of the β-aspartyl phosphoenzyme intermediate; (ii) the E_1–E_2 conformational change; and (iii) H^+ transport. The Post-Albers scheme, based on more than three decades of investigation into the partial reactions of P-ATPases, describes the generally accepted relationship among these steps. In the top limb of the cycle, as diagrammed for the H^+-ATPase (Figure 1.3), ATP is bound with high affinity to the E_1 form of the enzyme. Intracellular H^+ (or a hydro-

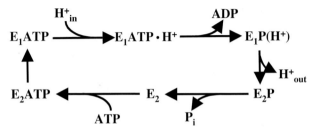

Figure 1.3 Reaction cycle of Pma1 H^+-ATPase. Based on the classical Post-Albers scheme for P-type ATPases.

nium ion; see below) then binds and stimulates the transfer of γ-phosphate from ATP to create the high-energy β-aspartyl phosphate intermediate; E_1P shifts to E_2P, accompanied by release of the transported H^+ to the extracellular medium; and ATP binds with low affinity to the E_2 form, accelerating the return to E_1.

1.3.2
ATP Binding and Phosphorylation

Although some of the amino acid residues that contribute to the ATP-binding site of P-type ATPases were first identified by affinity labeling and site-directed mutagenesis (reviewed in Ref. 47), a much more complete picture has emerged with the crystal structure of Ca^{2+}-ATPase, where TNP-ATP was used to visualize the nucleotide site directly [27]. As expected, key residues have remained virtually unchanged during evolution. They include F487 in Ca^{2+}-ATPase (F451 in yeast H^+-ATPase), K492 (K456), and K515 (K474), which lie deep in the pocket surrounding the adenosine moiety, and R560 (R509) near the mouth of the pocket, where it probably interacts with the β or γ phosphate of ATP. Consistent with this picture, mutational analysis has demonstrated the functional importance of K474 in the yeast Pma1 ATPase [48], and Pardo and co-workers have shown that K474 of the Neurospora enzyme is labeled by fluorescein 5-isothiocyanate in an ATP- and ADP-protectable way [49]. Overall, however, the N domain has been less well conserved than the rest of the cytoplasmic headpiece, with Ca^{2+}- and H^+-ATPases sharing only ca. 20% identity in this region. More work is needed to correlate the structural differences with known differences in catalytic properties. In particular, the yeast H^+-ATPase has a K_m for MgATP of ca.1.5 mM and K_DS that are too high to measure accurately; its relatively weak affinity may relate to the smaller size of the N domain (ca. 150 residues) than in Ca^{2+}-ATPase (ca. 240 residues).

By contrast, the P domain is similar in size (ca. 165 residues) and displays a very high degree of sequence identity (up to 40%) among P_2-type ATPases. This is not surprising in view of its central role in the reaction cycle. Functionally important P-domain motifs include 347-DKTGTLT in sarcoplasmic reticulum Ca^{2+}-ATPase (378-DKTGTLT in yeast H^+-ATPase) and 601-DPPR (534-DPPR), both located in or near the "hinge" region between the P and N domains; and 701-TGDGVNDAP (632-TGDGVNDAP), located at the interface of the P and A domains. The first of these motifs surrounds the Asp residue that forms the characteristic high-energy E_1P intermediate, while the second and third contain residues involved in phosphoryl group transfer [50–52]. Based on site-directed mutagenesis studies of yeast Pma1 H^+-ATPase, amino acid substitutions within these highly conserved parts of the P domain often lead to a pronounced defect in protein folding, which can be recognized by an elevated sensitivity of the ATPase to trypsin [53]. In severe cases, newly synthesized ATPase fails to leave the endoplasmic reticulum and is eventually degraded by the proteasome (see below). In milder cases, the ATPase manages to reach the plasma membrane, but is then recognized by a yet-to-be-identified quality control mechanism and sent to the vacuole for degradation [54]. Similar folding problems have been reported for Na^+,K^+-ATPase carrying mutations in the conserved TGDGVNDSP motif [55]. Thus, as might be expected from the crystal structures, the interfaces between the P, N, and A domains play a critical role in protein folding and stability.

Valiakhmetov and Perlin [56] have recently reported encouraging progress towards mapping the P domain of yeast Pma1 H^+-ATPase. In this study, oxidative cleavage by Fe^{2+} was used to identify parts of the 100 kDa polypeptide that lie close to bound Mg^{2+} and ATP during phosphoryl group transfer. Of the cleavage sites described, three are near conserved residues (Thr-558 and Lys-615) known to take part in the phosphorylation mechanism; the fourth cleavage site (tentatively located within 335-PVGLPA) provides intriguing evidence that the cytoplasmic end of membrane segment 4 may play a hitherto-unsuspected role in this part of the reaction cycle.

1.3.3
E1–E2 Conformational Change

Profound conformational changes accompany ATP hydrolysis by P-type ATPases [29], and are intricately involved in the transport mechanism. For simplicity, the reaction cycle is usually drawn as an alternation between two major states, E_1 and E_2 (Figure 1.3), even though the shift from one to the other presumably involves a carefully orchestrated sequence of smaller movements. Early evidence for structural differences between E_1 and E_2 came from proteolytic studies of Na^+,K^+- and Ca^{2+}-ATPases, in which distinctive cleavage patterns could be seen depending upon the step of the reaction cycle (reviewed in Ref. 57). Differences were also reported in intrinsic tryptophan fluorescence and in the accessibility of individual amino acid residues to group-specific reagents [57]. Several years ago, Karlish and co-workers introduced a versatile way to probe conformational changes in Na^+,K^+-ATPase by mapping and comparing the sites of Fe^{2+}- and Cu^{2+}-catalyzed oxidative cleavage in the presence of ligands that pull the ATPase into the E_1 or E_2 state. To date, the results fit well with the E_1 and E_2 structures of Ca^{2+}-ATPase, and indeed serve to extend those structures by providing information on the conformation-dependent binding of ATP and Mg^{2+} (reviewed in Ref. 31).

As expected, Pma1 H^+-ATPase also displays major conformational changes during its reaction cycle. Early work by Scarborough [17] and by our laboratory [18] showed that the tryptic fragmentation pattern of the Neurospora enzyme could be altered by ligands that pull the ATPase into the E_1 state (ADP) or E_2 state (vanadate). Further insight into the nature of the E_1–E_2 conformational change has come from a novel group of yeast H^+-ATPase mutants. Haber and co-workers selected the first example of this group (S368F) based on its resistance to hygromycin B [58], an antibiotic later shown to require a negative membrane potential for uptake into the cell [59]. Not only was H^+-ATPase activity reduced in S368F, but the mutant enzyme also had three kinetic abnormalities: a large increase in IC_{50} for inhibition by vanadate, a several-fold lowering of the $K_{1/2}$ for MgATP, and an alkaline shift in pH optimum [60]. Soon afterwards, three similar mutants were identified by our laboratory as part of a scanning mutagenesis study of membrane segment 4 (I332A, V336A, V341A; Ref. 61). In this case, the amino acid substitutions were located in the middle of M4, making it difficult to imagine that vanadate binding was affected directly. Much more likely was a shift in the conformational equilibrium of the ATPase from E_2 (where vanadate binds tightly as a P_i analogue) towards E_1 (which is expected to have low affinity for vanadate but high affinity for ATP and H^+; see reaction cycle in Figure 1.3). Mutants with similar kinetic changes have since been described for Na^+,K^+-ATPase [62, 63], gastric H^+,K^+-ATPase [64], and sarcoplasmic reticulum Ca^{2+}-ATPase [65, 66] and interpreted in the same way.

The discovery of the vanadate-resistant phenotype has provided a convenient way to map regions of the yeast H^+-ATPase polypeptide that work together to affect the E_1-E_2 equilibrium. Not surprisingly, given the global nature of the conformational change [29], "E_1–E_2" mutations are scattered throughout the 100 kDa polypeptide (see Figure 5 in Ref. 67). Against this general background, however, a remarkable cluster has recently been discovered in a scanning mutagenesis study of the stalk region (S4) that links the cytoplasmic end of membrane-spanning segment 4 to the phosphorylation site, D378. As reported by Ambesi and co-workers [68], mutants at thirteen successive positions from I359 through G371 (including S368F) have undergone a 10- to 200-fold increase in the IC_{50} for vanadate; most of them also show a parallel decrease in the apparent affinity for MgATP and an alkaline shift in pH optimum. In the Ca^{2+}-ATPase-based homology model of the H^+-ATPase [32], the I359–G371 stretch forms a short α-helix (P1) that is oriented transversely between the membrane and the phosphorylation site (D378). Based on the kinetic behavior of the E_1E_2 mutants, it seems likely that this helix plays a central role in the conformational shift that lies at the heart of the transport mechanism.

1.3.4
H⁺ Pumping

Sequence identity drops to less than 20 % in the M domain of P_2-type ATPases, making it difficult to recognize functional features based on sequence analysis alone. Even before the Toyoshima E_1Ca structure appeared, however, there was considerable evidence from mutagenesis studies that membrane-spanning segments 4, 5, 6, and 8 (M4, M5, M6, and M8) define the actual transport pathway of P_2-ATPases [69–72]. The way in which these four segments cooperate to form the two Ca^{2+}-binding sites of Ca^{2+}-ATPase is now clear from the 2.6 Å crystal structure of that enzyme [27]. As summarized above, side-chain oxygen atoms (D, E, N, T) and main-chain carbonyl oxygen atoms (A, V, I) are involved, and the actual coordination geometry is made possible by local unwinding of M4 and M6.

Detailed structures are not yet available for other P_2-ATPases, but site-directed mutagenesis has begun to shed light on the residues involved in cation binding and transport. For the yeast Pma1 H^+-ATPase, scanning mutagenesis has been performed along the entire length of M4 [61], M5 [73], M6 and M8 (Petrov et al. and Guerra et al., manuscripts in preparation and throughout much of M1 and M2 [74]. Among the mutants that have been constructed and analyzed, four types of defects have been noted:

(i) *Protein misfolding, leading to ER retention and eventual degradation.* Mutations at 16 intramembrane positions fall into this category, presumably because they disrupt the precise helix packing needed to stabilize the M domain of the H^+-ATPase. Not surprisingly, changes in charged residues are poorly tolerated in M5 (R695, H701) and M6 (D730, D739) of the H^+-ATPase (Petrov et al., manuscript in preparation). Kaplan and co-workers have shown that the M5–M6 hairpin of Na^+,K^+-ATPase is only tenuously embedded in the lipid bilayer, requiring occluded K^+ ions to keep it from being lost into the medium after cytoplasmic portions of the polypeptide are removed by proteolysis [75, 76]. M8 appears to play an even more important role in proper maturation of the H^+-ATPase, since alanine substitutions at seven positions along this segment (I794, F796, L797, Q798, I799, L801, I807) lead to misfolding and retention in the ER (Guerra et al., manuscript in preparation).

(ii) *Decrease in ATPase activity.* Alanine substitutions in M4 (at G333, L338, V341), M5 (Y691, S699), M6 (L721, V723, F724, I727, F728, L734, Y738) cause H^+-ATPase activity to fall below 20 % of the value seen in the wild-type control [61, 73; Petrov et al., manuscript in preparation]. The preponderance of such mutants in M6 again calls attention to the special nature of this membrane segment and points to the importance of bulky hydrophobic residues (Phe, Tyr, Leu, Val, Ile) for its proper functioning.

(iii) *Shift in conformational equilibrium from E_1 to E_2.* The "E_1–E_2" phenotype, in which the H^+-ATPase becomes strongly resistant to vanadate, has already been discussed (see above). Six such mutants have been discovered in the membrane domain, and the fact that three of them are located in M4 (I332A, V341A, M346A) points to a possible link between S4 and M4 in mediating the E_1–E_2 conformational change (Petrov et al., manuscript in preparation). Similar mutants are also found in M1 (M128C; Ref. 74), M5 (V692A; Ref. 73), and M6 (V723A; Petrov et al., manuscript in preparation).

(iv) *Mutational change in coupling between ATP hydrolysis and H^+ transport.* H^+-ATPases pose a special challenge in studies of transport mechanisms since there are no radioisotopic methods to assay unidirectional proton fluxes or to detect protons occluded within the membrane. Instead, experimenters generally rely upon a fluorescent probe such as acridine orange or 9-amino-6-chloro-2-methoxyacridine (ACMA) to measure the initial rate of proton uptake by ATPase that has been expressed in inside-out secretory vesicles [77] or purified and reconstituted into proteoliposomes [9–11]. Such measurements gain in reliability if the MgATP concentration is varied over a broad range (say, from 0 to 3.0 mM) and the initial rate of fluorescence quenching plotted as a function of the initial rate of ATP hydrolysis. The resulting plots are typically linear, giving evidence for constant coupling across a wide range of pump velocities [61].

To date, more than 80 mutant H^+-ATPases have been analyzed in this way. For most amino acid substitutions, little or no change is seen in the slope of the quenching vs. hydrolysis plot. However, membrane-spanning segments 5, 6, and 8 contain twelve positions at which mutations have caused the slope (or "coupling ratio") to fall below 50 % of that seen in the wild-type control [73, Petrov et al., and Guerra et al., manuscripts in preparation]. Based on these results, both hydrophobic residues (I712, L700, L708, L713, M791) and polar residues (T733, N792, E703, E803) contribute directly or indirectly to the transport pathway, as do two alanines (A726 and A732) and a glycine (G793).

An unexpected effect on coupling has come from mutations at two positions in M8 of the yeast Pma1 ATPase. In the first case (E803, homologous to E908 of Ca^{2+}-ATPase), replacement by Asn or Ala caused pumping to fall to barely detectable levels, but replacement by Gln led to a remarkable two-fold *increase* in the slope of the quenching vs. hydrolysis plot [78]. Approximately one helical turn away, substitution of S800 by Ala also produced a ca. 2-fold increase in slope (Guerra, L., et al., manuscript in preparation). Control measurements, in which the rate of decay of ΔpH was tracked after the addition of vanadate to turn off the pump, showed no significant change in the passive permeability of the membrane [78]. Thus, the enhanced transport seen in S800A and E803Q appears to reflect an actual improvement in the effective stoichiometry of the pump. Further work is required to understand the relationship between this finding and earlier reports of variable stoichiometries in the Neurospora [79] and yeast [80] Pma1 ATPases.

Efforts are already under way to model the transport pathway of Pma1 ATPase, with the longer-range goal of synthesizing the results of site-directed mutagenesis into a clear mechanistic picture of how proton transport actually works. On the one hand, it is tempting to draw analogies with better-understood proton pumps such as bacteriorhodopsin [81] and F_oF_1 ATPase [82], where the carboxyl group of an essential Asp or Glu residue is alternately protonated and deprotonated as H^+ ions move through the membrane. On the other hand, the clearcut resemblance between Pma1 H^+-ATPase and P_2-type Na^+,K^+, H^+,K^+, and Ca^{2+}-ATPases suggests that hydronium ions (H_3O^+) may be the actual transported species, crossing the membrane via one or more binding sites analogous to those seen in the crystal structure of sarcoplasmic reticulum Ca^{2+}-ATPase.

Both possibilities are currently being pursued. Based on homology modeling, Bukrinsky and co-workers have proposed that D730 (M6) and I331, I332, and V334 (M4) of yeast Pma1 ATPase may form a binding site for H_3O^+, analogous to site II of sarcoplasmic reticulum Ca^{2+}-ATPase [83]. In a similar model of Neurospora Pma1 ATPase, however, Radresa and his colleagues have pointed to the relative shortage of negatively charged groups in the region corresponding to Ca^{2+} sites I and II [84]. Rather, these authors emphasize a bacteriorhodopsin-like string of polar cavities, starting at D378 (the catalytic phosphorylation site), that may conduct protons from the cytoplasmic surface into the interior of Pma1 ATPase. It is unclear whether site-directed mutagenesis will be able to distinguish unambiguously between the two types of mechanistic models or whether other kinds of data, including a long-hoped-for high-resolution Pma1 structure, will be required.

1.4
Biogenesis

Alongside studies of reaction mechanism, there has been great interest in the bio-genesis of P_2-type ATPases which, as large polypeptides anchored in the lipid bi-layer by multiple hydrophobic helices, pose a special challenge to the membrane insertion and trafficking machinery. The yeast H^+-ATPase has proved to be an ex-cellent model system. Unlike Na^+,K^+-ATPase, which has been equally well studied (reviewed in 85, 86), the yeast ATPase lacks the complication of an auxiliary β-sub-unit. Work in yeast also benefits from powerful genetic tools that are not yet avail-able in mammalian systems.

1.4.1
Pma1 Mutants with Defects in Folding and Biogenesis

As expected, yeast Pma1 ATPase is synthesized and inserted into the membrane in the rough endoplasmic reticulum [reviewed in 87]. Studies by Addison on the clo-sely related Neurospora H^+-ATPase have defined the way in which alternating sig-nal anchor (SA) and stop-transfer (ST) sequences serve to thread the 100 kDa poly-peptide into the bilayer, helped by hairpin formation between neighboring mem-brane segments [88, 89]. At the earliest time points in a pulse-chase experiment, the ATPase can already be protected against trypsinolysis by low concentrations of MgADP, MgATP, and vanadate, indicating that its cytoplasmic domain has folded to form recognizable binding sites for nucleotides and inhibitors [90]. It then travels along the well-known secretory pathway to the plasma membrane.

The availability of nearly 300 point mutants of the yeast H^+-ATPase has provided a wealth of material to explore the structural requirements for ATPase folding and trafficking [reviewed in 67]. At one end of the spectrum are mutants like D378N, D378S, and D378A, which are severely misfolded [91], become trapped in the ER [91, 92], and are eventually ubiquitinated and degraded in the proteasome [93]. Such mutants act in a dominant lethal fashion to prevent growth when co-ex-pressed with wild-type Pma1-ATPase [91, 92]; this behavior has recently been traced to the formation of mixed oligomers between mutant and wild-type polypep-tides [38].

At the other end of the spectrum are mutants that display a modest folding de-fect and a relatively mild genetic phenotype. By means of confocal microscopy, it has been possible to visualize a transient arrest of one such ATPase (G381A) in prominent punctate structures, which resemble the vesicular-tubular complexes (VTCs) that serve as a normal ER-Golgi intermediate in mammalian cells [94]. Thus, the yeast structures may correspond to specific ER exit ports that become amplified as the mutant ATPase is synthesized. The G381A protein eventually es-capes the VTCs and travels to the plasma membrane, still in a misfolded, trypsin-sensitive state. There, its behavior points to an additional, yet-to-be-characterized quality control step that recognizes the ATPase as defective and sends it to the va-cuole for degradation [94]. A separate series of experiments has shown that G381A can impose its phenotype on co-expressed wild-type ATPase, transiently retarding the wild-type protein in the ER and later stimulating its degradation in the vacuole [94]. Both effects serve to lower the steady-state amount of wild-type ATPase in the plasma membrane and can thus explain the co-dominant genetic behavior of the G381A mutation.

Between the two extremes are numerous genetically intermediate mutants, which display a recessive lethal phenotype but have not yet been studied biochemically. All three kinds of mutations cluster in regions of the H⁺-ATPase that are likely to be important for protein folding: in the ^{231}TGES and ^{632}TGDGVNDAP motifs, located at the interface between the A and P domains; in the ^{535}DPPR and ^{378}DKTGTLT motifs, situated within or immediately adjacent to the "hinge" regions between the N and P domains; and in the membrane-spanning segments, known to play a critical role in protein folding and stability (reviewed in Ref. 67).

1.4.2
Use of Pma1 Mutants to Screen for Other Genes that Play a Role in Biogenesis and Quality Control

Because Pma1 ATPase is delivered to the cell surface by way of the secretory pathway, its biogenesis can be probed with the classical collection of temperature-sensitive "sec" mutants that define the basic architecture of that pathway. For example, *sec18*^{ts}, *sec7*^{ts}, and *sec6*^{ts} strains have made it possible to detect stepwise Ser/Thr phosphorylation of the ATPase between the endoplasmic reticulum and the plasma membrane [95]. The *sec6*^{ts} strain has also permitted the expression of mutant Pma1 ATPases in secretory vesicles, where ATP hydrolysis and ATP-dependent proton transport can be examined quantitatively, free of contamination by wild-type ATPase [77].

More recently, genetic screens have been invented to uncover genes that play a specialized role in ATPase biogenesis (Table 1.2). In one such approach, Haber and his colleagues [96] began with the *pma1-114* mutant, which is moderately resistant to hygromycin B due to a low membrane potential; they then used higher concentrations of the same drug to select "*mop*" (modifier of *pma1*) mutations leading to increased resistance. Among the isolates were several allelic strains (*mop2*) with a reduced amount of Pma1 ATPase at the cell surface. The new mutations did not affect transcription of the *PMA1* gene, nor did they cause Pma1 ATPase to be arrested intracellularly. Rather, the 108 kDa Mop2 protein (also known as End4 or Sla2) appears to play a role in setting the amount and/or stability of the ATPase at the plasma membrane.

In parallel, Chang and co-workers have used misfolded Pma1 mutants to pinpoint other genes involved in ATPase biogenesis and quality control. One novel gene identified in this way is *EPS1*, which, when disrupted, suppresses the dominant lethal phenotype of *pma1-D378N* and re-routes the mutant ATPase to the plasma membrane [93]. Interestingly, *EPS1* encodes a protein disulfide isomerase that may act as a membrane-bound chaperone. Other genes, when overexpressed [*AST1*; 97] or disrupted [*SOP*; 98], allow a temperature-sensitive mutant ATPase

Table 1.2 Proteins governing the biogenesis of Pma1 H$^+$-ATPase.

Protein	Subcellular location	Function	Effect on Pma1 ATPase	Ref.
Eps1	ER (integral)	Protein disulfide isomerase	Deletion of *EPS1* prevents the proteasomal degradation of Pma1-D378N ATPase and reroutes it to the plasma membrane	93
Lst1	ER (peripheral)	Component of COPII vesicle coat	Deletion of *LST1* inhibits the delivery of Pma1 ATPase to the cell surface	99–100
Ast1	Multiple membranes (peripheral)	Targeting to the plasma membrane	Overexpression of *AST1* reroutes a temperature-sensitive Pma1 ATPase mutant from the vacuole to the plasma membrane	97
Mop2 (End4, Sla2)	Cytoskeleton	Actin filament organization Cell wall organization and biogenesis Endocytosis, exocytosis Polar budding	Mutation of *MOP2* reduces the amount of Pma1 ATPase in the plasma membrane	96
Sop1-16	Various	Not known	Deletion of each *SOP* gene perturbs the endosomal trafficking and re-routes a temperature-sensitive Pma1 ATPase mutant from the vacuole to the plasma membrane	98

(*pma*1-114) to escape quality control and move to the plasma membrane. The nature of these gene products and the specificity of their role in biogenesis are currently under active study.

Kaiser and his co-workers [99] have focused genetically on the formation and activity of COPII vesicles, which carry newly synthesized proteins from the ER to the Golgi. Among ten *LST* (*l*ethal with *sec t*hirteen) genes identified by a method known as "synthetic lethality", the product of one (Lst1p) helps to package Pma1 ATPase into the vesicles [99, 100]; significantly, it is homologous to a known COPII coat protein, Sec24p. More work will be required to understand the exact function of Lst1p, but its discovery hints at differentiation of the secretory machinery to handle large, physiologically important cargo molecules such as the ATPase.

1.4.3
Role of Lipid Rafts

In addition to the protein components of the secretory pathway, membrane lipids also play a vital role in the biogenesis of Pma1 ATPase. Recent studies by Simons and his colleagues have shown that newly synthesized ATPase and other proteins destined for the plasma membrane are sorted preferentially into lipid rafts, which form an ordered, sphingolipid- and ergosterol-rich phase within the bilayer [101]. If a mutation such as *pma1-7* impairs the ability of the 100 kDa ATPase polypeptide to associate with the rafts, the ATPase is mistargeted to the vacuole rather than the plasma membrane [36]. Thus, the rafts are an essential part of the biogenesis process; at the plasma membrane, their high sphingolipid and sterol content may also protect against the stresses of the external environment [36, 101].

1.5
Regulation

Although P$_2$-ATPases are structurally and functionally similar in most respects, very different mechanisms have evolved to regulate their activity; this is perhaps not surprising in view of the wide range of physiological roles played by members of the family. In some cases, regulation is mediated by an interacting protein: for example, phospholamban for sarcoplasmic reticulum Ca^{2+}-ATPase [102] or 14-3-3 protein for plant plasma-membrane H$^+$-ATPase [103–105]. Small proteolipids are also associated with many P$_2$-ATPases and appear to modulate activity (see above). In other cases, regulation is accomplished intramolecularly by post-translational modification – usually phosphorylation of one or more Ser/Thr residues (reviewed in Ref. 106).

The latter mechanism is clearly involved in the well-documented posttranslational regulation of yeast H$^+$-ATPase activity by glucose. It has been known since 1983 that, when yeast cells are placed in carbon-free medium, there is a rapid 5- to 10-fold decrease in H$^+$-ATPase activity, and when glucose is added back, activity rebounds completely in less than 5 minutes [107]. Although the mechanism is not yet fully understood, there is growing evidence to implicate the C-terminus of the ATPase, acting as an autoinhibitory domain [108]. Mutations at potential phosphorylation sites near the C-terminus affect the ability of glucose to stimulate ATPase activity [109], and in thermolysin digests of the ATPase, two (as yet unidentified) phosphopeptides decrease in amount during carbon starvation and increase again upon glucose addition [95]. Thus, it has been proposed that the C-terminus becomes dephosphorylated during carbon starvation, allowing it to interact in an inhibitory way with one or more catalytically important parts of the ATPase; upon addition of glucose, the C-terminus is rephosphorylated and the inhibition is released.

Recent evidence points to a role for stalk segment 5 (the cytoplasmic extension of M5) in this regulatory process. Cys substitutions at seven positions along one face of S5 lead to strong constitutive activation of the ATPase [110]. Furthermore, labeling studies with a membrane-impermeant fluorescent maleimide have pinpointed three S5 cysteines that fail to react in ATPase from glucose-starved cells but become reactive in ATPase from glucose-metabolizing cells, pointing to a significant conformational change between the two states of the protein [111]. Work is now needed to identify the Ser/Thr residues that undergo glucose-dependent phosphorylation, and, by means of biochemical and biophysical approaches, to probe the interaction of the C-terminus with other parts of the ATPase including S5. Progress is also being made towards defining the elements of the signal transduction pathway. In particular, Portillo and co-workers have recently identified two kinases (Ptk2 and Hrk1) that may mediate Ser phosphorylation of the ATPase in response to glucose [112].

1.6
Emerging Knowledge of Other Yeast P-type ATPases

The sequence of the yeast genome, published in 1996 [113], allowed the first identification of the entire set of P-type ATPases that support the growth and physiological functioning of a simple eukaryotic cell. In all, there are 16 P-ATPases in *Saccharomyces cerevisiae* [114; Table 1.3]. The plasma membrane contains Pma and Ena ATPases that mediate the efflux of H^+ and Na^+ (or K^+) from the cell. Of them, only Pma1 ATPase (the subject of this chapter) is well expressed and highly active under normal laboratory conditions. Ena1 ATPase is induced during growth at elevated

Table 1.3 P-type ATPases in *Saccharomyces cerevisiae*.

Specificity	Protein	Type	Subcellular location	Size	Ref.
H^+	Pma1	IIIA	Plasma membrane	918 aa	This chapter
H^+	Pma2	IIIA	Plasma membrane	947 aa	115,116
Na^+ K^+	Ena1 (Hor6, Pmr2) to Ena5[a]	IID	Plasma membrane	1091 aa	117–119
Ca^{2+}, Mn^{2+}	Pmr1 (Ldb1)	IIA	Golgi	950 aa	120–125
Ca^{2+}	Pmc1	IIB	Vacuolar membrane	1173 aa	126,127
Cu^{2+}	Ccc2	IB	Golgi	1004 aa	128,129
	Pca1 (Cad2, Pay2)	IB	Unknown	1216 aa	130
PL	Drs2 (Fun38, Swa3)	IV	Golgi	1355 aa	131–133
	Neo1	IV	Unknown	1151 aa	133
	Dnf1	IV	Plasma membrane	1571 aa	133
	Dnf2	IV	Plasma membrane	1612 aa	133
	Dnf3	IV	Golgi	1656 aa	133
Unknown	Spf1 (Cod1, Pio1, Per9)	V	ER	1215 aa	134,135
	YOR291w	V	Unknown	1472 aa	136

[a] Various strains of *Saccharomyces cerevisiae* contain a tandem cluster of up to 5 nearly identical *ENA* genes. PL: Aminophospholipid.

Na$^+$ or Li$^+$ concentrations and serves to protect the cell against excessive accumulation of Na$^+$, Li$^+$, or K$^+$ [117–119].

Ca^{2+} homeostasis in yeast relies upon a complex system of transporters and signaling pathways including two endomembrane P-type ATPases: Pmr1 and Pmc1, which are homologues of mammalian SPCA and PMCA ATPases [120–127]. Both work together to maintain a low Ca^{2+} concentration in the cytoplasm. In addition, by pumping Ca^{2+} into the Golgi, Pmr1 creates a lumenal Ca^{2+} concentration (10 µM) that permits the proper functioning of the secretory pathway [123]. Interestingly, Pmr1 ATPase can also transport Mn^{2+} ions, and it plays an essential role in the resistance of yeast to millimolar Mn^{2+} [124]. By site-directed mutagenesis of amino acid residues with oxygen-containing side chains throughout M4, M5, M6, M7, and M8, followed by screening for altered sensitivity to Mn^{2+} and ion-chelating agents, Rao and co-workers have identified two residues in M6 (Asn-774 and Asp-778) that are likely to be involved in divalent cation binding [124]. Even more intriguing is a third mutant in M6 (Q783A) that transports Ca^{2+} normally but displays a 60-fold reduction in affinity for Mn^{2+} [124]; results of this kind promise to be useful in modeling the molecular basis for cation selectivity.

Studies have barely begun on the remaining yeast P-type ATPases. Two belong to the Type IB subfamily and are related to the Cu^{2+}-ATPases known to underlie Menkes and Wilson's diseases; they are Ccc2, located in a late-Golgi or post-Golgi compartment [128, 129], and Pca1, whose location and function have not yet been established [130]. Five (Drs2/Fun38/Swa3, Neo1, and Dnf1-3) are Type IV ATPases, postulated to "flip" aminophosholipids between leaflets of the membrane bilayer [131–133]. One Type V ATPase (Spf1/Cod1/Pio1/Per9), located in the endoplasmic reticulum, has been implicated in Ca^{2+} homeostasis [134, 135], while another (YOR291W) is known only as an open reading frame [136]. Thus, it seems certain that the P-type ATPases of yeast will provide rich material for studies of subcellular targeting, molecular mechanism, and physiological regulation for many years to come.

Acknowledgments

The authors are grateful to members of the Slayman laboratory for helpful comments on the manuscript. Work in the laboratory has been supported by research grant GM15761 from the National Institute of General Medical Sciences.

References

1. Serrano R., Kielland-Brandt M. C., and Fink G. R., *Nature,* **1986,** *319,* 689–693.
2. Gradmann D. Hansen U. P., Long W. S., Slayman C. L., and Warncke J., *J. Membr. Biol.,* **1978,** *39,* 333–367.
3. Slayman C. L., Lu C. Y. H., and Shane L., *J. Membr. Biol.* **1973,** *14,* 305–338.
4. Slayman C. L., *Am. Zool.,* **1970,** *10,* 377–392.
5. Dufour J. P., and Goffeau A., *J. Biol. Chem.,* **1978,** *253,* 7026–7032.
6. Malpartida F., and Serrano R., *FEBS Lett.,* **1980,** *111,* 69–72.
7. Bowman B. J., Blasco F., and Slayman C. W., *J. Biol. Chem.,* **1981,** *256,* 12343–12349.
8. Addison R. and Scarborough G. A., *J. Biol. Chem.,* **1981,***256,* 13165–13171.
9. Villalobo A., Boutry M. and Goffeau A., *J. Biol. Chem.,* **1981,***256,* 12081–12087.
10. Malpartida F., and Serrano R., *J. Biol. Chem.,* **1981,** *256,* 4175–4177.
11. Perlin D. S., Kasamo K., Brooker R. J., and Slayman C. W., *J. Biol. Chem.,* **1984,** *259,* 7884–7892.
12. Perlin D. S., San Francisco M. J., Slayman C. W., and Rosen B. P., *Arch. Biochem. Biophys.* **1986,** *248,* 53–61.
13. Amory A., Foury F., and Goffeau A., *J. Biol. Chem.,* **1980,** *255,* 9353–9357.
14. Dame J. B. and Scarborough G. A., *Biochemistry,* **1980,** *19,* 2931–2937.
15. Dame J. B. and Scarborough G. A., *J. Biol. Chem.,* **1981,** *256,* 10724–10730.
16. Amory A., Goffeau A., McIntosh D. B., and Boyer P. D., *J. Biol. Chem.,* **1982,** *257,* 12509–12516.
17. Addison R., and Scarborough G. A., *J. Biol. Chem.,* **1982,** *257,* 10421–10426.
18. Mandala S. M., and Slayman C. W., *J. Biol. Chem.,* **1988,** *263,* 15122–15128.
19. Hager K. M., Mandala S. M., Davenport J. W., Speicher D. W., Benz E.J Jr., and Slayman C. W., *Proc. Natl. Acad. Sci. U. S.A.,* **1986,** *83,* 7693–7697.
20. Addison R., *J. Biol. Chem.,* **1986,** *261,* 14896–14901.
21. Hesse JE., Wieczorek L., Altendorf K., Reicin A. S., Dorus E., and Epstein W., *Proc. Natl. Acad. Sci. U. S. A.,* **1984,** *81,* 4746–4750.
22. Kawakami K., Noguchi S., Noda M., Takahashi H., Ohta T., Kawamura M., Nojima H., Nagano K., Hirose T., Inayama S. Hayashida H., Miyata T., and Numa S., *Nature,* **1985,** *316,* 733–773.
23. Shull G. E., Schwartz A., and Lingrel J. B., *Nature,* **1985,** *316,* 691–669.
24. MacLennan D. H., Brandl C. J., Korczak B., Green N. M., *Nature,* **1985,** *316,* 696–700.
25. Lutsenko S., and Kaplan J. H., *Biochemistry,* **1995,** *34,* 15607–156013.
26. Axelsen K. B., and Palmgren M. G., *J. Mol. Evol.,* **1998,** *46,* 84–101.
27. Toyoshima C., Nakasako M., Nomura H., and Ogawa H., *Nature,* **2000,** *405,* 647–655.
28. Zhang P., Toyoshima C., Yonekura K., Green N. M., and Stokes D. L., *Nature,* **1998,** *392,* 835–839.
29. Toyoshima C., and Nomura H., *Nature,* **2002,** *418,* 605–611.
30. Sweadner K. J., and Donnet C., *Biochem. J.,* **2001,** *356,* 685–704.
31. Karlish S. J.D., *Ann. New York Acad Sci.,* **2003,** *986,* 39–49.
32. Kühlbrandt W., Zeelen J., and Dietrich J., *Science,* **2002,** *297,* 1692–1696.
33. Auer M., Scarborough G. A., and Kühlbrandt W., *Nature,* **1998,** *392,* 840–843.
34. Goormaghtigh E., Chadwick C., and Scarborough G. A., *J. Biol. Chem.,* **1986,** *261,* 7466–7471.
35. Chadwick C. C., Goormaghtigh E., and Scarborough G. A., *Arch. Biochem. Biophys.,* **1987,** *252,* 348–356.
36. Bagnat M., Chang A., and Simons K., *Mol. Biol. Cell,* **2001,** *12,* 4129–4138.
37. Lee M. C., Hamamoto S., and Schekman R., *J. Biol. Chem.,* **2002,** *277,* 22395–22401.
38. Wang Q., and Chang A., *Proc. Natl. Acad. Sci. U. S.A.,* **2002,** *99,* 12853–12858.
39. Mandala S. M., and Slayman C. W., *J. Biol. Chem.,* **1989,** *264,* 16276–81621.

40. Navarre C., Ghislain M., Leterme S., Ferroud C., Dufour J. P., and Goffeau A., *J. Biol. Chem.,* **1992,** *267,* 6425–6428.

41. Navarre C., Catty P., Leterme S., Dietrich F. and Goffeau, A. *J. Biol. Chem.,* **1994,** *269,* 21262–21268.

42. Crambert G, and Geering K., *Sci. STKE 166,* **2003,** RE1.

43. MacLennan D. H., Abu-Abed M., and Kang C., *J. Mol. Cell. Cardiol.,* **2002,** *8,* 897–918.

44. MacLennan D. H., Asahi M., Tupling A. R.., *Ann. N. Y. Acad. Sci.,* **2003,** *986,* 472–480.

45. Beswick V., Roux M., Navarre C., Coic Y. M., Huynh-Dinh T., Goffeau A, Sanson A., and Neumann J. M., *Biochimie,* **1998,** *80,* 451–459.

46. Mousson F., Coic Y. M., Baleux F., Beswick V., Sanson A., and Neumann J. M., *Biochemistry,* **2002,** *41,* 13611–13616.

47. Jorgensen P. L., and Pedersen P. A., *Biochim. Biophys. Acta,* **2001,** *1505,* 57–74.

48. Maldonado A. M., and Portillo F., *J. Biol. Chem.,* **1995,** *270,* 8655–8659.

49. Pardo J. P., and Slayman C. W., *J. Biol. Chem.,* **1988,** *263,* 18664–18668

50. Farley R. A., Elquza E., Muller-Ehmsen J., Kane D. J., Nagy A. K., Kasho V. N., Faller L. D., *Biochemistry,* **2001,** *40,* 6361–6370.

51. Pedersen P. A., Jorgensen J. R., Jorgensen P. L., *J. Biol. Chem.,* **2000,** *275,* 37588–37595

52. Jorgensen P. L., Jorgensen J. R., Pedersen P. A., *J. Bioenerg. Biomembr.,* **2001,** *33,* 367–377.

53. DeWitt N. D., Turinho dos Santos C. F., Allen K. E., and Slayman C. W., *J. Biol. Chem.,* **2002,** *277,* 21027–21040.

54. Ferreira T., Mason A. B., Pypaert M., Allen K. E., and Slayman C. W., *J. Biol. Chem.,* **1988,** *277,* 21027–21040

55. Jorgensen J. R., and Petersen P. A., *Biochemistry,* **2001,** *40,* 7301–7308

56. Valiakmetov A., and Perlin D. S., *J. Biol. Chem.,* **2003,** *278,* 6330–6336.

57. Jorgensen P. L., and Andersen J. P., *J. Membr. Biol.,* **1988,** *103,* 95–120.

58. McCusker J. H., Perlin D. S., and Haber, J. E., *Mol. Cell. Biol.,* **1987,** *7,* 4082–4088.

59. Perlin D. S., Brown C. L., and Haber J. E., *J. Biol. Chem.,* **1988,** *263,* 18118–18122.

60. Perlin D. S., Harris S. L., Seto-Young D., and Haber J. E., *J. Biol. Chem.,* **1989,** *264,* 21857–21864.

61. Ambesi A., Pan R. L. and Slayman C. W., *J. Biol. Chem.,* **1996,** *271,* 22999–23005.

62. Boxenbaum N., Daly S. E., Javaid Z. Z., Lane L. K., Blostein R., *J. Biol. Chem.,* **1998,** *273,* 23086–23092.

63. Segall L., Lane L. K., Blostein R., *J. Biol. Chem.,* **2002,** *277,* 35202–35209.

64. De Pont J. J., Swarts H. G., Willems P. H., Koenderink J. B., *Ann. New York Acad. Sci.,* **2003,** *86,* 175–182.

65. Vilsen B., *Biochemistry,* **1997,** *36,* 13312–13324.

66. Toustrup-Jensen M., Hauge M., Vilsen B., *Biochemistry,* **2001,** *40,* 5521–5532.

67. Morsomme P., Slayman C. W., Goffeau A., *Biochim. Biophys. Acta,* **2000,** *1469,* 133–157.

68. Ambesi A., Miranda M., Allen K. E., and Slayman C. W., *J. Biol. Chem.,* **2000,** *275,* 20545–20550.

69. Clarke D. M., Loo T. W., Inesi G., MacLennan D. H., *Nature,* **1989,** *339,* 476–478.

70. Clarke D. M., Loo T. W., MacLennan D. H., *J. Biol. Chem.,* **1990,** *265,* 22223–22227.

71. Nielsen J. M., Pedersen P. A., Karlish S. J., Jorgensen P. L., *Biochemistry,* **1998,** *37,* 1961–1968.

72. Pedersen P. A., Nielsen J. M., Rasmussen J. H., Jorgensen P. L., *Biochemistry,* **1998,** *37,* 17818–17827.

73. Dutra M. B., Ambesi A., Slayman C. W., *J. Biol. Chem.,* **1998,** *273,* 17411–17417.

74. Seto-Young D., Hall M. J., Na S., Haber J. E., and Perlin D. S., *J. Biol. Chem.,* **1996,** *271,* 581–587.

75. Lutsenko S., Anderko R., Kaplan J. H., *Proc. Natl. Acad. Sci. U. S.A.,* **1995,** *92,* 7936–7940.

76. Gatto C., Lutsenko S., Shin J. M., Sachs G., Kaplan J. H., *J. Biol. Chem.,* **1999,** *274,* 13737–13740.

77. Nakamoto R. K., Rao R. and Slayman C. W., *J. Biol. Chem.*, **1991**, *266*, 7940–7949.

78. Petrov V. V., Padmanabha K. P., Nakamoto R. K., Allen K. E., and Slayman C. W., *J. Biol. Chem.*, **2000**, *275*, 15709–15716.

79. Warncke J., Slayman C. L., *Biochim. Biophys. Acta*, **1980**, *591*, 224–233.

80. Venema K., and Palmgren M. G., *J. Biol. Chem.*, **1995**, *270*, 19659–19667.

81. Lanyi J. K., *J. Biol. Chem.*, **1997**, *272*, 31209–31212.

82. Fillingame R. H., and Dimitriev O. Y., *Biochim. Biophys. Acta*, **2002**, *1565*, 232–245.

83. Bukrinsky J. T., Buch-Pedersen M. J., Larsen S., and Palmgren M. G., *FEBS Lett.*, **2001**, *494*, 6–10.

84. Radresa O., Ogata K., Wodak S., Ruysschaert J. M., and Goormatigh E., *Eur. J. Biochem.*, **2002**, *269*, 5246–5258.

85. Geering K., *J. Membr. Biol.*, **2000**, *174*, 181–190.

86. Geering K., *J. Bioenerg. Biomembr.*, **2001**, *33*, 425–438.

87. Ferreira T, Mason A. B., Slayman C. W., *J. Biol. Chem.*, **2001**, *276*, 29613–29616.

88. Lin J., and Addison R., *J. Biol. Chem.*, **1995**, *270*, 6935–6941.

89. Lin J., and Addison R., *J. Biol. Chem.*, **1995**, *270*, 6942–6948.

90. Chang A., Rose M. D., and Slayman C. W., *Proc. Natl. Acad. Sci. U. S.A.*, **1993**, *90*, 5808–5812.

91. DeWitt N. D., dos Santos C. F., Allen K. E., and Slayman C. W., *J. Biol. Chem.*, **1998**, *273*, 21744–21751.

92. Harris S. L., Na S., Zhu X., Seto-Young D., Perlin D. S., Teem J. H. and Haber J. E., *Proc. Natl. Acad. Sci. U. S.A.*, **1994**, *91*, 10531–10535.

93. Wang Q., and Chang A., *EMBO J.*, **1999**, *18*, 5972–5982.

94. Ferreira T., Mason A. B., Pypaert M., Allen K. E., Slayman C. W., *J. Biol. Chem.*, **2002**, *277*, 21027–21040.

95. Chang, A., and Slayman, C. W. *J. Cell Biol.*, **1991**, *115*, 289–295

96. Na S. Hincapie M., McCusker J. H., Haber J. E., *J. Biol. Chem.*, **1995**, *270*, 6815–6823.

97. Chang A., Fink G. R., *J. Cell Biol.*, **1995**, *128*, 39–49.

98. Luo W., Chang A., *J. Cell Biol.*, **1997**, *138*, 731–746.

99. Roberg K. J., Crotwell M., Espenshade P., Gimeno R. and. Kaiser C. A., *J. Cell Biol.*, **1999**, *145*, 659–672.

100. Shimoni Y., Kurihara T., Ravazzola M., Amherdt M., Orci L., Schekman R., *J. Cell Biol.*, **2000**, *151*, 973–984.

101. Bagnat M., Keranen S., Shevchenko A., Shevchenko A., Simons K., *Proc. Natl. Acad. Sci. U. S.A.*, **2000**, *97*, 3254–3259.

102. Asahi M., Kimura Y., Kurzydlowski K., Tada M., MacLennan D. H., *J. Biol. Chem.*, **1999**, *274*, 32855–32862.

103. Jahn T., Fuglsang A. T., Olsson A, Bruntrup I. M., Collinge D. B. Volkmann D., Sommarin M., Palmgren M. G., Larsson C., *Plant Cell*, **1997**, *9*, 1805–1814.

104. Fuglsang A. T., Visconti S., Drumm K., Jahn T., Stensballe A., Mattei B., Jensen O. N., Aducci P., Palmgren M. G., *J. Biol. Chem.*, **1999**, *274*, 36774–36780.

105. Maudoux O., Batoko H., Oecking C., Gevaert K., Vandekerckhove J., Boutry M., Morsomme P., *J. Biol. Chem.*, **2000**, *275*, 17762–17770.

106. Therien A. G., Blostein R., *Am. J. Physiol. Cell Physiol.*, **2000**, *279*, C541–566.

107. Serrano R. *FEBS Lett.*, *156*, **1983**, 11–14.

108. Portillo F., *Biochim. Biophys. Acta*, **2000**, *1469*, 31–42.

109. Portillo F., Eraso P. and Serrano R., *FEBS Lett.*, **1991**, *287*, 71–74.

110. Miranda M., Pardo J. P., Allen K. E., Slayman C. W., *J. Biol. Chem.*, **2002**, *277*, 40981–40988.

111. Miranda M., Allen K. E., Pardo J. P., Slayman C. W., *J. Biol. Chem.*, **2001**, *276*, 22485–22490.

112. Goossens A., de La Fuente N., Forment J., Serrano R., Portillo F., *Mol. Cell Biol.*, **2000**, *20*, 7654–7661.

113. Goffeau A. et al., *Science*, **1996**, *274*, 546, 563–567.

114. Catty P, de Kerchove d'Exaerde A, Goffeau A., *FEBS Lett.*, **1997**, *409*, 325–332.

115. Supply P., Wach A., Goffeau A., *J. Biol. Chem.*, **1993**, *268*, 19753–19759

116. Fernandes A. R., Sa-Correia I., *Yeast*, **2003**, *20*, 207–219.

117. Haro R., Garciadeblas B., Rodriguez-Navarro A., *FEBS Lett.*, **1991**, *291*, 189–191.

118. Benito B., Quintero F. J., Rodriguez-Navarro A., *Biochim. Biophys. Acta*, **1997**, *1328*, 214–226.

119. Benito B., Garciadeblas B., Rodriguez-Navarro A., *Microbiology*, **2002**, *148*, 933–941.

120. Antebi A, Fink G. R., *Mol. Biol. Cell*, **1992**, *3*, 633–654.

121. Sorin A., Rosas G., and Rao R., *J. Biol. Chem.*, **1997**, *272*, 9895–9901.

122. Durr G., Strayle J., Plemper R., Elbs S., Klee S. K., Catty P., Wolf D. H., Rudolph H. K., *Mol. Biol. Cell*, **1998**, *9*, 1149–1162.

123. Strayle J., Pozzan T., and Rudolph H. K., *EMBO J.*, **1999**, *18*, 4733–4743.

124. Wei Y., Chen J., Rosas G., Tompkins D. A., Holt P. A., Rao R., *J. Biol. Chem.*, **2000**, *275*, 23927–23932.

125. Mandal D., Rulli S. J., Rao R., *J. Biol. Chem.*, **2003**, *278*, 35292–35298.

126. Cunningham K. W., and Fink G. R., *J. Exp. Biol.*, **1994**, *.196*, 157–166.

127. Cunningham K. W., and Fink G. R., **1994**, *J. Cell Biol. 124*, 351–363.

128. Fu D, Beeler T. J., Dunn TM., *Yeast*, **1995**, *11*, 283–292.

129. Huffman D. L., and O'Halloran T. V., *Annu. Rev. Biochem.*, **2001**, *70*, 677–701

130. Rad M. R., Kirchrath L., Hollenberg C. P., et al., *Yeast*, **1994**, *10*, 1217–1225.

131. Tang X., Halleck M. S., Schlegel R. A., Williamson P., *Science*, **1996**, *272*, 1495–1497.

132. Chen C. Y., Ingram M. F., Rosal P. H., Graham T. R., *J. Cell Biol.*, **1999**, *147*, 1223–1236.

133. Hua Z., Fatheddin P., Graham T. R., *Mol. Biol. Cell*, **2002**, *13*, 3162–3177.

134. Cronin S. R., Rao R., Hampton R. Y., *J. Cell Biol.*, **2002**, *157*, 1017–1028.

135. Vashist S., Frank C. G., Jakob C. A., and Davis T. W., *Mol. Biol. Cell*, **2002**, *13*, 3955–3966.

136. Poirey R., Cziepluch C., Tobiasch E., Pujol A., Kordes E., Jauniaux J. C., *Yeast*, **1997**, *13*, 479–482.

2

Regulation of the Sarco(endo)plasmic Reticulum Ca^{2+}-ATPase by Phospholamban and Sarcolipin

David H. MacLennan and *Evangelia G. Kranias*

2.1
Introduction

2.1.1
Background to Ca^{2+} Signaling

In the evolution of metabolism based on phosphates and organic acids, it was critical that cytosolic Ca^{2+} concentrations be maintained below about 10 µM. Thus Ca^{2+} pumps, known as plasma membrane Ca^{2+} ATPases (PMCAs) or sarco(endo)-plasmic reticulum Ca^{2+} ATPases (SERCAs), which lower resting Ca^{2+} concentrations to <100 nM and establish a 10,000- fold gradient across the cell membrane play an important role in shaping cellular metabolism [1]. The Ca^{2+} gradient underlies the use of Ca^{2+} as a cellular signaling molecule. Brief openings of plasma membrane or organellar Ca^{2+} channels create rapid elevations of cytosolic Ca^{2+} to 1–10 µM, a range in which Ca^{2+} binds to and activates specific proteins (the 'on' mechanism); Ca^{2+} pumps are among the first proteins activated, to eject Ca^{2+} (the 'off' mechanism), assuring that the elevation of Ca^{2+} will be transient.

The trigger for muscle contraction is the entry of Ca^{2+} into the myoplasm (Figure 2.1), mediated by Ca^{2+} release channels (ryanodine receptors) that tap the Ca^{2+} store in the lumen of the sarcoplasmic reticulum (SR), or plasma membrane Ca^{2+} channels (dihydropyridine receptors) that tap the high concentrations of Ca^{2+} in the extracellular space. For cardiac and skeletal muscle, the key Ca^{2+} binding protein is troponin C and the key event is a Ca^{2+}-triggered conformational change in troponin C, which removes a steric block to the interaction between actin in thin filaments and myosin in thick filaments, permitting contraction [2]. The steric block is reinstated when myoplasmic Ca^{2+} levels are reduced, so that Ca^{2+} dissociates from troponin C. The trigger for relaxation is removal of cytosolic Ca^{2+} by the combined activity of SERCAs, PMCAs and Na$^+$/Ca^{2+} exchangers (NCX) [3]. In the hearts of humans and larger mammals, SERCA2a activity determines the rate of removal of >70 % of cytosolic Ca^{2+}, thereby determining the rate of relaxation of the heart, and influencing cardiac contractility by determining the size of the lume-

Handbook of ATPases. Edited by M. Futai, Y. Wada, J. H. Kaplan
Copyright © 2004 WILEY-VCH Verlag GmbH & Co. KGaA, Weinheim
ISBN 3-527-30689-7

nal Ca^{2+} store that is available for release in the next beat. In fast-twitch skeletal muscle, SERCA1a accounts for a virtually all of the Ca^{2+} removal from the myoplasm.

In this chapter, we discuss two proteins, phospholamban (PLN) and sarcolipin (SLN) that regulate the activity of SERCAs in cardiac and skeletal muscle (Figure 2.2). We will discuss advances in the understanding of molecular aspects of the in-

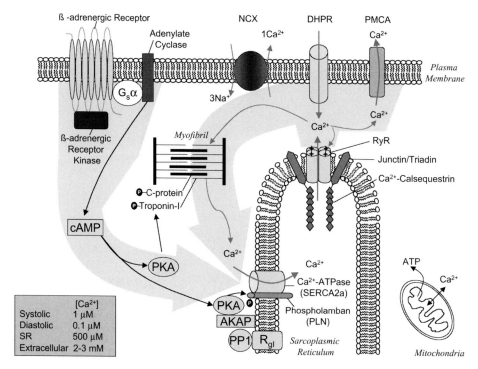

Figure 2.1 Interactions between cardiac signaling pathways. The heart provides an example of how two signaling pathways that involve the elevation of two intracellular second messengers, cAMP and Ca^{2+}, can interact physiologically. In response to depolarization, Ca^{2+} enters the cytoplasm through Ca^{2+} channels in the plasma membrane (dihydropyridine receptors; DHPR). This 'trigger' Ca^{2+} then binds to the Ca^{2+} release channels (ryanodine receptors; RyRs) to stimulate Ca^{2+} release from the SR. After activating muscle contraction by binding to troponin C in the thin filament, Ca^{2+} is removed from the myoplasm by plasma-membrane Ca^{2+} pumps (PMCA) or Na^{+}/Ca^{2+} exchangers (NCX), located in the plasma membrane, or by SERCA2a, located in the SR. Since SERCA2a activity accounts for removal of >70% of myoplasmic Ca^{2+} in humans, it determines both the rate of Ca^{2+} removal (and, consequently, the rate of cardiac muscle relaxation) and the size of the Ca^{2+} store (which affects cardiac contractility in the subsequent beat). SERCA2a activity is regulated by its interaction with PLN, which is a target for phosphorylation by PKA through the second signaling pathway – the β-adrenergic receptor pathway (see text). In its dephosphorylated form, PLN is an inhibitor of SERCA2a, but, when phosphorylated by PKA (or CaM kinase), PLN dissociates from SERCA2a, activating this Ca^{2+} pump. As a result, the rate of cardiac relaxation is increased and, on subsequent beats, contractility is increased in proportion to the elevation of Ca^{2+} release from the SR. PLN is dephosphorylated by a protein phosphatase (PP1), terminating the stimulation phase.

teraction between PLN and SERCA that have arisen from extensive mutagenesis and structural modeling [4]. We describe gain and loss of inhibitory function mutations in PLN [5], which led to investigation of PLN mutations as a potential cause of cardiomyopathy, and the studies carried out in transgenic mice overexpressing superinhibitory forms of PLN that have confirmed this potential [6–8]. The creation and analysis of PLN-null mice [9] and of mice overexpressing PLN [10] have elucidated the remarkable role that PLN actually plays in the regulation of the kinetics of cardiac contractility [11]. The fact that ablation of PLN or suppression of the inhibitory function of PLN can intervene to prevent the progression of dilated cardiomyopathy in well-characterized animal models opens the door to investigation of both the diverse pathways that lead to end-stage heart failure and to potential interventions [12–14]. Finally, we discuss mutations in PLN that cause dilated cardiomyopathy in humans and identify differences in the response of humans and mice to alterations in PLN function [15, 16].

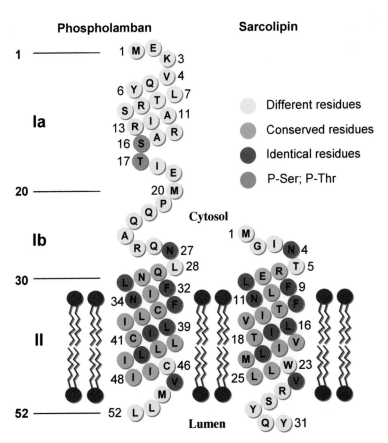

Figure 2.2 Comparison of the amino acid sequence, topology and folding of PLN and SLN. The two proteins share a high degree of amino acid sequence identity and conservation in their transmembrane helices. However, their cytosolic and lumenal sequences are divergent.

Although sarcolipin [17] was discovered before phospholamban [18], its homology to PLN [19] and its PLN-like role as a regulator of SERCA function [20] was demonstrated much later. The patterns of SLN and PLN expression are complementary, so that SLN is expected to play a more important role in regulation of skeletal muscle contractility, while PLN plays a critical role in cardiac muscle contractility. Nevertheless, the finding that SLN and PLN are supcrinhibitory when co-expressed [20, 21], and that a favored PLN–SLN binary complex forms a strongly inhibited ternary complex with SERCA, raises the possibility that SLN plays a role in cardiac contractility, since SLN is expressed in the atrium of the heart [22].

2.1.2
β-Adrenergic Signaling in the Heart

The Ca^{2+} signal transduction pathway is one of two major signaling pathways in the heart (Figure 2.1). Cardiac function is regulated on a beat to beat basis through the sympathetic nervous system. When the demand arises, the heart may respond to stress and increase blood flow to peripheral tissues within seconds. This is based on the large cardiac reserve in humans; the slow heart beat rate and submaximal contractility at rest are increased dramatically following the release of adrenaline into the blood [23]. Adrenaline and other β-agonists initiate a second important signal transduction pathway in the heart by binding to and activating β-adrenergic receptors in the cell membrane (Figure 2.1). The signal proceeds through G$_s$-proteins to stimulate the formation of 3′,5′cyclic AMP (cAMP) by adenylyl cyclase [24]. Elevations in cAMP concentration activate cAMP-dependent protein kinase (PKA), which then phosphorylates and alters the function of a few cardiac proteins, including PLN [25, 26], that have key effects on overall cardiac function.

In early studies, the addition of cAMP and PKA to cardiac SR vesicles was found to increase both the rate and extent of Ca^{2+} uptake [25, 26]. This stimulatory effect was shown to be mediated by the phosphorylation of a reversibly phosphorylated, transmembrane protein co-localized with SERCA2a in the cardiac SR that was named phospholamban [25–28]. Depending on its phosphorylation state, PLN binds to and regulates the activity of SERCA2a.

A useful assay of the regulatory interaction between PLN and SERCA molecules involves measurement of the Ca^{2+}-dependence of Ca^{2+} transport in isolated SR vesicles (Figure 2.3) [26]. By varying the Ca^{2+} concentration over four orders of magnitude and calculating the rate of ATP-dependent Ca^{2+} transport relative to the maximal rate (V_{max}) observed at 10 μM Ca^{2+}, the apparent affinity of SERCA for Ca^{2+} can be determined. Ca^{2+} affinity is expressed as K_{Ca} in pCa units. Typically, the Ca^{2+} affinity of SERCA1a or SERCA2a is about 6.4–6.6 pCa units and it is reduced to 6.0–6.2 pCa units in the presence of PLN. This change is often expressed as ΔK_{Ca} and is about −0.35 pCa units.

Kinetic analysis showed that the major effect of the association of dephosphorylated PLN with SERCA2a is to diminish the apparent affinity of SERCA2a for Ca^{2+}, with little or no effect on V_{max} at saturating Ca^{2+} and ATP [29]. Equilibrium measurements of Ca^{2+} binding to the ATPase, in the absence of ATP, are unchanged by

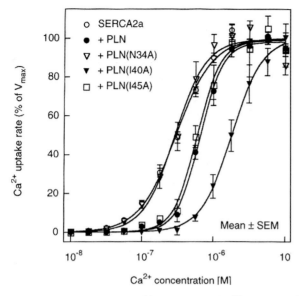

Figure 2.3 Measurement of Ca^{2+}-dependence of Ca^{2+} transport. The affinity of SERCA for Ca^{2+} is measured in Ca^{2+} uptake assays. In the presence of PLN, SERCA2a has a diminished affinity for Ca^{2+}: in the absence of PLN (PLN-null), or following phosphorylation of PLN, apparent Ca^{2+} affinity is increased. Mutations in PLN domain II affect the affinity of SERCA2a for Ca^{2+}: some (e. g. N34A) decrease affinity, suggesting that the amino acid mutated normally interacts with SERCA; some (e. g. I45A) have no effect; some (e. g. I40A) have superinhibitory effects on Ca^{2+} affinity.

PLN. This might suggest that binding of Ca^{2+} and PLN to SERCA are not competitive, so that PLN can bind to the Ca^{2+}-bound form of SERCA. This seems highly unlikely in light of the structural considerations described below. In fact, the effect of PLN is to reduce the bidirectional rate constants of a rate-limiting, Ca^{2+}-dependent conformational change. Due to the second order nature of that step, the effect of PLN is reflected as a displacement of the Ca^{2+} concentration curve required for ATPase activation [29, 30].

In the physiologically relevant Ca^{2+} concentration range of 100 nM to 1 µM, diminished Ca^{2+} affinity diminishes SERCA2a activity. However, SERCA2a activity is increased two- to three-fold by phosphorylation of PLN by either PKA or Ca^{2+}-calmodulin-dependent protein kinase (Ca^{2+}-CaM kinase). *In vivo*, PLN is phosphorylated by both cAMP-dependent and Ca^{2+}-CaM-dependent protein kinases during β-adrenergic stimulation [31–36]. Relief of PLN inhibitory effects on SERCA2a is the major contributor to the positive inotropic (affecting the force of cardiac contractions) and lusitropic (affecting cardiac relaxation) effects of β-agonists [32–36].

2.2
Phospholamban–SERCA Interactions

2.2.1
SERCA Structure and Function

Cloning of SERCA molecules revealed that three *ATP2A* genes encode at least seven SERCA isoforms. *ATP2A1*, encoding SERCA1 (OMIM 108730), is located on human chromosome 16p12.1, *ATP2A2*, encoding SERCA2 (OMIM 108740), is on 12q23-24.1 and *ATP2A3*, encoding SERCA3 (OMIM 601929), is on 17p13.3 [37]. The genes contain 22–23 exons and alternative splicing occurs among the 3′ exons.

SERCA1a is expressed almost exclusively in the sarcoplasmic reticulum of adult fast-twitch skeletal muscle fibers. SERCA2a is expressed in the sarcoplasmic reticulum of cardiac and slow-twitch skeletal muscle fibers, while SERCA2b is expressed ubiquitously and, in this sense, is the "housekeeping" Ca^{2+} pump of the body. SERCA3 is expressed at highest levels (together with SERCA2b) in endothelial cells, in epithelial cells of the trachea, intestine and salivary glands and in platelets, mast cells and lymphocytes. In salivary gland acinar and duct cells, SERCA3 is expressed at the basal pole, while SERCA2b is expressed at the luminal pole. All three SERCA3 isoforms have a lower apparent Ca^{2+} affinity, a higher sensitivity to vanadate and a higher pH optimum than SERCA2b [38]. These unusual properties and cell type distribution suggest that SERCA3 might have a specialized function.

Since PLN inhibits SERCA1a and SERCA2a equally, these proteins have been used interchangeably in both mutagenic [39] and structural [4] analysis of the PLN–SERCA interaction. By contrast, PLN is much less effective in altering the low Ca^{2+} affinity of SERCA3 [40, 41]. In studies of chimeric SERCA2a/SERCA3 molecules designed to identify the sites of PLN–SERCA interaction [40], cytosolic amino acids 467–762 from SERCA2a were associated with high Ca^{2+} affinity and amino acids 336–412 from SERCA2a were associated with PLN interaction. This result appears counterintuitive, since Ca^{2+} is bound in the transmembrane domain and it is these Ca^{2+} binding sites that determine the true affinity of SERCA pumps for Ca^{2+}. The effects of cytoplasmic sequences on Ca^{2+} affinity can be explained by the fact that PLN asserts its inhibitory effect by binding to SERCA2a in its E$_2$ conformation (Figure 2.4). PLN alters the kinetics of Ca^{2+} transport by prolonging the time SERCA2a spends in the E$_2$ conformation [29]. Thus, changes in Ca^{2+} affinity are described as apparent rather than real. It is probable that the high conservation of amino acids in critical transmembrane helices obviated the use of chimeric molecules to identify PLN interaction sites in the membrane domain.

Biochemical analysis [42] and crystal structures of SERCA1a in two different conformations [43, 44] serve as a paradigm for the structure–function relationships that drive Ca^{2+} transport in all three SERCA enzymes and, indeed, for cation transport in all P-type ATPases. In the cycle of Ca^{2+} transport catalyzed by SERCAs at least four interconvertible phosphorylated and unphosphorylated conformations of the enzyme have been defined: E$_1$ATP.2Ca^{2+} is formed in the Ca^{2+} binding

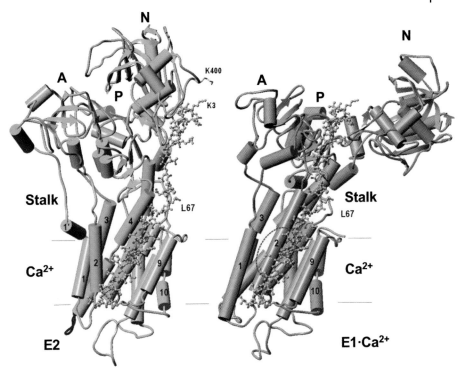

Figure 2.4 Structural model of the sites of interaction between PLN and SERCA. The domains of SERCA1a are indicated as: (Ca^{2+}), Ca^{2+} binding domain; (Stalk), the stalk domain connecting cytosolic domains to the transmembrane domain; (A) actuator domain; (P) phosphorylation domain; and (N) nucleotide binding domain. Two conformations of SERCA1a are shown to move around a single, modeled conformation of PLN. In the E_2 conformation of SERCA1a, a groove is formed in the lipid-facing surface of the transmembrane domain [4]. The carboxy-terminal transmembrane helix of PLN can fit into this groove and form interaction sites with amino acids in transmembrane helices M2, M4, M6 and M9 of SERCA1a. PLN is prevented from continuing as a helix into the cytoplasm by the bulk of M4. Accordingly, it unwinds, but forms a second, amino-terminal helix that fits into a groove in the cytosolic nucleotide-binding (N) domain of SERCA1a, where it is stabilized by hydrophobic interactions and a salt bridge. The unstructured segment between the two helices is exposed, so that interactions with kinases are feasible. In fact, further unwinding of the carboxy-terminal end of the amino-terminal helix is needed to accommodate the access of kinases to the phosphorylation sites at Ser16 and Thr17. As SERCA1a moves to the $E_1 \cdot Ca^{2+}$ conformation, there is a major movement of M2, which narrows the transmembrane groove. The extensive PLN–SERCA1a bonding in this region is gone and PLN sits forlornly on the surface of the SERCA1a transmembrane domain. This disruption of the transmembrane PLN–SERCA1a interaction would terminate the ability of PLN to inhibit the enzyme. The large movement of the N-domain and changes in its conformation would also tend to disrupt PLN–SERCA1a interactions. Although PLN is represented here as remaining static while SERCA1a moves around it, PLN is likely to move away from SERCA1a once its interaction sites are disrupted, perhaps forming homopentamers.

stage; $E_1P(Ca^{2+})_2$, the high energy phosphoenzyme conformation, is formed in the phosphorylation (Ca^{2+} occlusion) stage; E_2P, the low energy phosphoenzyme, is formed in the stage of Ca^{2+} release to the lumen; and E_2 is formed after hydrolysis of the acylphosphate and release of Pi [44].

SERCA1a has cytoplasmic nucleotide binding (N), phosphorylation (P) and actuator (A) domains, a stalk domain and ten transmembrane helices, of which four make up the Ca^{2+} binding domain (Figure 2.4). Extensive conformational changes are powered overall by ATP hydrolysis, but are induced as discrete steps by Ca^{2+} binding, phosphoenzyme formation, Ca^{2+} release and dephosphorylation. In the E_2 conformation [44], Ca^{2+} has access to amino acids near the center of the bilayer, but these amino acids are not yet oriented as high affinity Ca^{2+} binding sites and there is no pathway for Ca^{2+} entry into the lumen. In this conformation, the cytosolic domains are clustered together. In the E_1Ca^{2+} conformation [43], the cytosolic domains are dispersed, but the transmembrane domain is ordered so that two molecules of Ca^{2+} are bound, side-by-side in sites formed by the precise juxtaposition of amino acids in four transmembrane helices. The cytosolic access pathway is now closed, but the outline of a lumenal access pathway emerges. Thus Ca^{2+} transport, energized by the formation of a phosphoenzyme from ATP, involves: access of cytosolic Ca^{2+} to potential Ca^{2+} binding sites in the transmembrane domain, while access of Ca^{2+} to the lumen is obviated; formation of high affinity Ca^{2+} binding sites under conditions where access of Ca^{2+} to the lumen is obviated; and disruption of the high affinity Ca^{2+} binding sites under conditions where access of Ca^{2+} to the lumen is gained, while access of Ca^{2+} from the cytosol is obviated.

Because Ca^{2+} transport requires such extensive movements of both cytosolic and transmembrane domains, inhibitory proteins such as PLN and SLN act by preventing these conformational movements. In its dephospho-form, PLN binds to the E_2 configuration of SERCA2a and inhibits Ca^{2+} pump activity: the conformational changes in SERCA, which accompany Ca^{2+} elevation, force PLN out of its binding site in SERCA (Figure 2.4). Phosphorylation of PLN alters the conformation of PLN and, as a consequence, disrupts PLN–SERCA2a interactions. Both of these mechanisms relieve Ca^{2+} pump inhibition [45]. An antibody against residues 7–16 of PLN mimics phosphorylation of PLN in its activation of the Ca^{2+} ATPase [46, 47].

2.2.2
PLN Structure and Function

Analysis of a cDNA clone of PLN revealed a 52 amino acid protein of 6.1 kDa, that forms a homopentamer [48]. This accounts for the original observation that phosphorylated PLN has an apparent mass of 22 kDa [26]. On the basis of sequence analysis it was suggested that the protein is organized in three domains: cytosolic domain Ia, amino acids 1–20, contains Ser16, the site of phosphorylation by PKA, and Thr17, the site of phosphorylation by Ca^{2+}-Cam kinase; cytosolic domain Ib, amino acids 21–30, is rich in amidated amino acids; amino acids 31–52, which traverse the membrane, make up domain II.

Three NMR studies of PLN structure have been published [49–51], but the only full length structure is that of the C41F-substituted protein in chloroform–methanolone [51]. The conclusion that PLN consists of two helices, spanning amino acids 4–16 and 21–49, and connected by a short β turn is supported by full and partial structures.

2.2.3
Approaches to the Study of PLN–SERCA Interactions

James *et al.* [52] provided the first clear evidence for a physical interaction between PLN and SERCA2a by cross-linking Lys3 in PLN domain Ia to Lys397 and Lys400 in the cytosolic nucleotide-binding (N) domain of SERCA2a (Figure 2.4). Cross-linking was disrupted by elevated Ca^{2+} or by phosphorylation of PLN.

Several investigators have attempted to reconstitute the PLN–SERCA interaction by removing detergent from homogeneous mixtures of purified PLN [53], purified SERCA2a, and phospholipid [54–58]. Reconstitution has also been used to study the interaction of PLN domains with SERCA2 [57–59]. In general, high concentrations of PLN or PLN fragments, ranging from 8- to 100-fold molar excess, have been required to achieve functional interaction. Such reconstitution experiments are controversial since they are difficult to perform and reproduce and the ratio between the interacting proteins is far from physiological.

A second approach to the study of PLN–SERCA2a or the equivalent PLN–SERCA1a interactions [40] has been to co-express the two proteins in heterologous cell culture and to compare the properties of wild-type and mutant forms of these proteins. The Ca^{2+} dependence of Ca^{2+} transport in microsomes isolated from transfected cells has been used to measure the effects of mutations in either SERCA or PLN molecules on Ca^{2+} affinity (functional interaction) and PLN–SERCA2a co-immunoprecipitation (physical interaction). Loss and gain of functional or physical interaction has identified potential interacting amino acids.

2.2.4
SERCA Residues Essential for Cytoplasmic Interaction with PLN

In attempts to localize the regions in SERCA2 and SERCA3 responsible for the different apparent Ca^{2+} affinities, chimeric SERCA2/SERCA3 molecules have been created and tested for Ca^{2+} affinity and for alterations in Ca^{2+} affinity that could be attributed to an interaction with PLN [60]. SERCA3 chimeras containing cytosolic aa 467–762 from SERCA2 had high Ca^{2+} affinity, and SERCA3 chimeras with aa 336–412 and 467–762 of SERCA2 had high Ca^{2+} affinity and interacted functionally with PLN. The PLN interacting region could be narrowed further to include only those sequences that were different between SERCA2 and SERCA3. Within the likely PLN interacting sequence in SERCA2a (aa 365–407), only mutations to Lys-Asp-Asp-Lys-Pro-Val402 affected the functional interaction of SERCA2a with PLN [61]. A SERCA2a chimera, which contained amino acids 334–464 of SERCA3 and could not interact with PLN, gained the ability to interact with

PLN when the SERCA3 sequence Gln-Gly-Glu-Gln-Leu-Val402 was substituted with the SERCA2 sequence Lys-Asp-Asp-Lys-Pro-Val402. Such studies confirmed that a cytosolic site in SERCA2a for interaction with PLN lies between amino acids Lys397 and Leu402 [61].

2.2.5
PLN Residues Essential for Cytoplasmic Interaction with SERCA

Co-expression and mutagenesis has also been used to evaluate the roles of amino acids in PLN domain Ia in PLN–SERCA2 interaction [62]. Mutation of positively charged residues, Lys3, Arg9, Arg13 and Arg14, of the negatively charged residue, Glu2, and of hydrophobic residues, Val4, Leu7, Ile12 and Ile18, all affected functional interaction with SERCA2. Thus, there was evidence, both in the SERCA2 interaction site sequence and in the PLN domain 1a interaction site sequence, that electrostatic and hydrophobic interactions are critical and that mutations in either hydrophilic or hydrophobic amino acids in PLN cytosolic domains can alter PLN–SERCA functional interactions [62].

Coexpression and mutagenesis has also been used to evaluate the roles of amino acids in PLN domain Ib in PLN–SERCA2 interaction [62, 63]. Mutation of Gln22, Gln23 and Ala24 had unaltered function, mutation of Pro21, Arg25A, Gln26 and Leu28A lost inhibitory function, and mutants N27A, Q29A and N30A gained inhibitory function. For the mutant N27A, the change in apparent Ca²⁺ affinity induced in SERCA2a was about 1 pCa unit. From these experiments it is apparent that domain 1b is also involved in PLN–SERCA interactions.

2.2.6
PLN Residues Essential for Transmembrane Interactions with SERCA

To determine whether transmembrane interaction sites were important in PLN–SERCA2a interactions, SERCA2 was expressed with various PLN constructs in which the N-terminal cytoplasmic domains were either deleted or replaced [41]. PLN deletions included domain Ia, domain Ib, or both domains Ia and Ib and replacements of domains Ia and Ib were with HA, Myc or Flag epitopes. While PLN lowered K_{Ca} by 0.35 pCa units, a construct containing only amino acids 28–52 (mostly domain II), lowered K_{Ca} by about 0.17 pCa units. All three epitope-labeled domain II constructs lowered K_{Ca}, but the HA-domain II construct lowered K_{Ca} by 0.6 pCa units and the domain Ib-deleted construct lowered K_{Ca} by at least 1 pCa unit (a phenomenon sometimes referred to as "supershifting" of Ca²⁺ affinity). These constructs did not uncouple Ca²⁺ ATPase from Ca²⁺ transport and inhibition was reversible by appropriate antibodies. All of the PLN constructs in which domain II was intact, but domains Ia and Ib were modified, inhibited SERCA3, presumably because SERCA3 contains an appropriate transmembrane sequence interaction site. This was in contrast to the diminished interaction with native PLN, where incompatible cytoplasmic interaction sites in SERCA3 might act as steric blockers to PLN interaction. In this interpretation, when the cytosolic domain of PLN is removed, the transmembrane interaction site in SERCA3 is unmasked.

These studies led to the proposal that there are PLN–SERCA interaction sites in both cytoplasmic domains and transmembrane domains and to the proposal that these two sites regulate each other through long-range interactions that are analogous to the long-range interactions that occur within SERCA1 during Ca^{2+} transport [41].

Scanning mutagenesis of the PLN transmembrane sequence was carried out first to investigate the phenomenon of pentamer formation by PLN [64, 65]. Mutation of the linear sequence of the transmembrane domain led to cyclical destabilization of the pentamer. Destabilization was associated with mutation of leucines and isoleucines spaced approximately every 3.5 residues, indicating that the transmembrane sequence was helical. It was proposed that the PLN pentamer is held together by "leucine zippers" located in two close parallel columns on one face of each transmembrane helix [65, 66].

When PLN inhibitory function was analyzed, two faces of the monomer could be distinguished [5, 67]. Mutation of amino acids on one face of the helix had little effect on pentamer formation, but diminished the ability of PLN to interact with and inhibit SERCA. Mutation of the other face, on which the leucine zipper was located, not only disrupted the ability of PLN to form pentamers, but, in most

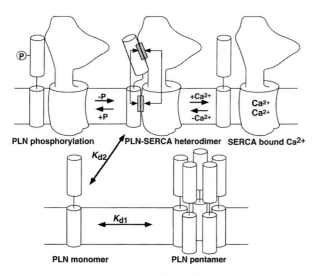

Figure 2.5 Regulatory features of PLN–SERCA interaction. Two steps can be dissected in the reversible inhibition of SERCA activity by PLN: the association–dissociation of pentameric PLN (K_{d1}); and the association–dissociation of monomeric PLN/SERCA (K_{d2}). PLN mutations can, potentially, alter both dissociation constants, leading to gain or loss of function, and SERCA mutations can affect K_{d2}, leading to gain or loss of function. Phosphorylation of PLN and Ca^{2+} binding to SERCA are driving forces for dissociation of the PLN–SERCA2a complex that activates SERCA. Phosphorylation of PLN dissociates functional interactions, but is less effective than Ca^{2+} binding in breaking up physical interactions.

cases, led to superinhibition of SERCA. By contrast, mutation of Asn27 and Asn30 in domain Ib, which also led to superinhibition, did not affect pentamer stability [63].

These observations demonstrate that the PLN monomer must be the active species, while the PLN pentamer must act as an inactive or less active reservoir (Figure 2.5) [5, 67]. They also suggested that superinhibition could be achieved in two ways. One mechanism would involve mass action, resulting from the 3- to 4-fold increase in the concentration of the inhibitory monomeric species, due to a change in the dissociation constant for the PLN pentamer. The other mechanism would involve diminished dissociation of the PLN–SERCA inhibited complex through a change in the affinity of the mutant PLN monomer for SERCA.

2.2.7
SERCA Residues Essential for Transmembrane Interactions with PLN

The discovery of a face on the PLN transmembrane helix that would interact with SERCA stimulated investigation of the complementary interacting face in transmembrane helices in SERCA. Since PLN inhibits SERCA1a and SERCA2a equally [40], mutagenesis of the amino acids in transmembrane helices M4, M5, M6 and M8 of SERCA1a that were likely to lie on the lipid-facing surface of the molecule were tested for their ability to interact with PLN [39]. Mutation of Leu321 in transmembrane helix M4 in SERCA1a and of Val795, Leu802, Thr805 and Phe809 in transmembrane helix M6 diminished the ability of PLN to inhibit Ca^{2+} transport activity [39], suggesting that these were critical residues in SERCA for interaction with PLN. Transmembrane helices M4 and M6 contribute strongly to Ca^{2+} binding [42, 43], but the helical faces involved in Ca^{2+} binding are different from those involved in interactions with PLN.

Attempts were then made to determine whether amino acids in the extension of M6 into the cytosolic loop between M6 and M7 might be involved in interaction with PLN [68]. Mutagenesis of Asp813 in SERCA1a led to a spectrum of results that were consistent with a site of interaction between Asp813 and Asn27 or Asn30 in PLN. However, the crystal structures of SERCA1a in the E1 and E2 conformation showed that interaction at this site is unlikely to occur [43, 44]. The Asp813 to Ala mutation disrupts hydrogen bonding to Asn755 in M5 and to Ser917 in L89. The extensive rearrangements in the stalk region that result are likely to be an indirect, rather than a direct cause of loss of the ability of PLN to interact with SERCA1a.

The crystal structure of SERCA1a in the E2 conformation also showed a close association of the cytosolic end of M4 with the cytosolic end of M6, leading to the possibility that Asn27 in PLN might interact with Leu321 in M4. This possibility was tested by disulfide cross-linking studies in which the PLN mutant N27C was co-expressed with the SERCA1a mutant L321C. The formation of a direct S–S cross-link, in the absence of added oxidant, demonstrated that these amino acids were directly apposed and likely to interact. In other experiments, cross-linking between the PLN mutant N30C and the natural Cys^{318} in SERCA2a was shown

[69]. In these experiments the cross-linker, bis-maleimidohexane, was sufficiently long to bridge about 10 Å between Cys side-chains, so it was not as readily inter-pretable as the N27C/L321C cross-link.

2.2.8
Structural Modeling of the PLN–SERCA Inhibitory Interaction

The finding of the N27C/L321C cross-link permitted the initiation of modeling of the transmembrane interaction between PLN and SERCA1a [4]. Modeling led di-rectly to the prediction that a cross-link would also form between V49C in PLN and V89C in SERCA1a. This prediction was confirmed by cross linking between PLN V49C and SERCA1a V89C, in the absence of added oxidant. These amino acids provided reference points from which structural modeling of the PLN/SERCA1a complex could proceed with confidence. In the model that was devel-oped (Figure 2.4), PLN domain II fits into a groove formed by M2, M4, M6, and M9 helices, on the lipid-facing surface of SERCA1a in the E_2 conformation. When SERCA1a moves to the $E_1.2Ca^{2+}$ conformation, [4, 43, 44], the groove be-comes much narrower due to a large movement of transmembrane helix M2, which prizes PLN away from its snug contact with SERCA.

The N27–L321 interaction site is also a potential collision point between the PLN helix and the M4 helix. As a result, the PLN helix must unwind upstream of this point in domain Ib. In this model, unwinding extends the length of the cytosolic domain of PLN so that it can fit into another groove in the cytosolic domain of SERCA1a, accounting for hydrophobic interactions between the PLN domain Ia helix and SERCA1a. In this cytosolic groove in SERCA1a, a salt bridge forms be-tween Lys3 in PLN and Asp399 in SERCA1a.

Although there is no clear evidence, it seems likely that the sites of phosphoryla-tion in PLN domain 1a would fit into the active sites of PKA or Ca^{2+}-CaM kinase when PLN is bound to SERCA, since phosphorylation disrupts the inhibitory PLN–SERCA interaction. In the model, there is a space between SERCA1a and much of domain 1b and part of domain 1a where kinases could fit. The crystal structure of PKA bound to a 20-amino-acid inhibitory peptide [70] is informative as to how PLN domain 1a must interact with PKA. By analogy, the domain Ia helix would have to unwind between Thr8 and Glu19 to fit into the active site of PKA.

In the transmembrane domains, PLN binds to amino acids in both transmem-brane helices M4 and M6, which are key helices in Ca^{2+} binding. If they cannot move, they cannot bind Ca^{2+}. PLN then extends into the cytosol to bind to the N-domain, presumably blocking its ability to participate in the clustering and dis-persion of the cytosolic domains, which power conformational changes in the transmembrane domain. This structural view of PLN, tying transmembrane he-lices together and extending into the cytosol to immobilize the N domain of SERCA, provides a visual illustration of the mechanism by which PLN inhibits SERCA in its E_2 conformation. Movement of M2 is induced by Ca^{2+} binding, so that M2 narrows the transmembrane groove in SERCA1a by occupying part of the space previously occupied by PLN. Thus, PLN is literally squeezed out of the transmembrane site to become non-inhibitory. It is not immediately obvious how PLN phosphorylation breaks up the PLN–SERCA interaction. Indeed, phosphory-lated PLN retains some physical interactions with SERCA [71].

An alternate structural model of PLN–SERCA interactions [72] is based on the premise that domain Ib and domain II of PLN form a continuous α-helix, as they do in an NMR structure [51]. This model is not very satisfactory because the distance constraints introduced by adherence to the NMR structure left the hy-drophobic portion of PLN suspended partially in and partially out of the mem-brane, in contrast to the seamless boundaries between water-accessible and water-inaccessible domains seen in the model of the PLN–SERCA1a complex shown in Figure 2.4.

2.3
Physiological Role of PLN in Basal Cardiac Function

2.3.1
Alterations in PLN Levels and Function by Transcription and Phosphorylation

PLN is expressed in high levels in mouse ventricles, at intermediate levels in atria and at low levels in smooth and slow-twitch skeletal muscles. The different levels of PLN expression in ventricular and atrial compartments correlate with differ-ences in the contractile parameters of these muscles [73].

PLN expression is also regulated during development. In mouse hearts, PLN transcript levels at birth were 30% of adult levels, approaching adult levels by day 15. PLN and SERCA2a transcript levels increased in a parallel manner, main-taining a constant PLN:SERCA2a stoichiometric ratio [74]. PLN protein levels are also regulated by the thyroid status of the heart, increasing in hypothyroidism and decreasing in hyperthyroidism. These alterations in PLN protein levels reflect par-allel alterations in the affinity of SERCA2 for Ca^{2+}, which, in turn, correlate with contractile parameters in hypothyroid and hyperthyroid hearts [75, 76].

PLN is phosphorylated *in vivo* during β-adrenergic stimulation of the heart. This typical "fight-or-flight" reaction is one of the main regulatory mechanisms by which cardiac output is increased. Studies in intact beating hearts and isolated car-diac myocytes show that both Ser16 and Thr17 in PLN become phosphorylated during isoproterenol stimulation. Phosphorylation of PLN increases the Ca^{2+}-affi-nity of SR Ca^{2+} transport. As a result, the amplitude of succeeding Ca^{2+} transients and the rate of their rise and fall are increased, leading to enhanced contractility. Dephosphorylation of PLN occurs mainly by a SR-associated type 1 phosphatase, which reverses the stimulatory effects of the β-agonists [77].

2.3.2
Targeting of PLN

The role of PLN in the regulation of cardiac function has been elucidated through targeting the PLN gene in embryonic stem cells and generating heterozygous and homozygous PLN-targeted mice. The heterozygotes expressed 60% of wild-type PLN levels, while the homozygotes expressed no PLN [9, 78]. These mice were indistinguishable from wild type at morphological and histological levels, indicating that ablation of the PLN gene was not associated with any cytoarchitectural abnormalities in the mouse heart.

The cardiac phenotype was then analyzed utilizing an integrative approach which combined studies at subcellular, organ and whole animal levels. Biochemical studies indicated that the affinity of SERCA2 for Ca^{2+} was increased significantly in the hearts of PLN-null mice. Physiological studies in single isolated cardiomyocytes revealed increased cell mechanics and Ca^{2+} kinetics, which agreed with the higher SR Ca^{2+} load and enhanced L-type Ca^{2+} channel currents observed in these PLN-deficient myocytes. The hyperdynamic cardiac function was also apparent at the whole organ level, when work-performing heart preparations were utilized under identical preload, afterload and heart rate (Figure 2.6). The enhanced cardiac function was confirmed in intact animals, using left ventricular catheter techniques and echocardiography. Thus, ablation of PLN was associated with significant increases in contractile parameters.

Modulation of cardiac function by PLN exhibited a gene–dosage relationship. The reduction in PLN levels was associated with linear increases in the affinity of SERCA2a for Ca^{2+} [9], and the rates of myocyte shortening and relengthening. Changes in contractility reflected similar increases in Ca^{2+} kinetics, such as the rates of increase and decrease of Ca^{2+} transients [79, 80]. A linear correlation was also observed between PLN levels and rates of contraction and relaxation in

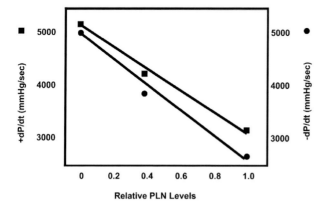

Figure 2.6 Linearity between cardiac PLN levels and contractility. Levels of PLN in the heart can be altered by ablation of PLN in one allele (heterozygote) or two alleles (homozygote). Contractility, measured as +dP/dt or −dP/dt, is increased linearly as PLN levels are decreased.

work-performing heart preparations [78] or intact mice (Figure 2.6) [81]. These results demonstrated that the ratio of PLN to SERCA2 is a critical determinant of cardiac SR function and basal contractility.

To examine the contribution of PLN in the response of the heart to β-agonists, PLN-null hearts or isolated myocytes were stimulated by isoproterenol and studied in parallel with wild-type animals. PLN-null myocytes, which were already operating at almost maximal rates of contraction and relaxation, were only minimally stimulated by isoproterenol [9, 82]. The attenuated responses were not due to alterations in the β-agonist pathway, including β-adrenergic receptor density, adenylyl cyclase activity, cAMP levels or phosphorylation levels of troponin I, C-protein or the 15 kDa sarcolemmal protein [82]. Thus, PLN is a major mediator of the β-adrenergic signaling pathway in the mammalian heart.

PLN ablation and the resultant hyperdynamic cardiac function were not associated with any alterations in the levels of SERCA2a, calsequestrin, Na^+/Ca^{2+} exchanger, myosin, actin, troponin I or troponin T [83]. However, Ca^{2+} release channel (ryanodine receptor) protein levels were decreased by 25 %, an important compensatory mechanism that would attenuate Ca^{2+} release from the SR, since the SR Ca^{2+} store was increased significantly in PLN-null hearts [80, 84]. These findings suggest that cross-talk between cardiac SR Ca^{2+}-uptake and Ca^{2+}-release occurred in an attempt to maintain appropriate resting Ca^{2+} levels and appropriate dynamics of Ca^{2+} transients in the hyperdynamic PLN knockout mice.

Another compensatory mechanism was an increase in the inactivation kinetics of the L-type Ca^{2+} channel current, since the SR Ca^{2+} store was significantly higher in the hyperdynamic PLN knockout hearts [85]. The levels of ATP, creatine kinase activity or creatine kinase reaction velocity were not altered, but ADP and AMP levels and the active fraction of mitochondrial pyruvate dehydrogenase were increased [83]. Thus, metabolic adaptations established a new energetic steady-state to meet the increased ATP demands in the hyperdynamic PLN-null hearts.

PLN deficiency was associated with significant enhancement of the rates of contraction and relaxation, as well as attenuation of contractile responses to β-adrenergic stimulation, due mainly to the "uninhibited" SR Ca-ATPase activity in these hearts. The phenotype of the PLN knockout hearts was preserved through aging, stress and other pathophysiological conditions. To determine whether the absence of PLN also reduced the reserve capacity of the murine cardiovascular system to respond to stress, the heart rate (HR), blood pressure, and metabolic responses of PLN-deficient mice to graded treadmill exercise (GTE) were evaluated. The exercise capacities of these mice, as measured by duration of exercise and peak oxygen consumption (VO_2), were normal. The oxygen pulse (VO_2/HR) curve was also normal in PLN mice, suggesting that an appropriate ability to increase stroke volume and oxygen extraction was maintained during exercise, despite an inability to increase cardiac contractility by β-adrenergic stimulation. Thus ablation of PLN, although resulting in diminished β-adrenergic inotropic reserve, does not compromise cardiac performance during maximal cardiovascular stress provided by graded treadmill exercise [86].

To determine whether this hyperdynamic function persists through the aging process, a longitudinal examination of age-matched PLN-deficient and wild-type mice was employed. Kaplan–Meier survival curves indicated no significant differences between PLN-deficient and wild-type mice over the first year (Figure 2.7). Examination of cardiac function revealed significant increases in the rates of contraction $(+dP/dt)$ and relaxation $(-dP/dt)$ in PLN-deficient hearts, compared with their wild-type counterparts at various time points over 24 months. Thus, the hyperdynamic cardiac function of the PLN-null hearts persisted throughout the process, without any alterations in heart-to-body mass ratio, cardiac cell length, or sarcomere length, and PLN ablation did not shorten life span [87]. These findings on the persistence of hyperdynamic cardiac function over the long term suggested that PLN may constitute an important target for treatment in heart disease.

PLN ablation did not compromise the ability of the heart to compensate for sustained aortic stenosis [88]. To determine whether the hyperdynamic PLN-null hearts could withstand a chronic aortic stenosis, the transverse section of the aorta was banded in PLN-null mice and their isogenic wild-type counterparts. Cardiac performance was followed with echocardiography in parallel in these animals and in sham-operated mice, before and at 2.5, 5 and 10 weeks after surgery. Cardiac decompensation was evidenced by the presence of lung congestion in some banded PLN-null and wild-type animals. The incidence of heart failure was not genotype-dependent. The development of left ventricular hypertrophy was similar between PLN-null and wild-type animals and longitudinal assessment of end-diastolic dimension indicated progressive increases after banding, with a greater dilation in failing mice. Fractional shortening was reduced in failing PLN-null and wild-type animals to a similar degree. The incidence of heart failure was similar between PLN-null and wild-type "banded" mice, although an increased pressure gradient across the band of the transverse aorta was observed in PLN-null animals

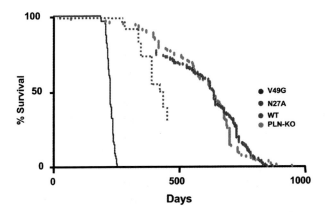

Figure 2.7 Longevity of mice with genetically altered PLN. PLN ablation, resulting in hyperdynamic cardiac function, does not compromise the life span of mice. In contrast, superinhibition of SERCA2 and cardiac contractility by the transgenic overexpression of PLN mutants N27A and V49G results in early mortality, which is more pronounced in V49G transgenic mice.

[88]. However, PLN ablation increased cardiac susceptibility to ischemic injury, with females being less susceptible than males [89]. This gender-specific protective effect appeared to be mediated by the ability of females to produce more nitric oxide [89], a potent vasodilator.

Collectively, these studies of *PLN* targeted animals indicate that PLN is a critical regulator of basal cardiac Ca^{2+} cycling and contractile parameters. The studies also show that PLN is a critical determinant of β-adrenergic agonist responses *in vivo*.

2.3.3
Role of PLN in Smooth and Skeletal Muscles

PLN is also expressed in very low levels in smooth and slow-twitch skeletal muscles [90, 91]. Ablation of PLN in mouse aorta was associated with significant decreases in sensitivity to potassium chloride and phenylephrine stimulation compared to wild types [90]. PLN is also present in vascular endothelium and it appears to modulate endothelium-dependent relaxation of the aorta [92]. In soleus muscles, ablation of PLN was associated with significant increases in relaxation rates, without any effects on contraction rates. Isoproterenol had no effect on contraction rates but stimulated the relaxation rate of wild-type soleus to the high basal level observed in PLN-deficient soleus [91]. Thus, PLN is a regulator of contractility in smooth and slow-twitch skeletal muscles [90, 91, 93, 94].

As indicated above, *in vitro* studies have shown that PLN can also interact and regulate SERCA1, the fast-twitch skeletal muscle isoform. Consistent with these observations, ectopic expression of PLN in mouse fast-twitch skeletal muscle decreased the Ca^{2+}-affinity of SERCA1 and depressed rates of skeletal muscle relaxation. Furthermore, the prolongation of muscle relaxation correlated with the expression levels of PLN in transgenic fast-twitch skeletal muscle [95]. These studies confirm that SERCA1 activity can be regulated by PLN *in vivo*, just as it has been shown to be regulated by PLN *in vitro*.

2.3.4
Overexpression of PLN

PLN gene targeting studies revealed that the stoichiometric ratio of PLN:SERCA is a critical determinant of the regulation of Ca^{2+}-cycling and contractility in cardiac muscle. Therefore, an important question in evaluating PLN function is the functional stoichiometry between PLN and SERCA2a. To determine this stoichiometry and to determine whether all of the SR Ca^{2+}-pumps are functionally regulated by PLN *in vivo*, transgenic mice were generated which overexpress wild-type PLN specifically in the heart, under the control of the α-myosin heavy chain promoter [10]. Transgenic mice expressing PLN at two-fold wild-type levels in the heart showed no morphological abnormalities and no alteration in heart:body weight, heart:lung weight or cardiomyocyte size. However, the Ca^{2+}-affinity of the SR Ca^{2+} transport system was diminished significantly, without an alteration in the V_{max} of Ca^{2+} transport. When the relative levels of PLN to SERCA were plotted against the

$EC_{50}s$ of SERCA for Ca^{2+} in PLN-overexpressing, wild-type, PLN-heterozygous and PLN-homozygous hearts, a close linear correlation was obtained, indicating that the overexpressed PLN was coupled functionally to SERCA [11].

The reduced Ca^{2+}-affinity of SERCA resulted in decreases in the amplitude of the Ca^{2+} transient and prolongation in the rate of decay of the Ca^{2+}-transient in Fura-2 loaded transgenic cardiomyocytes. The depressed Ca^{2+}-cycling parameters were associated with diminished contractile parameters in cardiomyocytes (Figure 2.8). The decreased contractility in isolated myocytes was consistent with attenuated function in intact mice, assessed by echocardiography [10]. However, the PLN inhibitory effects associated with depressed contractile parameters were abolished upon isoproterenol stimulation (Figure 2.8). These findings suggest that a fraction of SERCA pumps in native SR is not regulated by PLN.

To determine the PLN levels under which maximal functional inhibition of SERCA occurs, transgenic mice with 4-fold PLN levels were generated [96]. Isolated myocytes from these transgenic hearts exhibited a robust depression of intrinsic contractile parameters and Ca^{2+} cycling, which were abolished by administration of the β-adrenergic receptor agonist, isoproterenol. In contrast, cardiac function *in vivo* was depressed substantially by administration of the β-adrenergic receptor antagonist, propranolol, suggesting that enhanced sympathetic drive acted as a compensatory mechanism. Assessment of plasma norepinephrine levels revealed a significant elevation, which was associated with enhanced PLN phosphorylation, relieving PLN inhibitory function. However, upon aging, there was desensitization of the β-adrenergic pathway and PLN phosphorylation was reduced. This

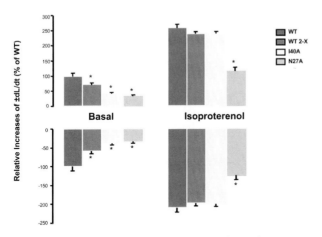

Figure 2.8 Effect of PLN levels and isoproterenol stimulation on mouse cardiac myocyte contractility. The two-fold overexpression of wild-type PLN or I40A or N27A PLN mutants results in significant depression of myocyte contractile parameters (+dL/dt: rate of shortening; −dL/dt: rate of relengthening). β-Agonists, such as isoproterenol, increase the contractility of wild-type cardiac myocytes, measured as +dL/dt or −dL/dt. The depressed contractile parameters in myocytes from transgenic mice overexpressing wild-type PLN or superinhibitory I40A PLN are fully relieved by isoproterenol stimulation. However, the superinhibitory N27A PLN mutant does not respond fully to isoproterenol.

resulted in contractile dysfunction and remodeling, which progressed to heart failure and early mortality [96]. Early mortality was gender-dependent and appeared to be regulated by the levels of p38 and the MKP-1 phosphatase [97].

These findings suggest that increased inhibition of SERCA2a by PLN results in depressed SR Ca^{2+}-cycling and contractility, which serves as a stimulus for increased adrenergic drive. Initially, this serves as an important compensatory mechanism, but ultimately leads to cardiac remodeling and congestive heart failure. The elevated catecholamines, which attenuated the PLN inhibitory effects in wild-type PLN overexpressing hearts, did not allow assessment of the functional PLN:SERCA2a stoichiometry.

2.3.5
Physiological Role of PLN in β-Adrenergic Stimulation

A different approach to the assessment of functional PLN:SERCA2a ratios was to overexpress different levels of a nonphosphorylatable form of PLN in which both phosphorylation sites Ser16 and Thr17 were mutated to Ala and could not be phosphorylated [98]. The apparent affinity of SERCA2a for Ca^{2+} was maximally inhibited in this model at a functional stoichiometry of 2.6 PLN/1.0 SERCA2a, indicating that ca. 40% of the SR Ca^{2+} pumps are regulated functionally by PLN under normal conditions in mouse hearts[98]. The measured PLN:SERCA2a molar ratio in canine hearts lies between 2 and 2.5:1[99]. On the assumption that PLN exceeds SERCA2a by 2.5-fold and that only 40% of SERCA2a is normally inhibited, then for every PLN monomer that interacts with SERCA2a about five monomers do not. At least 90% of these are likely to form into homopentamers [5].

The functional significance and interplay of dual site phosphorylation of PLN was also determined through the generation and characterization of transgenic mice expressing phosphorylation-site specific PLN mutants in the null background. The S16A mutant hearts exhibited a depressed response to isoproterenol and lack of Thr17 phosphorylation. The T17A hearts exhibited Ser16 phosphorylation and a response to isoproterenol similar to that observed with wild types [100, 101]. These findings suggested, first, that Ser16 phosphorylation may be a prerequisite for Thr17 phosphorylation; and, second, that Ser16 phosphorylation may be sufficient to mediate the contractile responses of the heart to β-agonist stimulation. However, Thr17 phosphorylation has been observed to be independent of Ser16 phosphorylation in hearts perfused with elevated Ca^{2+} (in the presence of okadaic acid to inhibit phosphatase activity) [36], in hearts recovering from an ischemic insult [102] and in isolated cardiomyocytes subjected to frequency-dependent increases of contractile parameters [103]. Thus, phosphorylation of Thr17 in PLN may be important when stress is applied directly to the heart, while Ser16 phosphorylation may mediate the β-agonist responses *in vivo*.

2.3.6
Superinhibitory PLN Mutants

The question of whether superinhibitory mutations in PLN might impair cardiac function was tested in a series of transgenic mice. Overexpression of the superinhibitory PLN mutants, N27A, L37A, I40A and V49G, was associated with increased inhibition of the affinity of SERCA2a for Ca^{2+} [6–8, 15]. Depressed SERCA2a function resulted in significant decreases in the temporal pattern and amplitude of the Ca^{2+} transients in cardiac myocytes and in the extent and temporal pattern of contraction and relaxation of the myocytes (Figure 2.8).

The depression of myocyte Ca^{2+} kinetics and mechanics, observed with the monomeric L37A and I40A PLN mutants, was reversed by isoproterenol, in line with the view that superinhibition with these mutants is mainly due to mass action resulting from increases in the concentration of the active PLN monomer. Periodic reversal of the superinhibitory effects of these mutants by the elevation of endogenous β-agonists probably allows the mice to survive. Although life span did not appear to be shortened for these animals, significant left ventricular hypertrophy was noted by echocardiography and female mice often died with greatly enlarged hearts during delivery of second or third litters.

The strongest superinhibitory mutant yet noted in heterologous cell culture is N27A [63]. Yet neither N27A nor another superinhibitory mutant, V49G [8], alter the monomer: pentamer ratio. These observations indicate that their efficacy must be due to an enhanced affinity for SERCA and this is confirmed by structural modeling studies [4]. Interactions between Asn27 in PLN and Leu321 in SERCA1a and between Val49 in PLN and Val89 in SERCA1a have been shown to be very close, and mutation of any of these residues has significant effects on PLN–SERCA affinity [4]. With N27A and V49G mutants, impaired functions could not be reversed fully with the experimental application of isoproterenol. Accordingly, PLN becomes a chronic inhibitor and SERCA 2a no longer has the potential to be fully functional. Mice that overexpress the N27A mutation on the PLN-null background, survive for less than one year, dying of dilated cardiomyopathy (Figure 2.7) [104], while transgenic males overexpressing V49G die of dilated cardiomyopathy at 6 months (Figure 2.7) [8].

Analysis of the various PLN mutants described above suggests that PLN can be viewed as an unnecessary encumbrance to a healthy functional heart, since mice live happily without it. It becomes a problem, however, if it is overexpressed or mutated to a superinhibitory form. If the inhibitory function of PLN can be reversed by endogenous β-agonists, life is not threatened, but dilated cardiomyopathy occurs if the mutation is highly inhibitory and is not reversible by endogenous β-agonists.

2.4
Phospholamban in Heart Failure

2.4.1
Introduction

The abnormal Ca^{2+}-cycling in animal models of heart failure and human failing hearts has been suggested to reflect, at least in part, impaired Ca^{2+}-uptake by the SR [105–107]. Investigations of muscle strips using myothermal studies have revealed that the amplitude of the Ca^{2+} transient and the rate of Ca^{2+} cycling are diminished in the failing human myocardium [108]. Accordingly, studies in isolated ventricular strips and myocytes have revealed that abnormal Ca^{2+} transients are present in failing cardiac tissue. During systole, the amplitude of the Ca^{2+} transient is reduced and, during diastole, free Ca^{2+} levels are elevated [105, 109]. The duration of the Ca^{2+} transient is also prolonged in failing human hearts [110].

Not surprisingly, abnormalities in Ca^{2+} cycling correlate well with the degree to which contractile force is diminished [111]. Failing hearts also exhibit a diminished or flat response to increasing frequency of stimulation [109], unlike normal hearts, which continue to produce greater force as the frequency of stimulation is increased; this phenomenon is known as a positive "Treppe" or a positive force–frequency relationship. Failing hearts do not exhibit this phenomenon, primarily because the reserve of SR function is very limited. Indeed, under increasing frequency of stimulation, this diminished SR functional reserve is reflected in a depressed ability to enhance intracellular Ca^{2+} cycling and, consequently, a decreased ability to generate increasing force with increased frequency of stimulation [109, 112].

In accordance with these findings, biochemical studies have indicated that both the V_{max} of Ca^{2+} uptake and the Ca^{2+}-affinity of SERCA are diminished in failing human hearts, compared to healthy donor hearts [113, 114]. Numerous studies have attempted to link this decrease in SR Ca^{2+} transport activity to changes in levels of key SR proteins. The level of SERCA2a mRNA is reduced in failing compared to non-failing hearts [115, 116]. Most findings at the protein level indicate that the level of PLN protein remains unchanged, whereas the level of SERCA2a protein decreases by up to 40 % in end-stage dilated cardiomyopathy or ischemic cardiomyopathy [107, 114, 117–119]. There appears to be a significant correlation between the decrease in SERCA2a protein and the extent of myocardial dysfunction [119]. A decrease in the level of SERCA relative to PLN would be expected to lead to an increased functional stoichiometry of PLN to SERCA, increased inhibition of the Ca^{2+} affinity of SERCA2a, and prolonged relaxation. Indeed, findings in transgenic mice demonstrate clearly that the functional stoichiometry of PLN to SERCA is a key determinant of cardiac function [9], and that increased inhibition of SERCA by PLN reflects a major SR defect in human heart failure (Figure 2.9) [96]. In addition, the phosphorylation status of PLN at Ser16 and Thr17 is decreased in failing hearts [114, 120], indicating an increased inhibition by PLN.

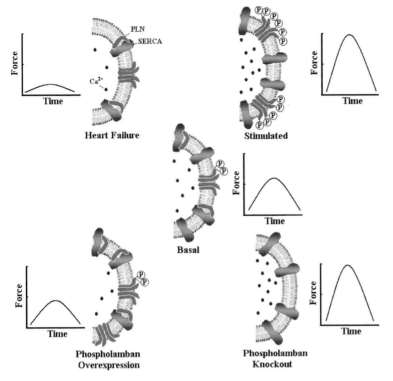

Figure 2.9 Model of the ways that PLN–SERCA2a interactions are altered in mutant mice and in heart failure. Under basal conditions (Basal), dephosphorylated PLN interacts with SERCA2a to lower its affinity for Ca^{2+}, thereby lowering Ca^{2+} pump activity at low Ca^{2+} concentrations, and lowering the SR Ca^{2+}-load, which is available for release during cardiac muscle contraction. Phosphorylation of PLN (Stimulated) dissociates the inhibited PLN–SERCA complex, stimulating Ca^{2+} transport into the lumen of the SR and increasing the rate of relaxation. Since more Ca^{2+} is accumulated in the SR lumen, a greater Ca^{2+}-store is available for release in a subsequent beat, resulting in enhanced contractile force. PLN ablation (Phospholamban knockout) has stimulatory effects on Ca^{2+} transport that are similar to PLN phosphorylation. In contrast, higher levels of the active PLN monomer that accompany the overexpression of wild-type PLN (Phospholamban overexpression) are more inhibitory to SERCA activity and contractility. In heart failure (Heart failure), SERCA2a expression, but not PLN expression, is diminished, altering the functional PLN:SERCA ratio and resulting in higher inhibition. PLN is also less phosphorylated in heart failure, becoming more inhibitory overall. Phosphorylation, in dissociating the monomer from SERCA, may enhance pentamer formation.

2.4.2
Potential Therapies

Decreased phosphorylation of PLN can be attributed to the combined effect of decrease in PKA activity and a ~2.5- fold increase in SR-associated protein phosphatase 1 (PP1) activity in human heart failure [121]. Mice overexpressing the catalytic subunit of PP1 have a marked decrease in PLN phosphorylation that recapitulates the phenotype observed in heart failure [122]. Moreover, the adenovirus-mediated expression of a constitutively active protein phosphatase 1 inhibitor-1 (I-1) in failing human cardiomyocytes increased the contractility of failing human cardiomyocytes significantly following administration of isoproterenol [122]. Infection of adult and neonatal rat cardiomyocytes with an adenovirus encoding the full-length inhibitor-1 was also associated with a marked increase in PLN phosphorylation at Ser16 and an increase in cardiac contractility [123]. Thus, increasing the phosphorylation of PLN at Ser16 may represent a promising therapeutic target in the treatment of heart failure.

Another potential therapeutic strategy to restore perturbed Ca^{2+} uptake into the SR is to increase SERCA2a levels or to attenuate the inhibitory effect of PLN on SERCA2a. Overexpression of SERCA2a, using recombinant adenovirus-mediated gene transfer, led to enhancement of contractility in isolated failing human cardiomyocytes [124, 125] and improved cardiac function, metabolism and survival in a rat model of heart failure [126, 127]. Transgenic overexpression of SERCA2a was also associated with increases in both SR Ca^{2+} uptake and contractility [128, 129]. These studies support the notion that an enhanced expression of SERCA2a restores disturbed intracellular Ca^{2+} handling by decreasing the relative ratio of PLN to SERCA2a.

In addition, the overexpression of SERCA1a in transgenic mice [130] led to markedly enhanced cardiac contractility, lending further support to the hypothesis that increasing SERCA levels and activity can be beneficial to the heart. The enhanced cardiac function of SERCA overexpressing mice persisted as the mice aged [131], indicating that, at least in murine species, the increased energy consumption of hypercontractile hearts do not have long-term deleterious effects.

Other potential approaches to influence the PLN–SERCA2a complex may utilize such techniques as expression of PLN antisense RNA or expression of PLN mutants to decrease PLN levels or activity. Such studies, utilizing adenoviral-mediated gene transfer in isolated cardiac myocytes, demonstrated increased rates of Ca^{2+} transient decay [128, 132, 133]. Infection of failing human cardiomyocytes with PLN anti-sense mRNA resulted in contractile properties similar to cardiomyocytes infected with SERCA2a [133]. In addition, overexpression of the PLN-dominant negative mutants, K3E/R14E or V49A, in cardiomyocytes improved SR function and myocyte contractility [12, 13]. Though these findings are promising, the notion that targeting PLN at the cellular level or expressing dominant negative forms of PLN can enhance cardiac contractility has yet to be tested in larger mammalian species.

In some murine models of heart failure rescue, PLN ablation has been shown to rescue the animal from the disease phenotype. PLN ablation prevented the functional, structural and histological abnormalities in cardiomyopathic mice lacking muscle Lim protein (MLP), which exhibit heart failure [12]. Through restoration of appropriate SR Ca^{2+} cycling, a multitude of structural and functional defects in the MLP-null mice were prevented. It was proposed that enhanced SR Ca^{2+} cycling in the PLN-null/MLP-null model decreases wall stress on the mouse hearts and mitigates wall stress-associated activation of MAP-kinase pathways, which ultimately results in cardiac hypertrophy and dilated cardiomyopathy [134]. These studies have a counterpart in experiments in which inhibition of PLN activity by cardiac-specific expression of the PLN S16E mutant protein (which might mimic the phosphorylated, non-inhibitory form of PLN) resulted in improved function and delayed progression to heart failure significantly in a hamster cardiomyopathic model [14]. PLN ablation has also been shown to rescue the depressed function of calsequestrin-overexpressing hearts [135] and to restore the basal contractility and exercise performance of a model with cardiac overexpression of a mutant myosin heavy chain [136].

Success in the rescue of heart failure by PLN ablation in murine models has not always been achieved. Studies aimed at correcting depressed SR Ca^{2+} cycling and contractility by PLN ablation in a mouse overexpressing both a G-protein involved in β-adrenergic signaling, Gαq, and a mutant myosin binding protein C that causes familial hypertrophic cardiomyopathy in humans, were less successful; cardiomyocyte contractility was rescued, but no improvement was observed in whole-heart contractile function or in cardiac dilation [137]. These findings provide the very important insight that PLN ablation may be beneficial only selectively as a therapy for heart failure and may even depend on the etiology of heart failure. Moreover, it has become clear that rescuing cardiomyocyte contractility does not necessarily rescue heart failure.

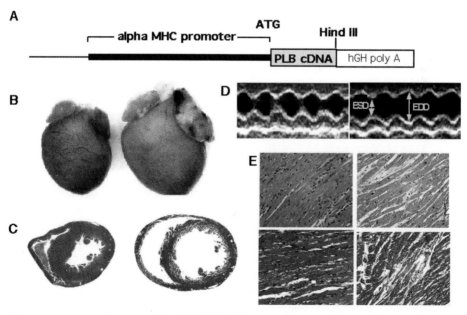

Figure 2.10 Effects of overexpression of the R9C PLN mutant. Chronic inhibition of SERCA2a by overexpression of the PLN mutant, R9C, results in: (A) cardiac hypertrophy; (B) cardiac function (monitored by M-mode echocardiography), which is more impaired than that observed when similar levels of wild-type PLN are overexpressed; (C) dilated cardiomyopathy and (D) increased wall thickness and interstitial fibrosis observed in longitudinal sections of hearts stained with Masson's trichrome. (E) EDD: end diastolic dimension; ESD: end systolic dimension.

Thus, some studies have provided a clear demonstration of the beneficial effects of improved SR Ca^{2+}-cycling through PLN ablation on cardiac function and remodeling, while others have shown that normalization of myocyte Ca^{2+}-handling may not translate into improved cardiac function *in vivo* or into reversal of remodeling. However, these findings should be interpreted cautiously, since they are based on results in genetically altered mouse models where compensatory mechanisms, arising throughout early developmental stages, may contribute to or mask the observed phenotypes.

2.5
Human PLN Mutations as a Cause of Cardiomyopathy

2.5.1
PLN R9C Mutant

In recent years, human hypertrophic or dilated cardiomyopathy has often been associated with a variety of naturally occurring heritable mutations. In attempts to identify the genetic variation in SR proteins that, alone or in combination with environmental influences, may modulate cardiac function and progression to heart failure, the *PLN* gene was examined in human patients with dilated cardiomyopathy. Two naturally occurring mutations in the coding region of the human *PLN* gene have been identified to date.

The inheritance of the PLN mutation, R9C, was linked to the dominant inheritance of dilated cardiomyopathy in a large American family with a probability of 10,000 to 1 that the mutation was causal of the disease (lod score over 4 at $\Theta = 0.00$) [15].

The effects of the PLN R9C mutation were characterized by expression in heterologous cell culture, by the creation of a transgenic mouse and by analysis of cardiac tissue obtained from an explant (Figure 2.9). In all cases, the level of PLN phosphorylation was reduced dramatically. In heterologous cell culture, the R9C mutant PLN had reduced inhibitory properties in the homozygous state, but, when expressed in the heterozygous state, did not act in a dominant fashion to prevent the inhibition of SERCA2a by wild-type PLN. The key effect of the mutation was enhancement of the affinity of R9C mutant PLN for PKA. In attempting to phosphorylate mutant PLN, PKA becomes trapped in a stabilized mutant-PLN/PKA complex and can no longer dissociate to phosphorylate wild-type PLN molecules. The effect appears to be local and to be restricted to the SR, perhaps because a specific fraction of PKA is associated with this membrane through A-kinase anchoring proteins (AKAPs) [138]. These results explain the dominant effects of the mutation. Affected individuals must go through life with chronically inhibited SERCA2a and can never draw upon their full cardiac reserve. The long-term effects of the chronic, specific inhibition of PLN phosphorylation on Ca^{2+} transients are clearly sufficiently deleterious to cause the onset of dilated cardiomyopathy in humans in their teenage years.

While these findings were recapitulated in a PLN R9C overexpressing transgenic line (Figure 2.10, see p. 49) [15], analysis of the transgenic expression of a PLN S16A-T17A double mutant in mice, which should also lead to chronic PLN inhibition provided different results. This mutant PLN produced the same shift in Ca^{2+} affinity as unphosphorylated, wild-type PLN, but could not be phosphorylated. Measures of basal contractility and Ca^{2+} transients were similar in wild-type and mutant myocytes, but isoproterenol did not increase the rate of Ca^{2+} removal in the mutant. Increased L-type Ca^{2+} current (I_{Ca}) density, with unaltered characteristics, turned out to be the major compensatory response in these mutant myocytes. The modulation of I_{Ca} may compensate partially for the loss in SERCA2a responsiveness, thereby providing partial normalization of β-adrenergic inotropy in S16A-T17A double mutant mice, permitting them to survive without cardiomyopathy [139].

2.5.2
PLN L39stop Mutant

In the second PLN mutation that has been associated with dilated cardiomyopathy, a termination codon is substituted for the Leu39 codon (L39stop) [16]. This mutation, discovered in two large Greek families, truncates the 52 amino acid PLN protein near the middle of the highly conserved transmembrane domain II.

In the first family, the heterozygous inheritance of the L39stop mutation led to left ventricular hypertrophy in one-third of the older members of the family, without diminished contractile performance. However, the recessive inheritance of two copies of the mutant *PLN* gene led to dilated cardiomyopathy and heart failure in two teenage siblings. In heterologous expression in HEK-293 cells, the L39stop mutant protein was unstable and was undetectable in the ER. As a result, there was no effect of the mutant protein, in either the homozygous or heterozygous state, on the Ca^{2+} affinity of SERCA2a. A very small amount of the truncated protein survived in non-ER fractions and was shown by confocal microscopy to be mislocated. In further experiments, myocyte contractility and Ca^{2+} cycling were not altered in isolated adult rat myocytes infected with adenoviral vectors containing L39stop PLN cDNAs.

In a cardiac explant, no PLN protein was detected. Nevertheless, *PLN* mRNA was readily detectable in the myocardium of PLN L39stop homozygotes, although at lower levels than those measured in normal or idiopathic cardiomyopathy hearts. Histopathologic analysis of both explanted hearts revealed significant interstitial fibrosis and myofibrillar disarrangement. Clinical history and physical examinations of the other family members failed to elicit symptoms or signs of heart failure. Taken altogether, these two individuals can be considered to be equivalent to a PLN-null genotype with a phenotype of dilated cardiomyopathy. Thus, in contrast to the benefits of PLN ablation in mouse, humans lacking PLN develop lethal cardiomyopathy. A caveat in these studies is that the number of affected individuals is very low so that the lod score for linkage vs. non-linkage cannot be used to support the causal nature of the L39 stop mutation in dilated cardiomyopathy.

A second family carrying the PLN L39stop mutation contained several heterozygotes, one of whom had left ventricular hypertrophy. Within the family, a father of unknown genotype died prematurely of dilated cardiomyopathy and two of his children, with heterozygous genotypes, exhibited dilated cardiomyopathy with an ejection fraction of 20 25 %. The mother and the mother's family did not carry any detectable mutation in the *PLN* gene. The variable expression of the clinical phenotype, elicited by a single copy of the L39stop mutation, was associated with hypertrophy in some members of both families and with overt dilated cardiomyopathy in another. This indicates that environmental perturbations, an undetected mutation in a second gene or an additional stress that is, as yet, undetected can contribute to the mechanism by which a reduction in the content of functional PLN can induce cardiac dysfunction. The findings with humans suggest that a fine balance between the degree of SERCA2a inhibition and augmentation by PLN is essential for normal cardiac function.

So, how can this discrepancy between the cardiac phenotypes in mice and humans be explained? In contrast to humans, mice have relatively little cardiac reserve and differ in the balance of myocyte Ca^{2+} fluxes (Figure 2.11). With heart beat rates ranging up to 800 bpm, mice are normally at $\frac{2}{3}$ of the theoretical maximum and the SR Ca^{2+} store is nearly full at normal heart rates. Humans, with

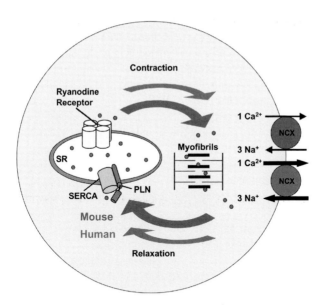

Figure 2.11 Differences in the regulation of Ca^{2+} transients between mice and humans. With heart beat rates ranging up to 800 bpm, mice are normally at $\frac{2}{3}$ of the theoretical maximum and the SR Ca^{2+} store is nearly full at normal heart rates. Mice cycle 90 % of their cardiac Ca^{2+} through the SR and only 10 % through extracellular spaces in a very short temporal cycle. Humans, with heart beat rates ranging between 60 and 180, have a huge cardiac reserve. A major part of this reserve is the potential for a gain in SR Ca^{2+} content. Normally humans cycle only 70 % of their Ca^{2+} through the SR in a longer temporal cycle. Darker arrows and NCX, mouse; lighter arrows and NCX, humans; red dots, Ca^{2+} ions.

heart beat rates ranging between 60 and 180, have a huge cardiac reserve [23]. A major part of this reserve is the potential for a gain in SR Ca^{2+} content [140]. If defective PLN regulation limits the normal gain in SR Ca^{2+} content with increasing heart rate, cardiac reserve will be compromised, as observed in human heart failure [140]. Thus, PLN modulation may be of more paramount importance in humans, which have a life span 40 times that of a mouse.

2.6
Sarcolipin

2.6.1
Introduction

Sarcolipin was discovered as a protein that copurifies with SERCA1 [17]. It was first isolated as a proteolipid by its extraction into acidified chloroform–methanol and shown to have a mobility in SDS-PAGE corresponding to a mass of about 6 kDa. Sequencing of the rabbit proteolipid [141] and subsequent cloning of cDNA and genomic DNA from different sources [19] showed it to be a 31 amino acid, transmembrane protein, which was named sarcolipin (SLN). The 7 amino acid, hydrophilic, cytosolic N-terminal sequence is poorly conserved among species, but the 19 amino acid, hydrophobic, transmembrane sequence and a more hydrophilic, 5 amino acid, C-terminal sequence are well conserved. Alignment of the transmembrane sequences of PLN and SLN showed considerable identity, indicating that they are homologous proteins (Figure 2.2) [20]. This view was supported by the finding that *SLN* and *PLN* genes have similar structures and that expression of both genes is muscle-specific [20].

Early attempts to determine a function for the proteolipid were contradictory. When purified proteolipid was reconstituted with preparations of SERCA1a that were deficient in proteolipid, the ratio of Ca^{2+} translocation to ATP hydrolysis increased over 5-fold [142]. However, the proteolipid abolished Ca^{2+} accumulation, acting like an ionophore when added to the exterior of reconstituted liposomes. In another investigation of the role of sarcolipin in coupling processes, however, these results were not confirmed and no effects on coupling ratios were observed [142].

In more recent studies [143], when pure SLN was reconstituted with SERCA1a in defined lipid bilayers, it had no significant effect on the rate of ATP hydrolysis or on the Ca^{2+}-dependence of ATP hydrolysis, even at a 50:1 molar ratio of SLN:SERCA1a. However, at molar ratios of 2:1 or higher, SLN decreased Ca^{2+} accumulation by reconstituted vesicles. These effects could be simulated by assuming that SLN increases the rate of slippage on the ATPase and the rate of passive leak of Ca^{2+} mediated by the ATPase. It was suggested that the presence of SLN could be important in non-shivering thermogenesis in which heat is generated by hydrolysis of ATP by SERCA1a. In other studies using reconstitution, however, both SLN and PLN were shown to be inhibitors of SERCA1a [144].

To circumvent the controversy generated by the differing results obtained through reconstitution of SLN and SERCA in lipid bilayers, interactions between the two proteins have been studied by coexpression with SERCA and measurement of altered SERCA function in isolated microsomes. These studies have shown that SLN, like PLN, is a reversible inhibitor of SERCA activity and that its affect, like that of PLN, is to diminish Ca^{2+} affinity of SERCA [20, 21].

Because it has proven difficult to prepare antibodies against SLN, most functional studies have been carried out with rabbit SLN fused N-terminally with the Flag epitope (NF-SLN) [20]. Co-expression of SERCA1a in HEK-293 cells with either native rabbit SLN or NF-SLN led to decreases in the apparent affinity of SERCA1 for Ca^{2+} that were similar to those observed by co-expression with PLN. Anti-Flag antibody reversed the inhibition of Ca^{2+} uptake by NF-SLN at low Ca^{2+} concentrations. Although early experiments indicated that SLN also increased V_{max} for Ca^{2+} uptake [20], this is likely to be incorrect and to arise from difficulty in measuring levels of expression of SLN [21]. NF-SLN was found to be a more effective inhibitor of SERCA2a than of SERCA1a: It was not only equal to PLN in its ability to decrease apparent Ca^{2+} affinity, but it also decreased V_{max} at pCa 5 [21].

When SERCA1a or SERCA2a were co-expressed with both NF-SLN and PLN, inhibition was synergistic, reducing ΔK_{Ca} by about 1 pCa unit and V_{max} at pCa 5 by 20–45 % [20, 21]. Co-immunoprecipitation of a ternary complex showed that NF-SLN increased the binding of PLN to SERCA dramatically. Co-immunoprecipitation of PLN and SLN without SERCA showed that SLN inhibits the formation of PLN pentamers and that SLN binds directly to PLN in a binary complex [21]. Since SLN inhibits the formation of PLN pentamers, it is likely that the PLN–SLN binary complex is highly stable and is likely to predominate when these two proteins are brought together at equimolar concentrations under equilibrium conditions. Co-immunoprecipitation also showed that elevated Ca^{2+} dissociates both PLN and SLN from their binary complexes with both SERCA1a and SERCA2a, but is much less effective in dissociation of the ternary SLN/PLN/ SERCA complex.

2.7
Physiological Role of SLN

2.7.1
SLN Expression

SLN expression in humans is predominantly in muscle and is largely complementary to PLN expression [19]. SLN is expressed most abundantly in fast-twitch skeletal muscle, to a lower extent in slow-twitch skeletal muscle, and to an even lower extent in cardiac muscle, but it is also expressed in trace amounts in pancreas and prostate [19]. In rat, however, the pattern of expression is different [145], since SLN is expressed to a low extent in fast-twitch EDL muscle and nearly 18-fold higher in both diaphragm (mixed fast and slow) and cardiac muscles. In mouse and human heart, SLN expression is restricted to the atrium and, in the mouse model, SLN expression is increased during embryonic development and persists throughout the life span [22]. By contrast, PLN is largely restricted to the ventricle. Thus the atrial-specific expression of SLN may be related to functional differences between atria and ventricles and SLN may play an important role in regulation of Ca^{2+} cycling in the atrium. Although SLN expression in skeletal muscle seems to mimic the expression pattern of SERCA1a, this is not true in the heart, where there is abundant expression of SLN in the atrium, but very low expression of SERCA1.

2.7.2
Overexpression of SLN

NF-SLN was overexpressed in rat soleus muscle, by intramuscular injection and electrotransfer of rabbit NF-SLN cDNA, to explore the possibility that SLN could regulate SERCA2a activity and slow-twitch soleus muscle contractility, just as PLN can regulate cardiac contractility. NF-SLN reduced peak isometric force, slowed the rates of contraction and relaxation and increased susceptibility to fatigue. The V_{max} of Ca^{2+} uptake in postnuclear homogenates from these muscles was also reduced. Thus NF-SLN seems to impair muscle contractile function indirectly by inhibiting SERCA function and thus lowering basal Ca^{2+} stores in the sarcoplasmic reticulum.

2.7.2.1 Response of the SLN Gene to Chronic Stimulation

Ca^{2+} ATPase and Ca^{2+} uptake activities are decreased by 30–50%, without an alteration in the amount of Ca^{2+} ATPase protein, after 3–4 days of chronic low-frequency stimulation of rabbit fast-twitch skeletal muscle [146]. After chronic low-frequency stimulation for 6 to 7 weeks, fast-twitch, fast-fatiguable muscles transform into slow-twitch, fatigue-resistant muscles [147]. Enhanced fatigue-resistance, absence of twitch potentiation and prolonged contraction and relaxation times are associated with significant increases in Na^+,K^+ ATPase, PLN and SERCA2a concentrations, and significant decreases in SERCA1, ryanodine receptor, dihydropyridine receptor and triadin concentrations. In the first 3–4 days, however, SERCA1 mRNA and protein levels were unaltered after stimulation. In contrast, SLN mRNA was decreased by 15% and SLN protein was reduced by 40% [20]. At the time of these studies, it appeared that SLN might increase SERCA1a activity, and that loss of SERCA1a activity might be associated with diminished SLN content. However, further study has shown that SLN functions only as an inhibitor [21], so this explanation is no longer valid. Nevertheless, the observation that the *SLN* gene is much more transcriptionally responsive than the *ATP2A1* gene suggests that changes in SLN expression might represent an early functional adaptation to chronic low-frequency stimulation.

2.7.3
Inhibition of SERCA Function by SLN Plus PLN

SLN co-expressed with PLN becomes superinhibitory [20, 21]. SLN has a higher affinity for PLN than PLN itself, so that it can depolymerize the PLN pentamer. At equal concentrations and at equilibrium, the predominant species in the membrane seems to be a PLN–SLN dimer. A ternary PLN–SLN–SERCA complex can be dissolved in non-ionic detergents and immunoprecipitated by antibodies against either PLN or SLN, suggesting that the ternary complex, rather than any of the three possible binary complexes, is the most stable complex, when all three proteins are together [148]. The amount of the complex is strongly determined by PLN mutant sequences known to alter PLN–SERCA interactions.

The sequence homology between PLN and SLN (Figure 2.2) makes it possible to compare the effects of mutations in SLN and in PLN on the ability of either molecule to interact with wild-type or mutant SERCA molecules. As described above, such studies with PLN have led to the prediction of which amino acids in both molecules might interact in the PLN–SERCA complex. Comparison of the effects on SERCA binding of corresponding mutations in PLN and SLN led to the conclusion that PLN and SLN bind to the same transmembrane binding site in SERCA1a [148].

2.7.4
Modeling of the SLN-SERCA and SLN-PLN–SERCA Interactions

The successful modeling of the PLN–SERCA interaction led to modeling of the binary SLN-SERCA interaction and of the ternary PLN–SLN–SERCA interaction. Modeling of the interaction of SLN with SERCA1a, based on the atomic structures of SERCA1a in the E2 conformation [44] and SLN in a lipid bilayer [149] showed that SLN could fit into the same groove, formed by transmembrane helices M2, M4, M6 and M9, that binds PLN. Conformational changes would force SLN out of this groove as movement of M2 narrows the groove. Thus the PLN–SERCA and SLN–SERCA binary complexes are comparable, except that the C-terminal end of SLN extends to the lumenal loop connecting M1 and M2 of SERCA1 where Tyr29 and Tyr31 interact with aromatic residues in SERCA1a.

Further modeling of the PLN–SLN binary complex shows that it can also fit snugly into the PLN-binding groove in the E$_2$ conformation of SERCA1a to form a PLN–SLN–SERCA1a ternary complex [148] (Figure 2.12). The tighter fit creates a higher affinity, while the unique C-terminus of SLN adds to the stability by binding to aromatic residues in the loop connecting M1 and M2 helices of SERCA1a. Although a PLN–SLN complex fits snugly into this site, a PLN–PLN complex would not because of differences in the backbone of the PLN and SLN helices and the different interactions that occur in PLN–PLN and PLN–SLN dimers. The superinhibition that occurs with the PLN–SLN complex can be related directly to its higher affinity for SERCA1a than either PLN or SLN alone. These studies hint at another level of regulation of SERCA2a by a high affinity PLN–SLN complex that requires much more investigation.

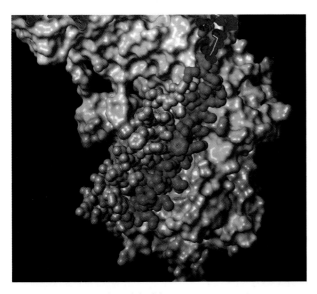

Figure 2.12 Structural model of the SLN–PLN–SERCA1a ternary complex. The SLN–PLN binary complex can fit into the same groove in the E_2 conformation of SERCA1a as PLN alone (Figure 2.3). Additional SLN–SERCA1a binding sites make the ternary complex more stable than the binary PLN–SERCA1a complex, leading to superinhibition.

Acknowledgments

Original studies described in this review were supported by Heart and Stroke Foundation of Ontario grant T-5042, Canadian Institutes of Health Research grants MT-12545 and MOP49493 to D.H.M, the Neuromuscular Research Program, and by National Institutes of Health (USA) grants HL-26057, HL-64018, and HL-52318 to E.G.K.

References

1. D. H. MacLennan and R. A. Reithmeier, *Nat. Struct. Biol.*, **1998**, *5*, 409–411.
2. A. S. Zot and J. D. Potter, *Annu. Rev. Biophys. Biophys. Chem.*, **1987**, *16*, 535–559.
3. D. M. Bers, *Circ. Res.*, **2002**, *90*, 14–17.
4. C. Toyoshima, M. Asahi, Y. Sugita, R. Khanna, T. Tsuda, and D. H. Mac-Lennan, *Proc. Natl Acad Sci. U. S.A,* **2003**, *100*, 467–472.
5. Y. Kimura, K. Kurzydlowski, M. Tada, and D. H. MacLennan, *J. Biol. Chem.*, **1997**, *272*, 15061–15064.
6. J. Zhai, A. G. Schmidt, B. D. Hoit, Y. Kimura, D. H. MacLennan, and E. G. Kranias, *J. Biol. Chem.*, **2000**, *275*, 10538–10544.
7. E. Zvaritch, P. H. Backx, F. Jirik, Y. Kimura, S. Leon de, A. G. Schmidt, B. D. Hoit, J. W. Lester, E. G. Kranias, and D. H. MacLennan, *J. Biol. Chem.*, **2000**, *275*, 14985–14991.
8. K. Haghighi, A. G. Schmidt, B. D. Hoit, A. G. Brittsan, A. Yatani, J. W. Lester, J. Zhai, Y. Kimura, G. W. Dorn, and D. H. 2nd, MacLennan, and

E. G. Kranias, *J. Biol. Chem.*, 2001, *276*, 24145–24152.

9. W. Luo, I. L. Grupp, J. Harrer, S. Ponniah, G. Grupp, J. J. Duffy, T. Doetschman, and E. G. Kranias, *Circ. Res.*, 1994, *75*, 401–409.

10. V. J. Kadambi, S. Ponniah, J. M. Harrer, B. D. Hoit, G. W. Dorn, II, R. A. Walsh, and E. G. Kranias, *J. Clin. Invest.*, 1996, *97*, 533–539.

11. K. L. Koss and E. G. Kranias, *Circ. Res.*, 1996, *79*, 1059–1063.

12. S. Minamisawa, M. Hoshijima, G. Chu, C. A. Ward, K. Frank, Y. Gu, M. E. Martone, Y. Wang, J. Ross, Jr., E. G. Kranias, W. R. Giles, and K. R. Chien, *Cell*, 1999, *99*, 313–322.

13. H. He, M. Meyer, J. L. Martin, P. M. McDonough, P. Ho, X. Lou, W. Y. Lew, R. Hilal-Dandan, and W. H. Dillmann, *Circulation*, 1999, *100*, 974–980.

14. M. Hoshijima, Y. Ikeda, Y. Iwanaga, S. Minamisawa, M. O. Date, Y. Gu, M. Iwatate, M. Li, L. Wang, J. M. Wilson, Y. Wang, J. Ross, Jr., and K. R. Chien, *Nat. Med.*, 2002, *8*, 864–871.

15. J. P. Schmitt, M. Kamisago, M. Asahi, G. H. Li, F. Ahmad, U. Mende, E. G. Kranias, D. H. MacLennan, J. G. Seidman, and C. E. Seidman, *Science*, 2003, *299*, 1410–1413.

16. K. Haghighi, F. Kolokathis, L. Pater, S. Adamopoulos, R. A. Linch, M. Asahi, A. O. Gramolini, G. Fan, S. B. Liggett, G. Dorn, II, D. H. MacLennan, D. T. Kremastinos, and E. G. Kranias, *J. Clin. Invest.*, 2003, *111*, 869–876.

17. D. H. MacLennan, C. C. Yip, G. H. Iles, and P. Seeman, *Cold Spring Harbor Symp. Quant. Biol.*, 1972, *37*, 469–478.

18. M. A. Kirchberber, M. Tada, and A. M. Katz, *Rec. Adv. Stud. Cardiac Struct. Metab.*, 1975, *5*, 103–115.

19. A. Odermatt, P. E. Taschner, S. W. Scherer, B. Beatty, V. K. Khanna, D. R. Cornblath, V. Chaudhry, W. C. Yee, B. Schrank, G. Karpati, M. H. Breuning, N. Knoers, and D. H. MacLennan, *Genomics*, 1997, *45*, 541–553.

20. A. Odermatt, S. Becker, V. K. Khanna, K. Kurzydlowski, E. Leisner, D. Pette, and D. H. MacLennan, *J. Biol. Chem.*, 1998, *273*, 12360–12369.

21. M. Asahi, K. Kurzydlowski, M. Tada, and D. H. MacLennan, *J. Biol. Chem.*, 2002, *277*, 26725–26728.

22. S. Minamisawa, Y. Wang, J. Chen, Y. Ishikawa, K. R. Chien, and R. Matsuoka, *J. Biol. Chem.*, 2003, *278*, 9570–9575.

23. D. A. Kass, J. M. Hare, and D. Georgakopoulos, *Circ. Res.*, 1998, *82*, 519–522.

24. A. Levitzki, *Science*, 1988, *241*, 800–806.

25. A. M. Katz, *Ann. New York Acad. Sci.*, 1998, *853*, 9–19.

26. M. A. Kirchberger, M. Tada, D. I. Repke, and A. M. Katz, *J. Mol. Cell. Cardiol.*, 1972, *4*, 673–680.

27. M. Tada, M. A. Kirchberger, and A. M. Katz, *J. Biol. Chem.*, 1975, *250*, 2640–2647.

28. P. J. Raia La and E. Morkin, *Circ. Res.*, 1974, *35*, 298–306.

29. T. Cantilina, Y. Sagara, G. Inesi, and L. R. Jones, *J. Biol. Chem.*, 1993, *268*, 17018–17025.

30. G. Hughes, A. P. Starling, R. P. Sharma, J. M. East, and A. G. Lee, *Biochem. J.*, 1996, *318*, 973–979.

31. E. G. Kranias and R. J. Solaro, *Nature*, 1982, *298*, 182–184.

32. A. D. Wegener, H. K. Simmerman, J. P. Lindemann, and L. R. Jones, *J. Biol. Chem.*, 1989, *264*, 11468–11474.

33. L. Talosi, I. Edes, and E. G. Kranias, *Am. J. Physiol.*, 1993, *264*, 791–797.

34. J. P. Lindemann, L. R. Jones, D. R. Hathaway, B. G. Henry, and A. M. Watanabe, *J. Biol. Chem.*, 1983, *258*, 464–471.

35. J. L. Garvey, E. G. Kranias, and R. J. Solaro, *Biochem. J.*, 1988, *249*, 709–714.

36. C. Mundina-Weilenmann, L. Vittone, M. Ortale, G. C. Cingolani de, and A. Mattiazzi, *J. Biol. Chem.*, 1996, *271*, 33561–33567.

37. F. Wuytack, L. Raeymaekers, and L. Missiaen, *Cell Calcium*, 2002, *32*, 279–305.

38. K. D. Wu, W. S. Lee, J. Wey, D. Bungard, and J. Lytton, *Am. J. Physiol.*, 1995, *269*, 775–784.

39. M. Asahi, Y. Kimura, K. Kurzydlowski, M. Tada, and D. H. MacLennan, *J. Biol. Chem.*, 1999, *274*, 32885–32862.

40. T. Toyofuku, K. Kurzydlowski, M. Tada, and D. H. MacLennan, *J. Biol. Chem.*, **1993**, *268*, 2809–2815.

41. Y. Kimura, K. Kurzydlowski, M. Tada, and D. H. MacLennan, *J. Biol. Chem.*, **1996**, *271*, 21726–21731.

42. D. H. MacLennan, W. J. Rice, and N. M. Green, *J. Biol. Chem.*, **1997**, *272*, 28815–28818.

43. C. Toyoshima, M. Nakasako, H. Nomura, and H. Ogawa, *Nature*, **2000**, *405*, 647–655.

44. C. Toyoshima and H. Nomura, *Nature*, **2002**, *418*, 605–611.

45. H. K. Simmerman and L. R. Jones, *Physiol. Rev.*, **1998**, *78*, 921–947.

46. F. N. Briggs, K. F. Lee, A. W. Wechsler, and L. R. Jones, *J. Biol. Chem.*, **1992**, *267*, 26056–26061.

47. G. L. Morris, H. C. Cheng, J. Colyer, and J. H. Wang, *J. Biol. Chem.*, **1991**, *266*, 11270–11275.

48. J. Fujii, A. Ueno, K. Kitano, S. Tanaka, M. Kadoma, and M. Tada, *J. Clin. Invest.*, **1987**, *79*, 301–304.

49. R. J. Mortishire-Smith, S. M. Pitzenberger, C. J. Burke, C. R. Middaugh, V. M. Garsky, and R. G. Johnson, *Biochemistry*, **1995**, *34*, 7603–7613.

50. P. Pollesello, A. Annila, and M. Ovaska, *Biophys. J.*, **1999**, *76*, 1784–1795.

51. S. Lamberth, H. Schmid, M. Muenchbach, T. Vorherr, J. Krebs, E. Carafoli, and C. Griesinger, *Helv. Chim. Acta*, **2000**, *83*, 2141–2152.

52. P. James, M. Inui, M. Tada, M. Chiesi, and E. Carafoli, *Nature*, **1989**, *342*, 90–92.

53. L. R. Jones, H. K. Simmerman, W. W. Wilson, F. R. Gurd, and A. D. Wegener, *J. Biol. Chem.*, **1985**, *260*, 7721–7730.

54. M. Inui, B. K. Chamberlain, A. Saito, and S. Fleischer, *J. Biol. Chem.*, **1986**, *261*, 1794–1800.

55. H. W. Kim, N. A. Steenaart, D. G. Ferguson, and E. G. Kranias, *J. Biol. Chem.*, **1990**, *265*, 1702–1709.

56. G. Szymanska, H. W. Kim, J. Cuppoletti, and E. G. Kranias, *Mol. Cell. Biochem.*, **1992**, *114*, 65–71.

57. T. Sasaki, M. Inui, Y. Kimura, T. Kuzuya, and M. Tada, *J. Biol. Chem.*, **1992**, *267*, 1674–1679.

58. L. G. Reddy, L. R. Jones, S. E. Cala, J. J. O'Brian, S. A. Tatulian, and D. L. Stokes, *J. Biol. Chem.*, **1995**, *270*, 9390–9397.

59. G. Hughes, J. M. East, and A. G. Lee, *Biochem. J.*, **1994**, *303*, 511–516.

60. T. Toyofuku, K. Kurzydlowski, J. Lytton, and D. H. MacLennan, *J. Biol. Chem.*, **1992**, *267*, 14490–14496.

61. T. Toyofuku, K. Kurzydlowski, M. Tada, and D. H. MacLennan, *J. Biol. Chem.*, **1994**, *269*, 22929–22932.

62. T. Toyofuku, K. Kurzydlowski, M. Tada, and D. H. MacLennan, *J. Biol. Chem.*, **1994**, *269*, 3088–3094.

63. Y. Kimura, M. Asahi, K. Kurzydlowski, M. Tada, and D. H. MacLennan, *J. Biol. Chem.*, **1998**, *273*, 14238–14241.

64. I. T. Arkin, P. D. Adams, K. R. MacKenzie, M. A. Lemmon, A. T. Brunger, and D. M. Engelman, *Embo J.*, **1994**, *13*, 4757–4764.

65. H. K. Simmerman, Y. M. Kobayashi, J. M. Autry, and L. R. Jones, *J. Biol. Chem.*, **1996**, *271*, 5941–5946.

66. P. D. Adams, I. T. Arkin, D. M. Engelman, and A. T. Brunger, *Nat. Struct. Biol.*, **1995**, *2*, 154–162.

67. J. M. Autry and L. R. Jones, *J. Biol. Chem.*, **1997**, *272*, 15872–15880.

68. M. Asahi, N. M. Green, K. Kurzydlowski, M. Tada, and D. H. MacLennan, *Proc. Natl. Acad. Sci. U. S.A*, **2001**, 98, 10061–10066.

69. L. R. Jones, R. L. Cornea, and Z. Chen, *J. Biol. Chem.*, **2002**, *277*, 28319–28329.

70. D. R. Knighton, J. H. Zheng, L. F. Eyck Ten, N. H. Xuong, S. S. Taylor, and J. M. Sowadski, *Science*, **1991**, *253*, 414–420.

71. M. Asahi, E. McKenna, K. Kurzydlowski, M. Tada, and D. H. MacLennan, *J. Biol. Chem.*, **2000**, *275*, 15034–15038.

72. M. C. Hutter, J. Krebs, J. Meiler, C. Griesinger, E. Carafoli, and V. Helms, *Chem. Biochem.*, **2002**, *3*, 1200–1208.

73. K. L. Koss, S. Ponniah, W. K. Jones, I. L. Grupp, and E. G. Kranias, *Circ. Res.*, **1995**, *77*, 342–353.

74. J. M. Harrer, K. Haghighi, H. W. Kim, D. G. Ferguson, and E. G. Kranias, *Am. J. Physiol.*, **1997**, *272*, 57–66.

75. E. Kiss, G. Jakab, E. G. Kranias, and
I. Edes, *Circ. Res.*, **1994**, *75*, 245–251.

76. A. G. Brittsan, E. Kiss, I. Edes, I. L.
Grupp, G. Grupp, and E. G. Kranias,
J. Mol. Cell. Cardiol., **1999**, *31*, 1725–
1737.

77. E. G. Kranias and J. Di Salvo, *J. Biol.
Chem.*, **1986**, *261*, 10029–10032.

78. W. Luo, B. M. Wolska, I. L. Grupp, J. M.
Harrer, K. Haghighi, D. G. Ferguson,
J. P. Slack, G. Grupp, T. Doetschman,
R. J. Solaro, and E. G. Kranias, *Circ.
Res.*, **1996**, *78*, 839–847.

79. B. M. Wolska, M. O. Stojanovic, W. Luo,
E. G. Kranias, and R. J. Solaro, *Am.
J. Physiol.*, **1996**, *271*, 391–397.

80. L. Li, G. Chu, E. G. Kranias, and
D. M. Bers, *Am. J. Physiol.*, **1998**, *274*,
1335–1347.

81. J. N. Lorenz and E. G. Kranias, *Am.
J. Physiol.*, **1997**, *273*, 2826–2831.

82. E. Kiss, I. Edes, Y. Sato, W. Luo, S. B.
Liggett, and E. G. Kranias, *Am. J.
Physiol.*, **1997**, *272*, 785–790.

83. G. Chu, W. Luo, J. P. Slack, C. Tilgmann, W. E. Sweet, M. Spindler, K. W.
Saupe, G. P. Boivin, C. S. Moravec,
M. A. Matlib, I. L. Grupp, J. S. Ingwall,
and E. G. Kranias, *Circ. Res*, **1996**, *79*,
1064–1076.

84. L. F. Santana, A. M. Gomez, E. G.
Kranias, and W. J. Lederer, *Heart
Vessels*, **1997**, *suppl. 12*, 44–49.

85. H. Masaki, Y. Sato, W. Luo, E. G.
Kranias, and A. Yatani, *Am. J. Physiol.*,
1997, *272*, H606–H612.

86. K. H. Desai, E. Schauble, W. Luo, E.
Kranias, and D. Bernstein, *Am. J.
Physiol.*, **1999**, *276*, H1172–H1177.

87. J. P. Slack, I. L. Grupp, R. Dash,
D. Holder, A. Schmidt, M. J. Gerst,
T. Tamura, C. Tilgmann, P. F. James,
R. Johnson, A. M. Gerdes, and E. G.
Kranias, *J. Mol. Cell. Cardiol.*, **2001**, *33*,
1031–1040.

88. H. Kiriazis, Y. Sato, V. J. Kadambi,
A. G. Schmidt, M. J. Gerst, B. D. Hoit,
and E. G. Kranias, *Cardiovasc. Res.*,
2002, *53*, 372–381.

89. H. R. Cross, E. G. Kranias, E. Murphy,
and C. Steenbergen, *Am. J. Physiol.
Heart Circ. Physiol.*, **2003**, *284*, H683–
H690.

90. J. Lalli, J. M. Harrer, W. Luo, E. G.
Kranias, and R. J. Paul, *Circ. Res.*, **1997**,
80, 506–513.

91. J. P. Slack, I. L. Grupp, W. Luo, and
E. G. Kranias, *Am. J. Physiol.*, **1997**,
273, C1–C6.

92. R. L. Sutliff, J. B. Hoying, V. J. Kadambi, E. G. Kranias, and R. J. Paul,
Circ. Res., **1999**, *84*, 360–364.

93. K. Nobe, R. L. Sutliff, E. G. Kranias,
and R. J. Paul, *J. Physiol.*, **2001**, *535*,
867–878.

94. M. J. Lalli, S. Shimizu, R. L. Sutliff,
E. G. Kranias, and R. J. Paul, *Am. J.
Physiol.*, **1999**, *277*, H963–H970.

95. J. P. Slack, I. L. Grupp, D. G. Ferguson,
N. Rosenthal, and E. G. Kranias, *J. Biol.
Chem.*, **1997**, *272*, 18862–18868.

96. R. Dash, V. Kadambi, A. G. Schmidt,
N. M. Tepe, D. Biniakiewicz, M. J.
Gerst, A. M. Canning, W. T. Abraham,
B. D. Hoit, S. B. Liggett, J. N. Lorenz,
G. W. Dorn II, and E. G. Kranias,
Circulation, **2001**, *103*, 889–896.

97. R. Dash, A. G. Schmidt, A. Pathak,
M. J. Gerst, D. Biniakiewicz, V. J.
Kadambi, B. D. Hoit, W. T. Abraham,
and E. G. Kranias, *Cardiovasc. Res.*,
2003, *57*, 704–714.

98. A. G. Brittsan, A. N. Carr, A. G.
Schmidt, and E. G. Kranias, *J. Biol.
Chem.*, **2000**, *275*, 12129–12135.

99. J. Colyer and J. H. Wang, *J. Biol. Chem.*,
1991, *266*, 17486–17493.

100. W. Luo, G. Chu, Y. Sato, Z. Zhou, V. J.
Kadambi, and E. G. Kranias, *J. Biol.
Chem.*, **1998**, *273*, 4734–4739.

101. G. Chu, J. W. Lester, K. B. Young, W.
Luo, J. Zhai, and E. G. Kranias, *J. Biol.
Chem.*, **2000**, *275*, 38938–38943.

102. L. Vittone, C. Mundina-Weilenmann,
M. Said, P. Ferrero, and A. Mattiazzi,
J. Mol. Cell. Cardiol., **2002**, *34*, 39–50.

103. D. Hagemann, M. Kuschel, T. Kuramochi, W. Zhu, H. Cheng, and R. P.
Xiao, *J. Biol. Chem.*, **2000**, *275*, 22532–
22536.

104. A. G. Schmidt, J. Zhai, A. N. Carr, M. J.
Gerst, J. N. Lorenz, P. Pollesello, A.
Annila, B. D. Hoit, and E. G. Kranias,
Cardiovasc. Res., **2002**, *56*, 248–259.

105. D. J. Beuckelmann, M. Nabauer, and
E. Erdmann, *Circulation*, **1992**, *85*,
1046–1055.

106. K. Dipla, J. A. Mattiello, K. B. Margu-
lies, V. Jeevanandam, and S. R. Houser,
Circ. Res., **1999**, *84*, 435–444.

107. G. Hasenfuss, H. Reinecke, R. Studer,
M. Meyer, B. Pieske, J. Holtz, C. Ho-
lubarsch, H. Posival, H. Just, and H.
Drexler, *Circ. Res.*, **1994**, *75*, 434–442.

108. G. Hasenfuss, L. A. Mulieri, B. J. Lea-
vitt, P. D. Allen, J. R. Haeberle, and
N. R. Alpert, *Circ. Res.*, **1992**, *70*, 1225–
1232.

109. B. Pieske, B. Kretschmann, M. Meyer,
C. Holubarsch, J. Weirich, H. Posival,
K. Minami, H. Just, and G. Hasenfuss,
Circulation, **1995**, *92*, 1169–1178.

110. J. K. Gwathmey, L. Copelas, R. Mac-
Kinnon, F. J. Schoen, M. D. Feldman,
W. Grossman, and J. P. Morgan, *Circ.
Res.*, **1987**, *61*, 70–76.

111. J. P. Morgan , *New Engl. J. Med.*, **1991**,
325, 625–632.

112. J. J. Mercadier, A. M. Lompre, P. Duc,
K. R. Boheler, J. B. Fraysse, C. Wis-
newsky, P. D. Allen, M. Komajda, and
K. Schwartz, *J. Clin. Invest.*, **1990**, *85*,
305–309.

113. C. Mittmann, T. Eschenhagen, and
H. Scholz, *Cardiovasc. Res.*, **1998**, *39*,
267–275.

114. R. Dash, K. F. Frank, A. N. Carr, C. S.
Moravec, and E. G. Kranias, *J. Mol.
Cell. Cardiol.*, **2001**, *33*, 1345–1353.

115. M. Arai, N. R. Alpert, D. H. MacLennan, P. Barton, and M. Periasamy, *Circ.
Res.*, **1993**, *72*, 463–469.

116. M. Arai, H. Matsui, and M. Periasamy,
Circ. Res., **1994**, *74*, 555–564.

117. R. H. Schwinger, M. Bohm, U.
Schmidt, P. Karczewski, U. Bavendiek,
M. Flesch, E. G. Krause, and E. Erdmann, *Circulation*, **1995**, *92*, 3220–
3228.

118. K. F. Frank, B. Bolck, K. Brixius, E. G.
Kranias, and R. H. Schwinger, *Basic
Res. Cardiol.*, **2002**, *97 (Suppl 1)*,
72–178.

119. M. Meyer, W. Schillinger, B. Pieske, C.
Holubarsch, C. Heilmann, H. Posival,
G. Kuwajima, K. Mikoshiba, H. Just,
G. Hasenfuss, et al., *Circulation*, **1995**,
92, 778–784.

120. R. H. Schwinger, G. Munch, B. Bolck,
P. Karczewski, E. G. Krause, and E.

Erdmann, *J. Mol. Cell. Cardiol.*, **1999**,
31, 479–491.

121. J. Neumann, T. Eschenhagen, L. R.
Jones, B. Linck, W. Schmitz, H. Scholz,
and N. Zimmermann, *J. Mol. Cell.
Cardiol.*, **1997**, *29*, 265–272.

122. A. N. Carr, A. G. Schmidt, Y. Suzuki,
F. Monte del, Y. Sato, C. Lanner,
K. Breeden, S. L. Jing, P. B. Allen,
P. Greengard, A. Yatani, B. D. Hoit,
I. L. Grupp, R. J. Hajjar, A. A. DePaoli
Roach, and E. G. Kranias, *Mol. Cell.
Biol.*, **2002**, *22*, 4124–4135.

123. A. El-Armouche, T. Rau, O. Zolk, D.
Ditz, T. Pamminger, W. H. Zimmermann, E. Jackel, S. E. Harding, P.
Boknik, J. Neumann, and T. Eschenhagen, *FASEB J.*, **2003**, *17*, 437–439.

124. F. J. Giordano, H. He, P. McDonough,
M. Meyer, M. R. Sayen, and W. H.
Dillmann, *Circulation*, **1997**, *96*,
400–403.

125. F. Monte del, S. E. Harding, U.
Schmidt, T. Matsui, Z. B. Kang, G. W.
Dec, J. K. Gwathmey, A. Rosenzweig,
and R. J. Hajjar, *Circulation*, **1999**, *100*,
2308–2311.

126. M. I. Miyamoto, F. Monte del, U.
Schmidt, T. S. DiSalvo, Z. B. Kang, T.
Matsui, J. L. Guerrero, J. K. Gwathmey,
A. Rosenzweig, and R. J. Hajjar, *Proc.
Natl. AcaD. Sci. U. S.A*, **2000**, *97*,
793–798.

127. F. Monte del, E. Williams, D. Lebeche,
U. Schmidt, A. Rosenzweig, J. K.
Gwathmey, E. D. Lewandowski, and
R. J. Hajjar, *Circulation*, **2001**, *104*,
1424–1429.

128. H. He, F. J. Giordano, R. Hilal-Dan-
dan, D. J. Choi, H. A. Rockman, P. M.
McDonough, W. F. Bluhm, M. Meyer,
M. R. Sayen, E. Swanson, and W. H.
Dillmann, *J. Clin. Invest.*, **1997**, *100*,
380–389.

129. D. L. Baker, K. Hashimoto, I. L. Grupp,
Y. Ji, T. Reed, E. Loukianov, G. Grupp,
A. Bhagwhat, B. Hoit, R. Walsh, E.
Marban, and M. Periasamy, *Circ. Res.*,
1998, *83*, 1205–1214.

130. E. Loukianov, Y. Ji, D. L. Baker, T. Reed,
J. Babu, T. Loukianova, A. Greene,
G. Shull, and M. Periasamy, *Ann. New
York Acad. Sci.*, **1998**, *853*, 251–259.

131. W. Zhao, K. F. Frank, G. Chu, M. J. Gerst, A. G. Schmidt, Y. Ji, M. Periasamy, and E. G. Kranias, *Cardiovasc. Res.*, **2003**, *57*, 71–81.

132. K. Eizema, H. Fechner, K. Bezstarosti, S. Schneider-Rasp, A. Laarse van der, H. Wang, H. P. Schultheiss, W. C. Poller, and J. M. Lamers, *Circulation*, **2000**, *101*, 2193–2199.

133. F. Monte del, S. E. Harding, G. W. Dec, J. K. Gwathmey, and R. J. Hajjar, *Circulation*, **2002**, *105*, 904–907.

134. R. Knoll, M. Hoshijima, H. M. Hoffman, V. Person, I. Lorenzen-Schmidt, M. L. Bang, T. Hayashi, N. Shiga, H. Yasukawa, W. Schaper, W. McKenna, M. Yokoyama, N. J. Schork, J. H. Omens, A. D. McCulloch, A. Kimura, C. C. Gregorio, W. Poller, J. Schaper, H. P. Schultheiss, and K. R. Chien, *Cell*, **2002**, *111*, 943–955.

135. Y. Sato, H. Kiriazis, A. Yatani, A. G. Schmidt, H. Hahn, D. G. Ferguson, H. Sako, S. Mitarai, R. Honda, L. Mesnard-Rouiller, K. F. Frank, B. Beyermann, G. Wu, K. Fujimori, G. W. Dorn II, and E. G. Kranias, *J. Biol. Chem.*, **2001**, *276*, 9392–9399.

136. K. Freeman, I. Lerman, E. G. Kranias, T. Bohlmeyer, M. R. Bristow, R. J. Lefkowitz, G. Iaccarino, W. J. Koch, and L. A. Leinwand, *J. Clin. Invest.*, **2001**, *107*, 967–974.

137. Q. Song, A. G. Schmidt, H. S. Hahn, A. N. Carr, B. Frank, L. Pater, M. J. Gerst, K. B. Young, B. D. Hoit, B. K. McConnell, C. E. Seidman, J. G. Seidman, G. W.I. Dorn, and E. G. Kranias, *J. Clin. Invest.*, **2003**, *111*, 859–867.

138. M. A. Fink, D. R. Zakhary, J. A. Mackey, R. W. Desnoyer, C. Apperson-Hansen, D. S. Damron, and M. Bond, *Circ. Res.*, **2001**, *88*, 291–297.

139. A. G. Brittsan, K. S. Ginsburg, G. Chu, A. Yatani, B. M. Wolska, A. G. Schmidt, M. Asahi, D. H. MacLennan, D. M. Bers, and E. G. Kranias, *Circ. Res.*, **2003**, *92*, 769–776.

140. B. Pieske, B. Beyermann, V. Breu, B. M. Loffler, K. Schlotthauer, L. S. Maier, S. Schmidt-Schweda, H. Just, and G. Hasenfuss, *Circulation*, **1999**, *99*, 1802–1809.

141. A. Wawrzynow, J. L. Theibert, C. Murphy, I. Jona, A. Martonosi, and J. H. Collins, *Arch. Biochem. Biophys.*, **1992**, *298*, 620–623.

142. E. Racker and E. Eytan, *J. Biol. Chem.*, **1975**, *250*, 7533–7534.

143. W. S. Smith, R. Broadbridge, J. M. East, and A. G. Lee, *Biochem. J.*, **2002**, *361*, 277–286.

144. S. Hellstern, S. Pegoraro, C. B. Karim, A. Lustig, D. D. Thomas, L. Moroder, and J. Engel, *J. Biol. Chem.*, **2001**, *276*, 30845–30852.

145. G. Gayan-Ramirez, L. Vanzeir, F. Wuytack, and M. Decramer, *J. Physiol.*, **2000**, *524*, 387–397.

146. E. Leberer, K. T. Hartner, and D. Pette, *Eur. J. Biochem.*, **1987**, *162*, 555–561.

147. D. Pette and G. Vrbova, *Rev. Physiol. Biochem. Pharmacol.*, **1992**, *120*, 115–202.

148. M. Asahi, K. Kurzydlowski, S. Leon De, M. Tada, C. Toyoshima, and D. H. MacLennan, *Proc. Natl. Acad. Sci. U. S.A.*, **2003**, *100*.

149. A. Mascioni, C. Karim, G. Barany, D. D. Thomas, and G. Veglia, *Biochemistry*, **2002**, *41*, 475–482.

3

Catalytic and Transport Mechanism of the Sarco-(Endo)Plasmic Reticulum Ca^{2+}-ATPase (SERCA)

Giuseppe Inesi and *Chikashi Toyoshima*

Summary

Transport of Ca^{2+} by the ATPase of intracellular membranes occurs with an optimal ratio of two Ca^{2+} per ATP, resulting in a three orders of magnitude Ca^{2+} gradient. The ATPase sequential reactions include high-affinity binding of two Ca^{2+}, a phosphorylated enzyme intermediate, vectorial dissociation of bound Ca^{2+}, and hydrolytic cleavage of Pi. The cycle is highly reversible. The catalytic mechanism includes general acid–base catalysis and metal-ion assistance by Mg^{2+}. The phosphorylated intermediate is formed through attack of the electrophilic ATP terminal phosphate by a nucleophilic carboxylate group within the catalytic site. The phosphoenzyme intermediate then reacts with water to yield Pi. The Ca^{2+} transport mechanism has been defined at the molecular and atomic level by chemical, mutational and crystallographic analysis. Locations of Ca^{2+}, ATP and phosphorylation sites within the ATPase protein are known, as well as the amino acid residues involved in binding and catalytic reactions. The transduction mechanism clearly relies on a long-range linkage triggered by high-affinity Ca^{2+} binding at one end, and ATP utilization at the other end of the ATPase molecule. The linkage includes rearrangement of transmembrane helices and large scale motions of cytosolic domains.

3.1
Introduction

The Ca^{2+} transport ATPase of intracellular membranes was first obtained with a microsomal fraction of rabbit skeletal muscle homogenates that prevented activation of muscle contractile proteins upon addition of ATP [1, 2]. This "relaxing effect" was attributed to sequestration of medium Ca^{2+} into the lumen of vesicular fragments of sarcoplasmic reticulum (SR), due to active transport by a membrane-bound ATPase. In fact, the presence of Ca^{2+} transport ATPase in the SR membrane is related to its functional role in muscle relaxation. Ca^{2+} ATPase was then found to

Handbook of ATPases. Edited by M. Futai, Y. Wada, J. H. Kaplan
Copyright © 2004 WILEY-VCH Verlag GmbH & Co. KGaA, Weinheim
ISBN 3-527-30689-7

be associated with intracellular membranes of various tissues, and therefore the enzyme is referred to as Sarco- Endoplasmic Reticulum Calcium ATPase ("SERCA"). Cloning of cDNA was accomplished [3] following partial amino acid sequencing [4], and three gene products (SERCA1, SERCA2 and SERCA3) with 2–3 splice variants for each primary transcript were subsequently revealed. The related isoforms display specific tissue distribution, functional roles and regulatory mechanisms. It is clear that the SERCA isoforms play a prominent role in numerous cytosolic signaling mechanisms, where Ca^{2+} sequestration into intracellular stores is required for subsequent release through passive channels upon various physiological stimuli [5]. SERCA gene defects have been related to muscle [6] and skin [7, 8] diseases of humans, and defects of SERCA function contribute to the development of cardiac failure [9]. Various aspects of the SERCA ATPase function and structure have been reviewed [10–20].

3.2
Experimental Systems

Vesicular fragments of longitudinal SR, obtained from differential centrifugation of rabbit skeletal muscle, provide a very useful experimental system for functional and structural studies. SR vesicles contain a high quantity of Ca^{2+} ATPase (SERCA1 isoform), which accounts for approximately 50% of the total protein, and is densely spaced within the plane of the membrane (Figure 3.1). The SR vesicles allow parallel measurements of ATPase activity and Ca^{2+} transport, and control of the ionic environment in two compartments delimited by the native mem-

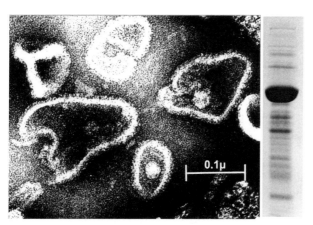

Figure 3.1 Structural characterization of SR vesicles. Left panel: Negatively stained SR vesicles, obtained from rabbit skeletal muscle, visualized by electron microcopy. Densely spaced granules on the surface of the membrane correspond to Ca^{2+} ATPase molecules. Right panel: Electrophoretic analysis demonstrating that the Ca^{2+} ATPase separates as a prominent band that accounts for the major portion of the membrane protein [107].

brane. For this reason, they have been used advantageously in studies of the mechanism of ATP utilization for Ca^{2+} transport.

The Ca^{2+} ATPase can be selectively solubilized from the native vesicles with detergents, and further purified by absorption in reactive red agarose columns and elution with nucleotide [21]. The purified protein has been used for reconstitution of proteo-liposomal vesicles of controlled lipid composition, exhibiting Ca^{2+} transport coupled to ATP utilization [22]. Protein purified by this method has been also used for crystallization and diffraction studies [23, 24]. In addition to the native ATPase obtained directly from muscle homogenates, recombinant ATPase has been obtained by transfection of cultured cells with SERCA cDNA [25]. Cultured cells exhibiting exogenous SERCA gene expression can then be homogenized to prepare microsomal vesicles for functional and structural studies, and mutational analysis. Recombinant viral vectors allow efficient transfer of SERCA gene into cultured cells, yielding microsomal preparations with a recombinant Ca^{2+} ATPase content that accounts for 7–8 % of the microsomal protein [26]. This relatively high content has been very helpful in functional and structural studies of wild-type and mutant enzyme.

3.3
Functional Characterization

Active transport of Ca^{2+} into SR vesicles is observed in the presence of ATP, with an optimal stoichiometric ratio of two Ca^{2+} per ATP undergoing hydrolytic cleavage. Acetyl phosphate [27] or *p*-nitrophenyl phosphate [28] can be used as substrate by the enzyme instead of ATP, although with lesser kinetic competence. The enzyme is activated by medium Ca^{2+} within the micromolar range and is inhibited by lumenal Ca^{2+} in the millimolar range, indicating that ATP is utilized to change the affinity and orientation of the Ca^{2+} binding sites. Thereby, a three orders of magnitude Ca^{2+} gradient is formed across the membrane. Furthermore, using reconstituted ATPase preparations deprived of pathways for passive leakage of ions (unavoidably inherent to native SR vesicles) , it was shown that Ca^{2+} transport is accompanied by H^+ countertransport with a 1:1 ratio, and the pump is electrogenic [29]. Therefore, the free energy required for active transport of Ca^{2+} into the vesicles can be defined as

$$\Delta G = RT\ln([Ca^{2+}_{in}]/[Ca^{2+}_{out}] + zF\Delta V \qquad (1)$$

where $[Ca^{2+}_{in}]$ and $[Ca^{2+}_{out}]$ refer to the Ca^{2+} concentration in the lumen of the vesicles and in the outer medium, z is the electrical charge of the transported species, R and F are the gas and Faraday constants, respectively, and ΔV is the transmembrane electrical gradient. The free energy requirement, estimated in this manner and based on experimentally observed gradients, shows that active transport of Ca^{2+} is compatible with the chemical potential of ATP.

Clarification of the intermediate reactions of the ATPase catalytic cycle began with the realization that the ATP terminal phosphate is transferred covalently to an aspartyl residue (Asp351), to form a phosphorylated enzyme intermediate prior to hydrolytic release of Pi [30]. Phosphoryl transfer shows the same Ca^{2+} concentration dependence as ATP hydrolysis, indicating that enzyme activation by Ca^{2+} must occur early in the catalytic cycle. In fact, cooperative binding of two Ca^{2+} per ATPase in the absence of ATP, with an apparent K_d in the micromolar Ca^{2+} range, was demonstrated by direct equilibrium measurements [31]. The sequence of partial reactions making up the catalytic cycle was further elucidated by transient state experiments [32], showing that addition of ATP to SR vesicles pre-incubated with micromolar Ca^{2+} is followed by rapid formation of phosphorylated intermediate and vectorial displacement of two bound Ca^{2+} per ATPase. These initial events are then followed, after a time lag, by hydrolytic cleavage of the acylphosphate intermediate (Figure 3.2).

The ATPase cycle is highly reversible, as demonstrated by ATP synthesis coupled to Ca^{2+} efflux from loaded vesicles [33]. The partial reactions of the reverse cycle

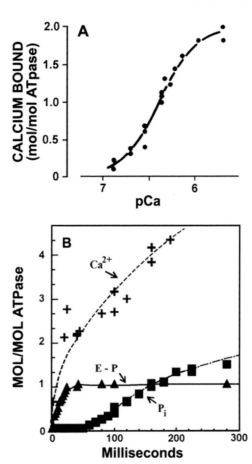

Figure 3.2 Functional characterization of SR vesicles. (A) Calcium binding at equilibrium in the absence of ATP, demonstrating cooperative binding of two Ca^{2+} per ATPase [31]. (B) Pre-steady state experiment showing rapid enzyme phosphorylation and inward displacement of the two Ca^{2+} upon addition of ATP. Hydrolytic cleavage of Pi occurs after a lag period [32].

were defined by the finding that, in the absence of bound Ca^{2+}, the enzyme can be phosphorylated with Pi to yield ADP-insensitive phosphoenzyme [34]. This phosphoenzyme can then be made ADP sensitive by addition of millimolar Ca^{2+}, leading to formation of ATP [35, 36]. These functional findings were interpreted with a mechanism based on interconversion of two states [37, 10]. One state (E1), stabilized by Ca^{2+} binding, would have the Ca^{2+} sites in high affinity and cytosolic orientation. The alternate state (E2), stabilized by enzyme phosphorylation, would have the Ca^{2+} sites in low affinity and lumenal orientation.

A more explicit view of the coupling mechanistic can be obtained by considering a minimal number of partial reactions, and their equilibrium constants $(K_1–K_7)$:

(Reaction 1) $E{\cdot}2H^+ + 2Ca^{2+}{}_{out} \leftrightarrow E{\cdot}2Ca^{2+} + 2H^+{}_{out}$ $\quad$ $(K_1 = 10^{12}\ M^{-2})$

(Reaction 2) $E{\cdot}2Ca^{2+} + ATP \leftrightarrow ATP{\cdot}E{\cdot}2Ca^{2+}$ $\quad$ $(K_2 = 10^5\ M^{-1})$

(Reaction 3) $ATP{\cdot}E{\cdot}2Ca^{2+} \leftrightarrow ADP{\cdot}E{\text-}P{\cdot}2Ca^{2+}$ $\quad$ $(K_3 = 1.0)$

(Reaction 4) $ADP{\cdot}E{\text-}P{\cdot}2Ca^{2+} \leftrightarrow E{\text-}P{\cdot}2Ca^{2+} + ADP$ $\quad$ $(K_4 = 10^{-4}\ M)$

(Reaction 5) $E{\text-}P{\cdot}Ca^{2+} + 2H^+{}_{in} \leftrightarrow E{\text-}P{\cdot}2H^+ + 2Ca^{2+}{}_{in}$ $\quad$ $(K_5 = 10^{-6}\ M^2)$

(Reaction 6) $E{\text-}P{\cdot}2H^+ \leftrightarrow E{\cdot}2H^+{\cdot}P_i$ $\quad$ $(K_6 = 1)$

(Reaction 7) $E{\cdot}2H^+{\cdot}Pi \leftrightarrow E{\cdot}2H^+ + P_i$ $\quad$ $(K_7 = 10^{-2}\ M)$

where the hyphens denote covalent binding and dots non-covalent binding, and the equilibrium constants relate to conditions allowing constant temperature (25°C) and pH (7.0). A detailed account of the equilibrium and kinetic characterization of these partial reactions can be found in Inesi et al. [38].

In the above reaction sequence, the initial high-affinity binding (reaction 1) of two Ca^{2+} activates the enzyme, permitting utilization of ATP and formation of a phosphorylated intermediate (reactions 2–4). In turn, enzyme phosphorylation lowers the affinity and changes the vectorial orientation of bound Ca^{2+}, thereby increasing the probability of its dissociation into the lumen of the vesicles (reaction 5). Note that the equilibrium constant for the enzyme phosphorylation by ATP is nearly 1, indicating that the free energy of ATP is conserved by the enzyme, and utilized to change the Ca^{2+} binding characteristics. Finally, the phosphoenzyme undergoes hydrolytic cleavage and releases Pi (reactions 6 and 7) before entering another cycle. The reaction sequence shows clearly that the direct mechanistic device for translocation of bound Ca^{2+} is enzyme phosphorylation, rather than hydrolytic cleavage of acylphosphate [39].

As the catalytic and transport cycle is likely to include conformational transitions in addition to the chemical reactions listed above, such transitions are, notably, coupled implicitly with the chemical reactions subjected to experimental measurement (25°C, neutral pH), and their influence is reflected by the equilibrium constants given above. In fact, the standard free energies $(-RT\ln K)$ of the partial reactions add up to the standard free energy of ATP hydrolysis (γ-phosphate), as expected. Most interestingly, the standard free energy diagram for the partial reactions reveals that the chemical potential of ATP does not manifest itself in the phosphoryl transfer or hydrolytic cleavage reactions (K_4 and K_6 near 1), but rather in the drastic reduction of the enzyme affinity for Ca^{2+} (compare K_1 to $1/K_5$). Therefore, under standard conditions, the free energy involved in active transport of n (2 in our case) calcium ions per cycle can be considered to be

$$\Delta G = nRT\ln(K_a^{CaEP}/K_a^{CaE}) \tag{2}$$

where K_a^{CaE} and K_a^{CaEP} are the *association* constants of the enzyme for Ca^{2+} in the ground state and following activation by ATP. With reference to the reaction scheme given above, the two relevant constants are K_1 and $1/K_5$.

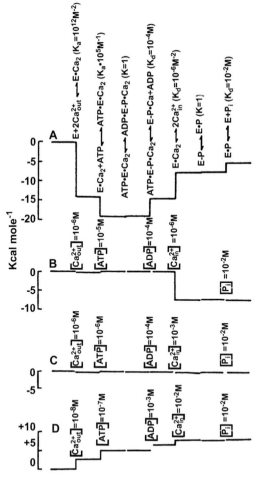

Figure 3.3 Free-energy changes for the partial reactions of the ATPase cycle. (A) Free energy changes under standard conditions. (B, C and D). Actual free energy changes in the presence of concentrations of ligands that may be present under various experimental conditions [adapted from Ref. 41].

Free-energy diagrams have been constructed for the Ca^{2+} ATPase [10, 41], calculating the standard free-energy change for each partial reaction from pertinent equilibrium constants. In the standard free-energy diagram (Figure 3.3A, see p. 68), the enzyme returns to its original energy level after a cycle, and the net drop of free energy between the first and the last partial reaction corresponds to the free energy of ATP hydrolysis, which is the free energy yield by the cycle. This energy may be recovered to a variable degree by the system, depending on the Ca^{2+} concentrations on the two sides of the membrane (i. e. the gradient). As the standard free energy applies only to cases in which the concentrations of reagents and products are all 1 M, corrections must be made to account for the concentration of ligands (i. e. Ca^{2+}), yielding the "actual" free energy operative under a specific set of conditions. At nearly physiological concentrations of ligands, the actual free energies of the binding and dissociation reactions are much smaller than their standard values. For instance, binding of $2Ca^{2+}$ to 1 mol of enzyme with an affinity of $\sim 10^{12}$ M^{-2} yields 16.4 kcal under standard conditions ($[Ca^{2+}_{out}] = 1M$), whereas the actual free energy is ~ 0 in the presence of 10^{-6} M Ca^{2+}_{out}. Conversely,

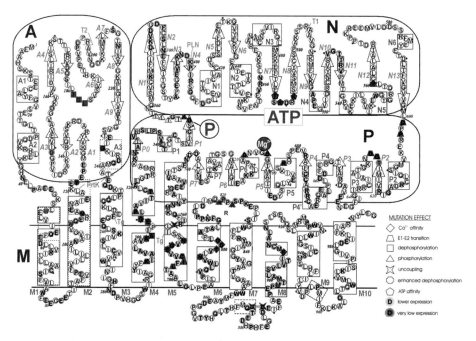

Figure 3.4 Two-dimensional diagram of the Ca^{2+} ATPase amino-acid sequence. The transmembrane region includes ten helical segments, while the cytosolic region includes a large loop yielding the phosphorylation (P) and nucleotide binding (N) domains, and a smaller loop yielding the A domain (see text). Secondary structure elements (boxes for helices; arrows for β-strands) refer to the enzyme conformation with Ca^{2+} bound. T1, T2 and PrtK refer to sites for early digestion by trypsin and proteinase K, respectively. Residues whose mutation produces inhibition of ligand binding or catalytic reactions are evidenced as indicated in the figure; red signifies strong inhibition. [18, 20].

dissociation of $2Ca^{2+}$ from 1 mol of enzyme with an affinity of $\sim 10^3$ M^{-2} requires 8.2 kcal under standard conditions, but yields 8.2 kcal of actual free energy (Figure 3.3B) in the presence of 10^{-6} M Ca^{2+}_{out} (no gradient), and nearly zero free energy (Figure 3.3C) in the presence of 10^{-3} M Ca^{2+}_{out} (maximal gradient). For this reason, utilization of the free energy derived from 1 mol of ATP for transport of 2 mol of Ca^{2+} (i.e. *efficiency*) is proportional to the Ca^{2+} gradient. Furthermore, depending on the concentrations of ligands, the cycle may yield free energy in the forward or even in the reverse direction (cf. Figure 3.3 parts B, C and D). It is then clear that correction of the standard free energy diagram for the actual concentrations of substrates and products (i.e., ATP, ADP, Pi, Ca^{2+}_{out}, Ca^{2+}_{in}) yields net free energy changes permitting net forward or reverse fluxes of Ca^{2+}.

Although the diagrams of Figure 3.3 are based on a sequence of partial reactions for productive coupling of catalysis and transport, the sequence may not be absolutely obligatory, but rather subjected to stochastic distribution. For instance, the phosphoenzyme may, conceivably, undergo hydrolysis even before dissociating Ca^{2+}, albeit at a very slow rate. Significant flow of "non-ideal" routes would reduce the 2:1 coupling ratio of Ca^{2+} transport to ATP hydrolysis ("slippage of the pump"). In fact, experimentally, a perfect 2:1 ratio is rarely obtained, and "slippage of the pump" can occur even when passive "leak" of transported Ca^{2+} through the lipid bilayer is not significant [42, 43]. Such slippage may be relevant to the physiological phenomenon of "thermogenesis" [44].

The free energy diagrams are also important from the kinetic point of view, because correction of the standard free energies by the ligand concentrations fills the "energy wells" corresponding to the binding reactions. A diagram of "actual" free energies under conditions permitting rapid flow of reaction does not display the large variations seen in the standard free-energies diagram. This prevents accumulation of intermediates in the low-energy wells. Another advantage of this analysis is to be mechanistically explicit, and to define the specifications of the transport "machine." It makes clear that, although the pump is in principle capable of establishing a transmembrane gradient commensurate with the free energy of the coupled substrate (as predicted by Eq. 1), it will in fact be limited by kinetic constraints to the Ca^{2+} concentration ranges dictated by the binding constants of the enzyme in the high- and low-affinity states. For instance, the same pump will operate ideally in raising a gradient from the micromolar to the millimolar range, but much less competently from the nanomolar to the micromolar range, or from the millimolar to the molar range.

It is then apparent that the basic coupling mechanism of catalysis and transport is based on mutual destabilization of Ca^{2+} and phosphorylation sites. This is accomplished by protein conformational changes, coupled to interconversion of cation binding and phosphorylation potentials. Recent progress in the characterization of the ATPase structure at high resolution allows consideration of the coupling mechanism in terms of structural and functional relationships.

3.4
Structural Characterization

Structural studies of the Ca^{2+} ATPase have been performed with the SERCA1 isoform of sarcoplasmic reticulum, a protein that includes 994 amino acids distributed into a membrane-bound region, an extra-membranous (cytosolic) region, and with a short stalk between them. Analysis of the primary sequence [3], electron microscopy of ordered ATPase arrays [23, 45–47] and X-ray diffraction from three-dimensional crystals [24, 48] indicate that the membrane-bound region includes ten transmembrane segments, mostly in helical conformation. The extramembranous region, or "headpiece", consists a small loop ("A" domain) between the cytosolic amino terminus and the loop between transmembrane segments M2 and M3, and a larger loop between transmembrane segments M4 and M5 (Figure 3.4). This larger loop includes the nucleotide binding ("N") domain corresponding to the central portion of the loop, and the phosphorylation ("P") domain formed by folding of the proximal and distal segments of the loop. The high-resolution (2.6 Å) structure, derived from 3D crystals of the ATPase obtained in the presence of Ca^{2+} (E1·2Ca^{2+}), can be outlined as follows.

3.4.1
Extramembranous Region and the Catalytic Domains of E1·2Ca^{2+}

The P, N and A domains are in an open conformation and appear quite separate in the image (Figure 3.5) derived from 3D crystals obtained in the presence of Ca^{2+}.

The P domain is formed by the proximal (Asn330 to Asn359) and the distal (Lys605 to Asp737) segments of the long cytosolic loop intervening between the M4 and M5 transmembrane segments. It folds into a seven-stranded parallel β-sheet flanked by eight short helices, to form a Rossmann fold. The catalytic site, including the D351 residue that undergoes phosphorylation and the highly conserved 351DKTGT segment, is in a central and solvent accessible position within the fold. Residues expected to play a catalytic role are arranged in analogy to dehalogenases [49, 50] and small phosphatases [51]. Significant interactions with the transmembrane region occur through direct connections to M4 and M5, and hydrogen bonding with the L67 loop.

The N domain corresponds to the sequence (Gln360 to Asp601) intervening between the two P domain segments. It folds into a seven-stranded antiparallel β-sheet, delimited by two helix bundles (Figure 3.5). Involvement of this domain in nucleotide binding was first suggested by chemical derivatization of K515 [52, 53] and K492 [54, 55], and mutational analysis of the Phe487–K492 segment [56]. The presence of nucleotide was demonstrated directly by Toyoshima et al. [24] in E1·2Ca^{2+} crystals soaked with TNP-AMP. The adenosine moiety of the nucleotide resides within a pocket formed between the β-sheet and a long α-helix containing T441 at one end. The pocket is delimited by residues F487, K492, M494, K515 and L565. A site for early trypsin digestion (T1), corresponding to R505, is on a solvent-exposed loop on the uppermost surface of the N domain.

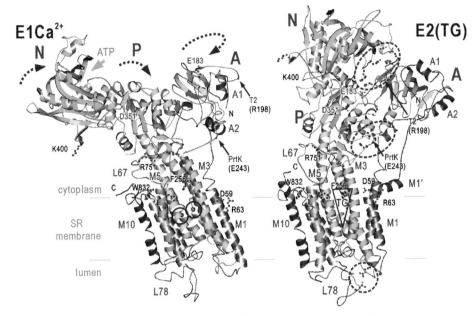

Figure 3.5 Ribbon representation of the Ca²⁺ ATPase structure in the Ca²⁺-bound form (E1·2Ca²⁺) and in the Ca²⁺-free form (E2·TG). The colors change gradually from the amino terminus (blue) to the carboxy terminus. The two purple circles in the E1·2Ca²⁺ transmembrane region identify the location of bound Ca²⁺, and the dotted arrows show the directions of N, P and A domains movements upon Ca²⁺ dissociation. The dashed circles in E2·TG indicate stabilization of headpiece domains by hydrogen bonding in the Ca²⁺ free conformation. [48].

The A domain includes the cytosolic loop (W107–A241) between the M1 and M2 transmembrane segments. It folds into a prevalently β-stranded rounded (deformed jelly roll) structure, attached to M2 and M3 through two rather long and unstructured loops. In addition, the A domain includes the N-terminal segment (M1–G46) of the ATPase, folded into two short helices adjacent to the β-stranded structure. The 181TGES motif, at the outer edge of the A domain, has been subjected to mutational analysis, suggesting a functional role of these residues in the Ca²⁺ as well as in the Na⁺,K⁺ -ATPase [57, 58]. Furthermore, sites for trypsin (T2 at R198) and proteinase K (L119 and T242) digestion are present on the outer surface of the A domain and its connections to M2 and M3, and have been the subjects of studies revealing catalytically relevant conformational changes [59–65].

3.4.2
Transmembrane region of E1·2Ca²⁺

The ten helical segments that make up the transmembrane domain present significant variation in their length and axial positioning (Figure 3.5). M5 is the longest segment of helical conformation, extending from the luminal side of the membrane to reach the P domain, in close proximity to the catalytic D351 residue. M4 is also rather long, but unwound in the middle portion to split into two helical segments; the upper segment is directly connected to the P domain, clamped with M5 by hydrogen bonds. Unwinding of helical structure is also found in M6, whose upper segment continues with the cytosolic loop L67 that wraps around M5. On the cytosolic side of the membrane, the N terminus, L67, and, of course, the L23 and L45 loops yielding the A and P/N domains are rather long. Conversely, L89 and the C-terminus are of very limited length. The lumenal loops are also short, with the exception of L78, which is substantially longer than the others.

A most important feature of the transmembrane region is the calcium binding domain. Involvement of E309 (M4), E771 (M5), N796, T799 and D800 (M6), and E908 (M8) in calcium binding, was originally suggested by mutational analysis revealing a lack of specific Ca²⁺ functional effects [66]. Mutational interference with calcium binding was later demonstrated directly by the use of high yield systems for recombinant protein expression [26].

Crystallographic analysis of E1·2Ca²⁺ [24] shows two high density peaks, 5.7 Å apart, surrounded by M4, M5, M6 and M8, and attributed to the two calcium ions known to bind cooperatively to the ATPase. The side chains of E771 (M5), N768 (M5), T799 (M6), D800 (M6) and E908 (M8) contribute side-chain oxygen atoms for stabilization of one Ca²⁺ (site I). Stabilization of the other Ca²⁺ (site II) is obtained with side-chain oxygen atoms of E309 (M4), N796 (M6) and D800 (M6), and main-chain carbonyl oxygen atoms of Val304, Ala305 and Ile307. Notably, participation of all these residues is rendered possible by the unwinding of the M4 and M6 helices, and D800 contributes its two acidic side-chain oxygen atoms to coordinate both Ca²⁺ (site I and II). Two water molecules participate in stabilization of site I. Interestingly, homology modeling of the α-subunit of the Na⁺,K⁺ ATPase, based on the atomic models of SERCA1 in the presence and in the absence (see below) of Ca²⁺, yields a satisfactory representation of the binding sites for Na⁺ and K⁺ in the corresponding transmembrane region [67].

3.4.3
Enzyme Structure in the Absence of Ca²⁺ (E2·TG)

The crystal structure (3.1 Å resolution) of the ATPase in the absence of Ca²⁺, and in the presence of the inhibitor thapsigargin (TG), was reported [48] following lower resolution maps derived from EM studies of tubular crystals [46, 47]. A most apparent feature of the E2-TG structure is the compact appearance of the headpiece, resulting from rotation and approximation of the A, N and P domains. The direction of domain movements is shown in Figure 3.5, indicating large rotation of the A domain on the axis normal to the membrane, and changes in the inclination of the N and P domains relative to the plane of the membrane and to each other. The compact conformation of the headpiece appears to be stabilized by hydrogen bonds at the A–N and A–P interfaces, suggesting an important role of the A domain in interfacing headpiece domains. Exposure of the trypsin (T2 at R198) and proteinase K (L199 and E243) sites to the solvent varies in the absence and in the presence of Ca^{2+}, due to the A domain rotation. Very little change in secondary structure is noted, in spite of the large domain movements.

The structure of the membrane-bound region is quite different in the absence, as compared to the presence, of Ca^{2+}. A prominent difference involves the arrangement of the first six (M1–M6) transmembrane helices. Most apparent is the tilting of the upper part of the long M5 segment, to follow the inclination of the P domain where it appears to be integrated with its upper portion, while also connected to the L67 loop at least through hydrogen bonding at R751. As a consequence of such tilting of the upper part, M5 bends in its middle portion, with G770 as a pivoting point. A consensual tilt of M4 is also observed, related to its direct connection to the inclined P domain. Furthermore, M4 and M3 are displaced by about 5.5 Å toward the lumenal surface of the membrane, while M3 presents a curvature in the opposite direction to M5. At the same time, M1 and M2 appear displaced toward the cytosolic surface of the membrane, while M1 is displaced laterally, and its top part is bent. Unwinding of M6 is very important, as it results in rotation of N796, T799 and D800 and displacement of their side chains from optimal positions for Ca^{2+} binding. Also important, in this regard, is the large rotation of the E309 (M4) side chain which faces a direction opposite to that observed in the presence of Ca^{2+}.

3.4.4
Thapsigargin-binding Domain

Thapsigargin (TG) is a plant-derived sesquiterpene lactone that inhibits the SR ATPase with extremely high affinity and specificity [68]. TG interferes with all partial reactions of the catalytic cycle, and induces stabilization of ordered ATPase arrays [69, 70], consistent with a general effect on the enzyme. Mutational studies suggested that the TG binding site is near the M3 segment, at the cytosolic membrane surface, with a prominent influence of the Phe256 residue [71, 72]. In fact, TG can be clearly identified in the E2·TG crystal structure, residing in a cavity delimited by the M3, M5 and M7 helices (Figure 3.6, see p. 76) near the cytosolic surface of the membrane [48]. It is likely that the presence of TG in this cavity confers structural stabilization to the enzyme, thereby preventing conformational responses to cation and substrate binding as required for progression of the catalytic cycle. Different density profiles of the L34 and L78 lumenal loops, detected by electron microscopy in the presence and in the absence of TG, may be related to slight deformation of the protein structure [73].

3.4.5
Interaction with Phospholamban

Phospholamban is a 22 kDa protein containing five identical subunits, and found prevalently in cardiac sarcoplasmic reticulum [74]. Association of phospholamban monomers with SERCA ATPase [75] produces ATPase inhibition, which is mostly observed at low Ca^{2+} concentrations. The inhibition is reversed by phospholamban phosphorylation, which provides a regulatory mechanism for cardiac and slow muscle contraction [76]. Crosslinking [77] and mutational studies [78, 79] have provided information and useful constraints that were used in modeling the interaction of a phospholamban atomic model with the crystal structure of SERCA1a in the absence of Ca^{2+} [80]. A satisfactory fit of phospholamban was found in the transmembrane region, within a groove formed by the juxtaposition of M2, and the upper parts of M4, M6 and M9. The model was extended into the ATPase cytoplasmic region so that its K3 residue can crosslink the K397 and K400 residues of the ATPase. The arrangement causes minor disruption of the phospholamban N-terminal helix near the cytosolic surface of the membrane. Notably, the groove occupied by phospholamban in the transmembrane region is closed in the structure of the enzyme with bound Ca^{2+}, consistent with relief of the phospholamban inhibition by Ca^{2+}.

3.5
Binding of Ligands, Catalytic Events and Conformational Changes

A conformational mechanism for cation transport ATPases was envisioned by early models, suggesting that interconversion of two states, E1 and E2, would explain vectorial displacement of bound Ca^{2+} against a concentration gradient [81, 10]. It is now clear that a conformational mechanism is indeed involved in Ca^{2+} transport by the ATPase, considering the long-range intramolecular linkage between catalytic and Ca^{2+} binding sites, and the evidence provided by functional and structural studies. As explained above, a more realistic view of the catalytic and transport mechanism may be obtained by considering each of the sequential steps of the catalytic and transport cycle (Figure 3.3), and relating free energy and conformational changes that occur with each step and through the entire cycle.

3.5.1
Ca^{2+} Binding and Catalytic Activation

Binding of two calcium ions per ATPase molecule is an absolute requirement for enzyme activation. Fitting the experimental equilibrium binding isotherm (Figure 3.2A) requires a cooperative binding equation, with two separate and different constants for binding of two calcium ions, indicating a sequential binding mechanism, including conformational changes that account for enzyme activation [31]. Spectroscopic studies provided early suggestions of Ca^{2+}-induced conformational effects

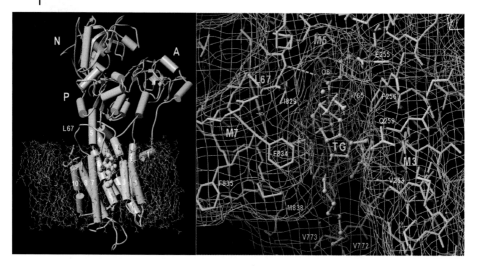

Figure 3.6 Thapsigargin-binding site in the Ca^{2+} ATPase. Note its presence in a cavity surrounded by the M3, M5 and M7 helices, near F256. Hydrophobic residues on the surfaces of these helices are complementary in shape to the TG molecule. [48].

[82, 83]. The recently determined crystal structures of E2·TG [48] and E1·2Ca^{2+} [24] provide a detailed representation of the initial and final conformations for the Ca^{2+} binding reactions, even though possibly influenced by the crystallization conditions (i.e., pH, TG, high Ca^{2+}). It is clear that, as a consequence of Ca^{2+} binding, the side chains of residues involved in binding undergo a remarkable change in orientation, especially that of E309, as well as those of N796, T799, D800 and N768. These changes are permitted by responsive displacement and reorganization of pertinent transmembrane helices, such as M4, M5 and M6. The resulting coordination geometry is different for the two sites (Figure 3.7). Site I includes five side-chain oxygen atoms (Asn768 and Glu771 from M5, Thr799 and Asp800 from M6, and Glu908 from M8) and two water oxygen atoms. Site II includes four side-chain oxygen atoms (Asn796 and both Asp800 side chain oxygens from M6, and Glu309 from M4) and three main chain carbonyl oxygen atoms (Val304, Ala305 and Ile307 from M4). The denomination I and II assumes that the two sites are occupied sequentially as predicted by the cooperative mechanism. In fact, single mutations of site I residues prevent binding on both sites I and II. Conversely, single mutations of site II residues interfere with binding on site II but not on site I [84].

The functional role of the Ca^{2+}-induced conformation lies in its absolute requirement for enzyme activation, transmitted to the extramembranous domains through a long-range intramolecular linkage. The requirement for Ca^{2+} activation involves ATP utilization for phosphoenzyme formation in the forward direction of the cycle, as well as formation of ATP upon addition of ADP to phosphoenzyme

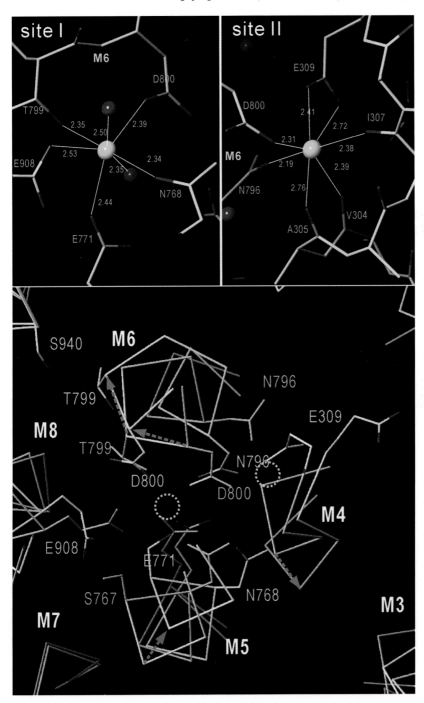

Figure 3.7 Coordination geometry of the two Ca^{2+} sites. Site I includes six side-chain oxygen atoms and two water molecules. Site II has four side-chain and three main chain oxygen atoms. The movements of participating residues following Ca^{2+} dissociation is shown in the lower panel with magenta for E1-Ca2, and atom color for E2-TG. [24, 48].

formed with Pi in the reverse direction of the cycle [85]. Activation is not obtained by Ca^{2+} occupancy of site I, but requires engagement of Glu309 and Asn 796 by Ca^{2+} in site II, which may then be considered a "trigger point" for enzyme activation. Ca^{2+} binding affects directly M4, M5 and M6, and is then extended to other transmembrane helices. This includes straightening of the M5 bent (as well as those of M1 and M3), reversal of the M4 and M5 tilt, rotation of the unwound part of M6, and displacement of M3 and M4 upward, and of M1 and M2 downward [48]. The rearrangement of the transmembrane helices affects the extramembranous region mostly through connections with the P domain. The long M5 helix impinges into the P domain Rossmann fold, and M4 has also a relationship of continuity with the P domain, whereby the upper end of M4 and M5 seem to be constrained by P domain anti-parallel β-strands. The L67 loop is hydrogen bonded to M5 as well as to the P domain, and the upper end of M3 is also connected by hydrogen bonding to the bottom of the P domain. Therefore, the Ca^{2+}-induced rearrangement of transmembrane helices translates into an approximately 30° inclination of the P domain, and a 90° inclination of the N domain with respect to the plane of the membrane. Simultaneously, the A domain undergoes a ca. 110° rotation about an axis normal to the plane of the membrane [24, 48], as the interaction between the conserved [181]TGES (A domain) and the [703]DGVND (P domain) is disrupted. Consequently, the three extra-membranous domains become free to move and, in some instances, are widely separated. As mentioned above, such large movements of the extra-membranous domains are not accompanied by large changes in secondary structure. Nevertheless, single mutations of several residues in the segments connecting the Ca^{2+} binding region with the phosphorylation domain, such as M4 [86], M5 [87] and the L67 loop [88], interfere with the phosphorylation reactions. Notably, the presence of TG, wedged in a cavity delimited by the M3, M5 and M7 helices near the cytosolic surface of the membrane, is highly inhibitory. Therefore, a responsive and highly precise arrangement of transmembrane helices and extra-membranous domains is required for catalytic activation.

3.5.2
Nucleotide Binding and Substrate-induced Conformational Fit

Studies of crystals soaked with TNP-AMP [24] revealed that the adenosine moiety of this inactive nucleotide analog is located within a flap of the N domain, where nucleotide binding is interfered with by chemical derivatization or mutation of F487, K492 and K515 and R560 [13, 89–91]. Accordingly, a recent model (Figure 3.8, see p. 80) based on ATP·Fe^{2+} (replacing Mg^{2+}) catalyzed oxidation, and on the crystal structure of the ATPase in the E2·TG state, places the adenosine moiety in a pocket delimited by K492 and K515, with stabilization provided by R560 and F487 [92]. In the model, the ATP·Mg^{2+} complex is in a folded configuration, with Mg^{2+} stabilized by γ- and β-phosphate oxygen atoms and by the Thr441 side-chain (N domain), while the β-phosphate approximates Thr353 (P domain), and the γ-phosphate approximates Asn359 and Asp601 (hinge region). This arrangement results from early collision of the ATP-Mg complex with the ATPase in the open conformation, since Ca^{2+} is required for the Fe^{2+}-catalyzed oxidation. However, the N domain must then approximate the P domain, and the geometry of ATP be rearranged to place its terminal phosphate near D351 (P domain), i. e. the residue undergoing catalytic phosphorylation.

Useful information was provided in this regard by ATPase protection from digestion with proteinase K, suggesting a nucleotide-dependent conformational change that includes repositioning of the A domain concomitant with approximation of the N and P domains [62]. Mutational analysis of the nucleotide-dependent protection [93] indicates that R560 and E439 (N domain), as well as D351, K352, T353, K684, D703, N706 and D707 (P domain) are involved in this effect. Such a participation of N and P residues demonstrates that approximation of the two domains does indeed occur, as required to accommodate the nucleotide molecule by means of adenosine moiety interaction with the N domain, as well as phosphate interaction with the P domain. The A domain must also reposition through interaction with the N domain so as to protect its loop containing the site for proteinase K digestion. Most importantly, the Ca^{2+} requirement for nucleotide protection indicates that even though a compact arrangement of the headpiece is favored by nucleotide binding, the transmembrane domain retains bound Ca^{2+}. The compact headpiece conformation obtained under these conditions is not identical to that observed in the absence of Ca^{2+}, but evidently represents an additional specific state produced by substrate induced conformational fit of the three cytoplasmic domains on the Ca^{2+} activated enzyme. An apparently similar approximation of nucleotide binding ("fingers") and catalytic ("palm") domains is known to occur in DNA polymerases [94].

Although ATP is likely to form an initial complex with Mg^{2+} due to the cation binding property of the nucleotide, Mg^{2+} is not required for nucleotide protection of ATPase digestion by proteinase K [93]. This suggests that even though initial binding of ATP-Mg^{2+} complex may occur, ultimately ATP and Mg^{2+} reach the catalytic site through a random mechanism [95]. Accordingly, nucleotide binding (and its conformational effect) can be obtained either in the presence or in the absence of Mg^{2+}, even though the subsequent phosphoryl transfer and hydrolytic reactions require Mg^{2+}. Interestingly, structural snapshots [96] of the reaction cycle of phosphoserine phosphatase (PSPase, another member of the haloacid dehalogenase superfamily structurally analogous to the Ca^{2+} ATPase) show that the initial substrate–enzyme complex does not include Mg^{2+}, while interaction with Mg^{2+} occurs concomitantly with the phosphoryl transfer and hydrolytic reactions. The additional stabilization provided by Mg^{2+} to the transition state, relative to that of the enzyme–substrate complex, is of definite kinetic advantage.

It is then apparent that the adenosine moiety of the nucleotide substrate resides in the N domain, within a positively charged pocket delimited by K492 and K515, and is stabilized by F487 through ring stacking and by R560 through interaction with phosphate oxygens. Conversely, the phosphate moiety interacts with P domain residues [93], with direct stabilization of γ-phosphate oxygen by K684 in analogy to stabilization of phosphoserine by the corresponding K144 in the PSPase

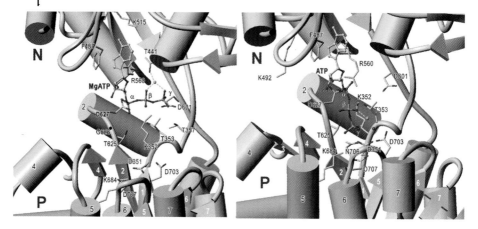

Figure 3.8 Models of ATP binding to the Ca²⁺ ATPase following substrate-induced approximation of phosphorylation (P) and nucleotide (N) domains. Left: Initial collision of ATP-Mg complex, with folded configuration of the phosphate moiety, and engagement of T441 [92]. Right: ATP bound at substrate site, with γ-phosphate close to reactive residue Asp351. Note that Mg²⁺ bound to ATP (green sphere) would not match the Mg²⁺ position expected in the phosphoenzyme intermediate (purple sphere), indicating that ATP and Mg²⁺ bind independently at the catalytic site [93]. The close homology to the structure of the phosphorylserine substrate site (step II in Figure 3.9) is notable. Modeling was based on the structures of the E2·TG domain [48] and of the ATPase containing the MgFx complex (93).

[96]. The geometry of ATP bound at the substrate site, and the role of additional P domain residues, are illustrated by a model based on the highly homologous atomic model of the phosphoserine phosphatase (PSPase) with its substrate bound. The ATPase model was obtained by combining that for the N-domain taken from the E2 (TG) form and that for the P-domain from the MgFx complex (considered to be an E-P analog). ATP in an extended form was placed so that the γ-phosphate comes exactly to the same position as that of the phosphate in phosphoserine. This conformation is stabilized by K352, T353, G626, K684 and N706, in close homology with the PSPase model and consistent with ATPase mutational analysis [93].

3.5.3
Phosphoryl Transfer and Hydrolytic Cleavage

The Ca²⁺ ATPase mechanism includes a covalent (aspartylphosphate) intermediate, general acid–base catalysis and metal-ion assistance by Mg²⁺. The covalent intermediate is formed through attack of the electrophilic ATP terminal phosphate by a nucleophilic carboxylate group within the catalytic site. The phosphoenzyme intermediate reacts then with water to yield inorganic phosphate.

It is useful to consider the Ca²⁺ ATPase mechanism in the light of the PSPase (Figure 3.9), due to a remarkable structural and catalytic analogy, and the high-resolution structures available for the PSPase catalytic intermediates [96]. D11 is the residue undergoing phosphorylation in the PSPase, corresponding to D351 in the ATPase. In the PSPase, D13 serves as a general acid for proton donation to the leaving group (i. e., serine), a role that may be sustained by the corresponding T353 in the Ca^{2+} ATPase with regard to leaving ADP. The same residue would then act as a general base, extracting a proton from nucleophilic water involved in hydrolytic cleavage of the phosphoenzyme. Possibly related to this function, mutation of T353 reduces ATPase activity by 40 %. Furthermore, mutation of T353 interferes with the conformational change produced by nucleotide binding [93], suggesting involvement of T353 in hydrogen bonding and substrate stabilization. Upon headpiece closure, the ATP terminal phosphate gains an optimal position relative to the D351 nucleophile (II in the PSPase scheme), through interactions with K684, K352, T625, G626, T353 and N706, whose mutations produce total or partial ATPase inhibition.

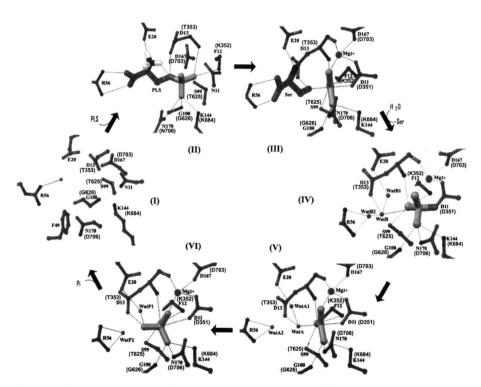

Figure 3.9 Structural snapshots of the phosphorylserine phosphatase (PSPase) reaction cycle as a reference to the homologous catalytic site of Ca^{2+} ATPase. (I) Apo-enzyme structure. (II) Substrate (phosphorylserine, bound structure, using mutant D11N. (III) Model of serine-bound transition state structural analog. (IV) Phospho-aspartyl intermediate structural analog: PSPase-BeF$_3$ complex. (V) Transition state structural analog: A1F$_3$-PSPase complex. (VI) Product, Pi, bound structure. The corresponding residues of the Ca^{2+} ATPase are shown in parentheses. (Adapted from Ref. 96).

In analogy with the PSPase mechanism (Figure 3.9), we consider that the subsequent ADP-bound transition state (III in the PSPase scheme) includes Mg^{2+}, which is coordinated by D703, T353, D351, and phosphate oxygen, providing conformational stabilization as well as electrophilic catalytic assistance. At this stage, the ATP terminal phosphate interacts with D351, while its oxygen atoms interact with Mg^{2+}, as well as with side or main chains of K352, T353, G626, K684 and N706, whose positions are now constrained by the pentacovalent transition state of the phosphate group. This conformation allows the ADP product to leave, aided by proton donation, possibly by T353, within the context of general acid catalysis. The consequent phospho-aspartyl intermediate (IV in the PSPase scheme) acquires bound water, whose nucleophilic character is enhanced through proton extraction by the same residue, leading to phosphoenzyme cleavage.

3.5.4
Interconversion of Phosphorylation and Ca²⁺ Binding Potentials

Due to remarkable structural and functional analogies at the active site, the PSPase cycle outlined in Figure 3.9 is an excellent term of comparison for the Ca^{2+} ATPase catalytic chemistry. However, a specific and important difference lies in the substrate utilized by the two enzymes, as the equilibrium constant for cleavage of the ATP phosphoanhydride bond is considerably higher than that for cleavage of the phosphoserine phosphoester bond [97]. Accordingly, a specific feature of the Ca^{2+} ATPase is its ability to harvest free energy derived from ATP utilization, and use it to displace bound Ca^{2+} through a long-range intramolecular linkage.

The specific step related to interconversion of phosphorylation and Ca^{2+} binding potentials is often referred to as $E1\text{-}P\cdot2Ca^{2+} \leftrightarrow E2\text{-}P\cdot2Ca^{2+}$, to indicate the reversible transition of an intermediate of high phosphorylation potential and high Ca^{2+} affinity, to an intermediate of low phosphorylation potential and low Ca^{2+} affinity (Figure 3.3). It is clear that the high or low phosphorylation potential of the phophoenzyme intermediate is related to whether the Ca^{2+} binding domain is in the high- or low-affinity configuration, respectively. In view of the functional and structural information described so far, it is apparent that small conformational changes occurring concomitantly with ATP utilization at the catalytic site are amplified and transmitted to the Ca^{2+} binding domain through a long-range relay mechanism. Inversion of phosphoryl oxygen atoms during the transition state of the phosphorylation reaction, and displacement of residues interacting with these oxygen atoms, commence the triggering perturbation [98]. In fact, formation of a stable transition state analog with vanadate reduces the ATPase affinity for Ca^{2+} [99] and interferes with the Ca^{2+}-induced rise of ATPase intrinsic fluorescence [100].

Considering the residues assisting catalytic chemistry (Figure 3.9), N706 appears to have an interesting role, as the corresponding N170 of the PSPase undergoes significant displacement when the transition state for the phosphoryl transfer reaction is formed following substrate binding. In fact, this residue provides a nitrogen atom for coordination of the same phosphate oxygen as K684. D703 (corresponding to D167 of the PSPase) is also interesting, as it is engaged in Mg^{2+} chelation in concomitance with the transition state conformational change.

The interest in N706 and D703 engagement lies in the reversible interaction of the conserved 703DGVND sequence (P domain) with the 181TGES sequence (A domain), which plays a prominent role in stabilizing the relative positions of A, P and N domains. Notably, Ca^{2+} binding releases and allows the A domain to rotate, as indicated by changes in the susceptibility of an A domain loop to proteolyitc digestion [64, 65], and therefore the 703DGVND and TGES184 are separated. Nucleotide binding, however, reverses this effect [62], with further enhancement by the phosphorylation reaction, whereby the stabilizing interaction of the two conserved P and A segments is again favored. Noteworthy is that the A domain is linked directly to M1, M2 and M3, and its rotation must be coupled to displacement of these transmembrane helices that have functionally important hydrogen bond connections to M4 near the Ca^{2+} binding domain, and to the L67 loop that is in turn hydrogen bonded to the P domain. This explains how positioning of the A domain can be influenced by events at the phosphorylation site, which may be considered the "trigger point" for long-range displacement of bound Ca^{2+}. The important functional role of this connection is demonstrated by the complete inhibition of ATPase activity resulting from proteolytic cut of the link between the A domain and M3 [64, 101]. In addition to the 181TGES and 703DGVND interaction, the A and P domain interface includes the P6 helix and the 239MAATEQ loop connecting the A domain to M3. Mutagenesis of 181TGES [13] and excision of 239MAATE with proteinase K [65] provide further evidence for the importance of this interface for the energy transduction mechanism. An important functional role of the conserved 703DGVND was also demonstrated for the Na^+,K^+ ATPase [102].

It is important to recognize that rotation of the A domain allows a large hinge-bending motion of the N domain, most prominently upon Ca^{2+} binding and dissociation [24, 48, 103], but also upon nucleotide binding and enzyme phosphorylation [62]. Bending of the N domain, and its interaction with the P domain, is related to specific and Ca^{2+}-dependent functional events, i.e. ATP binding to the N domain and phosphoryl transfer within the P domain. The functional relevance of N and P domain displacement, relative to each other, was demonstrated by experiments on crosslinking R678 (P domain) and K492 (N domain) with gluteraldehyde, which is influenced by the occurrence of catalytic reactions [104].

An interesting question relates to the pathway for vectorial translation of Ca^{2+} through the protein. It seems likely that Ca^{2+} access to the Ca^{2+}-deprived enzyme (E2) occurs through an aqueous channel, lined with negative charge provided by M1 and A domain residues near the cytosolic membrane interface, and leading to the binding sites in the transmembrane region. Cooperative binding of two Ca^{2+} then occurs in exchange for H^+, with a concomitant conformational change. Engagement of E309 on site II serves as a gating device for occlusion of bound Ca^{2+} [105], and as the trigger for long-range activation of the phosphorylation reaction [85]. Therefore, in the E2 state, the Ca^{2+} sites are open to the cytosolic medium and become occluded following Ca^{2+} binding (E1·$2Ca^{2+}$ state). It is then the phosphorylation reaction, and the concomitant conformational transition, that opens an access to the luminal side, permitting dissociation of Ca^{2+} and binding of H^+. The effect of enzyme phosphorylation on the calcium sites begins with small displacements of catalytic site residues, amplified to produce headpiece domain motions, including A domain rotation. Stress is consequently placed on M1, M2 and M3, as well as M4, M5 and the L67 loop, resulting in perturbation of the calcium binding sites between M4, M5, M6 and M8. Release of calcium is then followed by hydrolytic cleavage of the phosphorylated intermediate, as in the absence of bound Ca^{2+} the enzyme manifests catalytic inclusion of water concomitant with the highly reversible phosphorylation reaction with Pi [34, 106].

In conclusion, the transduction mechanism relies on a long-range linkage of transmembrane helices and cytosolic domains, whereby large-scale and mutually exclusive motions are triggered by high-affinity Ca^{2+} binding at one end, and ATP utilization at the other. This is rendered possible by the ability of the protein to utilize free energy derived from binding and phosphorylation reactions. Thereby, the cycle proceeds forward to Ca^{2+} release, or in reverse to ATP synthesis, depending on the concentrations of ligands, substrate and products.

References

1. S. Ebashi and F. Lippman, *J. Cell. Biol.*, **1962**, *14*, 389–400.
2. W. Hasselbach and M. Makinose, *Biochem. 2.* **1961**, *333*, 518–528.
3. D. H. MacLennan, C. J. Brandl, B. Korczak, and N. M. Green, *Nature*, **1985**, *316*, 696–700.
4. G. Allen, B. J. Trinnaman, and N. M. Green, *Biochem. J.*, **1980**, *187*, 591–616.
5. E. Carafoli, L. Santella, D. Branca, and M. Brini, *Crit. Rev. Biochem. Mol. Biol.*, **2001**, *36*, 107–260.
6. A. Odermatt, P. E. Taschner, V. K. Khanna, H. F. Busch, G. Karpati, C. K. Jablecki, M. H. Breuning, and D. H. MacLennan, *Nat. Gen.*, **1996**, *14*, 191–194.
7. A. Sakuntabhai, V. Ruiz-Perez, S. Carter, N. Jacobsen, S. Burge, S. Monk, M. Smith, C. S. Munro, M. O'Donovan, N. Craddock, R. Kucherlapati, J. L. Rees, M. Owen, G. M. Lathrop, A. P. Monaco, T. Strachan, and A. Hovnanian, *Nat. Gen.*, **1999**, *21*, 271–277.
8. Z. Hu, J. M. Bonifas, J. Beech, G. Bench, T. Shigihara, H. Ogawa, S. Ikeda, T. Mauro, and E. H. Epstein, Jr., *Nat. Gen.*, **2000**, *24*, 61–65.
9. J. K. Gwathmey, L. Copelas, R. MacKinnon, F. J. Schoen, M. D. Feldman, W. Grossman, and J. P. Morgan, *Circ. Res.*, **1987**, *61*, 70–76.
10. L. Meis de and A. L. Vianna, *Annu. Rev. Biochem.*, **1979**, *48*, 275–292.
11. D. J. Bigelow and G. Inesi, *Biochim. Biophys. Acta*, **1992**, *1113*, 323–338.
12. G. Inesi, D. Lewis, D. Nikic, A. Hussain, and M. E. Kirtley, *Adv. Enzymol. Relat. Areas Mol. Biol.*, **1992**, *65*, 185–215.
13. J. P. Andersen, *Biosci. Rep.*, **1995**, *15*, 243–261.
14. J. V. Moller, B. Juul, and M. le Maire, *Biochim. Biophys. Acta*, **1996**, *1286*, 1–51.
15. D. H. MacLennan, W. J. Rice, and N. M. Green, *J. Biol. Chem.*, **1997**, *272*, 28815–28818.
16. E. Mintz and F. Guillain, *Biochem. Biophys. AcTp* **1997**, *1318*, 52–70.
17. J. M. East, *Mol. Membr. Biol.*, **2000**, *17*, 189–200.

18. F. Wuytack, L. Raeymaekers, and L. Missiaen, *Cell Calcium*, **2002**, *32*, 279–305.

19. N. M. Green and D. L. Stokes, *Annu. Rev. Biophys. Biomol. Struct.*, **2003**, *32*, 445–468.

20. C. Toyoshima, *Handbook of Metalloproteins*, **2003**, John Wiley and Sons.

21. R. J. Coll and A. J. Murphy, *J. Biol. Chem.*, **1984**, *259*, 14249–14254.

22. D. Levy, A. Gulik, A. Bluzat, and J. L. Rigaud, *Biochim. Biophys. Acta*, **1992**, *1107*, 283–298.

23. D. L. Stokes and N. M. Green, *Biophys. J.*, **1990**, *57*, 1–14.

24. C. Toyoshima, M. Nakasako, H. Nomura, and H. Ogawa, *Nature*, **2000**, *405*, 647–655.

25. K. Maruyama and D. H. MacLennan, *Proc. Natl. Acad. Sci. U. S.A.*, **1988**, *85*, 3314–3318.

26. C. Strock, M. Cavagna, W. E. Peiffer, C. Sumbilla, D. Lewis, and G. Inesi, *J. Biol. Chem.*, **1998**, *273*.

27. L. de Meis , *Biochim. Biophys. Acta*, **1969**, *172*, 343–344.

28. G. Inesi, *Science*, **1971**, *171*, 901–903.

29. X. Yu, S. Carroll, J. L. Rigaud, and G. Inesi, *Biophys. J.*, **1993**, *64*, 1232–1242.

30. T. Yamamoto and Y. Tonomura, *J. Biochem. (Tokyo)*, **1967**, *62*, 558–575.

31. G. Inesi, M. Kurzmack, C. Coan, and D. Lewis, *J. Biol. Chem.*, **1980**, *255*, 3025–3031.

32. G. Inesi, C. Coan, S. Verjovksi-Almeida, M. Kurzmack, and D. And Lewis, *Ann. New York Acad. Sci.*, **1978**, *11*, 1212–1219.

33. M. Makinose and W. Hasselbach, *FEBS Lett.*, **1971**, *12*, 271–272.

34. H. Masuda and L. Meis de, *Biochemistry*, **1973**, *12*, 4581–4585.

35. A. F. Knowles and E. Racker, *J. Biol. Chem.*, **1975**, *250*, 3538–3544.

36. L. Meis de and R. K. Tume, *Biochemistry*, **1977**, *16*, 4455–4463.

37. M. Makinose, *FEBS Lett.*, **1973**, *37*, 140–143.

38. G. Inesi, M. Kurzmack, and D. Lewis, *Methods Enzymol.*, **1988**, *157*, 154–190.

39. G. Inesi, T. Watanabe, C. Coan, and A. Murphy, *Ann. New York Acad. Sci.*, **1982**, *402*, 515–534.

40. C. M. Pickart and W. P. Jencks, *J. Biol. Chem.*, **1984**, *259*, 1629–1643.

41. G. Inesi, *Annu. Rev. Physiol.*, **1985**, *47*, 573–601.

42. X. Yu and G. Inesi, *J. Biol. Chem.*, **1995**, *270*, 4361–4367.

43. W. S. Smith, R. Broadbridge, J. M. East, and A. G. Lee, *Biochem. J.*, **2002**, *361*, 277–286.

44. L. de Meis, *J. Membr. Biol.*, **2002**, *111*, 1–9.

45. L. Dux and A. Martonosi, *J. Biol. Chem.*, **1983**, *258*, 2599–2603.

46. C. Toyoshima, H. Sasabe, and D. L. Stokes, *Nature*, **1993**, *362*, 467–471.

47. P. Zhang, C. Toyoshima, K. Yonekura, N. M. Green, and D. L. Stokes, *Nature*, **1998**, *392*, 835–839.

48. C. Toyoshima and H. Nomura, *Nature*, **2002**, *418*, 605–611.

49. T. Hisano, Y. Hata, T. Fujii, J. Q. Liu, T. Kurihara, N. Esaki, and K. Soda, *Proteins*, **1996**, *24*, 520–522.

50. L. Aravind, M. Y. Galperin, and E. V. Koonin, *Trends Biochem. Sci.*, **1998**, *23*, 127–129.

51. W. Wang, R. Kim, J. Jancarik, H. Yokota, and S. H. Kim, *Structure (Camb.)*, **2001**, *9*, 65–71.

52. U. Pick, *Eur. J. Biochem.*, **1981**, *121*, 187–195.

53. C. Mitchinson, A. F. Wilderspin, B. J. Trinnaman, and N. M. Green, *FEBS Lett.*, **1982**, *146*, 87–92.

54. H. Yamamoto, M. Tagaya, T. Fukui, and M. Kawakita, *J. Biochem. (Tokyo)*, **1988**, *103*, 452–457.

55. D. B. McIntosh, D. G. Woolley, and M. C. Berman, *J. Biol. Chem.*, **1992**, *267*, 5301–5309.

56. D. B. McIntosh, D. G. Woolley, B. Vilsen, and J. P. Andersen, *J. Biol. Chem.*, **1996**, *271*, 25778–25789.

57. P. L. Jorgensen and J. Petersen, *Biochim. Biophys. Acta*, **1985**, *821*, 319–333.

58. P. L. Jorgensen and J. P. Andersen, *J. Membr. Biol.*, **1988**, *103*, 95–120.

59. Y. Imamura and M. Kawakita, *J. Biochem. (Tokyo)*, **1989**, *105*, 775–781.

60. J. P. Andersen, B. Vilsen, E. Leberer, and D. H. MacLennan, *J. Biol. Chem.*, **1989**, *264*, 21018–21023.

61. D. M. Clarke, T. W. Loo, and D. H. MacLennan, *J. Biol. Chem.*, **1990**, *265*, 22223–22227.

62. S. Danko, K. Yamasaki, T. Daiho, H. Suzuki, and C. Toyoshima, *FEBS Lett.*, **2001**, *505*, 129–135.

63. M. le Maire, S. Lund, A. Viel, P. Champeil, and J. V. Moller, *J. Biol. Chem.*, **1990**, *265*, 1111–1123.

64. B. Juul, H. Turc, M. L. Durand, Gomez, d. G., Denoroy, L., Moller, J. V, Champeil, P., and le Maire, M., *J. Biol. Chem.* **1995**, *270*, 20123–20134.

65. J. V. Moller, G. Lenoir, C. Marchand, Montigny, C., le Maire, M., Toyoshima, C., Juul, B. S., and Champeil, P., *J. Biol. Chem.*, **2002**, *277*, 38647–38659.

66. D. M. Clarke, T. W. Loo, G. Inesi, and D. H. MacLennan, *Nature*, **1989**, *339*, 476–478.

67. H. Ogawa and C. Toyoshima, *Proc. Natl. Acad. Sci. U. S. A.*, **2002**, *99*, 15977–15982.

68. Y. Sagara and G. Inesi, *J. Biol. Chem.*, **1991**, *266*, 13503–13506.

69. Y. Sagara, J. B. Wade, and G. Inesi, *J. Biol. Chem.*, **1992**, *267*, 1286–1292.

70. Y. Sagara, F. Fernandez-Belda, L. Meis de, and G. Inesi, *J. Biol. Chem.*, **1992**, *267*, 12606–12613.

71. M. Yu, L. Zhong, A. K. Rishi, M. Khadeer, G. Inesi, A. Hussain, and L. Zhang, *J. Biol. Chem.*, **1998**, *273*, 3542–3546.

72. L. Zhong and G. Inesi, *J. Biol. Chem.*, **1998**, *273*, 12994–12998.

73. H. S. Young, C. Xu, P. Zhang, and D. L. Stokes, *J. Mol. Biol.*, **2001**, *308*, 231–240.

74. M. Tada, M. A. Kirchberger, and A. M. Katz, *J. Biol. Chem.*, **1975**, *250*, 2640–2647.

75. A. D. Wegener and L. R. Jones, *J. Biol. Chem.*, **1984**, *259*, 1834–1841.

76. M. A. Kirchberger and M. Tada, *J. Biol. Chem.*, **1976**, *251*, 725–729.

77. P. James, M. Inui, M. Tada, M. Chiesi, and E. Carafoli, *Nature*, **1989**, *342*, 90–92.

78. T. Toyofuku, K. Kurzydlowski, M. Tada, and D. H. MacLennan, *J. Biol. Chem.*, **1994**, *269*, 3088–3094.

79. T. Toyofuku, K. Kurzydlowski, M. Tada, and D. H. MacLennan, *J. Biol. Chem.*, **1994**, *269*, 22929–22932.

80. C. Toyoshima, M. Asahi, Y. Sugita, R. Khanna, T. Tsuda, and D. H. MacLennan, *Proc. Natl. Acad. Sci. U. S.A.*, **2003**, *100*, 467–472.

81. R. Albers, *Annu. Rev. Biochem.*, **1967**, *36*, 727–756.

82. Y. Dupont, *Biochem. Biophys. Res. Commun.*, **1976**, *71*, 544–550.

83. C. Coan, S. Verjovski-Almeida, and G. Inesi, *J. Biol. Chem.*, **1979**, *254*, 2968–2974.

84. Z. Zhang, D. Lewis, C. Strock, G. Inesi, M. Nakasako, H. Nomura, and C. Toyoshima, *Biochemistry*, **2000**, *39*, 8758–8767.

85. G. Inesi, Z. Zhang, and D. Lewis, *Biophys. J.*, **2002**, *83*, 2327–2332.

86. Z. Zhang, C. Sumbilla, D. Lewis, S. Summers, M. G. Klein, and G. Inesi, *J. Biol. Chem.*, **1995**, *270*, 16283–16290.

87. T. L. Sorensen and J. P. Andersen, *J. Biol. Chem.*, **2000**, *275*, 28954–28961.

88. Z. Zhang, D. Lewis, C. Sumbilla, G. Inesi, and C. Toyoshima, *J. Biol. Chem.*, **2001**, *276*, 15232–15239.

89. A. J. Murphy, *Arch. Biochem. Biophys.*, **1977**, *180*, 114–120.

90. K. Yamagata, T. Daiho, and T. Kanazawa, *J. Biol. Chem.*, **1993**, *268*, 20930–20936.

91. S. Hua, H. Ma, D. Lewis, G. Inesi, and C. Toyoshima, *Biochemistry*, **2002**, *41*, 2264–2272.

92. S. Hua, G. Inesi, H. Nomura, and C. Toyoshima, *Biochemistry*, **2002**, *41*, 11405–11410.

93. H. Ma, G. Inesi, and C. Toyoshima, *J. Biol. Chem.*, **2003**, *278* (31), 28938–28943.

94. T. A. Steitz, *J. Biol. Chem.*, **1999**, *274*, 17359–17398.

95. J. Reinstein and W. P. Jencks, *Biochemistry*, **1993**, *32*, 6632–6642.

96. W. Wang, H. S. Cho, R. Kim, J. Jancarik, H. Yokota, H. H. Nguyen, I. V. Grigoriev, D. E. Wemmer, and S. H. Kim, *J. Mol. Biol.*, **2002**, *319*, 421–431.

97. P. J. Romero and L. Meis de, *J. Biol. Chem.*, **1989**, *264*, 7869–7873.

98. G. Inesi, C. Sumbilla, and M. E. Kirtley, *Physiol. Rev.*, **1990**, *70*, 749–760.

99. G. Inesi, D. Lewis, and A. J. Murphy, *J. Biol. Chem.*, **1984**, *259*, 996–1003.

100. B. F. Fernandez, D. A. Garcia, and G. Inesi, *Biochim. Biophys. Acta*, **1986**, *854*, 257–264.

101. P. L. Jorgensen and J. H. Collins, *Biochim. Biophys. Acta*, **1986**, *860*, 570–576.

102. P. A. Pedersen, J. R. Jorgensen, and P. L. Jorgensen, *J. Biol. Chem.*, **2000**, *275*, 37588–37595.

103. C. Xu, W. J. Rice, W. He, and D. L. Stokes, *J. Mol. Biol.*, **2002**, *316*, 201–211.

104. D. C. Ross, G. A. Davidson, and D. B. McIntosh, *J. Biol. Chem.*, **1991**, *266*, 4613–4621.

105. B. Vilsen and J. P. Andersen, *Biochemistry*, **1998**, *37*, 10961–10971.

106. D. B. McIntosh and P. D. Boyer, *Biochemistry*, **1983**, *22*, 2867–2875.

107. D. Scales and G. Inesi, *Membr. Biophys. J.*, **1976**, *16*, 735–751.

4
The Na,K-ATPase: A Current Overview

Jack H. Kaplan

4.1
Introduction

In contrast to other P-type ATPase chapters in this volume, it was decided to include an overview, rather than a detailed, complete review of this protein and its properties. The major reason for this is that there have appeared many authoritative and thorough reviews of the system and its various properties during the last several years.

This overview will summarize the various areas of research activity and point the reader to the most recent reviews of the large literature dedicated to the Na,K-ATPase. In instances where the area has not yet received a thorough review, recent articles will be cited. One reason for the widespread interest and voluminous literature on this particular protein lies in its central physiological role in higher erukaryotes, and especially humans.

The Na,K-ATPase is central to the functioning of almost all cells in humans for the generalized purposes of volume control and electrolyte balance. The uptake of nutrients and removal of other substrates often occur via mechanisms that are dependent on an inwardly directed electrochemical potential gradient for Na ions. This gradient is maintained by the operation of the Na,K-ATPase. In addition certain organs such as kidney and intestine, are essential for the maintenance of electrolyte, fluid and metabolite balance in the intact organism. These epithelial systems are able to carry out their roles because of the controlled and specific tissue and cellular localizations of the Na,K-ATPase. At the cellular level the protein is only located at the basolateral surface of the epithelial cells of the kidney and gastrointestinal system, and in the kidney, the nephron shows different and characteristic Na,K-ATPase densities along its length.

The Na, K-ATPase is most closely related to the gastric H, K-ATPase that is the subject of another Chapter in this volume, and together they represent the only P-type ATPases that are composed of more than one essential subunit. There are many close similarities in the structures and properties of these two heterodimeric ion pumps.

Handbook of ATPases. Edited by M. Futai, Y. Wada, J. H. Kaplan
Copyright © 2004 WILEY-VCH Verlag GmbH & Co. KGaA, Weinheim
ISBN 3-527-30689-7

In this overview I present a summary of the areas of interest and references to recent authoritative reviews by leading workers in this field. Where recent reviews are not available, references to recent key papers are provided. Most of the contemporary work in the field has been influenced by the application of mutagenic analysis in heterologous expression systems, the application of refined methods of protein chemistry, and most recently by the incorporation of the high resolution structures obtained from the closely related (but not identical), SERCA, the Ca-ATPase from the sarcoplasmic reticulum of muscle, that is reviewed elsewhere in this volume.

4.2
Structure-Function and Reaction Mechanism

The relation between structure and function, and the mechanism of energy transduction or coupling has received attention in several recent review articles [1, 2, 3]. These articles have made use of the new SERCA structures from the group of Toyoshima, and a useful review article appeared quite soon after the appearance of these structures that systematically compared the structure of SERCA with what could be deduced about the structure of the α-subunit of the Na,K-ATPase [4]. A detailed analysis of structural changes taking place in the M5M6 loop that accompany the E1 to E2 conformational transition provides a good example of the use of modeling of Na,K-ATPase onto the high resolution SERCA structure and deducing possible details of conformational transitions.[5]. The essential involvement of this region in cation-dependent coupling in P-type pumps was pointed out earlier, when the inherent instability or mobility of these transmembrane segments was reported [6, 7].

4.3
The Roles of the β-Subunit

The α-subunit is the major catalytic subunit and shows great similarity with SERCA. The second essential subunit of the Na,K-ATPase, and of the closely related H,K-ATPase, is structurally unrelated to the α-subunit. It is a smaller type II membrane protein, having around 300 aminoacids and is extensively glycosylated [8, 9]. Prior to the appearance of heterologous expression and cloning, it was extremely difficult to investigate its function since any methods used to separate the two subunits caused enzyme inactivation. Several studies showed that reduction of the disulfide bonds in the extracellular domain of the β-subunit resulted in the loss of activity [10, 11]. The presence of K ions protected against this reduction and associated loss in activity. This led to the suggestion that the β-subunit was essentially involved in stabilizing the K occluded state [11]. There was also evidence that the β-subunit alters its conformation during the reaction cycle, as the α-subunit binds the various pump ligands [12]. This implies that conformational changes in the β-subunit may play a functional role in pump mechanism. This possibility has not been rigorously examined.

The successful cloning and possibility of heterologously expressing the isolated subunits has enabled an experimental approach to be used in investigating the role of the β-subunit. It is clear that in order to achieve successful delivery of a functional Na,K-ATPase to the plasma membrane both subunits are required and that a major role of the β-subunit is in acting as a chaperone in allowing exit of the assembled αβ-heterodimer from the endoplasmic reticulum [13, 14]. That this is so is supported by the observation that when β alone is expressed, it is delivered to the plasma membrane [14, 15]. Expression of α in the absence of β results in retention in the endoplasmic reticulum [14, 15]. The roles of the β-subunit in assembly and delivery have recently been reviewed [16]. Much of the work has been carried out in *Xenopus* oocytes, a system that has proven to be very useful in such studies, but one that does have certain limitations. Recent studies in insect cells, that lack endogenous Na,K-ATPase, have enabled a more complete analysis of the steps involved and of the molecular determinants that may be important in the separable processes of assembly, exit from the endoplasmic reticulum and delivery to the plasma membrane. It seems clear that in addition to important interactions in the extracellular domain of the β-subunit and the M7M8 loop of the α-subunit, other interactions involving the intramembrane segments also play a role [15].

4.4
A Third Subunit

In 1978, from the results of a study of photolabelling of the renal Na,K-ATPase with a photoaffinity analog of ouabain, it was proposed that a small additional peptide component formed a part of the enzyme complex [17]. This proposal remained largely untested for several years. More recently it became clear that this small peptide was clearly associated with the two major subunits of the Na,K-ATPase in renal tissue [18, 19]. Evidence has been accumulated that this peptide may have a physiologically important role. There have been several reviews of this area, and now it seems clear that the γ-subunit appears in several forms and is but one member of the family of FXYD protein regulators, that include CHIF, phospholemman etc. [20, 21]. The current picture emerging is that substrate affinities seem to be altered by the presence of γ, although it is not yet certain that this is the major or only function of this subunit. In a very interesting recent development an inherited form of hypomagnesia has been associated with a mutation in the γ-subunit [22]. Thus while the details of the physiological role of this subunit have not yet been resolved, it is clear that its presence in the kidney is of significance to human electrolyte balance.

4.5
Regulation of Na,K-ATPase Function

In most cases, regulation of the function of the Na,K-ATPase in cells occurs by altering the number of pump molecules present in the plasma membrane. A very thorough and complete review of the many aspects of the regulation of Na,K-ATPase function appeared recently [23]. The number of Na pump molecules in the plasma membrane in a cell is the resultant of insertion processes and retrieval processes (endocytosis and exocytosis). Both of these pathways are under regulation by complex sets of cellular proteins, and probably include in addition adaptor proteins and cytoskeletal elements among others. These pathways are central to many cell biological processes. In order to influence the rates of insertion or retrieval for a particular membrane protein it is necessary to have some kind of signal that may modulate the protein:protein interactions that underly these processes. During the last few years it has become apparent that such signals, in response to the actions of extracellular effectors, such as dopamine, are produced by phosphorylation of the α-subunit of the Na,K-ATPase by specific protein kinases at a number of sites. The short-term regulation of this type, involving PKC isoforms has recently been reviewed [24]. An added complexity in one notable case is that the response of the pump may be affected by the intracellular Na concentration [25]. If confirmed this points to the presence of signaling pathways that are sensitive to cellular ionic composition or to the conformational equilibrium poise of the target protein. An interesting short review, that expands on earlier work by its authors, involves both small molecule and lipid interactions, and also considers the possible basis for cross-talk between PKA and PKC signaling pathways [26].

4.6
The Importance of Subunit Isoforms

There have been four different isoforms of the α-subunit reported, and there are several reviews that compare their structures, which show a very high level of identity [28]. There is a chapter in this volume that discusses the physiological functions that such isoform diversity may subserve. The most widely expressed isoform, in mammals, is the renal or "housekeeping" form ($\alpha1$), the most selectively expressed form is $\alpha4$, found only in testes [30, 31], and the $\alpha2$, and $\alpha3$ forms have less specialized tissue distributions, but clearly have either special locations within the cell membrane, compared with the more uniformly expressed $\alpha1$, or may exhibit cell specificity within an organ such as brain, or heart A short review of functional aspects of isoform diversity appeared recently [29].

All β-subunits are glycosylated at N-linked consensus sites in multiple locations, and all reported β-subunits have three highly conserved disulfide bridges in their extracellular domain [8]. The β-subunit isoforms and their physiological significance have been less well documented. The most common, the $\beta1$ form, that accompanies $\alpha1$, in the ubiquitous kidney enzyme, has three glycosylation sites, and interestingly it has been amply documented that there is little effect of preventing glycosylation [15, 32]. It may be that an entirely unglycosylated β-subunit is more susceptible to cellular proteases [15]. Strikingly, the β2 isoform has multiple glycosylation sites (six or seven). It has been shown that the homologous β-subunit in the H,K-ATPase, which has seven glycosylation sites, fails to assemble and form a functional enzyme with H,K-ATPase α-subunit if glycosylation at a key site is prevented [33]. It has yet to be shown whether or not the same holds for the Na,K-ATPase assembly utilizing β2. The Na,K-ATPase β2 subunit may also serve other physiological functions as it has been recognized as an adhesion factor in the brain, where it was shown to be immunologically identical to AMOG, the adhesion molecule of glia [34]. Since that time a great deal of work, much emanating from the Schachner group suggests that this β-subunit isoform may play a special role in neuronal cell outgrowth and development. Interestingly, this isoform has recently been implicated in the occurrence of polycystic kidney disease, where mislocation of functional Na,K-ATPase subunits to the apical membrane of tubular cells has been reported [35].

4.7
Heterologous Expression of the Na,K-ATPase

The particular problem that arises with a ubiquitously expressed protein such as the Na,K-ATPase is that many of the most commonly employed expression systems have high levels of endogenous Na,K-ATPase. This holds for almost all mammalian cells and also *Xenopus* oocytes. In the case of mammalian cells, this difficulty was overcome in a series of studies from the Lingrel group, by converting, via several specific residue substitutions, a normally ouabain-sensitive form into a form that was ouabain-resistant. Functional studies could then be carried out in the presence of low levels of ouabain, so that the endogenous activity was inhibited, and the heterologous activity revealed. There are two systems that have been reported so far where there is either negligible or zero endogenous Na,K-ATPase activity. These are the yeast, *S. cerevisiae* [36], or sf9 or High Five insect cells that are infected with baculovirus containing the cDNA for the protein subunits [37]. This latter system has had the advantage, for assembly studies, that cell fractionation protocols to obtain endoplasmic reticulum, golgi and plasma membrane fractions have been developed [14, 15], but the disadvantage that many reagents that are available for proteins in mammalian cells are not available for insect cells.

4.8
Recent Highlights

There have been several novel developments and advances in recent years that have not yet gained the attention of the more generally appreciated areas discussed above. In some cases it is not yet evident how widely they will be utilized or how this new knowledge will affect our ideas about the biological functions of the Na pump. These include both newer methods as well as new areas of study.

The use of optical methods in establishing fine structural changes occurring at specific locations has been pioneered in studies of ion channels. In these systems site-directed modifications and the attachment of fluorescent probes at these sites has enabled detailed models to be proposed about the movements of specific aminoacid residues during channel opening or gating [38, 39]. This approach has been termed "voltage clamp fluorometry". The approach was recently introduced to the transport area in studies of the Na,K-ATPase. Here specific fluorescent labeling of site directed cys residue substitutions enabled the fluorescence to be recorded under voltage clamp conditions using heterologously expressed Na pump in *Xenopus* oocytes [40]. The extent to which electrical signals, that were simultaneously monitored, overlapped the fluorescent traces suggests that this will be a very valuable method. It was possible to detect small fluorescent changes in a precise location (in the M5M6 extracellular loop) during pump turnover under physiological conditions. The changes were interpreted as reflections of movements of residue 790, at the extracellular end of M5, in the E1P to E2P transition. It appears as if this residue moves from a more sheltered to a more exposed environment during the transition. Since it has been shown that there are no essential cys residues in the Na,K-ATPase α–subunit, cys-directed labeling becomes a powerful method to introduce reporter chromophores at almost any location in the protein [41]. Similar studies have now been carried out on the gastric H,K-ATPase [42].

It has been known for many years that a site of action of palytoxin, an extremely venomous marine toxin, is the Na,K-ATPase. The essential element of the action of palytoxin is to produce a large membrane conductance that is proximal to or through the Na,K-ATPase protein. It has been an open question as to whether or not the channel represented a modification of pre-existing ion permeation pathways through the pump protein. In recent studies it has been shown that these channels are modulated by addition of Na pump ligands. Different pump ligands promote opening or closing of the channels [43]. It is possible that with judicious use of this property we will learn more about the details of the occluded cation state and the entry and exit of ions into this location.

One extremely controversial subject has been the state of oligomerization of the Na,K-ATPase in native membranes [1]. The essential issue is whether or not the native structure is the $\alpha\beta$-heterodimer, or a more highly associated form such as $(\alpha\beta)2$, $(\alpha\beta)4$, etc. There have been many studies where kinetic data have been interpreted as a consequence of the involvement of such higher order aggregates. The case in favor of such aggregates comes from cross-linking studies, inhibitor studies and associated phosphorylation and dephosphorylation kinetics, and recently the appearance of single particle images [27]. The counter argument arises from a lack of ability to cross-link low density ATPase molecules in their native membrane state, from analytical ultracentrifuge studies where monomeric protein, subsequently assayed, has normal activity. Although it must be admitted that any rapid monomer-oligomer associations would render this last data less than definitive. A cogent summary of much of this work, that was (admittedly) a personalized view concluded that the situation was far from resolved and the case was not proven that the functional unit of the Na,K-ATPase was an αβ-heterodimer [43]. Data has also been presented from coimmunoprecipitation studies using isoform-specific antibodies in insect cells that α,α associations take place in the native membranes [44].

Recent studies by the author of this overview (manuscript in preparation) confirm and extend such findings. The major questions that remain are, (i) what will constitute acceptable and convincing data that such associations to higher oligomeric forms exist in native membranes, and (ii) do such associations play a role in the activity of the protein?

In a recent article on α-subunit mutations, and their developmental consequences in *Drosophila*, an interpretation involving a dominant negative effect was proposed [45]. Interpretations of such genetic experiments are usually taken as evidence of the existence of dimeric associations of the specific gene product. If this interpretation is correct, it supplies strong biological evidence for the presence of active αα forms of the Na, K-ATPase in native membranes. In summary, we still lack a definitive and objective assessment of the strengths and weaknesses of data on both sides of the argument.

In its natural environment the Na,K-ATPase is surrounded by and doubtless interacts with an array of cellular proteins. Some of these protein partners have been known for some time, while the possible involvement of others in modulating the intracellular activity of the pump are only just being considered. It has been suggested for a number of years that interactions with ankyrin, a cytoskeletal component, may play an important role in Na pump trafficking [46]. It seems likely that such cytoskeletal interactions may play an important role in stabilizing the Na,K-ATPase at the plasma membrane, although direct evidence of such interactions with the intact Na pump protein is lacking. Recently a number of putative protein partners for the Na,K-ATPase have been proposed that emerged from a yeast two-hybrid screen [47]. Many of these are plausible candidates for functional cellular interactions and may play important roles. The investigation of the significance of these interactions will be an active area of research for the coming several years.

The recent realization that the Na,K-ATPase is not merely an isolated transporter working in the membrane, but has the potential to interact with a wide array of cellular components took on an exciting new dimension when it became apparent that the pump may interact with well known signaling cascades. The initial findings emerged from effects of ouabain that were observed prior to any change in cellular cation composition. It soon became evident that a battery of intracellular molecules might be affected by changes in the pump protein. Such partial inhibition of the pump caused apparent changes in Ras, MAPK's, protein tyrosine phosphorylationand other signaling pathways. These multiplicity of effects, resulting from modulation of a well-studied transporter were quite unexpected by most workers in the field. Their substantiation and an assessment of the importance of their roles in cell signaling represents a very exciting prospect for the future [43, 48, 49].

Acknowledgments

I would like to thank the many colleagues, students and collaborators who have contributed to the findings and ideas discussed in this short overview. The literature cited is by no means complete, but primary reports can be found in the review articles cited. The work that is referred to that was performed in the authors' laboratory was supported by NIH grants HL30315, and GM 39500.

References

1. Kaplan JH, Biochemistry of Na,K-ATPase, *Ann Rev Biochem*, **2002**, *71*, 511–535.
2. Jorgensen PL, Hakansson KO, and Karlish SJD, *Ann Rev Physiol*, **2003**, *65*, 817–849.
3. Kaplan JH, *J Bioenerg Biomembr*, **2001**, *33*, 365–448.
4. Sweadner KJ and Donnet C, *Biochem J*, **2001**, *356*, 685–704.
5. Jorgensen, PL, *Ann NY Acad Sci*, **2003**, *986*, 22–30.
6. Lutsenko S, Anderko R. and Kaplan JH, *Proc Nat Acad Sci USA*, **1995**, *92*, 7936–7940.
7. Gatto C, Lutsenko S, Shin JM, Sachs G, and Kaplan JH, *J Biol Chem*, **1999**, *274*, 13737–13740.
8. Horisberger JD, *The Na,K-ATPase: Structure-Function Relationship*, RG Landes, Austin, **1994**, 21–50.
9. Geering K, *J Membrane Biol*, **1990**, *115*, 109–121.
10. Kirley TL, *J Biol Chem*, **1990**, *265*, 4227–4232.
11. Lutsenko S and Kaplan JH, *Biochemistry*, **1993**, *32*, 6737–6743.
12. Lutsenko S and Kaplan JH, *J Biol Chem*, **1994**, *269*, 4555–4564.
13. Geering K, *J Bioenerg Biomembr*, **2001**, *33*, 425–438.
14. Gatto C, McCloud SM, and Kaplan JH, *Am J Physiol Cell Physiol*, **2001**, *281*, C982–C992.
15. Laughery MD, Todd ML, and Kaplan JH, *J Biol Chem*, **2003**, *278*, 34794–34803.
16. Geering K, *J Membr Biol*, **2000**, *174*, 181–190.
17. Forbush B, Kaplan JH, and Hoffman JF, *Biochemistry*, **1978**, *17*, 3667–3676.
18. Pu HX, Scanzano R, and Blostein R, *J Biol Chem*, **2002**, *277*, 20270–20276.
19. Therien AG, Karlish SJD, and Blostein R, *J Biol Chem*, **1999**, *274*, 12252–12256.
20. Sweadner K, Arystarkhova E, Donnet C, and Wetzel R, *Ann NY Acad Sci*, **2003**, *986*, 382–387.
21. Geering K, Beguin P, Garty H, Karlish S, Fuzesi M, Horisberger J, and Crambert G, *Ann NY Acad Sci*, **2003**, *986*, 388–394.
22. Meij IC, Koenderink JB, De Jong JC, De Pont JJHHM, Monnens LAH, Van Den Heuvel LPWJ, and Knoers NVAM, *Ann NY Acad Sci*, **2003**, *986*, 437–443.
23. Therien AG and Blostein R, *Am. J. Physiol Cell Physiol*, **2000**, *279*, C541–C566.
24. Pedemonte, CH and Bertorelli, AM, *J Bioenerg Biomembr*, **2001**, *33*, 439–447.
25. Efendiev R, Budu CE, Cinelli AR, Bertorello AM, and Pedemonte CH, *J Biol Chem*, **2003**, *278*, 28719–28726.
26. Cornelius F, Mahmmoud YA, and Christensen HRZ, *J Bioenerg Biomembr*, **2001**, *33*, 415–423.
27. Taniguchi K, Kaya S, Abe K, and Mardh S, *J Biochem*, **2001**, *129*, 335–342.
28. Blanco G and Mercer RW, *Am J Physiol Renal Physiol*, **1998**, *275*, F633–F650.
29. Lingrel J, Moseley A, Dostanic I, Cougnon M, He S, James P, Woo A, O'Connor K, and Neumann J, *Ann NY Acad Sci*, **2003**, *986*, 354–359.

30. Woo AL, James PF, and Lingrel JB, *J Biol Chem*, **2000**, *275*, 20693–20699.

31. Blanco G, Melton RJ, Sanchez G, and Mercer RW, *Biochemistry*, **1999**, *38*, 13661–13669.

32. Zamofing Z, Rossier BC, and Geering, K, *Amer J Physiol Cell Physiol*, **1989**, *256*, C958–C966.

33. Vagin O, Denevich S, and Sachs G, *Am J Physiol Cell Physiol*, **2003**, *285*, C968–C976.

34. Gloor S, Antonicek H, Sweadner KJ, Pagliusi S, Frank R, Moos M, and Schachner M, *J Cell Biol*, **1990**, *110*, 165–174.

35. Wilson PD, Devuyst O, Li X, Gati L, Falkenstein D, Robinson S, Fambrough D, and Burrow CR, *Amer. J Path*, **2000**, *156*, 253–268.

36. Pedersen PA, Rasmussen JH, and Jorgensen PL, *Biochemistry*, **1996**, *35*, 16085–16093.

37. De Thomaso AW, Xie ZJ, Liu G, and Mercer RW, *J Biol Chem*, **1993**, *268*, 1470–1478.

38. Mannuzzu LM, Moronne MM, and Isaccoff EY, *Science*, **1996**, *271*, 213–216.

39. Cha A and Bezanilla F, *Neuron*, **1997**, *19*, 1127–1140.

40. Geibel S, Kaplan JH, Bamberg E, and Friedrich T, *Proc Nat Acad Sci USA*, **2003**, *100*, 964–969.

41. Hu YK, Eisses JE, and Kaplan JH, *J Biol Chem*, **2000**, *275*, 30734–30739.

42. Geibel S, Zimmerman D, Zifarelli G, Becker A, Koenderink JB, HuY-H., Kaplan JH, Friedrich T, and Bamberg E, *Ann NY Acad Sci*, **2003**, *986*, 31–38.

43. Artigas P and Gadsby DC, *Proc Nat Acad Sci USA*, **2003**, *100*, 501–505.

43. Askari A, *Na/K-ATPase and Related ATPases*, Taniguchi K, Kaya S, eds., Elsevier Tokyo, **2000**, 17–26.

44. Blanco G, Koster JC, and Mercer RW, *Proc Nat Acad Sci*, **1994**, *91*, 8542–8546.

45. Palladino MJ, Bower JE, Kreber R, and Ganetzky B, *J Neurosci*, **2003**, *15*, 1276–1286.

46. De Matteis MA and Morrow JS, *Curr Opin Cell Biol*, **1998**, *10*, 542–549.

47. Pagel P, Zatti A, Kimura T, Duffield A, Chauvet V, Rajendran V, and Caplan MJ, *Ann NY Acad Sci*, **2003**, *986*, 360–368.

48. Xie Z, Kometiani P, Liu J, Haas M, and Tian J, *Na/K-ATPase and Related ATPases*, Taniguchi K, Kaya S, eds., Elsevier, Tokyo, **2000**, 617–624.

49. Xie Z, *Ann NY Acad Sci*, **2003**, *986*, 497–503.

5
Copper-Transporting ATPases: Key Regulators of Intracellular Copper Concentration

Ruslan Tsivkovskii, Tina Purnat, and *Svetlana Lutsenko*

5.1
Introduction

5.1.1
Role of Copper in Cell Physiology

Copper is essential for the growth and development of all organisms. Its role in cell metabolism is tightly linked to the chemical properties of this metal. In biological systems, copper exists in two forms, Cu^+ and Cu^{2+}. The high redox potential of the Cu^{2+}/Cu^+ pair is effectively utilized for electron-transfer reactions, especially by oxidative enzymes [1–3]. In mammalian cells, these enzymes include dopamine-β-monooxygenase, which is involved in neurotransmitter biosynthesis, tyrosinase, which is required for pigmentation, cytochrome c oxidase, a key enzyme of the respiratory chain, ceruloplasmin, a copper-dependent ferroxidase, and other important enzymes. In photosynthetic organisms, the copper cofactor is also used for dinitrification reactions [4]. Copper-dependent enzymes also, importantly, participate in the degradation of side products of O_2 metabolism such as O_2^- radicals. A well-known protein that utilizes copper to carry out this type of reaction is copper, zinc-dependent superoxide dismutase (SOD).

The ability of copper to activate molecular oxygen explains another important biological effect of copper – its toxicity at concentrations that exceed the cell's chelating capacity for this metal. Therefore, cells have to balance their requirements for copper with the need for its efficient removal when its concentration becomes too high. The key role in regulation of intracellular copper concentration belongs to the copper-transporting P-type ATPases (CTAs). These enzymes utilize the energy of ATP hydrolysis to transport copper from the cytosol across various cell membranes [5–7]. Genetic studies have identified putative CTAs in all phila; however, very few of these proteins have been functionally characterized. Although limited, this initial characterization uncovered many interesting and distinct properties of the copper-transporting ATPases. Mammalian CTAs are particularly fascinating due to the complex regulation of their function by copper and their important

Handbook of ATPases. Edited by M. Futai, Y. Wada, J. H. Kaplan
Copyright © 2004 WILEY-VCH Verlag GmbH & Co. KGaA, Weinheim
ISBN 3-527-30689-7

role in human health (see below). In this chapter, we focus primarily on the human copper-transporting ATPases, and analyze their structure, function, and regulation in comparison with CTAs from other organisms.

5.1.2
Specific Functions of Copper-transporting ATPases in Various Organisms

A database search has predicted more than 50 proteins as potential copper-transporting ATPases. For only four of them was copper transport directly demonstrated [8–11], although for a larger number of ATPases the consequences of inactivation of the corresponding genes for the organism were investigated, and the complementation studies (see below) performed. The functions of the characterized CTA are described briefly below.

5.1.2.1 Bacterial Copper-transporting ATPases

The first bacterial system for ATP-dependent transport of copper was described by Solioz and colleagues who characterized a *copABZY* operon associated with copper resistance in *Enterococcus hirae* [12–14]. They found that the P-type ATPase CopB, encoded in the copABZY operon, was required for copper resistance displayed by this organism. Subsequently, using inside-out membrane vesicles isolated from *E. hirae*, Solioz and Odermatt demonstrated the ATP-dependent accumulation of copper or silver catalyzed by CopB [9]. CopB is concluded to be a pump for the extrusion of the monovalent copper and silver ions, with copper probably being the natural substrate. The precise function of another P-type ATPase encoded by the copABZY operon, CopA, is less clear. Expression of CopA is induced when copper concentration is low, and the disruption of the *copA* gene renders cells copper-dependent [14]. Therefore, it was proposed that CopA is involved in copper uptake [12, 14]. Direct comparison of copper transport characteristics for CopA and CopB would be very interesting.

In *E. coli*, the resistance to copper is linked to the *copA* gene product [10, 15]. Studies of the ^{64}Cu uptake and efflux by wild-type *E. coli* cells and by cells in which *copA* was disrupted by mutagenesis yielded three lines of evidence implicating CopA in copper efflux *in vivo*: (i) increased sensitivity of the *copA* mutant to copper, (ii) hyper-accumulation of ^{64}Cu in the *copA* mutant compared to wild-type cells, and (iii) copper-dependent up-regulation of the *copA* promoter. These results complemented the experiments by Rensing and co-authors, who demonstrated the ability of the CopA protein to transport ^{64}Cu into inverted membrane vesicles in an ATP-dependent manner [10]. In recent work, Fan and Rosen expressed and purified the product of the *copA* gene and provided direct evidence that *copA* encodes a copper-transporting ATPase [16] (for details on functional characterization, see Section 5.3). Altogether, these studies concluded that the copper concentration in *E. coli* is determined primarily by CopA-mediated efflux, while copper uptake is a constitutive process [17].

The gene encoding protein homologous to *E. coli* CopA was found in the *copYAZ* operon from the oral bacterium *Streptococcus mutans JH1005* [18] S. *mutans* cells that lack the *copYAZ* operon are sensitive to a 4-fold lower copper concentration than the wild-type cells (200 versus 800 µM, respectively). At the same time, the wild-type and mutant cells show no differences in their susceptibility to other heavy metals, confirming that CopA is involved in regulation of intracellular copper [18]. Similar conclusions were reached for the CopA product of the *copAP* operons of *Helicobacter pylori* and *Helicobacter felis* [19]. Thus, it appears that the primary function of copper-transporting ATPases in bacteria is copper detoxification.

5.1.2.2 Cyanobacteria

These organisms contain internal thylakoid membranes that house protein complexes involved in electron-transfer reactions. In *Synechosystis PCC 6803* electron transfer between the complexes is mediated via copper bound to plastocyanin PetE. PetE is located in the lumen of thylakoid, so that copper has to be transported from the cell cytosol across the thylakoid membrane for incorporation into PetE. Cyanobacteria contain two genes encoding copper-transporting ATPase, PacS and CtaA [20–23]. Both copper-transporting ATPases, PacS and CtaA, are required for the delivery of copper to thylakoids. CtaA is thought to reside in the plasma membrane and is important for copper uptake into cytosol, while PacS is located in the thylakoid membranes and transports copper from the cytosol into the lumen of thylakoids [23]. The physiological role of PacS and CtaA was clarified only following analysis of the photooxidation properties of cells lacking either PacS or CtaA, because simple measurements of metal resistance were unrevealing. *Synechosystis PCC 6803* with a disrupted *pacS* gene has reduced tolerance to copper, while a mutant with disrupted *ctaA* has an unaltered metal tolerance but reduced copper accumulation [22]. Simultaneously, disruption of *ctaA* from *Synechococcus PCC7942* resulted in increased copper tolerance compared with the wild-type [21]. It seems that in the organisms where CTAs work to deliver copper to intracellular organelles the resistance to copper is not a reliable criterion of their function.

5.1.2.3 Archeobacteria

Enzymes from thermophilic organisms have attracted significant attention due to their remarkable stability and potential use for structural studies. Thermophilic, sulfur-metabolizing *Archaeoglobus fulgidus* contains a P-type ATPase, CopA (NCB accession number AAB90763), which shows specificity for copper and silver [24]. The functional characterization of this protein suggests that CopA of *A. fulgidus* drives the outward movement of the metal [24]. The availability of this protein in purified and active form makes it an excellent candidate for structural studies. Analysis of the *A. fulgidus* also revealed another copper-transporting ATPase, CopB. Unlike *A. fulgidus* CopA, which is Cu^+ and Ag^+ specific, CopB has a higher turnover rate in the presence of Cu^{2+} [25]. Comparative analysis of CopA and CopB will help to identify the structural determinants that allow the ATPases to discriminate between Cu^+ and Cu^{2+}.

5.1.2.4 Yeast

Yeast has served as an excellent model system for dissecting the intracellular mechanisms of copper distribution. However, the functional properties of the yeast copper-transporting ATPases remain essentially uncharacterized. In *Saccharomyces cerevisiae*, the *ccc2* gene encodes a P-type ATPase Ccc2 located in the late Golgi or post-Golgi compartment of the secretory pathway [26, 27]. Current data suggest that the major function of Ccc2 is to transport copper into the lumen of the secretory pathway, where copper is incorporated into the extracellular domain of Fet3, a copper-dependent ferroxidase. The ferroxidase activity of Fet3 is necessary for efficient iron uptake by yeast cells [28]. Disruption of *ccc2* prevents copper incorporation into Fet3, resulting in inactive Fet3 and iron deficiency. The effect of the *ccc2* deletion can be offset by expression of Ccc2 present on a separate plasmid or by heterologous expression of another copper-transporting ATPases [29–31]. This property forms the basis of a convenient complementation assay that has been widely used to characterize several CTAs (see below).

In contrast to bacterial cells, Ccc2 seems to play no significant role in copper export from the cell and/or in copper resistance. While in *E. coli* the copper concentration is determined primarily by the CopA-mediated efflux (see above), in *S. cerevisiae* the determining role in controlling intracellular copper belongs to the metal-uptake systems (CTR1 and analogues), which are tightly regulated by copper at the transcriptional and posttranslational levels [32–34].

Interestingly, the physiological role of the copper-transporting ATPase CaCRP1 from the yeast *Candida albicans* seems very different from that of Ccc2. *C. albicans*, which lives in the digestive tract of humans and animals, can tolerate a 10-fold higher concentration of copper than *S. cerevisiae*. The key role in *Candida*'s higher copper tolerance belongs to the copper-transporting ATPase CaCRP1 [35]. Significantly, CaCRP1 does not complement the iron-deficient phenotype of the *Δccc2* strain (see above), suggesting that it does not function in the Golgi compartment. In agreement with this prediction, recent studies demonstrated that CaCRP1-GFP displayed a plasma membrane localization rather than the intracellular punctate localization characteristic of Ccc2 [36]; the authors also reported that the *C. albicans* genome contained a second putative copper-transporting P-type ATPase with higher homology to Ccc2. This observation raised the interesting possibility that two copper-transporting ATPases of *C. albicans* have functional specialization, with the Ccc2 homologue working in the secretory pathway and contributing to the intracellular iron transport, and CaCRP1 functioning in the plasma membrane in copper detoxification. Consistent with this hypothesis, the deletion studies of the *C. albicans* Ccc2 homologue demonstrated that CaCcc2 is required for the high-affinity iron import [36]. Structural and functional comparison of CaCRP1 and CaCcc2 would be very interesting since it may identify structural determinants essential for predicted differential localization of these proteins. In addition, it would be useful to determine the metal-specificity of CaCcc2 and compare it with the specificity of CaCrp1. The CaCrp1 deletion strain exhibits increased sensitivity to silver and cadmium in addition to copper, indicating that this CTA is somewhat promiscuous with respect to the metal ion [37].

5.1.2.5 Plants

An unexpected role of a copper-transporting ATPase for plant homeostasis was revealed in a series of experiments designed to identify the genes involved in ethylene signaling in *Arabidopsis thiliana*. These studies found mutants named "responsive-to-antagonist1" (RAN1), which had a phenotype similar to the loss-of-function ethylene receptor mutants. This result suggested that RAN1 was necessary to form functional ethylene receptors. Sequence analysis of RAN1 cDNA demonstrated that it encodes a P-type ATPase with homology to yeast Ccc2 [38]. Expression of RAN1 complemented the defects of a *Δccc2* mutant, demonstrating its function as a copper transporter. Further studies confirmed that binding of ethylene by ethylene receptor requires copper cofactor, which is apparently delivered to the receptor by RAN1 similarly to the Ccc2-mediated delivery of copper to Fet3 in yeast [39]. It was also proposed that, in plants, this pathway may facilitate retrieval of copper from senescing leaf tissue [40].

5.1.2.6 *Caenorhabditis elegans*

The cDNA coding for a putative copper-transporting P-type ATPase (CUA-1) was cloned from *C. elegans* and was shown to provide functional complementation of the *S. cerevisiae Δccc2* mutant. Subsequent studies revealed that CUA-1 is expressed in intestinal cells of adult worm and in hypodermal cells in the larvae [41, 42], suggesting a possible role of CUA1 in copper export from the cells. The precise functional properties of CUA-1 and its intracellular localization in *C. elegans* remain to be determined.

5.1.2.7 Human Copper-transporting ATPases

These proteins hold a special position within the CTA family, for several reasons. The human copper-transporting ATPases ATP7A and ATP7B were the first copper transporters to be discovered. Identification of the *ATP7A* and *ATP7B* gene products as ATP-driven active copper transporters supplied the first information regarding molecular mechanisms regulating copper concentration in human cells [43–47]. Association of ATP7A and ATP7B with severe genetic disorders in humans (Menkes disease and Wilson's disease, respectively) sparked interest in understanding how copper is distributed in a cell, and led to subsequent identification and characterization of other copper-transporting ATPases described in earlier sections. The fascinating structural organization and regulatory mechanism acting on ATP7A and ATP7B are discussed in detail in subsequent sections. Here we summarize what is known about specific functions of ATP7A and ATP7B.

The function of human copper-transporting ATPases can be discerned from the phenotypes of the diseases associated with mutations or deletions of the respective genes in humans. Defects in *ATP7A* result in a severe neurodegenerative disorder known as Menkes disease [48, 49]. In Menkes disease, export of copper from intestinal cells is dramatically reduced [50]. As a result, copper accumulates in intestinal cells, while other tissues experience severe copper deficiency. The major symptoms of Menkes disease include growth and developmental delays, vascular and connective tissue abnormalities, and poor temperature control [48]. Kinetic studies with ^{64}Cu demonstrated reduced copper efflux from cells isolated from Menkes patients [51], suggesting that ATP7A plays a key role in transport of copper across plasma membrane.

In addition, the lack of functional ATP7A severely compromises the activity of such copper-dependent enzymes as tyrosinase and peptidylglycine alpha-amidating monooxygenase (PAM). These enzymes acquire copper within the secretory pathway, suggesting that another important function of ATP7A is to transport copper from the cytosol into the lumen of secretory vesicles. Indeed, ATP7A has been shown to be required for the formation of functional tyrosinase [52]. In this work, Petris et al. demonstrated that recombinant tyrosinase expressed in immortalized fibroblasts from Menkes disease patients (Menkes fibroblasts) was inactive, whereas normal fibroblasts that express ATP7A showed substantial tyrosinase activity. Co-expression of recombinant ATP7A and tyrosinase in Menkes fibroblasts led to activation of tyrosinase and melanogenesis [52].

In a recent study, Hansel and colleagues established that ATP7A is present in the endocrine and brain tissues that have a high expression of PAM, a copper-dependent monooxygenase [53]. The ATP7A levels are particularly high in the pituitary, where it is largely localized to the *trans*-Golgi network. These findings pointed to the role of ATP7A in delivery of copper to PAM. To test this hypothesis, the authors used *brindle* mice, bearing an inactivating mutation in *ATP7A* (for description of available mouse models, see Section 5.1.3), and measured the levels of amidated peptides, the products of PAM activity. In *brindle* mice, the amounts of several amidated peptides were reduced in pituitary and brain extracts despite normal levels of PAM protein, demonstrating that PAM function is compromised when ATP7A is inactive [53]. Based on these results and complementary investigations in yeast [54, 55], the authors concluded that the function of PAM depends on ATP7A. Furthermore, they proposed that the reduction in the ability of PAM to produce bioactive peptides involved in neuronal growth and development may contribute to neurodegeneration symptoms associated with Menkes disease.

Thus, in human cells, the copper-transporting ATPase ATP7A (Menkes disease protein, MNKP) has a dual function: to deliver copper to the secreted copper-dependent enzymes and to export copper out of the cell. MNKP can perform both functions because it is located in two different cell compartments: the *trans*-Golgi network (TGN) and plasma membrane. The localization of MNKP and its precise function (copper delivery into TGN versus export of copper across plasma membrane) depend on the copper concentration in a cell. At low copper concentrations, MNKP is located in TGN. When the concentration of copper is elevated,

MNKP is found predominantly in the plasma membrane [56]. Regulation of MNKP function and intracellular localization are described in detail in Section 5.6.

The function of another copper-transporting ATPase, ATP7B (the Wilson's disease protein, WNDP) is, in general, very similar to MNKP. In Wilson's disease, mutations in *ATP7B* result in marked reduction of copper export from the liver (where WNDP is primarily expressed) into the bile, pointing to the involvement of WNDP in a transport of copper across the canallicular plasma membrane. In addition, the lack of functional WNDP is associated with the absence of copper incorporation into cerulloplasmin, a secreted copper-dependent ferroxidase [57]. Similarly to MNKP, the function and intracellular localization of WNDP are regulated by copper. However, the distribution pattern of WNDP in a cell is less clear. At low intracellular copper, WNDP resides in TGN, while elevated copper induces re-localization of WNDP to a vesicular compartment, the identity of which is currently not known [58, 59]. Identification of the nature of this compartment and understanding its precise role in copper homeostasis would be extremely interesting. It is likely that from vesicles WNDP traffics to the plasma membrane [59, 60]. Whether this step occurs in a regulated fashion or it is a constitutive process remains unclear.

MNKP, which is more ubiquitously expressed than WNDP, has a housekeeping function, while the role of WNDP is more specialized. In several tissues (brain, kidneys, placenta) both mammalian CTAs are present [5, 61–63]. This raises important questions about specific roles of MNKP and WNDP in these tissues. Currently, it is unknown whether endogenous MNKP and WNDP are present and perform complementary functions in the same type of cells or their expression and functional roles are more cell specific. Heterologous expression of WNDP in cells lacking functional MNKP (Menkes fibroblasts) prevents accumulation of copper in these cells [64], suggesting that the general function of these transporters is very similar.

To better understand the role of WNDP and MNKP in tissues, Moore and Cox employed in situ hybridization and immunohistochemistry and localized these two CTAs in mouse kidneys [63]. In kidneys, the defects in either *ATP7A* or *ATP7B* result in copper accumulation, suggesting that in this organ the respective copper-transporting ATPases have complementary functions. Consistent with this hypothesis, Moore and Cox found that both MNKP and WNDP are expressed in glomeruli, a network of blood capillaries. In this location, both ATPases may work together to regulate copper levels in the filtrate. Interestingly, WNDP but not MNKP was also seen in the kidney medulla [63], suggesting the specific role of WNDP in copper re-absorption. Surprisingly, these findings differ from the results of earlier work by Grimes et al., who demonstrated expression of ATP7A in the proximal and distal tubules with only slight expression in the glomeruli [65]. Moore and Cox suggested that these differences could be due to the use of different strains of mice, different antibody, and possibly different diet of experimental animals [63]. More experimental data on localization of WNDP and MNKP in tissues are clearly needed. The specific functions of MNKP and WNDP, their transport characteristics and regulation also remain to be elucidated. Recent biochemical investigations suggest that the *in vitro* properties of MNKP and WNDP are not identical (see Section 5.3).

5.1.3
Animal Models for Menkes Disease and Wilson's Disease

Several mammalian orthologues of MNKP and WNDP have been isolated and their primary sequences have been characterized. Sheep, unlike other mammals, do not efficiently excrete copper via bile and often accumulate copper in the liver. The reason for this phenomenon is not clear, but it is not due to any mutations in the primary sequence of the WNDP sheep orthologue [66]. Characterization of sheep WNDP also revealed that, in addition to the full-length product, the sheep liver contains a minor (10%) alternatively spliced transcript. This transcript encodes the WNDP variant, which is 1505 amino acids long (the "normal" size is 1444 residues), due to alternative splicing of the first and second exons. The alternative splicing generates an extended N-terminal region, rich in proline and acidic residues [66]. A database search did not find any significant homology of this region to known proteins, so the functional significance of the alternative form is unclear. Both forms of ovine ATP7B were expressed in Menkes fibroblasts and were shown to correct the copper-retention phenotype of these cells, confirming the ability of sheep ATP7B orthologues to act as copper-transporters [64]. Similarly, the alternatively-spliced product was localized to TGN in CHO cells and traffics in response to copper, as expected for the major WNDP product [67].

The murine orthologues of both MNKP and WNDP were also identified and were shown to be approximately 85% identical to the human proteins [68–70]. The major sequence variations between the human and murine homologues are clustered within the N-terminal portion of these proteins, while the rest of the sequences are essentially identical. Interestingly, spontaneous mutations in *ATP7B* were reported in mice. These mutations result in a so-called toxic milk phenotype, which resembles Wilson's disease. Two distinct mutations causing this phenotype where described. The "original" toxic milk (*tx*) mice accumulate copper in the liver, kidney, brain, muscle and other tissues. Homozygous *tx* adults do not show overt signs of disease; however, litters born to such parents are deficient in copper and die at about 2 weeks of age [71]. Immunohistochemistry experiments using the lactating *tx* mouse mammary gland demonstrated that WNDP in this tissue was mislocalized, which would explain the inability of the *tx* mouse to secrete normal amounts of copper in milk [72]. In this original strain, the *tx* mutation was identified as a single base A4066>G substitution in *ATP7B*. This substitution causes the change of Met1356 to Val in the membrane portion of WNDP [69] and disrupts the functional activity of WNDP in transfected CHO cells [11].

Another autosomal recessive mutation resulting in toxic milk phenotype arose in the C3H/HeJ animal colony at the Jackson Laboratory (*txj*) and was recently characterized by Cox and colleagues. This group identified the causative mutation in *txj* mouse as a glycine to aspartate substitution (Gly712Asp) in the second putative transmembrane segment of ATP7B [73]. The phenotype has not been described in detail, but appears to be similar to the *tx* mice. Lastly, in Long Evans Cinnamon (LEC) rats, extensive deletion in the 3′ region of the *ATP7B* gene [74] is associated with accumulation of copper in the liver and severe hepatic pathologies (for a review see Ref. 75). The mutant mice and LEC rats may serve as animal models to study Wilson's disease pathology. In fact, LEC rats have been extensively characterized [75, 76]. In addition to toxic milk mice and LEC rat, genetically engineered *ATP7B* (−/−) mice have become recently available. These mice accumulate copper in the liver and other tissues and develop morphologic abnormalities similar to Wilson's disease [77]. The *ATP7B* (−/−) mice are particularly useful for analysis of copper-induced pathologies because the original strain is available and can be used as a control for comparative investigations.

Animal models are also available for analysis of MNKP function. In mice, the *mottled* locus encodes a homologue of *ATP7A* gene. Mutations in this locus lead to symptoms similar to Menkes disease. Phenotypically, the affected mottled males were divided into three categories: (1) those that die before birth; (2) those that die in the early postnatal period; and (3) animals in which affected males survive to adulthood [78]. These phenotypic variations are associated with various *ATP7A* gene lesions. The most severely affected mottled alleles were shown to carry a deletion of exons 11–14 and an insertion in exon 10 leading to missplicing [78]. Two of the less severe mottled mutants are *brindled* and *blotchy*. The causative mutation in the *brindled* mouse is an in-frame deletion of 6 bp that encode Ala799 and Leu800, located in the cytoplasmic loop between the fourth and fifth transmembrane domains of MNKP [65]. In addition, these mice have substitution of Ala514 for Thr; however, the functional significance of this mutation remains unclear [79]. Curiously, the A3189C nucleotide change, which results in Lys>Thr substitution in the highly conserved DKTG motif (for CTA sequence motifs, see Section 5.2), was found in the viable, class 3, animals [78, 79]. In the *blotchy* mouse, the splice site mutations result in abnormal mRNA splicing and skipping of exon 11 [68, 80, 81]. Finally, a point mutation resulting in Pro>Ser substitution in the eighth transmembrane domain of MNKP was identified, further illustrating the variety of genetic abnormalities in *ATP7A* locus [82]. Genetically engineered animals lacking *ATP7A* are not yet available.

In summary, characterization of the copper-transporting ATPases in various organisms indicates that the function of CTAs is two-fold. In cells lacking organelles, the major role of the copper-transporting ATPases is to regulate intracellular copper concentration by transporting copper across the plasma membrane. In organisms that have intracellular compartments, the ATPases have an additional function – to transport copper into this compartment for incorporation into the copper-dependent enzymes. In certain organisms the dual function is apparently performed by two distinct CTAs, which are specialized for either copper export or copper delivery to the intracellular organelles. In mammalian cells, MNKP and WNDP and their homologues carry out both functions, which are under tight control by intracellular copper.

5.2
Major Structural Features of Copper-transporting ATPases

5.2.1
Copper-transporting ATPases as Members of the P-type ATPase Family

Copper-transporting ATPases belong to the P_1(CPX-, P1B-)-subfamily of P-type ATPases [83, 84]. The members of this subfamily are involved in the transport of various transition metals (Cu^+, Ag^+, Cu^{2+}, Zn^{2+}, Ni^{2+}, Cd^{2+}, Pb^{2+}) across cell membranes, in contrast to other P-type ATPases that transport non-heavy metals (Na^+, K^+, Ca^{2+}, Mg^{2+}) or protons. Current data suggest that the P_1-type ATPase subfamily can be further divided into sub-groups based on their specificity towards transported ion. There appears to be a considerable difference in ion selectivity between the ATPases transporting monovalent copper and the ATPases involved in the transport of divalent metals. The Cu^+-transporting ATPases seems highly selective for their substrate. They are also likely to transport silver [16, 24]; however, whether silver transport has any significant physiological role remains unclear. In contrast, the ATPase transporting divalent metals demonstrate a broad range of ion specificity. For example, ZntA from *E. coli* was shown to transport Zn^{2+}, Cd^{2+}, and Pb^{2+} [85, 86]. Although less selective, the ATPases transporting divalent

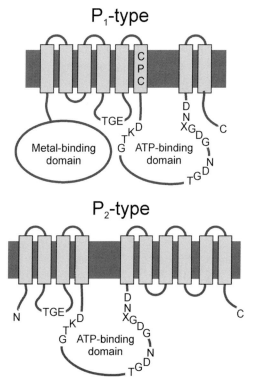

Figure 5.1 Comparison of the transmembrane organization for the P_1-type (top) and the P_2-type ATPases (bottom). TGE, DKTG, TGDN and GDGxND indicate the sequence motifs conserved in all P-type ATPases. CPC is a signature motif conserved in the P_1-type ATPases.

metals show a preference for certain ions (for example, ZntA is much less active in the presence of Ni^{2+}, Co^{2+}, or Cu^{2+} [87, 88]). As described below, metal selectivity correlates with the structural organization of the ATPases [89].

Currently, there is a considerable disparity between the amount of structural and functional information available for the P_1-type ATPases and P_2-type ATPase, the latter having been studied much longer and in greater detail. In particular, the recent determination of the high-resolution structure of Ca^{2+}-ATPase in Ca^{2+}-free and Ca^{2+}-bound form [90, 91] represents a major breakthrough in understanding the structural organization of the P_2-type ATPases. In addition, electron microscopy experiments have yielded low-resolution structures for proton-transporting P-type ATPase [92]. For H^+-ATPase and Na^+/K^+-ATPases, sequence similarity permitted homology modeling using the high-resolution structure of Ca^{2+}-ATPase [93–97]. In stark contrast, very little structural information is available for any of the P_1-type ATPases, including CTAs. High-resolution structures were determined only for isolated MBS (the small domains about 70 amino acids long), while the rest of the P_1-type ATPase structure remains essentially uncharacterized. The available structural information is summarized in the following sections.

5.2.1.1   Sequence Motifs in the Structure of CTAs

Several sequence motifs characterize the P-type ATPase superfamily; all these motifs can be found in CTAs (Figure 5.1). They include: DKTG, TGDN, and GDGxND sequences in the ATP-binding domain and the TGE sequence in the smaller cytosolic domain preceding the ATP-binding domain. In addition to common motifs, CTAs and other P_1-type ATPase have structural features that are distinct for the members of this sub-family. These features include Cys, His, or Met-rich sequences at the cytosolic N-terminal portion of the protein that form metal-binding sites, the signature sequences CPx in the transmembrane portion, and the conserved motif HP with an invariant His residue (Figure 5.2).

5.2.2
Topology and Organization of the Transmembrane Domain

The major role of the transmembrane domain is to provide a hydrophilic pathway for copper translocation across the membrane. The arrangement of the transmembrane segments with respect to the cytosolic domains is distinct for the P_1- and P_2-type ATPases and correlates well with the chemical nature of the transported ion [83]. P_1-type ATPases that transport transition metals were predicted to have eight transmembrane segments (TMS), while the P_2-type ATPase that translocate alkali and alkali-earth metals have 10 TMS (Figure 5.1). In the P_1-type ATPases, only one pair of TMS is located a C-terminal to the ATP-binding domain, in contrast to the P_2-type ATPases which have six TMS downstream the ATP-binding domain (Figure 5.1). Furthermore, all P_1-ATPases have four hydrophobic segments prior to the conserved TGEA/S motif, while the P_2-type ATPases possess only one pair of TMS in this region.

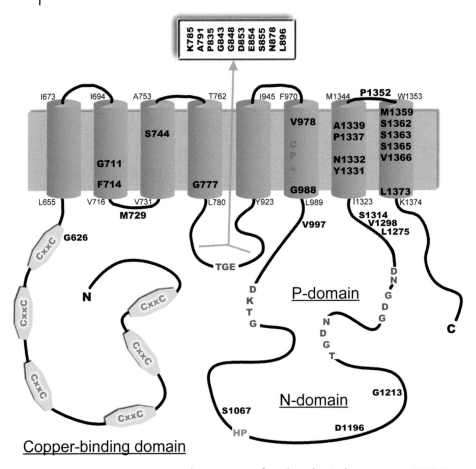

Figure 5.2 Domain structure and sequence motifs in the Wilson's disease protein (WNDP), a human copper-transporting a P1-type ATPase. The positions of the transmembrane segments (TMS) were predicted with GCG Wisconsin Package's TransMem program for human, mouse, rat, and sheep WNDP. The boundaries of WNDP TMS were proposed based on the comparison of the predicted TMS in the alignment. The residues characteristic of copper-transporting ATPases (bold) were identified based on sequence alignment of the P₁-type ATPases with experimentally proven specificity for copper. Residues invariant in all P-type ATPases are in red; residues conserved in the P₁-type ATPases are in orange.

The above transmembrane organization of the P₁-type ATPase was experimentally confirmed for the putative cadmium-transporting ATPase from *H. pylori* using two different techniques. Melchers et al. utilized the *in vitro* transcription/translation of cDNA constructs in which putative transmembrane segments were fused with the fragment of the β-subunit of the gastric H^+/K^+-ATPase [98]. This fragment of the β-subunit has glycosylation sites, consequently the orientation of the corresponding fusion proteins in the membrane can be determined by the presence or absence of glycosylation. Another experimental approach utilized the *in vivo* expression of the ATPase-alkaline phosphatase fusion proteins (phoA-fusions) in *E. coli* [99]. Both approaches confirmed the eight TMS topology, initially predicted based on hydropathy profiles. Later, the same eight TMS topology was demonstrated for the putative copper-transporting ATPase of *H. pylori* [19] and for CadA protein from *S. aureus* [100]. The hydropathy profiles of various members of the P_1-type subfamily are similar, therefore it is likely that all CTAs have he transmembrane organization shown in Figure 5.1. How the transmembrane segments of P_1-type ATPases are arranged with respect to each other is currently unknown.

Similarly, the structural determinants that define metal-specificity of the P_1-type ATPase remain poorly understood. Recent data suggest that the size of the transported ion is not a key factor in determining metal-specificity of the P_1-ATPases [87]. Rather, the charge and chemical properties of the transported ions appear to be more important in the process of selection. The Lewis acid–base theory classifies Cu^+ or Ag^+ as soft acids, i. e. more easy polarizable, better electron acceptors. Zn^{2+}, Ni^{2+}, Cd^{2+}, Co^{2+} reside on the border between hard and soft acids and are less polarizable than soft acids. The distinct properties of the transported metals suggest that the P_1-type ATPases may utilize different types and number of ligands to coordinate different transition metals within the membrane.

This hypothesis is supported by the results of a phylogenetic analysis of more than 150 known members of the P_1-subfamily (Figure 5.3), illustrating segregation of the copper-transporting ATPase into a separate group. Since the phylogenetic tree is based on the primary sequence comparison, this grouping of copper-transporting ATPase indirectly suggests characteristic structural motifs specific for CTAs. In fact, Arguello has recently identified sequence motifs that are characteristic of the P_1-type ATPases with different metal specificity [89], including copper. We carried out a similar analysis for the P_1-type ATPases with experimentally proven selectivity for copper and identified several residues conserved in the membrane portion of all copper-transporting ATPases (Figure 5.2).

The most noticeable among them is a CPx motif, which is present in all P_1-type ATPases and serves as a signature sequence for this subfamily [9]. This motif was proposed to form an intramembrane copper-binding site. In agreement with this hypothesis, mutation of the amino-acid residues in this motif abrogates functional activity of CTAs (e. g. [16, 101, 102]). In some bacterial CTAs, the second Cys residue in the CPC motif is replaced with His, while the first Cys and the Pro residue in the CPC motif are invariant. In contrast, in several non-copper-transporting P_1-ATPases the first Cys in CPC motif is replaced with Ser/Thr or Ala. It is tempting to speculate that formation of the intramembrane metal-binding site in CTAs requires at least two amino acid residues with strong avidity for copper (Cys or His), while for the non-copper-transporting P_1-type ATPases only one Cys is essential for formation of the metal-binding site. At the same time, the CPx motif is present in the P_1-type ATPases with different metal-specificity. Therefore, it is likely that the residues from other TMS contribute to the intramembrane metal-binding site(s) in addition to the CPx motif. Notably, neither the stoichiometry of copper transport nor the number of intramembrane sites have been determined for any CTA.

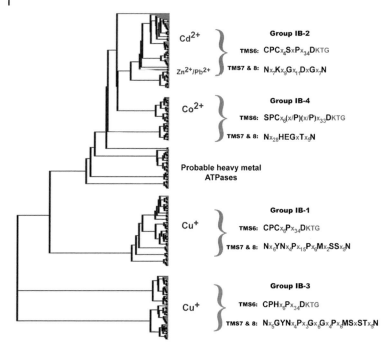

Figure 5.3 Phylogenetic tree of the P1-type ATPases. The search for the P_1-type ATPases was conducted in the NCBI Genebank, Protein databank and SwissProt database and cross-checked for duplicate sequences. The resulting 251 sequences were aligned with the ClustalW program using default settings, and the alignment was used to produce a phylogenetic tree with TREEVIEW. Assignment of metal specificity was based on available experimental information for individual members of each group. Group designations and the corresponding sequence motifs are as described by Arguello [89].

Sequence alignments of P_1-type ATPases with experimentally confirmed specificity for copper also revealed that TMS1–4 of CTAs show little sequence conservation. In contrast, the Asn residue in the third position of TMS7 (N1324 in WNDP) is invariant in the P_1-type subfamily, while the motif YN (Y1331N1332 in WNDP) located seven residues downstream this Asn is conserved in all CTAs (Figures 5.2 and 5.4). Interestingly, these conserved residues belong to a polar face of TMS7, and in WNDP the residues N1324 and Y1331 are located exactly on a top of each other (Figure 5.4). It is tempting to speculate that the invariant Asn in TMS7 could be a common metal-binding ligand in all P_1-type ATPases, while the YN motif is involved in providing specificity for copper.

Overall TMS6, TMS7, and TMS8 contain the largest number of conserved residues in comparison to other TMS. Furthermore, TMS6 and TMS7 are directly linked to the ATP-binding domain, which is the key domain involved in phosphorylation and energy transduction in the P-type ATPases (see below). The high degree of conservation and close position to the ATP-binding domain point to the essential role of TMS6–8 in copper transport and in coupling of transport to ATP-hydrolysis.

Accurate modeling of TMS arrangement in CTAs is difficult because there is no sequence similarity between TMS of CTAs and Ca^{2+}-ATPase for which the structure is known. However, one can analyze the distribution of polar and conserved residues at the α-helical surfaces and assume that these residues face the protein interior rather than the lipid bilayer.

Using sequence alignments of several WNDP orthologues and hydropathy profile predictions, we identified putative transmembrane segments of WNDP (Figure 5.2). Analysis of these TMS revealed that they all have polar residues. In all but two TMS (TMS5 and TMS6) the polar residues are clustered on one side, suggesting that this side of TMS is oriented towards the hydrophilic portion of the protein rather than lipid bilayer. TMS5 and TMS6 have polar residues on all sides, and could be surrounded by other TMS. Using these predictions and the published structure of Ca^{2+}-ATPase, we propose a possible spatial organization of the transmembrane segments of WNDP (Figure 5.4). In this model, TMS6, which contains the CPC motif, is positioned in the middle of the bundle, surrounded by a ring of TMS3–5, TMS7, and TMS8. The central location of TMS6 is consistent with the important role of the CPC motif for CTA function.

5.2.3
N-terminal Copper-binding Domain

The N-terminal domain is an extremely interesting structural feature of CTAs. Here, we review the most prevalent structure of the N-terminal domain, presumably associated with the specificity of CTAs for Cu^+. The N-terminal domain of these ATPases consists of 1 to 6 repetitive sequences, containing metal-binding sites (MBS). Each of these repeats is about 70 amino acids long and includes a conserved sequence motif GMxCxxC (Figure 5.5A, see p. 115). The cysteine residues in this motif are invariant and are involved in Cu^+ coordination [103–107]. The number of MBS in the N-terminal domain seems to correlate with the relative position of the organism on the evolutionary ladder. The bacterial and yeast P_1-type ATPases usually have only one or two MBS, while the number of MBS increases up to three or four for enzymes from insects and plants and then reaches five or six MBS for mammalian proteins. This implies that the "functional core" of CTA includes only one or two MBS, whereas other MBS are important for protein regulation. Interestingly, in the MBS4 of mouse WNDP the second cysteine of the GMxCxxC motif is absent, while MBS4 of rat WNDP lacks both cysteines. Although the length and presumably fold of these murine MBS are preserved, the Cys substitutions are likely to eliminate copper binding, further suggesting that not all MBS in mammalian CTAs are important for protein function. Current mutagenesis and deletion data from several laboratories support this conclusion [59, 108–110].

In addition to the GMxCxxC motif, alignment of MBS from various CTAs demonstrates several conserved hydrophobic residues (mostly Val, Leu, Ile and rarely Met or Phe) distributed over the entire domain sequence (Figure 5.5A). These residues are likely to be important for maintaining the core structure of MBS [111]. Also, there is a highly conserved Gly residue in the C-terminus of MBS, often

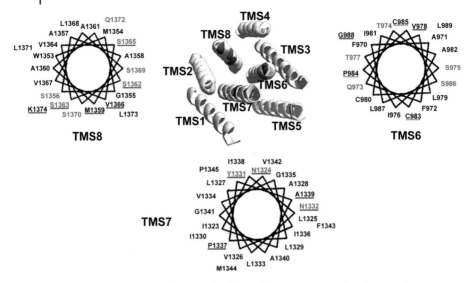

Figure 5.4 Transmembrane domain of Wilson's disease protein. In the absence of sequence similarity, the matching of WNDP and Ca^{2+}-ATPase TMS was performed based on the known functionally important motifs in both proteins. Specifically, it was assumed that the transmembrane hairpins connected to the common domains (the ATP-binding domain and the A-domain) are arranged similarly in both ATPases (i. e. TMS 1&2 in Ca^{2+}-ATPase → TMS 3&4 in WNDP, TMS 3&4 in Ca^{2+}-ATPase → TMS 5&6 in WNDP, TMS 5&6 in Ca^{2+}-ATPase → TMS 7&8 in WNDP). The positions of TMS1 and 2 of WNDP are uncertain, and they were arbitrarily placed at the positions equivalent to Ca^{2+}-ATPase TMS 7&8. The proposed spatial arrangement of the entire transmembrane Ca^{2+}-ATPase domain is in the center. The CPC motif is in green, and the invariant residues characteristic of CTAs are in dark pink. Helical wheel representations are shown for TMS 6, TMS 7, and TMS8, which are oriented in the same way as the corresponding helices in the center figure. In each helical wheel, the labels of polar residues are in blue, while other residues are labeled in black. Conserved residues are underlined in dark pink, and the residues in the CPC motif are underlined in green.

neighboring a hydrophobic residue (GlyPhe pair), which could provide flexibility to the corresponding region of MBS.

Multiple MBSs in the N-terminal domain of CTAs are connected by linkers of different length. In general, the linker connecting the two MBS closest to the membrane (the "core" MBS present in most CTAs) is very short (Figure 5.5B) regardless of a total number of MBS in the N-terminal domain. In contrast, a fairly long linker connects the core MBS and the remaining MBS. For example, in MNKP 92 amino acid residues are inserted between MBS1 and MBS2, while in WNDP the metal-binding domain MBS4 and MBS5 are separated by a stretch of 58 amino acid residues. The presence of long linkers suggests the need for conformational flexibility, which could be important for the regulatory role of the N-terminal domain (for details on the N-terminal domain function see Section 5.4).

NMR studies of recombinant MBS in their apo and metal-bound forms supplied the first detailed structural information on CTAs and other P$_1$-type ATPases. The

A

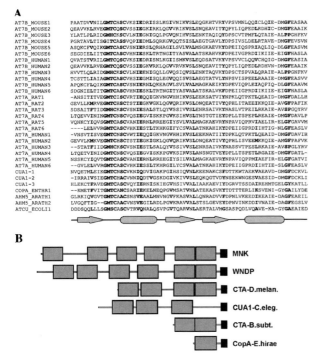

```
AT7B_MOUSE1   PAATDVVNILGMTCHSCVKSIEDRISSLKGIVNIKVSLEQGKHTVRYVPSVMNLQQICLQIE-DMGFEASAA
AT7B_MOUSE2   QEAVVKLRVEGMTCQSCVSSIEGKIRKLQGVVRIKVSLSNQEAVITYQPYLIQPEDLRDHIC-DMGFEAAIK
AT7B_MOUSE3   YLATLPLRIDGMHCKSCVLNIEGNIGQLPGVQNIHVSLENKTAQIQYDPSCVTPMFLQTAIE-ALPPGHFKV
AT7B_MOUSE4   PGRTAVLTISGITCASSVQPIEDMLSQRKGVQQTSISLAEGTGAVLYDPSIVSLDELRTAVE-DMGFEVSVN
AT7B_MOUSE5   ASQKCFVQIKGMTCASCVSNIERSLQRHAGILSVLVALMSGKAEVKYDPEIIQSPRIAQLIQ-DLGFEASVM
AT7B_MOUSE6   SEGDIELIITGMTCASCVHNIESKLTRTNGITYASVALATSKAHVKFDPEIVGPRDIIKIIE-EIGFHASLA
AT7B_HUMAN1   QVATSTVRILGMTCQSCVKSIEDRISNLKGIISMKVSLEQDSATVKYVPSVVCLQQVCHQIG-DMGFEASIA
AT7B_HUMAN2   QEAVVKLRVEGMTCQSCVSSIEGKVRKLQGVVRVKVSLSNQEAVITYQPYLIQPEDLRDHVN-DMGFEAAIK
AT7B_HUMAN3   HVVTLQLRIDGMHCKSCVLNIEENIGQLLGVQSIQVSLENKTAQVKYDPSCTSPVALQRAIE-ALPFGNFKV
AT7B_HUMAN4   TCSTTLIAIAGMTCASCVHSIEGMISQLEGVQQISVSLAEGTATVLYNPSVISPEELRAAIE-DMGFEASVV
AT7B_HUMAN5   APQKCFLQIKGMTCASCVSNIERNLQKEAGVLSVLVALMAGKAEIKYDPEVIQPLEIAQFIQ-DLGFEAAVM
AT7B_HUMAN6   SDGNIELTITGMTCASCVHNIESKLTRTNGITYASVALATSKALVKFDPEIIGPRDIIKIIE-EIGFHASLA
AT7A_RAT1     -ANSITITTVEGMTCISCVRTIEQQIGKVNGVHHIKVSLEEKSATVIYNPKLQTPKTLQEAID-DMGFDALLH
AT7A_RAT2     GEVLLKMRVEGMTCHSCTSTIEGKVGKLQGVQRIKVSLDNQEATIVKYPHLITAEEIKKQIE-AVGFPAFIK
AT7A_RAT3     SDSAITFTIDGMHCKSCVSNIESALSTLQYVSSIVVSLENRSAIVKYNASLVTPEILRKAIE-AVSPGQYRV
AT7A_RAT4     LTQEVVININGMTCNSCVQSIEGVISKKPGVKSIHVSLTNSTGTIEYDPLLTSPEPLREAIE-DMGFDAVLP
AT7A_RAT5     VQNKCYIQVSGMTCASCVANIERNLRREEGIYSVLVALMAGKAEVRYNPAVIQPRVIAELIR-ELGFGAVVM
AT7A_RAT6     ----LELVVRGMTCASCVHKIESTLTKHKGIFYCSVALATNKAHIKYDPEIIGPRDIIHTIG-NLGFEASLV
AT7A_HUMAN1   -VNSVTTISVEGMTCNSCVWTIEQQIGKVNGVHHIKVSLEEKNATIIYDPKLQTPKTLQEAID-DMGFDAVIH
AT7A_HUMAN2   GEVVLKMKVEGMTCHSCTSTIEGKIGKLQGVQRIKVSLDNQEATIVKYPHLIEVEEMKKQIE-AMGFPAFVK
AT7A_HUMAN3   --STATFIIDGMHCKSCVSNIESTLSALQYVSSIVVSLENRSAIVKYNASSVTPESLRKAIE-AVSPGLYRV
AT7A_HUMAN4   LTQETVINIDGMTCNSCVQSIEGVISKKPGVKSIHVSLANSNGTVEYDPLLTSPETLRGAIE-DMGFDATL-
AT7A_HUMAN5   NSSKCYIQVTGMTCASCVANIERNLRREEGIYSILVALMAGKAEVRYNPAVIQPPMIAEFIR-ELGFGATVI
AT7A_HUMAN6   --GVLELVVRGMTCASCVHKIESSLTKHRGILYCSVALATNKAHIKYDPEIIGPRDIIHTIE-SLGFEASLV
CUA1-1        NVQETMLEIKGMTCNSCVKNIQDVIGAKPGIHSIQVNLKEENAKCSFDTTKWTAEKVAEAVD-DMGFDCKVL
CUA1-2        -IRRAIVSIEGMTCHACVNNIQDTVGSKDGIVKIVVSLEQKQGTVDYNSEKWNGESVAESID-DMGFDCKLI
CUA1-3        HLEKCTFAVEGMTCASCVQYIERNISKIEGVHSIVVALIAAKAEVIYDGRVTSSDAIREHMTGELGYKATLL
COPA_ENTHR1   --KMETFVITGMTCANCSARIEKELNEQPGVMSATVNLATEKASVKYTDTTTERLIKSVENI-GYGAILYDE
AHM5_ARATH1   GLRKIQVGVTGMTCAACSNSVEAALMNVNGVFKASVALLQNRADVVFDPNLVKEDIKEAIE-DAGFEAEIL
AHM5_ARATH2   LVGQFTIG--GMTCAACVNSVEGILRDLPGVKRAVVALSTSLGEVEYDPNVINKDDIVNAIE-DAGFEGSLV
ATCU_ECOLI1   DDDSQQLLLSGMSCASCVTRVQNALQSVPGVTQARVNLAERTALVMGSASPQDLVQAVE-KA-GYGAEAIED
```

B

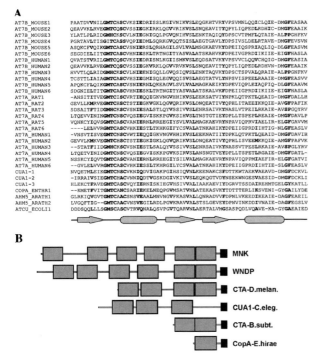

MNK

WNDP

CTA-D.melan.

CUA1-C.eleg.

CTA-B.subt.

CopA-E.hirae

Figure 5.5 Sequence alignment on the metal-binding repeats from different CTAs (A) and a modular organization of the repeats within the N-terminal domain (B). In the aligment, highly conserved residues are indicated in bold. Light gray boxes represent individual metal-binding repeats within the N-terminal domain of CTA; the dark gray box represents the rest of the CTA. WNDP – human Wilson's disease protein; MNKP – human Menkes disease protein; CTA-D. melan. – putative CTA from *D. melanogaster*; CUA1-C. eleg. – CTA from *C. elegans*; ATCU-B. subt. – CTA from *B. subtilis*; and CopA-E. hirae. – CTA from *E. hirae*.

structures have been solved for MBS4 of MNKP [107], MBS1 of yeast copper-transporting ATPase Ccc2a [106], and MBS2 of the potential copper-transporting ATPase from *B. subtilis* [112]. These studies demonstrated that all MBS have a very similar structural organization with a characteristic ferridoxin ($\beta\alpha\beta\beta\alpha\beta$) fold (Figure 5.6). Two cysteine residues in the GMxCxxC motif participating in copper coordination are located in the loop connecting the first β-sheet and α-helix, with the second cysteine positioned at the beginning of the α-helix (Figure 5.6). Both cysteines are exposed to the solvent, providing easy access for copper. The overall integrity of the MBS structure is maintained by a hydrophobic core, which is formed by hydrophobic residues, several of which are conserved in the structure of various CTAs.

Structural analysis also reveals noticeable clusters of positive and negative charges at the surface of all MBS. These charged clusters were proposed to interact with the corresponding but oppositely charged areas of the copper chaperones, the cytosolic proteins that serve as copper donors for CTA *in vivo* [113–115] (for further details, see Section 5.6.1). Sequence similarity between MBS of human copper-transporting ATPases (the pair-wise identity ranges from 20 to 60%) permitted

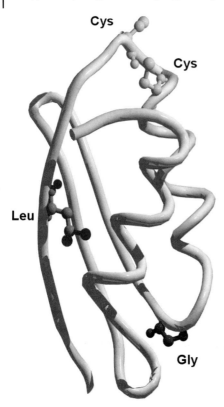

Figure 5.6 Structural model of the N-terminal metal-binding repeat 3 of Wilson's disease protein. The model was generated using available coordinates for MBS4 of MNKP (PDB accession # 1AWO) and "SpdbViewer" software. Cys residues from the GMxCxxC motif are at the top; the Leu residue homologous to the residue affected by the L492S mutation is mid-left, and the Gly85 and G591 affected by disease mutations Gly85Val and Gly591Asp are bottom-right.

homology modeling for all MBS using the NMR structure of MBS4 of MNKP [111]. As expected, the overall structure of all six MBS resembles MBS4. However, the surface distribution of charges looks quite distinct for different MBS. This diversity may reflect the difference in the structural and functional roles of metal-binding sub-domains. For example, while some MBS could be involved in interaction with the copper chaperone, the other may participate in the inter-domain or inter-MBS interactions. Such interactions could be important for regulation of either CTA function or, in mammalian ATPases, for regulation of intracellular trafficking or kinase-mediated phosphorylation (see Section 5.6 for details).

As discussed in later sections, binding of copper to the N-terminal domain of CTAs has a significant effect on conformation and possibly function of CTAs. Therefore, it is particularly instructive to compare the structures of MBS in the apo- and metal-bound forms. The solution structure of the MBS4 of MNKP was solved in the Ag^+-bound form [107], while MBS1 of yeast copper-transporting ATPase, ccc2a [106], and MBS2 of *B. subtilis* copper-transporting ATPase [112] were characterized in the Cu^+-bound state. Comparison of these structures reveals common consequences of metal binding. Importantly, binding of either copper or silver does not change overall fold of MBS. The hydrophobic core of MBS also remains essentially unaltered. At the same time, the first loop containing the metal-binding cysteines becomes more ordered compared with apo-MBS. Since copper-induced rearrangements are observed only in the flexible metal-binding loops of individual MBS, one may suggest that the significant structural changes observed in the N-terminal domains of human copper ATPases [105, 116] stem from interactions between MBS and/or from interactions of the flexible metal-binding loops with the other parts of the ATPase. Also, it seems likely that the long linkers connecting MBS in the N-terminal domain (Figure 5.5) may re-arrange as a result of copper-induced inter-protein interactions. These conformational rearrangements may play a key role in copper-dependent regulation of mammalian CTA function and intracellular localization (see more in Section 5.6).

How multiple MBS are arranged within the N-terminal domain is a tantalizing question, which remains to be addressed. So far, no structures of the full-length N-terminal domain or structures of two or three MBS together have been reported. In the absence of the high-resolution structures, coordination of copper in the N-terminal domain of MNKP (N-MNKP) and WNDP (N-WNDP) can been characterized by X-ray absorption spectroscopy (XAS). These experiments demonstrated that in either isolated MBS, or in the entire N-WNDP/N-MNKP, copper is coordinated by two cysteine residues serving as primary ligands [103, 105]. There is a slight distortion of linear geometry, presumably due to a third, distantly located, ligand. The nature of this third ligand is still unclear. Structural analysis of MBS revealed that the conserved Met residue in the GMxCxxC motif is not involved in metal coordination. Instead, it acts as a tether linking the metal-binding loop with the hydrophobic core of MBS [107]. The conserved Thr residue, adjacent to the first cysteine, is situated close to the copper ion and may serve as a third ligand [112]; however, this possibility is unconfirmed.

The maximum number of coordination sites for Cu^+ is six. Therefore, the two-coordinate state of copper in MBS leaves other orbitals vacant and available for interaction with additional protein ligands. Structural studies suggest that the conformation of the second Cys in the GMxCxxC motif is prearranged for metal binding, while additional protein ligand(s) alter their conformation when metal binds [112]. The exposure and conformational flexibility of copper-coordinating Cys provide a structural basis for easy exchange of copper. Such an exchange has been proposed to occur during copper delivery from the copper chaperones to MBS in the respective CTAs (see Section 5.6.1 for details).

The primary sequences of MBS of CTA and non-copper-transporting P_1-type ATPases, such as ZntA or CadA, are approximately 20% identical. In particular, the metal-binding motif CxxC is invariant in all P_1-type ATPases. This raises an interesting question as to whether MBS are metal-specific. Current data are somewhat controversial. It is known that Ag^+ binds to isolated MBS of CTA [107]. This is not surprising, considering the same oxidation state of Ag^+ and Cu^+ and their location within the same group of elements in the periodic table. Our group demonstrated that the isolated N-terminal domains of WNDP (N-WNDP) and MNKP (N-MNKP) bind copper with stoichiometry of one copper per MBS, bind zinc only slightly, and do not bind other metals [117]. These results suggest that MBS of human CTAs are copper-specific. At the same time, DiDonato et al. showed that N-WNDP could bind Cu^+, Zn^{2+} and other metals [118]. The discrepancy in the results could be due to differences in sample preparation. In the latter case, N-WNDP was subjected to denaturation and refolding, a procedure that could reduce metal specificity if such specificity is based on precise protein folding.

DiDonato et al. subsequently established that zinc binding to N-WNDP induced conformational changes that were different from those induced by copper and resulted in the eventual loss of secondary structure of N-WNDP [119]. Furthermore, zinc was found to be coordinated primarily by nitrogen atoms. Thus, it appears that the GMTCxxC motif in MBS is indeed specific for copper, while zinc binds elsewhere. The physiological significance of zinc binding to CTAs, although potentially important, remains to be established.

Recently, the solution structure of the isolated MBS of zinc-transporting ATPase ZntA was determined by NMR [112]. The structures of ZntA MBS in both the apo- and zinc-bound form are very similar to the structures of copper-bound MBS of CTAs with one exception. In addition to already known coordination of metal by two Cys residues, zinc is ligated by Asp residue adjacent to Cys in the GMDCxxC motif. Consequently, the authors hypothesized that the Asp residue could be important for providing specificity towards zinc. This conclusion is also supported by the invariance of the respective Asp residue in all ZntA-like sequences [111]. Significantly, in copper-transporting ATPases the equivalent position in the GMTCxxC motif is occupied by a highly conserved Thr, suggesting that the residue adjacent to the CxxC motif could be important for specificity of MBS for copper or zinc, respectively. This intriguing hypothesis awaits experimental verification.

5.2.4
ATP-binding Domain (ATP-BD)

This domain is located between TMS6 and TMS7 and contains the majority of the sequence motifs present in all P-type ATPases (Figure 5.2). Therefore, the overall fold of ATP-BD of CTAs is, probably, similar to the corresponding region of other P-type ATPases, such as Ca^{2+}-ATPase of the sarcoplasmic reticulum, for which the high-resolution crystal structure has been solved [90]. The crystal structure of Ca^{2+}-ATPase revealed that ATP-BD is composed of two sub-domains: the P-domain or phosphorylation domain, and the N-domain or nucleotide-binding domain (Figure 5.7). The primary sequence of the P-domain is not continuous. Instead, the domain is formed by the N- and C-terminal parts of ATP-BD, which come together in a folded protein. The P-domain of all CTA shares significant homology with the corresponding domains of other P-type ATPases and contains the signature motifs of P-type ATPases (Figures 5.2 and 5.7).

The DKTG sequence located in the P-domain is particularly important. The invariant Asp residue in this sequence (catalytic Asp) is a target of phosphorylation, a key step in the ATPase's catalytic cycle. Another motif, GDGxND, although separated from DKTG in the primary sequence by over 100 residues, spatially is very close and together with DKTG forms a site for ATP binding and phosphorylation (Figure 5.7). Current evidence suggests that in P-type ATPases the GDGxND motif

A

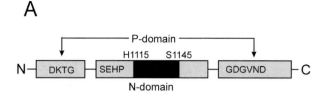

B

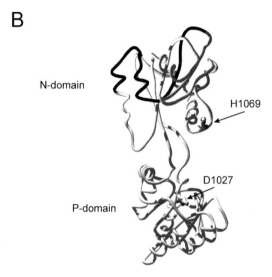

Figure 5.7 Relative positions of the conserved residues within the ATP-binding domain of WNDP (A) and molecular model of this domain (B). The N (nucleotide) and the P (phosphorylation) sub-domains are depicted. Letters and arrows (in B) indicate the position of conserved sequence motifs in each sub-domain, H1115 and S1145 are the borders of the sequence insert (see also Figure 5.8).

participates in coordination of Mg^{2+} from the Mg^{2+}-ATP complex during phosphorylation of catalytic Asp [120]. The third motif, TGDN, is upstream to GDGxND and is close to the flexible linker connecting the P- and N-domains. The TGDN sequence is thought to be involved in conformational transitions concomitant with ion translocation through the membrane [121] and may also contribute to coordination of ATP [122]. The role in conformational transitions was also proposed for the DxxK/R motif [123, 124], located in the linker connecting the N-domain and the P-domain, although the role of this site in coordination of Mg^{2+} [125] and in nucleotide binding [126] cannot be excluded. Overall, the highly conserved motifs in the P-domain of CTAs suggest that the basic chemistry of the phosphorylation process is common for all members of the P-type family. This has been confirmed recently for several CTAs (for details, see Section 5.3.2 on the functional characterization of CTAs).

The central portion of ATP-BD, which intercalates into the primary sequence of the P-domain, forms a separately folded N-domain (Figure 5.7). Unlike the P-domain, this region of CTAs shows no apparent sequence homology with the corresponding domains of other P-type ATPases and has no motifs common for all P-type ATPases. At the same time, sequence alignments of various P_1-type ATPases identify several amino-acid residues in the N-domain that are highly conserved among all members of this subfamily (Figures 5.2 and 5.8). The most prominent is the SxHP motif located 15–20 residues downstream DKTG. Histidine in this motif (His1069 in WNDP, Figure 5.2) is an invariant residue and is important for the CTAs' function, contributing directly or indirectly to ATP coordination and affecting protein function and stability (for details, see the Section 5.3).

Another motif, GxG, resides 10–15 residues downstream SxHP and is followed by a single invariant Gly residue (Figure 5.8). The distance between the conserved GxG and G residues varies significantly for different CTAs. For most bacterial and yeast enzymes, it is ca. 10–15 amino-acid residues, while for plant, insect, and mammalian CTA it is longer, extending up to 30 and 40 residues in WNDP and MNKP, respectively (Figure 5.8). The amino acid sequences of this extended loop are diverse in contrast to the rest of the N-domain, which shows considerable conservation. The variability of both the sequence and length of the extension suggests that this region is not important for the overall catalytic activity of copper-transporting ATPases. In the ATPases from higher eukaryotes this extended region could be involved in regulation through interaction with other regions of CTA or with other proteins [25, 127].

The function of the N-domain of CTAs is most likely similar to that of other P-type ATPases, i.e. to coordinate the adenine moiety of ATP and ADP during the catalytic cycle. For P_2-type ATPases, structural studies, site-directed mutagenesis, and chemical modification have identified several amino-acid residues within the N-domain that are important for coordination of ATP; for a review see Ref. 128. Interestingly, none of these residues is conserved in the N-domain of CTAs, suggesting that the details of ATP-binding in CTA may differ from that of the P_2-type ATPase subfamily. At present, the residues of the N-domain directly involved in ATP coordination remain unknown.

```
gi|17556548|cua-1        EALSEHP(x28)GHGVTCRIDSIRQSFSSLALSGSTCEI--PRLPDGQTITIPGTEVNLLQVSSKEVSQPNPDTANIVIGTERMMER
gi|12229551|AT7A_RAT     ESNSEHP(x28)GCGISCKVTNIEGLLHKSNLKIEENNIKNASLVQIDAINEQSSPSSSMIIDAHLSNAVNTQQYKVLIGNREWMIR
gi|12644462|AT7A_MOUSE   ESNSEHP(x28)GCGISCKVTNIEGLLHKSNLKIEENNIKNASLVQIDAINEQSSTSSSMIIDAHLSNAVNTQQYKVLIGNREWMIR
gi|1351993|AT7A_HUMAN    ESNSEHP(x28)GCGISCKVTNIEGLLHKNNWNIEDNNIKNASLVQIDASNEQSSTSSSMIIDAQISNALNAQQHKVLIGNREWMIR
gi|1351992|AT7A_CRIGR    ESNSEHP(x28)GCGISCKVTNIEGLLHKSNLKIEENNTKNASLVQIDAINEQSSTSSSMIIDAPLSNAVDTQQYKVLIGNREWMIR
gi|12229577|AT7B_MOUSE   EASSEHP(x28)GCGISCKVSNVEGILARSD----------------------LTAHPVGVGNPPTGEGAGPQTFSVLIGNREWMRR
gi|631354|Cu-tran        EASSEHP(x28)GCGIGCKVSNAEDILAHSERPLS----------------APASHLNEAGSLPAEKDAAPQTFSVLIGNREWLRR
gi|1703455|AT7B_HUMAN    EASSEHP(x28)GCGISCKVSNVEGILAHSERPLS----------------APASHLNEAGSLPAEKDAVPQTFSVLIGNREWLRR
gi|12643938|AT7B_SHEEP   EASSEHP(x28)GCGISCKVSVSESILAQGERLQG----------------PPTAHQNRVGSEPSETDAATQTFSVLIGNREWMRR
gi|3121725|AT7B_RAT      EASSEHP(x28)GCGISCKVSNVESILAHRG-------------------PTAHPIGVGNPPIGEGTGPQTFSVLIGNREWMRR
gi|416665|COPA_ENTHR     EHASEHP(x26)GAGISGTING--------V-------------------H------------------YFAGTRKRLAE
gi|15026756|heavy metal  EKASEHP(x26)GHGIEALIDS--------K-------------------R------------------VLLGNKKLMDN
gi|15636781|Cu-tran      EASSEHP(x45)GKGIQCLVDN--------K-------------------L------------------ILVGNRKLMSE
gi|12229667|AHM5_ARATH   EASSEHP(x45)GKGIQCLVNE--------K-------------------M------------------ILVGNRKLMSE
gi|24376802|Cu-tran      EALSEHP(x26)GLGLKGCVAD--------E-------------------T------------------LLVGNEKLMRQ
gi|2650152|pacS          ERRSEHP(x26)GEGVVA-----------D-------------------------------------GILVGNKRLMED
gi|28625435|CtaA         EARSEHP(x32)GAGVKAEIGLPGG-----K-------------------G---------QCTLFVGNARFILQ
gi|728935|ATU2_YEAST     ESISDHP(x29)GKGIVSKCQVNGN-------------------------------------TYDICIGNEALILE
gi|14132780|Cu-tran      ECHSEHP(x33)GLGIQANVTLANNP----Q-------------------T--------QFNVYVGNDKLIES
gi|2493005|COA2_HELPY    EKSSEHV(x26)GFGISAKTDYQG------A-------------------------------------KEVIKVGNSEFFNP
gi|10720391|COA1_HELPY   EKSSEHV(x26)GFGISAKTDYQG------T-------------------------------------KEIIKVGNSEFFNP
```

Figure 5.8 Sequence alignment of the N-domain region of various CTAs.

5.2.4.1 Homology Modeling of the ATP-binding Domain of Wilson's Disease Protein

The P-domain of CTAs shares approximately 30% sequence identity with the corresponding region of Ca^{2+}-ATPase and can be easily modeled using Ca^{2+}-ATPase coordinates. In contrast, the N-domain has no obvious sequence homology not only with Ca^{2+}-ATPase, but with any other proteins. This precludes modeling of the N-domain structure by conventional procedures. Consequently, to gain some insight into the structure of ATP-BD we carried out homology modeling experiments using the secondary structure predictions and threading algorithms [129]. This procedure revealed that the predicted fold of the entire ATP-BD of WNDP matches the fold of Ca^{2+}-ATPase ATP-BD in both the P-domain and the N-domains regions. The generated model is shown in Figure 5.7B.

The model of ATP-BD, although yet to be experimentally tested, is very useful for predicting positions of several conserved motifs and other regions of interest in the structure of the domain. For example, the model demonstrates that the extended and diverse segment of ATP-BD, which is characteristic of CTAs from higher eukaryotes (Figure 5.8), resides on the surface of the N-domain. In the model of ATP-BD, this segment, $H^{1115}-S^{1145}$, forms a groove that is well suited for possible interactions with other molecules, supporting the proposed regulatory role of this segment (Figure 5.7B). Also, the model predicts that the highly conserved motif SxHP with the invariant His residue is relatively close to the P-domain and, more importantly, points toward the catalytic Asp residue in the DKTG motif. The functional significance of this orientation will be discussed in Section 5.5.

5.2.5
A (actuator)-domain

This domain is located between TMS4 and TMS5 and is approximately 120 amino acids long. Its sequence is conserved among P_1-type ATPases and contains the TGE motif present in all P-type pumps (Figure 5.2). Threading algorithms predict that the overall fold of this domain is similar to the fold of the A-domain of Ca^{2+}-ATPase (our data, not shown). In Ca^{2+}-ATPase, the A-domain rotates during the catalytic cycle, facilitating conformational transitions between the so-called E1P and E2P phosphorylation states of the enzyme [91, 130]. Interestingly, mutations in the A-domain of MNKP, a human CTA, cause hyperphosphorylation of the enzyme after incubation with ATP [131]. This suggests that the E1P to E2P transition and subsequent dephosphorylation of MNKP is blocked by the mutations, in agreement with the predicted role of the A-domain.

5.2.6
C-terminal Tail

Most bacterial CTA do not have substantial protein sequences after TMS8. In contrast, mammalian, yeast, and plant CTAs may have a fairly long C-terminal tail, which can extend up to 100 amino acid residues. The primary sequence of this tail is rather diverse. However, the copper-transporting ATPase from *C. elegans*,

D. melanogaster, and mammalian MNKP and WNDP contain the LL motif. The di-leucine motif is important for trafficking of several membrane proteins, including MNKP [132]. CTAs from *C. elegans* and *D. melanogaster* possibly traffic within the cell similarly to mammalian ATPases; however, no experimental data are currently available on either trafficking or regulation of these CTAs.

5.3
Functional Properties of Copper-transporting ATPases

The conserved P-type ATPase motifs in the structure of CTAs suggests that the key steps of their catalytic cycle are the same as in all P-type ATPases. At the same time, dissimilarity in primary sequence and in transmembrane topology between CTAs and the P_2-type ATPases point to the distinct structural organization of the copper-translocation pathway in CTAs. The copper coordination environment is likely to be characteristic for this metal, and details of copper transfer into the membrane portion and subsequent release could be quite different from the corre-sponding steps for calcium, sodium, or other ions transported by the P_2-type ATPases. Detailed analysis of functional properties of CTAs is necessary to under-stand the functional consequences of structural differences between the P_1-type and P_2-type ATPases. Over the last several years, both the *in vivo* and *in vitro* assays for characterization of CTAs function have been developed. The following section will describe these functional assays and the information gained from functional studies.

5.3.1
Functional Complementation *In Vivo*

This procedure is based on the ability of cells with inactivated endogenous CTA to overcome certain metabolic defects following transformation with the plasmid en-coding the same CTA or CTA from another organism. As described in the intro-duction, inactivation of the yeast copper-transporting ATPase Ccc2 leads to (i) dis-ruption of copper incorporation into a multicopper oxidase Fet3p, (ii) significant decrease in Fet3p oxidase activity, and (iii) inability of this strain to grow on the iron-depleted media [27]. Gitlin and co-workers first demonstrated that this defect can be compensated by expression of the gene encoding another CTA, WNDP [29]. The authors also applied the yeast complementation assay to characterize several disease mutations found in WNDP. Subsequently, the assay has been widely used to evaluate functional activity of various MNKP and WNDP mutants [55, 58, 59, 108, 133] and to confirm the ability to transport copper by putative cop-per-transporting ATPases from different species (see Introduction).

Another modification of the *in vivo* functional complementation assay has been described for the characterization of the MNKP and WNDP mutants in more phy-siologically relevant mammalian cells [58]. The immortalized fibroblasts derived from the skin biopsies of Menkes disease patients (Menkes fibroblasts) accumulate copper due to decreased copper efflux. Expression of functional CTA in Menkes fibroblasts substantially increases copper efflux in the presence of elevated copper [58]. The amounts of ^{64}Cu remaining inside the cells can be quantified and compared with those from non-transfected Menkes fibroblasts; the difference serves as a measure of CTA's functional activity. This assay has also been successfully used to demonstrate the MNKP-mediated delivery of copper to tyrosinase within the secretory pathway [52].

Overall, the *in vivo* complementation assays provide convenient and robust experimental tools for initial characterization of CTA function. However, due to their indirect nature these assays do not permit detailed analysis of either catalytic or transport function of respective CTAs. Similarly, the complementation assays do not have sufficient resolution to dissect the precise changes in functional properties of proteins resulting from mutation or posttranslational modification. In the last five years, significant progress has been made in developing direct biochemical assays for CTAs.

5.3.2
Biochemical Characterization of CTAs

The functional characterization of several CTAs has been carried out using native membrane preparations from infected insect cells [134], vesicles preparations from mammalian cells [8, 110] or inside-out vesicles from bacterial cells [10]. The advantage of native membrane preparations is that the protein of interest remains embedded into the lipid bilayer and possible negative effects of solubilization are minimized. Conversely, certain biochemical assays, such as ATP hydrolysis or nucleotide binding, are difficult to perform in these preparations, because other proteins with similar activities often generate a very high background. Nevertheless, analysis of partial reactions is possible and produces valuable functional information [134, 135]. When protein expression is low (MNKP) immunoprecipitation has been employed to enrich the protein prior to subsequent functional measurements [110, 123]. To date, several bacterial CTAs have been purified to homogeneity and characterized [16, 24, 136, 137].

5.3.2.1 Transport Characteristics of CTAs
While CTA are important for various physiological processes, their own function is uniform. In all cells, these proteins couple a vectorial transport of copper ions across the cell membrane with hydrolysis of ATP. The putative catalytic scheme of CTA is shown in Figure 5.9. It includes the following key steps. (1) Binding of copper and ATP to the enzyme stabilizes the high-affinity state for copper (E1-state) and stimulates phosphorylation of Asp residue in the DKTG motif. (2) Phosphorylated protein undergoes conformational transition from the E1 state to the E2 state (low affinity for copper). (3) Copper dissociates from the low-affinity site at the opposite side of the membrane. (4) Copper dissociation is accompanied by protein dephosphorylation by a water molecule. (5) The enzyme converts from

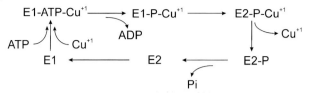

Figure 5.9 Basic catalytic cycle of the copper-transporting ATPases.

E2 into E1 state. At this point, the enzyme returns to its initial state and is ready for the next cycle. Current data on functional characterization of CTAs provide strong experimental support for this scheme.

Solioz and colleagues were first to demonstrate the ATP-dependent transport of radioactive copper using native inside-out vesicles containing CopB ATPase from *E. hirae* [9]. They also characterized the CopB-mediated transport in some detail (Table 5.1) by showing that the reaction obeys Michaelis–Menten kinetics with an apparent $V_{max} = 0.07$ nmol mg^{-1} min^{-1}, K_m for copper equal to 1 µM and K_m for ATP of 10 µM. Copper transport by CopB required 5 mM dithiotreitol, suggesting that the metal is transported in the reduced form. This conclusion is also supported by the ability of CopB to transport Ag$^+$ with very similar characteristics [9]. The authors also tested sensitivity of CopB to vanadate, a known potent inhibitor of the P-type ATPase. Noteworthy, the inhibitory concentration of vanadate was relatively high compared with other P-type ATPases: maximal inhibition was observed at 40 µM vanadate for copper transport and 60 µM for silver transport, respectively [9]. Half-maximum inhibition of activity in the presence of 40–50 µM vanadate was reported for mammalian CTAs [11].

Copper transport was also demonstrated for CopA ATPase of *E. coli* [10]. Everted membrane vesicles expressing this protein exhibited ATP-coupled accumulation of copper. Although, the authors did not carry out detailed characterization of transport, they showed that it was sensitive to vanadate and required 1 mM DTT. Interestingly, neither glutathione nor cysteine (that can also keep copper in the reduced form) at the same concentration could provide an appropriate environment for copper transport. Thus, it appears that for CopA of *E. coli* the presence of the reducing reagent is required to maintain the reduced state of both copper and the transporter.

ATP-driven copper transport by MNKP was shown using plasma membrane-enriched vesicles isolated from CHO cells overexpressing this protein [8]. The transport reaction also required DTT although at a much lower concentration, 25 µM. Kinetic parameters were similar to those of CopB from *E. hirae*: the apparent affinity for copper was in the low micromolar range and the K_m for ATP was ca. 17 µM [8]. Interestingly, V_{max} of MNKP appeared to be significantly higher than V_{max} of CopB (0.67 ± 0.38 nmol mg^{-1} min^{-1}). It is difficult to conclude whether the observed difference in V_{max} is significant, since the amounts of MNKP and CopB in the respective vesicular preparations have not been quantified and could be different. Later, Voskoboinik et al. characterized transport of copper by WNDP using conditions previously employed for analysis of MNKP [11]. The parameters of

Table 5.1 Major biochemical characteristics of copper-transporting ATPases from different sources.

Protein (Origin)	Vmax for copper transport, nmol/mg/min	Km for copper transport, µM	ATPase activity, nmol/min/mg	Vmax for ATPase activity, µmol/min/mg	ATP affinity, µM	$T_{1/2}$ for enzyme phosphorylation, s	Affinity for copper, µM	Source of the protein	Requirement for reducing reagent	Reference
MNKP (human)	1.15±0.33	4.0±2.3			17±7	10		Vesicles	DTT	(135)
MNKP (hamster)	0.67±0.38	7±3						Vesicles	DTT	(8)
WNDP (human)	1.4±0.2	8.4±1.6						Vesicles	DTT	(11)
WNDP (Human)					0.95±0.25	60	1.5±0.6	Vesicles	TCEP	(128)
CopA (E.coli)			130 (Purified)	0.19±0.06 (Purified)	520±50 (Purified)	15 (Vesicles)	1.5+0.5 (Vesicles)	Both	DTT	(17)
CopA (A.fulgidus)			225	0.062±0.002	250±40 (ATPase activity) 4.8±0.4 (phosphoryation)		2.1±0.3	Purified	Cysteine	(25)
CopA (E.hirae)	0.07 (Vesicles)		42	0.15	200	120	0.066 (from inhibition assay)	Purified		(132)
CopB (E.hirae)		1 (Vesicles)	1400±310	1.3±0.3	500±200			Both		(9, 98)

WNDP-mediated copper transport were similar to those of MNKP (Table 5.1). This similarity is not surprising considering the high level of homology between the two proteins.

In addition to characterization in mammalian cells, MNKP was heterologously expressed in oocytes, where it localized to the plasma membrane [138]. This system could provide a useful tool for electrophysiological analysis of copper transport. Unfortunately, no transport data have been reported for MNKP expressed in oocytes. Thus, it remains unclear whether ion transport by this CTA is electrogenic or not. Similarly, the nature of the counter-ion, if any, and the stoichiometry of copper transport (amount of copper translocated per enzyme cycle) remains to be determined.

5.3.2.2 ATPase Activity

During enzyme turnover, CTAs convert ATP into ADP and orthophosphate (Figure 5.9). As described above, analysis of ATPase activity for CTAs in membrane preparations is hindered by a low signal-to-background ratio. Consequently, measurements of the ATPase activity were reported only for bacterial CTAs, which were detergent-solubilized and purified. Bacterial CTAs that have been characterized in the purified form include CopA [137] and CopB [136] from *E. hirae*, CopA and CopB from *A. fulgidus* [24, 25] and CopA from *E. coli* [16]. The CopA proteins were expressed as His-tag fusions and purified using Ni-NTA resin. For CopB, the N-terminal His-rich sequence permitted purification of a non-tagged protein on a Ni-NTA resin. Although the detailed discussion of each purification protocol is beyond the scope of this chapter, it is worth mentioning that dodecylmaltoside (DDM) was consistently the most efficient detergent among several tested. DDM provided an optimal combination of a fairly high protein yield and low inhibition of protein activity. In contrast, Triton X-100 allowed an almost 100 % level of protein extraction from membranes, but the solubilized protein was completely inactive [24]. With CopA from *E. hirae*, the His-tag was removed by proteolytic cleavage after protein purification. CopA from *A. fulgidus* and CopA from *E. coli* were analyzed with the His-tag still present. Therefore, for these proteins the possible effect of the C-terminal fusion on protein activity still needs to be determined.

Interestingly, the ATPase activity of non-copper-transporting P_1-type ATPases, including ZntA from *E. coli* [88] and CadA from *L. monocytogenes* [139], can be detected even in crude membrane preparations. This difference between CTAs and non-copper-transporting P_1-type ATPases could be due to a higher turnover rate of the latter compared with CTAs. In fact, V_{max} for the *E. coli* ZntA is considerably higher than the corresponding value for CopA from the same organism (0.96–3.02 and 0.19 ± 0.06 μmol mg^{-1} min^{-1}, respectively), although the activity of CopB *E. hirae* (1.3 ± 0.3 μmol mg^{-1} ml^{-1}) is similar to ZntA. For ZntA, the activity varies depending on the metal present in the reaction, with maximum values observed in the presence of Pb^{2+} and minimal activity with Zn^{2+} (V_{max} of 3.02 ± 0.22 and 0.96 ± 0.02 μmol mg^{-1} min^{-1} [140], respectively).

Similarly to the transport assay, reducing reagents are required for measurements of the ATPase activity of CTAs. The role of the reducing reagent either for transport or for ATP hydrolysis is not entirely clear. Most likely, it is necessary to reduce Cu^{2+} to Cu^{+}, when Cu^{+} is a transported form of metal. In addition, the reducing reagent may prevent oxidation of the functionally important Cys residues in the protein. Interestingly, with CopA from *A. fulgidus* the addition of 20 mM cysteine was necessary to stimulate the CopA activity, while DTT and glutathione had no effect on the enzyme. It seems that in this case cysteine plays a role in stimulating catalytic activity by helping to deliver or remove copper from CopA, rather than maintaining a reducing environment. *A. fulgidus* is a sulfur-metabolizing thermophilic archeabacteria, which normally inhabits high temperature conditions. It would be interesting to know whether a stringent requirement of *A. fulgidus* for cysteine is related to the specific metabolism of these bacteria.

One of the initially puzzling properties of CopA and CopB from *E. hirae* was the apparent lack of a stimulatory effect of copper on the ATPase activity [136, 137]. However, treatment of these proteins with the specific copper chelator bathocuproine disulfonate (BCS) inhibited their catalytic activity, suggesting that copper was indeed required for catalysis. Furthermore, the inhibitory effect of BCS can be reversed by addition of extra copper. Altogether, these results suggest that, for certain CTAs, trace amounts of copper present in buffers are sufficient to sustain their enzymatic activity. This phenomenon was not observed for CopA from *A. fulgidus* and CopA from *E. coli*, which required additions of extra copper to the reaction buffers. It is possible that *in vivo* the affinity of these enzymes for copper is lower than that of *E. hirae* CTAs. Alternatively, isolation and purification steps may have different effects on the conformation and ligand-binding affinity of various CTAs.

5.3.2.3 Catalytic Phosphorylation

Transient phosphorylation of the invariant Asp residue in the DKTG motif is the signature event required for transmembrane ion-transport by all P-type ATPases. The acyl-phosphate intermediate can be distinguished from other phosphorylated products such as phosphoserine or phosphotyrosine by its transient nature, sensitivity to treatment with hydroxylamine at neutral pH, and instability at alkaline pH. The formation of the acylphosphate intermediate is usually detected using incubation with radioactive ATP followed by separation of the phosphorylated protein on an acidic polyacrylamide-gel and autoradiography. Analysis of catalytic phosphorylation has proven to be particularly useful for analysis of mammalian CTAs, for which ATPase activity measurements are currently difficult due to a lack of reliable purification procedures. Furthermore, by analyzing the effect of various ligands (copper, ATP, ADP, Pi, vanadate) several steps of the catalytic cycle can be dissected and the rates of phosphorylation, apparent affinities for nucleotides and copper, and the conformational state of the enzyme can be characterized.

The invariant DKTG motif, a predicted target of phosphorylation, is present in the structure of all CTAs. Consequently, efforts have been made to demonstrate that CTAs form a phosphorylated intermediate upon incubation with radioactive ATP. So far, the catalytic phosphorylation of Asp residue has been demonstrated for the following enzymes (Table 5.1): CopA and CopB ATPases from *E. hirae* [136, 137], CopA from *E. coli* [16], CopA from *A. fulgidus* [24] and human MNKP [141], and WNDP [134]. For these proteins, the authors verified the sensitivity of the phospho-intermediate to hydroxylamine and high pH. In addition, the transient nature of the phosphorylation product was confirmed by monitoring the disappearance of the radioactive intermediate after addition of cold ATP. The ATP-dependent phosphorylation was also shown to be reversible, as indicated by dephosphorylation of CTAs in the presence of ADP. As predicted, mutations of invariant Asp in the DKTG motif either to Glu or to Ala disrupt catalytic phosphorylation (e. g. [134]).

The kinetics of phosphoenzyme formation were measured for WNDP and CopA of *E. hirae*, and found to be considerably slower than those of other P-type ATPases [88, 142]. Interestingly, MNKP, which is highly homologous to WNDP, forms phosphoenzyme faster than WNDP [141]. This difference could be due to inherently distinct properties of these two enzymes. It is also possible that the important components necessary for efficient catalysis by WNDP are absent in insect cells, in which WNDP was expressed (phosphorylation of MNKP was analyzed using protein expressed in yeast). Interestingly, the transport characteristics of the same MNKP mutants expressed in mammalian cells and in yeast are different [110, 141], suggesting that the catalytic properties of CTAs may depend on expression system or on cell type. Further studies are necessary to identify the parameters that may affect the phosphorylation rates of CTAs.

5.3.2.3.1 Effect of Copper on ATP-dependent Phosphorylation

In general, the ion transported from the cytosol stimulates phosphorylation of the P-type ATPases by ATP. In agreement with this common property of the P-type ATPases, copper facilitates catalytic phosphorylation of several CTAs (Table 5.1). There are, however, some differences between CTAs with respect to copper activation. Addition of extra copper to the buffer is not necessary to observe maximum phosphorylation of *E. hirae* CopA and CopB [102, 137] or WNDP [134]. These ATPases do require copper for activity, as evidenced by their inhibition in the presence of specific chelators and subsequent reactivation by copper [134, 143] (see below). It appears that trace amounts of copper present in the buffers are sufficient for full activity of these CTAs. In contrast, extra copper should be added to the buffer to stimulate phosphorylation of MNKP (EC_{50} for copper activation of MNKP is close to 2 μM [141]) or CopA from *A. fulgidus* (apparent K_m of 1.5–2 μM [24]). The observed difference in copper-dependent activation of phosphorylation may reflect different affinities of CTAs for the metal. With human CTAs, this assumption is not unreasonable given that MNKP and WNDP are expressed in different types of tissue and may have distinct physiological functions.

Interestingly, the effect of copper on WNDP activity is cooperative, while for all bacterial enzymes and for MNKP, copper activation appears to follow a simple hyperbolic function. This suggests that with WNDP, which has six MBS in the

N-terminal region, more than one copper-binding sites are involved in stimulation of enzyme phosphorylation. In agreement with this hypothesis, our recent data demonstrate that site-directed mutagenesis of cysteines in either MBS5 or MBS6 of the N-terminal domain of WNDP eliminates the cooperative effect of copper without disrupting the ability of the protein to become phosphorylated in response to metal [109]. These results suggest MBS5 or MBS6 work together to facilitate copper binding to the intramembrane sites. How many copper ions bind in the membrane portion of CTAs and whether the intramembrane sites receive copper directly from the N-terminal domain remains to be determined.

5.3.2.3.2 Effect of Specific Copper Chelators on Activity of CTAs

Recent studies suggest that the effect of the copper chelators on CTAs is complex. As shown for CopA from *E. hirae* and WNDP, treatment with copper chelator inhibits catalytic phosphorylation [134, 143]. The inhibiting effect of chelator bathocuproine disulfonate (BCS) on WNDP is not due to simple chelation of copper in the reaction buffer, because very high concentrations of chelator (up to 500 µM) are required to decrease protein activity, and, particularly, because transferring the BCS-treated protein back to the original buffer does not restore the enzyme activity [134]. These results could be explained as follows. Prior to treatment with BCS, the enzyme exists in a conformation with high affinity for copper and can be activated by trace amounts of copper present in the buffer. The BCS treatment strips copper from the enzyme (probably from the N-terminal domain), shifting it to the low-affinity state. In this state, addition of extra copper is necessary for stimulation of phosphorylation. In fact, copper-dependent reactivation of CTAs treated with the chelator was directly demonstrated for CopA [143] and WNDP [134]. For WNDP, the effect of copper was shown to be pH dependent, with EC_{50} close to 1.5 µM at pH 6.0 and ca. 0.15–0.2 µM at pH 7.0 [134].

The reducing reagents have a distinct effect on the catalytic phosphorylation of WNDP. Prior to BCS treatment, the reducing reagent is not necessary for phosphorylation of WNDP, presumably because copper is already bound to the protein. In contrast, the reducing reagent is essential to observe copper-dependent reactivation of enzyme after its treatment with the chelator [134]. Strikingly, reactivation can only be achieved in the presence of a highly efficient phosphine-based reagent, TCEP, while DTT or glutathione are ineffective [134]. It is possible that, after removing copper from the N-terminal MBS or from the intramembrane site(s) of WNDP, the cysteine residues in the CXXC or CPC motifs become quickly oxidized, forming disulfide bonds. It is curious that oxidation, if it occurs, is very rapid (with WNDP, 15 min incubation on ice in the absence of the reducing reagent is sufficient to inactivate the enzyme, our observation) and that it is apparently prevented in the cell. Identification of the cytseine residues that become oxidized and cause inactivation of WNDP may contribute to understanding important steps in the copper transfer mechanism.

5.3.2.3.3 Phosphorylation by Inorganic Phosphate

The unique feature of all P-type ATPases is their ability to be phosphorylated in the presence of inorganic orthophosphate (P_i) and Mg^{2+}. This reaction occurs when the enzyme is stabilized in the E2 state, and is possibly due to reversibility of the dephosphorylation step in the enzyme catalytic cycle (Figure 5.9). To date, phosphorylation from P_i has been demonstrated only for WNDP [135]. As expected for the P-type ATPase, the efficiency of WNDP phosphorylation by P_i is affected by the presence of transported ion in a manner opposite to ATP-dependent phosphorylation. Specifically, copper chelator BCS increases the P_i-dependent formation of the phospho-intermediate, while copper inhibits phosphorylation from Pi [135].

5.4
Functional Role of the N-terminal Copper-binding Domain

The N-terminal metal-binding domain is a characteristic structural feature of the P_1-type ATPase subfamily. As described in earlier sections, the N-terminal domain of CTAs includes 1 to 6 repetitive elements, each containing the GMxCxxC sequence motif. When the human copper-transporting ATPases were discovered, one of the most frequently asked questions was: what is the role of the N-terminal metal-binding sites in protein function? A decade later, we know significantly more about structure and functional properties of N-terminal metal-binding sites (MBS); however, the precise function of the N-terminal domain is yet to be fully understood.

5.4.1
N-terminal Domain of Bacterial CTA

To examine the functional role of the N-terminal domain, it is useful to consider initially the bacterial CTAs because their structure is simpler and their regulation is likely to be less complex. Recently, Fan and Rosen characterized several variants of the E. coli CopA with modifications in the N-terminal domain. Specifically, these authors substituted the CxxC with AxxA in MBS1 and MBS2 of CopA, separately or together, or deleted the entire N-terminal domain containing both metal-binding sub-domains [16]. CopA variants with Cys>Ala substitutions did not impair the ability of CopA to transport copper in vivo as judged by normal copper resistance of cells expressing the mutant CopA. In contrast, the deletion of the entire N-terminal domain significantly decreased cell resistance to elevated copper. Consistent with these results, subsequent measurements of copper transport in vesicles showed that the mutations in MBS1 or in both MBS1 and MBS2 did not affect transport activity. Substitutions in MBS2 resulted in even higher transport activity. In contrast, the N-terminal deletion mutant did not exhibit any detectable copper transport. The changes in transport activity induced by the above mutations paralleled changes in catalytic phosphorylation of CopA. Specifically, the single or double CxxC to AxxA substitutions, as well as deletion of MBS1, did not abolish catalytic phosphorylation. However, no catalytic activity was detected for the deletion mutant lacking both MBS. These studies convincingly demonstrate that the physical presence of at least one N-terminal MBS is essential for the CopA activity, while the ability of MBS to bind copper is not.

Experiments on the copper dependence of CopA phosphorylation offer insight into the potential role of MBS. Both the double MBS substitution mutant and the MBS1 deletion mutant of CopA exhibited a slight increase in apparent affinity for copper, as indicated by K_m of 0.45 ± 0.1 and 0.84±0.2 µM, respectively, versus 1.5 ± 0.5 µM for wild-type CopA [16]. Thus, it seems possible that the N-terminal MBS may contribute to regulation of the affinity of the intramembrane sites for copper. Our recent studies of various N-terminal mutants of WNDP support this hypothesis. Specifically, mutations in MBS5 and MBS6 of WNDP (which are common for human and bacterial CTAs) have a clear effect on apparent affinity of the intramembrane sites for copper, as evidenced by a 7–8-fold decrease in copper EC_{50} in phosphorylation experiments, and by a markedly decreased ability of BCS to inhibit catalytic phosphorylation [109].

It is interesting to compare the results of the N-terminal domain analysis obtained for CTAs with the data for non-copper-transporting P_1-type ATPases. Most of these enzymes have only one metal binding site in their N-terminus. The consequences of deleting this N-terminal region in ZntA have been studied both *in vivo* and *in vitro* [87]. The deletion has no effect on cell resistance to lead, zinc, and cadmium *in vivo*, as well as on the affinity of the enzyme for ATP *in vitro*. Although the overall ATPase activity of the N-terminal deletion decreased 2–3-fold, the K_m for various metals (Cd, Zn, Pb) increased 2–5-fold. Importantly, the metal specificity is unchanged by deletion [87, 144]. Therefore, it was proposed that the N-terminal domain of ZntA is not essential for protein activity, but may be important for its regulation.

Similar experiments were performed for the cadmium-transporting ATPase CadA from *L. monocytogenes*, which is structurally similar to ZntA [139]. Deletion of the N-terminal metal-binding site in this enzyme did not significantly change its catalytic phosphorylation and did not affect the ATP-driven cadmium transport into the vesicles. At the same time, the affinity of CadA for cadmium, measured by cadmium dependence of ATPase activity, was increased three-fold. Furthermore, the cooperativity of the metal dependence curve for CadA was significantly diminished by the deletion of the N-terminus, pointing to functional interaction between the N-terminal and intramembrane sites. Later, the same group demonstrated that the N-terminus protects the intramembrane cadmium binding site against modification by the cysteine-directed reagent NEM, further supporting the inter-dependence of the two sites [139].

In summary, functional studies of bacterial enzymes, including CTAs, demonstrate that alterations of the N-terminal MBS affect the affinity of the intra-membrane site for the transported metal by several-fold. Therefore, it is possible that the N-terminal MBS regulate the affinity of the intramembrane site, depending on concentration of copper in the cell. Based on this model, at low intracellular copper the metal-binding sites of the N-terminal domain are predominantly in the apo form. In this state, the N-terminus interacts with the rest of CTA and keeps the intramembrane site(s) in the low-affinity state, preventing metal transport. After the copper concentration increases above a certain threshold, copper binds to the N-terminal domain. This alters the interdomain interactions, resulting in a conformational change of the CTA. The switch of the intramembrane site to the high-affinity state accompanies the conformation change and facilitates copper transfer to the intramembrane transport sites. Similarly, MBS in their apo form may support conversion of the enzyme into the low-affinity E2 state, facilitating copper release. In support of this hypothesis, recent studies of Cu^{2+}-transporting ATPase CopB from *A. fulgidus* concluded that the His-rich N-terminal domain of this CTA regulates the transport rate by controlling copper release and/or dephosphorylation rates [25]. The observed negative effect of the deletion of the entire N-terminal domain on CTA function indicates that some regions of the N-terminal domain contribute to the structural integrity of copper-transporting ATPases.

5.4.2
N-terminal Domain of Mammalian CTAs

The presence of as many as six metal-binding sites at the N-terminal domains of WNDP and MNKP (N-WNDP and N-MNKP, respectively) is a striking feature of human CTAs. The initial characterization of these domains focused on analysis of their biochemical properties. N-WNDP and N-MNKP were shown to bind copper *in vitro* and *in vivo* with a stoichiometry of one copper per MBS [104, 118]. Substantial conformational changes were detected during copper binding to N-MNKP [116] and N-WNDP [105]. Furthermore, binding of copper to N-MNKP was shown to be cooperative, with a Hill coefficient of 4 [116]. Thus, it was hypothesized that the N-terminal domain "senses" the changes in the intracellular copper concentration and regulate MNKP and WNDP in response to these changes.

Although this hypothesis is very attractive, very little work has been done to directly test the regulatory or "sensory" role of the N-terminal domain. N-WNDP was shown to interact with the ATP-binding domain of this protein in a copper-dependent manner, i. e. the interaction between the domains was shown to decrease after binding of copper to N-WNDP [127]. The decrease in domain–domain interactions was associated with an increased affinity of the ATP-binding domain for nucleotides [127]. This implies that the structural changes induced by copper binding to the N-terminal domain could be transmitted to other functional domains of WNDP. In agreement with this conclusion, the deletion of the first four MBS (which are unique for mammalian CTAs) was shown to be associated with an increased rate of catalytic phosphorylation, and possibly higher enzyme turnover [109]. The precise nature of the copper-induced conformational changes in CTAs remains to be determined.

Studies from several groups have demonstrated the functional non-equivalence of the N-terminal MBS in mammalian CTA. These experiments revealed that one or two MBS closest to the membrane portion of CTA are required for functional activity of the ATPases, while the more N-terminal MBS are not. Specifically,

Iida et al. introduced a series of sequential deletions into N-WNDP and studied their effect on the ability of mutant WNDP to complement *Δccc2* phenotype in yeast [30]. Only WNDP with intact MBS6 (closest to the membrane) was functional, i. e. could complement the *Δccc2* phenotype in yeast. Although different levels of protein expression made direct functional comparison of the mutants difficult, the study pointed to MBS6 as the site important for the transport function of WNDP. More detailed experiments using the same complementation assay were subsequently performed by Forbes and colleagues [108].

Along with the series of deletion constructs, these authors generated the WNDP mutants, in which the CxxC motif in one or several MBS was substituted with SxxS. Unlike deletions, the Cys>Ser substitution is thought to prevent copper binding to MBS without significant structural perturbations. Analysis of the substitution variants suggested that only one MBS was sufficient for WNDP function. Concurrently, inactivation of MBS6 did not have negative consequences on WNDP function if intact MBS4 and MBS5 were present. The MBS1–3 region did not have a compensatory effect. Altogether, the results pointed toward the important role of MBS6 in WNDP function. In agreement with this conclusion, the deletion of all MBSs but MBS6 produced a protein capable of copper transport.

Interestingly, different results were obtained for highly homologous of MNKP [58]. Mutation of CxxC to SxxS in MBS1 or MBS1–2 did not disrupt protein function as determined by the yeast complementation assay. However, further mutation of MBS from 3 to 6 completely abolished transport activity. This result contradicts another study, where similar mutations of either the first three MBS or all six MBS did not disrupt the protein ability to transport copper [110, 141]. Nevertheless, all experiments consistently point to unequal contributions of different MBS to MNKP and WNDP function. Our own experiments indicate that MBS5 and MBS6 of WNDP are important in copper-dependent activation of catalytic phosphorylation of this enzyme, while the first four MBS are not important for this process [109].

A somewhat different conclusion was reached by Voskoboinik et al., who mutated CxxC to SxxS in all six N-terminal MBS of MNKP and characterized transport properties of the mutant. By measuring copper transport using vesicles isolated from CHO cells, the authors showed that the mutation of all six MBS decreased the V_{max} of MNKP by approximately 4–5-fold; however, the protein remained active [110]. The authors concluded that the N-terminal MBS are not essential for overall MNKP activity. Later, this group reported expression of the same MNKP mutant in yeast [141]. Curiously, the transport data in the yeast system differed from the results obtained in the mammalian system, i. e. the MBS1–6 mutant showed no transport activity. Despite the loss of transport activity, the MBS1–6 mutant could form a phosphorylated intermediate in the presence of ATP. This phenomenon could be due to uncoupling of ATP-hydrolysis and copper transport in the MBS1–6 mutant expressed in yeast; however, why the uncoupling is observed in yeast but not in mammalian system is not yet clear.

Another interesting observation was a slower turnover of the MBS1–6 mutant and its higher sensitivity to treatment with BCS [141]. This latter result suggested that the affinity of the protein for copper was decreased. The authors concluded that the MBS1–6 mutation shifted the protein conformation from the E1 state toward the E2 state, i. e. the state with a low affinity for exported ion. Consistent with this interpretation, the MBS1–6 mutant was more sensitive to inactivation with vanadate, which is known to inhibit the P-type ATPases in the E2 state. The effect of the MBS1–6 mutation on MNKP is opposite to the consequences of the similar mutations in the N-terminal MBS of bacterial CTAs and in WNDP. In these latter cases, the apparent affinity of CTAs was increased as a result of mutations in MBS (see above). The discrepancy could be due to different types of introduced mutations. In bacterial CTAs and in WNDP, the Cys residues in the MBS were substituted with neutral and non-polar Ala. Cys>Ala substitution would abolish both the copper binding and potential protein–protein interactions. In contrast, MNKP Cys were replaced with Ser, which cannot bind copper, but can still form hydrogen bonds or contribute to the polarity of the environment. Comparing the effect of Cys>Ala and Cys>Ser mutation on function of either MNKP or WNDP may help in understanding the specific consequences of copper binding to MBS. Such studies are currently underway in our laboratory.

Despite remaining inconsistencies, current data suggest that the N-terminal domain of CTAs is involved in regulating copper translocation across the membrane. The regulation can be accomplished through shifting the enzyme conformation towards the E1 state following copper binding to the N-terminal MBS and can be mediated via interactions of the N-terminal domain with other regions of CTAs. The likely interaction partners of the N-terminal domain are the A-domain, which plays a critical role in the E1–E2 transitions, and the ATP-binding domain. In addition to our studies on the inter-domain interaction in WNDP, several recent experiments on P_2-type ATPases point to interaction of the N-terminus with other functional domains. The N-terminal tail of Ca^{2+}-ATPase was shown to interact with the A-domain [91]. A similar observation was made for the N-terminal segment of H^+-ATPase, which is longer than the corresponding region of Ca^{2+}-ATPase and which contacts both the A-domain and the ATP-binding domain [92]. An interaction between the N-terminal domain and the A-domain was proposed for Na^+/K^+-ATPase, based on functional characterization of various N-terminal deletion mutants [145]. In this study, the N-terminal domain was shown to regulate the conformational transitions of Na^+/K^+-ATPase between the E1 and E2 states [145].

Thus, it seems likely that, in the apo-form, one or two N-terminal MBS (which are common for all CTAs) may interact with the A-domain, which stabilizes the protein in the low-affinity E2-state. Binding of copper to MBS may alter interactions with the A-domain, inducing conformational changes in the latter. The conformational change is then transmitted to the rest of the protein, stabilizing it in a high-affinity E1 state. This would allow copper to bind to the intramembrane site(s) and initiate the catalytic cycle. Since the MBSs closest to the membrane are functionally important, it is likely that they interact with the A-domain. The more N-terminal MBS in human CTA may interact with the ATP-binding domain and provide additional regulation, as proposed [127].

5.5

Disease Mutations that Affect Function of MNKP and WNDP

The importance of CTA for cell metabolism is convincingly demonstrated by two severe genetic disorders in humans: Menkes disease and Wilson's disease. These diseases are caused by mutations in *ATP7A* and *ATP7B* genes, respectively. Identification and characterization of various mutations have significant medical and diagnostic value. The mutations can also serve as a powerful tool for understanding the functional role of various residues and structural domains of MNKP and WNDP. Some mutations affect amino acid residues conserved in all CTA from various species. In this case, characterization of the homologous mutation can be performed in bacterial or yeast CTA other than MNKP or WNDP, yielding useful information on the role of the conserved residue for all CTAs.

Currently, over 200 disease-causing mutations have been described for human CTAs, particularly for WNDP. Some of these mutations are listed in Tables 5.2 and 5.3. The mutations can be divided into the following groups: (1) missense mutations, causing single amino-acid substitutions in the protein sequence, (2) nonsense mutations, resulting in the insertion of a stop-codon and the premature termination of translation, (3) frame-shift mutations, usually caused by deletion or in-

Table 5.2 Mutations in WNDP documented in NCBI Protein database, excluding deletions and stop mutations.

Mutations of Invariantly Conserved Residues		Mutations Proximal to Conserved Residues (<4 AA away)		Mutations in other locations	
Location	*results of mutation*	*Location*	*results of mutation*	*Location*	*results of mutation*
MBS6	Gly^{626}>Ala	TMS2	Leu^{708}>Pro	MBS1	Gly^{85}>Val
TMS2	Gly^{711}>[Glu, Arg, Trp]		Gly^{710}>[Ser, Arg]	MBS5	Leu^{492}>Ser
TMS3	Ser^{744}>Pro		Tyr^{713}>Cys	MBS6	Phe^{608} Asp^{609} >Tyr
Loop 4-5	Asp^{918}>Asn	TMS3	Tyr^{741}>Cys		Arg^{616}>Gln
TMS6	Cys^{985}>Tyr		Ile^{747}>Phe		Asp^{642}>His
P-domain	Thr^{1031}>Ile	TMS4	Arg^{778}>[Gln, Leu, Gly Trp]		Met^{645}>Arg
	Gly^{1035}>Val	Loop 4-5	Pro^{840}>Leu	TMS1	Met^{665}>Ile
	Gly^{1266}>[Arg, Val]		Ile^{857}>Thr	Loop 1-2	Gly^{691}>Arg
	Asp^{1267}>Ala		Gly^{869}>[Arg, Val]	TMS4	Asp^{765}>Asn
	Asn^{1270}>Ser		Ala^{874}>Val		Met^{769}>[Arg, Val]
	Pro^{1273}>Leu		Arg^{919}>[Gly, Trp]	Loop 4-5	Leu^{795}>[Phe, Arg]

Table 5.2 Continued.

Mutations of Invariantly Conserved Residues		Mutations Proximal to Conserved Residues (<4 AA away)		Mutations in other locations	
Location	results of mutation	Location	results of mutation	Location	results of mutation
N-domain	Glu^{1064}>Ala		Ser^{921}>Asn		Gly^{891}>Val
	Glu^{1064}>Lys	**TMS6**	Thr^{977}>Met	**TMS5**	Thr^{933}>Pro
	His^{1069}>Gln	**P-domain**	Pro^{992}>Leu		Gly^{943}>[Asp, Ser]
	Gly^{1213}>Val		Ala^{1018}>Val	**Loop 5-6**	Ile^{967}>Phe
	Thr^{1220}>Met		Thr^{1033}>Ala		Arg^{969}>Gln
	Asp^{1222}>[Val, Tyr]		Arg^{1038}>Lys	**P-domain**	Ala^{1003}>[Thr, Val]
TMS8	Ser^{1363}>Phe		Ala^{1278}>Val		Arg^{1041}>[Pro, Trp]
			Arg^{1322}>Pro		Leu^{1043}>Pro
		N-domain	Gly^{1061}>Glu		Pro^{1052}>Leu
			Glu^{1068}>Gly		Val^{1239}>Gly
			Val^{1216}>Met		Val^{1262}>Phe
		TMS7	Leu^{1327}>Val		Ser^{1310}>Arg
			Gly^{1341}>Asp	**N-domain**	Leu^{1083}>Phe
					Gly^{1089}>[Glu, Val]
					Glu^{1095}>Pro
					Gly^{1101}>Arg
					Ile^{1102}>Thr
					Cys^{1104}>Phe
					Val^{1106}>Asp
					Glu^{1142}>His
					Val^{1146}>Met
					Ile^{1148}>Thr
					Arg^{1151}>His
					Trp^{1153}>[Cys, Arg]
					Met^{1169}>Thr
					Glu^{1173}>Lys
					Gly^{1176}>Arg
					Ala^{1183}>[Gly, Thr]
					Gly^{1186}>[Cys, Ser]
				TMS8	Trp^{1353}>Arg
					Gly^{1355}>Ser
					Ala^{1358}>Ser
				C-terminus	Thr^{1434}>Met

Table 5.3 Mutations in MNKD documented in NCBI Protein database, excluding deletions and stop mutations.

Mutations of Invariantly Conserved Residues		Mutations Proximal to Conserved Residues (<4 AA away)		Mutations in other locations	
Location	results of mutation	Location	results of mutation	Location	results of mutation
Loop 4-5	Gly^{876}>Glu	MBS6	Ala ^{629}> Pro	MBD	Ser^{637}> Leu
TMS6	Cys^{1000}>Arg	TMS2	Gly ^{727}> Arg	TMS3	Val^{767}> Leu
TMS6	Leu^{1006}>Pro	Loop 4-5	Leu^{873}> Arg	Loop 6-7	Gly ^{1019}>Asp
TGDN motif	Gly ^{1300}> Glu	TMS7	Ala ^{1362}> Val		
TGDN motif	Gly ^{1302}>Val				
P-domain	Asp^{1305}>Ala				

sertion of several nucleotides, and (4) splice mutations and large gene rearrangements that result in gross modification or entire loss of the transcript. Mutations of types 2–4 result mostly in the production of prematurely terminated products or proteins with addition of elongated non-relevant sequences (for instance, frame-shift mutation 3082AAGACT→AACT [146] and nonsense mutation 3424CAG→TAG [147] in WNDP). In general, the results of such a massive alteration of protein structure are devastating for the function, and useful structural or functional information cannot be extracted from analysis of these mutants. Consequently, we will only review the missense mutations found in MNKP and WNDP, since these mutations may help to identify the functionally important regions of CTAs.

The location of missense mutations within WNDP structure is described in Tables 5.2 and 5.3. Clearly, the mutations are found in all structural domains of WNDP, suggesting that all domains are likely to contribute to protein stability, activity, or regulation. Second, the distribution of mutations is not uniform over the protein sequence. The ATP-binding domain, the A-domain, and the transmembrane segments on average have more mutations per number of residues than the N-terminal domain. So far, only one mutation has been identified in the C-terminal tail. This observation is consistent with the fact that the three domains with a high mutation content form a "catalytic core" present in all P-type ATPases, and consequently play a crucial role in protein function. In contrast, the N-terminal region and the C-terminus are likely to be involved in regulation, rather than function, which may explain why the number of disease-causing mutations identified in these regions is much lower. Interestingly, several mutations modify residues conserved in all P-type ATPases, in such motifs as TGDN and GDGxDN (Figure 5.1). These mutations most likely affect the catalytic function of WNDP. Other mutations alter the residues that are invariant in P$_1$-ATPases only (see below). Characterization of these mutations may provide valuable information about the unique functional or structural properties of the P_1-ATPases. One such mutation is His1069Gln in WNDP, described in the following section.

Mutations in WNDP and MNKP occur with various frequencies and depend on the ethnicity of particular populations. Among various mutations identified in *ATP7B* gene, the substitution of His1069 for Gln (H1069Q) has attracted particular interest. The H1069Q replacement is the most frequent cause of Wilson's disease in the northern European population and one of the most frequent mutations in other populations [148]. Furthermore, His1069 resides in a conserved motif HP (Figure 5.2) within the ATP-binding domain, and is invariant in all P_1-type ATPases. This decidedly points to the critical role of this residue for CTA function. These observations stimulated efforts to determine the effect of H1069Q mutation on structure and functional characteristics of WNDP. The equivalent substitutions were also made in other P_1-type ATPase and the functional consequences examined.

Initial characterization of the H1069Q mutation in WNDP was carried out using yeast complementation assay. These studies produced inconsistent results. Iida et al. demonstrated that the mutant could rescue the *Δccc2* phenotype [30], suggesting that the H1069Q substitution did not affect protein function. In contrast, a complete loss of function for the same mutant was reported by another group [29]. Homologous mutation in MNKP resulted in a protein that was unable to restore the iron-deficient phenotype of yeast [58]. Importantly, in all three studies the mutation of His residue did not affect the protein expression levels in yeast.

Inconsistent results in yeast prompted analysis of the His1069Gln mutant in mammalian cells. For this purpose, Gitlin and co-workers utilized mouse fibroblast cells, which were defective in mouse homologue of MNKP [58]. Due to this defect, the cells accumulate significantly higher amounts of copper than the wild-type fibroblasts and show a considerable decrease of viability after incubation with increasing concentration of copper. Expression of wild-type WNDP cDNA in these cells restored both the copper efflux and cell viability almost to the level of control fibroblasts. In contrast, the His1069 mutants were unable to restore cell viability to the wild-type level, pointing to a disruption of the WNDP function [58]. These authors also demonstrated that the His1069Gln and His1069Ala mutants have lower stability in the cell and are localized mostly in the endoplasmic reticulum (ER) versus the *trans*-Golgi network observed for the wild-type WNDP. Interestingly, mislocalization of His1069 mutants can be corrected by growing cells at a lower temperature (28 versus 37°C), suggesting that the His replacement has a fairly mild effect on protein folding [58]. The mislocalization of the His1069Q mutant was later confirmed by analysis of endogenous WNDP in the biopsy samples from a Wilson's disease patients with this particular mutation [109]. These experiments provided the first direct demonstration of the intracellular consequences of the His1069 mutation; however, the role of His1069 in WNDP function remained unexplored.

To address this issue, the mutation of equivalent His was introduced into CopB ATPase of *E. hirae* [102], ZntA [88], MNKP [123], and WNDP [135]. In all these cases, a significant effect on CTA function was observed. The purified CopB

H>Q mutant had drastically reduced ATPase activity, and the level of catalytic phosphorylation of the mutant enzyme was decreased by 80 % [102]. Similarly, the MNKP H>Q mutant showed considerable loss of protein activity (80 % of wild type), and a 90 % decrease in catalytic phosphorylation. These effects appear to be due to the inability of mutated MNKP to bind ATP, as evidenced by a markedly increased K_m for ATP (2 mM compared with 16 µM for wild-type MNKP) [123].

For another P_1-type ATPase, ZntA, the H>Q mutant exhibited about a 60 % decrease of zinc-stimulated ATPase activity, and its phosphorylation from ATP and Pi was 32 % and 14 % of the wild-type level, respectively [88]. The authors concluded that the H>Q mutant could be trapped in the state intermediate between the E1 and E2 states due to impaired interaction between the N- and P-domains of the protein. Such an interpretation seems plausible, since the His residue belongs to the N-domain (Figure 5.7) and the N-domain is expected to come close to the P-domain for ATP to bind with high affinity and for phosphorylation to occur.

The most systematic study of the role of H1069 has been performed for WNDP expressed in insect cells [135]. For this purpose, the H1069 residue was substituted with Gln, Cys, and Ala. The structural and functional consequences of these mutations were then analyzed, using isolated ATP-BD and the full-length WNDP, to reveal that they did not markedly affect protein folding, as evidenced by the circular dichroism spectroscopy of isolated ATP-BD and unaltered proteolytic resistance of the full-length WNDP [135]. In contrast, functional analysis demonstrated the failure of the mutants to form a phosphorylated intermediate in the presence of radioactive ATP.

Interestingly, other catalytic steps were much less affected by the His replacement. Catalytic phosphorylation from Pi was decreased by 90 % for the H1069Q and H1069A mutants, but for the H1069C mutant it was indistinguishable from the wild-type. The affinity of the mutant H1069C for nucleotide showed a 4-fold decrease versus the wild-type WNDP, suggesting that the substitution of His residue had only a minor effect on nucleotide affinity. Overall, the data suggest that H1069 may contribute to orientation of ATP in the active site or may play a role in organizing the catalytic site by mediating domain–domain interactions. We hypothesized that His in the HP motif provides a precise orientation of ATP relative to catalytic Asp, allowing phosphorylation to occur. These conclusions are consistent with the predicted location of His1069 in the ATP-binding domain (Figure 5.6) in the exposed loop oriented towards the P-domain. The difference between the effects of His>Gln substitution on MNKP and WNDP affinity for ATP may reflect some variation in the structure of their nucleotide-binding sites.

5.5.1
Disease Mutations in the Transmembrane Domain

As described in earlier sections, the conserved motif CPx resides in the 6th transmembrane segment of all P_1-type ATPases (Figure 5.1). The effect of mutations in this motif on function of WNDP, MNKP and CUA-1 copper ATPase from *Caenor-*

habditis elegance was studied using a yeast complementation assay [31, 58, 101]. Cys to Ser mutants of WNDP or Pro to Ala of either WNDP [31] or MNKP [58] were unable to complement the *Δccc2* phenotype. A similar result was obtained for the CPC→CPA mutant of CUA-1 [101]. The CPC→APC mutation in CopB of *E. hirae* led to complete loss of ATPase activity and a marked decrease in enzyme phosphorylation [102]. Altogether, the data emphasize the importance of the CPC motif for the function of CTAs and are consistent with the proposed role of this motif in copper coordination during its translocation through the membrane.

Other mutations in predicted TMS have also been characterized. The Met1356-Val substitution in TMS8 of mouse WNDP [11] (Met1359 in human WNDP, Figure 5.4) and the analogous residue in TMS8 of MNKP completely abolished the ability of the proteins to transport copper into vesicles [141]. In addition, the Met1356Val substitution in MNKP caused the loss of catalytic phosphorylation [141]. Since the Met residue can serve as a ligand for copper and Met1356 is highly conserved in CTAs, the authors suggested that this residue could be a part of the intra-membrane copper-binding site(s). This hypothesis is consistent with the predicted location of this residue at the polar surface of the transmembrane segment (Figure 5.4).

5.5.2
Mutations in the ATP-binding Domain

Several mutations in ATP-BD affect the sequence motifs conserved in all P-type ATPases (DKTG, TGD, GDGxND – all in the P-domain) (Tables 5.2 and 5.3). The role of these motifs has been extensively characterized in the P_2-type ATPases (for example see Ref. 128) and will not be discussed in this section. It seems very likely that mutations within the highly conserved motifs would significantly impair the catalytic function of CTAs.

A very interesting mutation in the N-domain of MNKP, D1230A, has been recently described [123]. It was shown that the D1230A mutant had a normal level of phosphoenzyme formation, normal kinetics of phosphorylation and unaffected turnover. However, the copper transport activity of the mutant was decreased by 80%, while the apparent affinity for ATP was increased 20-fold. Another unusual property of this mutant was a complete lack of response to BCS treatment, i. e. the copper chelator did not decrease efficiency of ATP-dependent phosphorylation [123].

In the P-type ATPases, the transport of ions across the membrane and ATP hydrolysis are tightly linked. The marked decrease of copper transport in the absence of changes in catalytic phosphorylation suggests that the D1230A mutation uncouples these two events. Significantly, the equivalent Asp residue is conserved in all P_1-ATPases (Figure 5.2, in WNDP residue D1196). This residue is a part of the DxxK motif that resembles the well-known DPPR motif of the P_2-type ATPases located in the hinge region between the N- and P-domains (Figure 5.6). However, the effects of Asp substitution in DPPR and DxxK are not identical. Functional analysis of the DPPR motif in Ca^{2+}-ATPase revealed that the mutation of Asp residue abolished calcium transport by disrupting the E1P→E2P conformational transition, while the ability to form a phosphoenzyme from both ATP and Pi remained intact [121]. In contrast, the D1230A mutation in MNKP does not seem to alter the E1P→E2P transition, since the kinetics of dephosphorylation by cold ATP are identical to those of the wild-type protein. It was proposed that the D1230A mutant proceeds through the normal catalytic cycle (Figure 5.10), while the intramembrane site is trapped in the high-affinity copper E1-like state. Direct measurements of copper binding and analysis of the Pi-mediated phosphorylation would support this interesting hypothesis.

5.5.3
A-domain

In MNKP, the experimental mutations TGE→AAA and the disease mutation L873→R, located two residues upstream TGE motif, cause hyperphosphorylation [131] and an inability to complement the Δccc2 yeast phenotype. A similar effect of the mutations in the vicinity of TGE was previously reported for the P_2-type ATPases, where such mutations blocked ion-transport and enzyme turnover and led to accumulation of phosphorylated protein [121, 149]. Such behavior could be due to stabilization of the mutant protein in the E1-state due to a defect in the E1P→E2P transition. Mutations in the TGE region also have an interesting effect on the intracellular localization of MNKP. These effects are described in the Section 5.6.

5.5.4
Known Mutations in the N-terminal Domain of CTAs

Mutation of G173E in second MBS of copper-transporting ATPase RAN-1 of *Arabidopsis* was shown to abolish its function of supplying copper to the ethylene receptor [38]. The homologous Gly residue in the known structure of fourth MBS of MNKP is located inside the molecule and its replacement with bulkier and charged Glu may cause substantial changes in the structure of MBS and/or its interaction with copper chaperone.

A few disease mutations have been identified in the N-terminal domain of WNDP. Three of them were partially characterized [150]. Mutations G85V, L492S and G591D are located in the 1st, 5th and 6th MBS of WNDP, respectively. Both Gly85 and Gly591 residues occupy positions highly conserved in the sequence of MBS in various CTAs [111]. In the structure of MBS, Leu492 is predicted to reside inside the hydrophobic core. Therefore, its replacement with hydrophilic Ser may affect the stability and folding of MBS5. In contrast, Gly85 and Gly591 are on the surface of MBS, at the end opposite to the CxxC motif (Figure 5.6). By utilizing GST pull down assay, Hamza and Gitlin demonstrated that G85V, L492S and G591D mutations abolished interaction of mutant WNDP with the copper chaperone Atox1 [150]. The authors concluded that the disruption of interaction between

WNDP and Atox1 leads to the disease phenotype in patients with respective WNDP. Although very appealing, this hypothesis awaits further verification. In particular, the effect of these mutations on expression, stability, and function of WNDP needs to be elucidated.

5.6
Regulation of Copper-transporting ATPases

5.6.1
Copper Chaperones as Specific Regulators of CTAs

One of the unique features of copper homeostasis is an apparent lack of free copper in the cytosol, at least in eukaryotic cells [151]. Copper imported into cells quickly binds to proteins called copper chaperones, which then shuttle copper to various target proteins [152, 153]. The copper chaperones keep copper in the reduced Cu^+ state [154–156], stabilizing Cu^+ ion (which by itself is unstable in the aqueous environment), and minimizing redox reactions in the cell. Specific protein–protein interaction is thought to be necessary for delivery of copper from the chaperones to their respective protein targets. These targets include the cytosolic enzyme superoxide dismutase (SOD), the copper-transporting P-type ATPases located in the secretory pathway, and the mitochondria proteins involved in assembly of cytochrome C oxidase. Detailed descriptions of metallochaperones can be found in several excellent reviews [153, 157, 158]. Here, we will only discuss the metallochaperones that deliver copper to the copper-transporting ATPases.

The first copper chaperone Atx1 was originally discovered in yeast, *S. cerevisae*, as a multi-copy suppressor of oxidative damage in yeast cells lacking SOD [159]. Subsequently, it was shown that the primary role of Atx1 was to facilitate delivery of copper to the secretory pathway [160]. In *S. cerevisae*, the copper-transporting ATPase Ccc2 transports copper into the secretory pathway for incorporation into Fet3, the copper-dependent ferroxidase involved in iron uptake (see Introduction). The phenotype of the Δ*atx1* mutant is very similar to the Δ*ccc2* phenotype, i. e. the Δ*atx1* cells lack the functional Fet3 and are defective in the high-affinity iron uptake. However, the defects due to Atx1 inactivation can be suppressed by growing cells in elevated copper, while the defects in Ccc2 cannot be eliminated by high copper, suggesting that Atx1 functions upstream of Ccc2. Cytosolic localization of Atx1 led to the suggestion that it acts as a free-moving carrier of copper and shuttles the metal to Ccc2 [160].

These initial studies were followed by a series of experiments that yielded a wealth of information about the structure of Atx1 and its interactions with Ccc2. Atx1 is a soluble 7.5 kDa protein, which is likely to exist as a monomer. The metal-binding motif MxCxGC of Atx1 forms a copper-binding site, in which the metal is coordinated primarily by two cysteines [114, 161]. The crystal structure of Atx1 in the Hg(II) form was solved to 1.02 Å resolution by Rosenzweig et al. [152]. These studies revealed that the overall fold of Atx1 is identical to the fold of individual metal-binding domain of copper-transporting ATPases (Figure 5.6). Based on these observations and characterization of the functional properties of Atx1 [162, 163] it was proposed that the transfer of copper from Atx1 to Ccc2 occurs via a ligand exchange between structurally similar metal-binding sites of the chaperone and its target when they come together as a result of protein–protein interaction. The copper transfer is facilitated by (i) coordination of copper by only two cysteine residues in the metal-binding loop, (ii) position of the metal-binding site on the surface of the protein, and (iii) conformational flexibility of the metal-binding loop [163].

The proposed mechanism was tested using recombinant Atx1 and the first N-terminal metal-binding domain of Ccc2, Ccc2a. From NMR analysis of these proteins in solution, Arnesano et al. demonstrated that Atx1 and Ccc2a interact and that the changes in chemical shifts are consistent with formation of the Atx1–Ccc2 adduct [113]. Specifically, in NMR experiments apo-Atx1 displayed significant mobility in the metal-binding-site region, while this mobility was largely reduced in the Cu(I)-Atx1 complex and in the mixture of apo-Atox1 with apo-Ccc2a. The authors also estimated K_d for this interaction as $\sim 10^{-3}$ to 10^{-5} M. This fairly weak affinity pointed to the transient nature of the chaperone–target interactions. Structural studies also suggested how Atx1 recognizes Ccc2. The regions at the interface of Atx1 and Ccc2 were shown to contain many oppositely charged residues. This observation led to the hypothesis that the electrostatic interaction of complementary charges was a basis of specific Atx1–Cccc2a recognition (for a review see Ref. 111).

Direct evidence for copper transfer from Atx1 to Ccc2a was obtained by Huffman and O'Halloran [163] who demonstrated a reversible metal exchange between these two proteins. The authors determined that the delivery of copper by Atx1 to the isolated metal-binding site was not based on a higher copper affinity of the target domain and proposed that the physiological role of copper chaperones was to overcome the high copper-chelation capacity of the cytosol and to increase the rate of delivery of copper to the appropriate target proteins [158].

These insightful experiments formulated basic concepts of copper trafficking in eukaryotic cells and suggested a testable model for the molecular mechanism of copper transfer from chaperones to ATPases. However, several important questions remain. Given the reversible nature of copper exchange, it remains unclear what the driving force is for copper transfer from the chaperone to the ATPase. It has been suggested that the directionality of copper transfer is maintained by the subsequent ATP-dependent transport step leading to translocation of copper from the intramembrane sites across the membrane. How copper actually reaches the intramembrane sites is unknown. One possibility is that the N-terminal MBSs are directly involved in guiding copper towards the intramembrane site. Alternatively, the ATPases may contain additional low-affinity copper-binding sites, which form a pathway from the cytosolic portion of the protein to the higher affinity sites within the membrane. In this scenario, the N-terminal MBS would have a regulatory role, controlling enzyme conformation in response to change in copper concentration.

It seems important to complement the *in vitro* investigations of the copper-transfer mechanism with studies using live cells. Currently, the copper occupancy of chaperones in a cell at different copper concentrations remains uncharacterized. It is also unclear whether, in a cell, copper-bound Atx1 and its mammalian homologue Atox1 are present in excess to their targets, a condition that appears to be necessary for efficient transfer of copper *in vitro*. Finally, it would be important to understand the precise consequences of interactions between Atx1 and the ATPase for the functional activity of the latter. Recent experiments using mammalian equivalents of Atx1 and Ccc2 began to address some of these important questions.

Human copper chaperone Atox1 is very similar to Atx1. Initially identified by homology with Atx1 [164], Atox1 was later shown to play a very important role in mammalian copper homeostasis [165]. Genetically engineered mice with disruption of the Atox1 locus exhibit growth failure, skin laxity, hypopigmentation, and seizures because of perinatal copper deficiency. Studies using the Atox1-deficient cells derived from these mice revealed that the cells accumulated high levels of intracellular copper due to impaired copper efflux [165]. It was concluded that Atox1 plays an important role in delivering the intracellular copper to the copper-transporting ATPases, which then provide copper to proteins within the secretory pathway or export copper out of the cell. This conclusion is supported by studies showing protein–protein interactions between the metal-binding domains of MNKP/WNDP and Atox1 [150, 166, 167]. Furthermore, recent studies from our laboratory directly demonstrated saturable transfer of copper from Atox1 to the N-terminal domain of WNDP [168]. All six N-terminal copper-binding sites of WNDP can be metallated using the copper-Atox1 complex. The transfer of metal is associated with stimulation of catalytic activity of the full-length WNDP, as evidenced by a copper-Atox1 dependent increase in catalytic phosphorylation [168].

We hypothesized that in a cell the copper occupancy of chaperones depends on the intracellular copper concentration. At higher copper concentrations the Atox1: copper stoichiometry is likely to increase, resulting in delivery of larger amounts of copper to WNDP and stimulation of its activity. But what would happen with copper occupancy and activity of WNDP in the presence of large amounts of apo-chaperone, a situation presumably occurring in copper deficient cells? Our recent experiments demonstrate that incubation of the copper-bound N-terminal domain of WNDP or the full-length WNDP with the apo-chaperone results in removal of copper and inhibition of catalytic activity, respectively [168]. These results demonstrate that the copper chaperones may function not only in delivery of copper to the copper-transporting ATPases, but also in reversible inhibition of their activity under conditions of copper limitation. In mammalian cells, the Atox1-mediated changes in copper-occupancy of the copper-transporting ATPases could be tightly linked to other regulatory events, such as kinase-mediated phosphorylation and intracellular trafficking (see below)

Although in general the function of copper chaperone in yeast and mammalian cells seems to be very similar, there are interesting differences in specific details of copper transfer and its consequences. Mammalian copper-transporting ATPase

MNKP and WNDP have six homologous copper-binding sites, while there are only two copper-binding sites in Ccc2. This raises the question as to whether any MBS in CTAs can accept copper from the chaperone and whether copper migrates from one MBS to another before it reaches the membrane portion of the ATPase. Theoretically, each N-terminal MBS could be a target for interaction with the chaperone. Indeed, recent studies using purified Atox1 and recombinant MNKP fragments containing various MBS suggest that, *in vitro*, all metal-binding sub-domains of MNKP interact with Atox1 similarly [167]. All apo MBS interacted with Atox1 very weakly, if at all, while copper appears to stimulate the interaction [167].

The lack of specificity in interactions between Atxo1 and MBS *in vitro* is interesting considering that the molecular modeling of MBS structures demonstrated significant difference in the overall charge of MBS and in the distribution of charges at the surface of MBS [169]. In fact, these observations led to the suggestion that there could be a preferential binding site for Atox1, a conclusion apparently at odds with the above experimental data. However, the disagreement could be superficial. It is possible that individual domains interact with Atox1 predominantly via metal-binding sites and, therefore, similarly. Concurrently, in the full-length protein different MBS may interact through complementary surfaces and have their metal-binding sites differentially exposed for interactions with Atox1. Studies testing this model are currently underway in our laboratory.

Our recent studies revealed another interesting effect of the apo-Atox1 on WNDP. As described above, the transfer of copper by Atox1 is reversible; however, the forward and reverse reactions are not equivalent. Interaction of apo-Atox1 with the copper-bound N-terminal domain of WNDP results in removal of all but one copper, suggesting that one of the sites may either have much higher metal-binding affinity for copper or be less exposed. Significantly, inhibition of WNDP activity by apo-Atox1 is only partial. This indicates that occupation of all N-terminal MBS is not necessary for functional activity of WNDP, and that some copper remains bound to WNDP even in the presence of excess apo-Atox1. This effect of apo-Atox1 on WNDP copper occupancy and activity could be an important element of regulation of WNDP function in a cell (see also below).

5.6.2
Copper-dependent Protein Trafficking of MNKP and WNDP

Earlier sections briefly referred to the unique property of mammalian copper-transporting ATPases to relocate within the cell in response to a change in copper concentration. This phenomenon was first observed by Petris and colleagues [56], who found that in copper-resistant CHO cells MNKP was predominantly localized to the Golgi apparatus in basal medium, while in elevated copper MNKP was found in the plasma membrane. The shift in the steady-state distribution of MNKP was reversible and did not depend on new protein synthesis. Treatment of cells with reagents that block endosomal recycling resulted in accumulation of MNKP in cytoplasmic vesicles. The authors proposed that MNKP continuously cycled between the Golgi and the plasma membrane and that elevated copper shifted the steady-state distribution from the Golgi to the plasma membrane [56]. This conclusion was subsequently confirmed by several investigators [170–172]. Furthermore, it was demonstrated that WNDP also traffics in response to change in copper concentration [58, 60, 173]. Similarly to MNKP, WNDP resides in TGN in basal copper. In response to elevated copper WNDP relocates to a vesicular compartment (Figure 5.10) and then possibly to the plasma membrane. When the concentration of copper decreases, WNDP returns to TGN.

The recycling hypothesis suggested that the cytoplasmic regions of MNKP and WNDP contain the endocytic signal sequences. Indeed, the C-terminal tail of MNKP has a di-leucine motif L1487L1488, which is conserved in murine MNKP orthologues, and an acidic cluster. Deletion of the C-terminal segment downstream the di-leucine motif or substitution of the EDDD acidic cluster with AAAA did not alter the ability of MNKP to relocalize in response to copper [132]. In contrast, replacement of either individual or both leucines in the di-leucine motif resulted in the localization of the mutant MNKP in the plasma membrane even in the basal medium. These results demonstrated that the di-leucine motif L1487L1488 was essential for localization of MNKP within TGN [132]. This conclusion was subsequently confirmed by another group [174].

The work from Francis et al. demonstrated that the third transmembrane region of MNKP may contain structural information serving as a TGN retention signal [175]. Given that MNKP is a recycling protein, Petris and Mercer proposed that the TGN localization of MNKP occurs via two steps. The first involves direct retention of CTA in this compartment via the transmembrane region, and the second

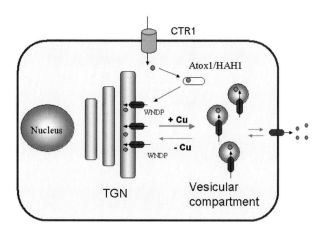

Figure 5.10 Regulated trafficking of human copper-transporting ATPases in a cell. Copper enters the mammalian cells through protein called CTR1 and is picked up by copper chaperones. The chaperone Atox1 delivers copper to the copper-transporting ATPases (WNDP is shown). At basal copper, WNDP is located in the TGN and transports copper to secreted copper-dependent enzymes. As the copper concentration becomes elevated, WNDP relocates to a vesicular compartment, where it sequesters excess copper. Subsequently WNDP reaches the plasma membrane; however, whether this step is regulated is unclear. When the concentration of copper decreases, WNDP returns to TGN.

step represents retrieval of the recycling MNKP from the plasma membrane using the di-leucine signal. The same group later demonstrated that retrieval occurs into the transferrin-containing endosomal pathway [176].

Two possible mechanisms for copper-induced localization have been considered. According to one, elevated copper stimulates exocytosis of MNKP from the TGN to the plasma membrane [56]. An alternative mechanism suggests that a constitutively recycling MNKP is trapped at the plasma membrane in the presence of elevated copper, which inhibits the retrieval of MNKP from the plasma membrane to the TGN, thereby increasing protein levels at the cell surface. To distinguish between these two scenarios Petris and Mercer introduced a c-myc tag between the first and second membrane-spanning regions of MNKP, to allow detection of surface MNKP in living cells [176]. Using cells expressing the myc-tagged MNKP constructs, these authors measured internalization of the anti-myc antibodies under basal and copper-loaded conditions. Under basal conditions, the time-dependent appearance of antibodies in the perinuclear space was detected, suggesting that the perinuclear pool of MNKP-tag was constitutively recycling via the cell surface and directing the internalization of anti-myc antibodies. Treatment of cells with copper resulted in a significant increase in the antibody staining at the plasma membrane and strong labeling of perinuclear region and vesicles throughout the cytoplasm. The authors concluded that in the presence of copper the MNKP-tag continued to be internalized and recycled through the endocytic compartments [176]. Therefore, inhibition of MNKP internalization is not the mechanism by which copper stimulates recruitment of the protein to the plasma membrane.

Interestingly, the time-course studies of antibody internalization in basal conditions did not reveal any staining of a cytosolic vesicular compartment, which would be expected if the return of MNKP-tag proceed through the endocytic pathway. At the same time, such vesicular staining was clearly observed in cells treated with copper. These results suggest that, although copper does not prevent endocytosis and recycling of MNKP to TGN, the kinetics of this process seem to be altered by copper.

5.6.3
Molecular Mechanism of Copper-dependent Trafficking

The above studies supply experimental evidence that the increased copper stimulates exocytosis of MNKP from TGN. Two approaches have been currently used to understand the molecular mechanism of this process. The first approach involves extensive site-directed and deletion mutagenesis in the copper-binding region of MNKP. Another approach utilized various mutations found in Menkes disease patients.

The role of the N-terminal copper-binding domain in trafficking of MNKP was investigated by Strausak et al. [177]. These authors demonstrated that the N-terminal region from amino acids 8 to 485, which included MBS1–4, was not essential for copper-induced trafficking of MNKP. In contrast, mutations of MBS5 and MBS6 disrupted the ability of MNKP to relocalize from TGN to the cell surface.

Further experiments revealed that only one of these MBS was necessary and sufficient for the copper-induced redistribution of MNKP [177]. Since earlier experiments demonstrated the role of MBS5 or MBS6 for the copper-transporting ATPase function, the trafficking studies raised an important issue as to whether the ability of mammalian copper-ATPases to traffic was coupled to their functional activity.

This issue has been recently addressed using MNKP and WNDP with single amino acid mutations. Several disease-causing mutations in MNKP and WNDP were shown to disrupt normal localization and trafficking of these CTAs [58, 109, 123, 172, 173, 178]. These mutations are located in various regions of the ATPases, suggesting that their effect on copper-induced trafficking is probably indirect. Consequently, it was hypothesized that the trafficking of MNKP and WNDP depends on copper transport activity. This conclusion was supported by finding that the mutations within the conserved CPC, SEHPL sequences, and DKTG sequences disrupted the copper-induced relocalization from the TGN [131]. Interestingly, the Leu873>Arg mutation in close proximity to the TGE motif resulted in the constitutive relocalization of MNKP to the plasma membrane even in the absence of additional copper [131].

A similar effect was observed when the entire TGE motif was substituted with AAA. As described earlier, the TGE motif of P-type ATPases is thought to facilitate conformational transitions during the catalytic cycle and dephosphorylation of the catalytic intermediate. In fact, an earlier work on P_2-ATPases demonstrated that mutations of the residues within the TGE motif trap proteins in the E1-P-like state [149, 179]. This is particularly interesting, since both the Leu873>Arg mutant and TGE>AAA mutant have a 2–3-fold higher level of catalytic phosphorylation compared with wild-type MNKP [131], which is consistent with the predicted stabilization in the E1-like state. Neither the Leu873>Arg nor TGE>AAA mutants can transport copper, suggesting that a certain conformational state rather than the ability to go through the entire cycle are necessary for relocalization of the copper-transporting ATPases.

The authors went somewhat further, suggesting that it is the formation of the phosphorylated intermediate during MNKP catalysis that triggers the exocytic trafficking of the transporter from the TGN. This conclusion was based on the observation that a double mutant containing both the 1044DKTG-EKTG and 875TGE-AAA mutations had an exclusively perinuclear localization and failed to undergo relocalization to the plasma membrane in cells exposed to elevated copper. It was concluded that the D1044E mutation suppressed the 875TGE-AAA mutation and therefore copper-induced trafficking of MNKP required formation of a phosphorylated catalytic intermediate [131]. The catalysis-dependent trafficking model implies that, in TGN, only a few MNKP and WNDP molecules are active, since each catalytic cycle is associated with MNKP and WNDP leaving the TGN. The redistribution of MNKP and WNDP from the TGN upon exposure to elevated copper is explained by an increase in the proportion of catalytically active proteins. This model seems inconsistent with the important role of MNKP and WNDP in delivering of copper to the enzymes localized in the secretory pathway. A more plausible explanation is that the copper-ATPases do need to adopt a certain conformation to become trafficking-competent; however, this conformation is not directly linked to the catalytic cycle. The above results with the mutants indicate that this trafficking-competent conformation resembles the E1 or E1P state of the enzyme. From this point of view, it seems quite interesting that the MBS1–6 mutant, which is catalytically competent but stabilized in the E2-state [110], does not traffic [177]. Recently, Kim et al. published an interesting paper that further demonstrates that the trafficking and catalytic activity can be uncoupled [180]

In this work, the authors characterized another disease mutation found in MNK patients, which resulted in the deletion of exon 8 in the MNKP gene. At the protein level, the deletion leads to an in-frame removal of 25 amino acid residues between Ser624 and Gln649 [180]. This preserves the MBS6 but shortens the linker connecting MBS6 to the membrane. The mutant MNKP is localized correctly to the TGN and can transport copper to tyrosinase, a copper-dependent enzyme within the secretory pathway [52]. However, the MNKP mutant failed to re-localize to the plasma membrane in response to elevated copper [180]. Thus, it seems likely that the important trafficking signal was altered or deleted in this mutant, although activity of the mutant was not impaired.

This work is very interesting in the context of recent studies from Mercer and colleagues, who generated chimera molecules using MNDP and WNDP cDNA to identify the regions of these proteins responsible for their differential trafficking (MNKP – to plasma membrane, WNDP – to cytosolic vesicles) [59]. Chimeras in which the N-terminal MBSs of MNKP were replaced with the corresponding MBSs of WNDP localized to the vesicular compartment, similarly to WNDP in elevated copper, pointing to the important role of the N-terminal domain in protein targeting. Further deletions of various MBSs of the chimera indicate that the targeting signal which directs the chimera to the vesicular compartment is present in the region of MBS6, and may involve the small segment close to the membrane [59]. Whether the signaling sequence works by itself or in the context of the entire protein sequence is still unclear.

Trafficking of WNDP has also been characterized in some detail. As described above, in basal copper WNDP is localized primarily to the TGN (Figure 5.10). WNDP shows somewhat more diffuse staining than MNKP, suggesting that copper may stimulate relocalization of WNDP at lower concentrations. The sequences of the WNDP homologues from mouse, rat, and sheep each include three consecutive leucines at a position corresponding to the dileucine motif of MNKP. These leucine residues might function in TGN targeting of WNDP homologues similarly to L1487-L1488 in MNKP. Ultrastructural studies showed that, with elevated copper levels, WNDP accumulated in large multi-vesicular structures resembling late endosomes, possibly a novel compartment for copper transport [181]. Identification of the molecular nature of this compartment could be extremely interesting and informative.

The effect of mutations in WNDP on both activity and trafficking was also investigated. Catalytically inactive mutants of WNDP Arg778Leu and CPC>SPS were unable to traffic in response to copper; and the Arg778Leu mutant was also mislocalized under basal conditions, presumably to the endoplasmic reticulum [173].

The WNDP mutant Gly943Ser, which has nearly normal function in the yeast complementation assay, was shown to localize normally to TGN under basal conditions, but was unable to redistribute in response to copper, further illustrating that the activity and trafficking of human CTAs can be uncoupled [173]. Altogether, the general mechanism of copper-stimulated trafficking appears to be similar for MNKP and WNDP. It is also interesting that both proteins undergo copper-dependent phosphorylation by a kinase (see below), a process that could be important for one of the trafficking steps.

5.6.4
Role of Copper Chaperone in CTA Trafficking

Recent studies suggest that Atox1 plays an important role in the regulation of trafficking of MNKP and possibly WNDP. Comparison of the dose- and time-dependent trafficking of MNKP in response to copper using quantitative confocal microscopy revealed a significant difference between MNKP expressed in the Atox1 (+/+) and Atox1 (−/−) cells. In cells containing endogenous Atox1 [Atox1(+/+) cells], MNKP is targeted to the Golgi compartment in the basal medium, as expected. In contrast, in the cells isolated from the transgenic mice lacking Atox1 [Atox1(−/−) cells], the localization of MNKP was more diffuse. This altered localization is probably due to elevated copper levels in the Atox1(−/−) cells [182].

This result is quite interesting. It suggests that either copper binds to MNKP in the absence of Atox1 and induces trafficking or that the copper-induced trafficking of MNKP is independent of copper-binding to the ATPase and is induced by other metal-dependent factors, for example by a kinase (see below). Further studies revealed that the situation is more complex. Although MNKP does relocalize in response to copper in the Atox1 (−/−) cells, larger amounts of copper are necessary to induce the trafficking step. Consequently, the authors proposed that Atox1 plays an essential role in establishing the threshold for copper-dependent movement of the copper-transporting ATPases within the secretory compartment [182]. This conclusion is consistent with the *in vitro* experiments, demonstrating the ability of Atox1 to regulate copper occupancy of WNDP [168].

5.6.5
Kinase-dependent Phosphorylation of Human-copper-transporting ATPases in Response to Copper

Studies using truncated, mutated, and chimera variants of MNKP and WNDP yielded important information about potential regions and conformations of CTAs essential for protein trafficking. At the same time, we still have less than a cursory knowledge of how changes in copper concentration are translated into relocalization of copper transporters. Work from DiDonato and colleagues [105] and our own studies indicate that binding of copper to the N-terminal domain of human CTAs results in significant conformational changes. Therefore it has been proposed that these copper-induced conformational changes serve as a signal for CTA trafficking. This hypothesis is quite attractive. If copper-binding to the N-terminal MBS leads to conformational changes and changes in domain–domain interactions, these events may result in exposure of new sites for interactions with protein targeting machinery or with other cell regulators. Recent data suggest that copper-induced kinase-mediated phosphorylation can be an important event associated with the re-distribution of MNKP and WNDP in a cell.

WNDP in a cell is phosphorylated by an uncharacterized kinase and the degree of phosphorylation is regulated by intracellular copper concentration [183]. There is also an interesting link between the phosphorylation level and intracellular localization of WNDP. In low copper, when WNDP is targeted predominantly to TGN, it has a basal level of phosphorylation. An increase in copper concentration results in a 2–3-fold increase in WNDP phosphorylation; under these conditions WNDP is found predominantly in the vesicular compartment. A decrease in copper concentration leads to dephosphorylation of WNDP to the basal level and return of protein to the TGN [183].

Recently, Voskoboinik et al. reported similar findings using MNKP [184]. In addition, this group carried out peptide mapping experiments and found that the majority of copper-dependent phosphorylation was on serine residues in two phosphopeptides. Importantly, there was no up-regulation of phosphorylation of a MNKP mutant with mutated N-terminal MBS. Further studies will have to determine what are the sites of basal and copper-induced phosphorylation in MNKP and WNDP and whether phosphorylation serves as a signal for trafficking or as a retention signal that keeps CTAs in a certain compartment at elevated copper. It is also possible that the copper-induced phosphorylation by a kinase is required for regulation of CTA activity in these compartments.

To further understand the mechanism of MNKP regulation, Cobbold et al. analyzed trafficking of the endogenous MNKP in HeLa cells. The copper-induced re-localization of MNKP was not sensitive to expression of a dominant-negative mutant protein kinase D, an enzyme implicated in regulating constitutive trafficking from the TGN to the plasma membrane. However, protein kinase A inhibitors blocked copper-stimulated relocalization of MNKP. Expression of constitutively active Rho GTPases such as Cdc42, Rac1 and RhoA revealed that Cdc42 was required for trafficking of MNKP to the cell surface; however, the precise details of this process remain to be elucidated [171].

The current model of regulation of mammalian CTAs in a cell can be summarized as follows (Figure 5.11). The activities of MNKP and WNDP are copper-dependent, and are likely to be regulated by their N-terminal domain. The copper chaperone Atox1 controls the copper-occupancy of the N-terminal domain through the reversible delivery of copper. At copper concentrations exceeding a certain threshold the N-terminal domain of mammalian CTAs binds copper and undergoes conformational transitions, which result in alterations of both the N-terminal domain and the ATP-binding domain and possibly other regions of the protein, such as the C-terminal tail. These events initiate the trafficking steps either directly or following phosphorylation of WNDP by a kinase. A decrease in copper concentration drives these reactions in reverse.

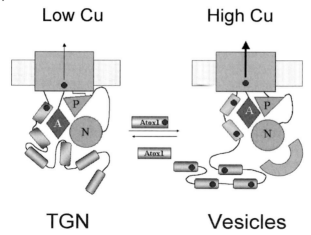

Figure 5.11 Model for copper-dependent regulation of the human copper-transporting ATPases. The model is based on experimental results for WNDP. At low copper, WNDP is located in the TGN and has a basal level of catalytic activity due to substoichiometric amounts of copper bound to N-WNDP. An increase in the intracellular copper concentration results in more Atox1-copper complexes and facilitates delivery of copper to WNDP. Saturation of N-WNDP with copper results in a conformational change in N-WNDP. This in turn weakens interactions between N-WNDP and ATP-BD, exposing sites for phosphorylation with a kinase or/and for interactions with the protein trafficking machinery. As a result of these copper-dependent events, WNDP is found in the vesicular compartment where it has a maximum transport activity. A decrease in copper concentration leads to removal of several coppers from N-WNDP by apo-Atox1, dephosphorylation of WNDP by a phosphatase, and its return to TGN.

Acknowledgments

We are grateful to Dr Jose Arguello for sharing his findings on the metal-specificity of the P1-type ATPase prior to publication and for helpful discussions. This work was supported by the National Institute of Health Grant 1-P01-GM067166-01 and National Science Foundation Grant MCB-0110057 to S. L. In addition, R. T. is a recipient of the American Heart Association postdoctoral fellowship (# 0120573Z).

References

1. R. K. Szilagyi and E. I. Solomon, *Curr. Opin. Chem. Biol.*, **2002**, *6*, 250–258.

2. S. T. Prigge, R. E. Mains, B. A. Eipper, and L. M. Amzel, *Cell. Mol. Life Sci.*, **2000**, *57*, 1236–1259.

3. P. Bielli and L. Calabrese, *Cell. Mol. Life Sci.*, **2002**, *59*, 1413–1427.

4. I. M. Wasser, S. Vries de, P. Moenne-Loccoz, I. Schroder, and K. D. Karlin, *Chem. Rev.*, **2002**, *102*, 1201–1234.

5. S. Lutsenko and M. J. Petris, *J. Membr. Biol.*, **2003**, *191*, 1–12.

6. I. Voskoboinik, J. Camakaris, and J. F. Mercer, *Adv. Protein Chem.*, **2002**, *60*, 123–150.

7. N. Fatemi and B. Sarkar, *Environ. Health Perspect*, **2002**, *110 (Suppl 5)*, 695–698.

8. I. Voskoboinik, H. Brooks, S. Smith, P. Shen, and J. Camakaris, *FEBS Lett.*, **1998**, *435*, 178–182.

9. M. Solioz and A. Odermatt, *J. Biol. Chem.*, **1995**, *270*, 9217–9221.

10. C. Rensing, B. Fan, R. Sharma, B. Mitra, and B. P. Rosen, *Proc. Natl. Acad. Sci. U. S. A.*, **2000**, *97*, 652–656.

11. I. Voskoboinik, M. Greenough, S. Fontaine La, J. F. Mercer, and J. Camakaris, *Biochem. Biophys. Res. Commun.*, **2001**, *281*, 966–970.

12. A. Odermatt, R. Krapf, and M. Solioz, *Biochem. Biophys. Res. Commun.*, **1994**, *202*, 44–48.

13. A. Odermatt and M. Solioz, *J. Biol. Chem.*, **1995**, *270*, 4349–4354.

14. A. Odermatt, H. Suter, R. Krapf, and M. Solioz, *J. Biol. Chem.*, **1993**, *268*, 12775–12779.

15. S. Franke, G. Grass, and D. H. Nies, *Microbiology*, **2001**, *147*, 965–972.

16. B. Fan and B. P. Rosen, *J. Biol. Chem.*, **2002**, *277*, 46987–46992.

17. C. Petersen and L. B. Moller, *Gene*, **2000**, *261*, 289–298.

18. N. Vats and S. F. Lee, *Microbiology*, **2001**, *147*, 653–662.

19. D. Bayle, S. Wangler, T. Weitzenegger, W. Steinhilber, J. Volz, M. Przybylski, K. P. Schafer, G. Sachs, and K. Melchers, *J. Bacteriol.*, **1998**, *180*, 317–329.

20. K. Kanamaru, S. Kashiwagi, and T. Mizuno, *Mol. Microbiol.*, **1994**, *13*, 369–377.

21. L. T. Phung, G. Ajlani, and R. Haselkorn, *Proc. Natl. Acad. Sci. U. S. A.*, **1994**, *91*, 9651–9654.

22. S. Tottey, P. R. Rich, S. A. Rondet, and N. J. Robinson, *J. Biol. Chem.*, **2001**, *276*, 19999–20004.

23. S. Tottey, S. A. Rondet, G. P. Borrelly, P. J. Robinson, P. R. Rich, and N. J. Robinson, *J. Biol. Chem.*, **2002**, *277*, 5490–5497.

24. A. K. Mandal, W. D. Cheung, and J. M. Arguello, *J. Biol. Chem.*, **2002**, *277*, 7201–7208.

25. S. Mana-Capelli, A. K. Mandal, and J. M. Arguello, *J. Biol. Chem.*, **2003**, *278*, 40534–40541.

26. D. S. Yuan, A. Dancis, and R. D. Klausner, *J. Biol. Chem.*, **1997**, *272*, 25787–25793.

27. D. S. Yuan, R. Stearman, A. Dancis, T. Dunn, T. Beeler, and R. D. Klausner, *Proc. Natl. Acad. Sci. U. S. A.*, **1995**, *92*, 2632–2636.

28. A. Dancis, D. S. Yuan, D. Haile, C. Askwith, D. Eide, C. Moehle, J. Kaplan, and R. D. Klausner, *Cell*, **1994**, *76*, 393–402.

29. I. H. Hung, M. Suzuki, Y. Yamaguchi, D. S. Yuan, R. D. Klausner, and J. D. Gitlin, *J. Biol. Chem.*, **1997**, *272*, 21461–21466.

30. M. Iida, K. Terada, Y. Sambongi, T. Wakabayashi, N. Miura, K. Koyama, M. Futai, and T. Sugiyama, *FEBS Lett.*, **1998**, *428*, 281–285.

31. J. R. Forbes and D. W. Cox, *Am. J. Hum. Gen.*, **1998**, *63*, 1663–1674.

32. A. Dancis, D. Haile, D. S. Yuan, and R. D. Klausner, *J. Biol. Chem.*, **1994**, *269*, 25660–25667.

33. M. M. Pena, S. Puig, and D. J. Thiele, *J. Biol. Chem.*, **2000**, *275*, 33244–33251.

34. M. M. O. Pena, K. A. Koch, and D. J. Thiele, *Mol. Cell. Biol.*, **1998**, *18*, 2514–2523.

35. Z. Weissman, I. Berdicevsky, B. Z. Cavari, and D. Kornitzer, *Proc. Natl. Acad. Sci. U. S.A.*, **2000**, *97*, 3520–3525.

36. Z. Weissman, R. Shemer, and D. Kornitzer, *Mol. Microbiol.*, **2002**, *44*, 1551–1560.

37. P. J. Riggle and C. A. Kumamoto, *J. Bacteriol.*, **2000**, *182*, 4899–4905.

38. T. Hirayama, J. J. Kieber, N. Hirayama, M. Kogan, P. Guzman, S. Nourizadeh, J. M. Alonso, W. P. Dailey, A. Dancis, and J. R. Ecker, *Cell*, **1999**, *97*, 383–393.

39. T. Hirayama and J. M. Alonso, *Plant Cell Physiol.*, **2000**, *41*, 548–555.

40. E. Himelblau and R. M. Amasino, *Curr.Opin. Plant. Biol.*, **2000**, *3*, 205–210.

41. T. Wakabayashi, N. Nakamura, Y. Sambongi, Y. Wada, T. Oka, and M. Futai, *FEBS Lett.*, **1998**, *440*, 141–146.

42. Y. Sambongi, T. Wakabayashi, T. Yoshimizu, H. Omote, T. Oka, and M. Futai, *J. Biochem. (Tokyo)*, **1997**, *121*, 1169–1175.

43. J. F. Mercer, J. Livingston, B. Hall, J. A. Paynter, C. Begy, S. Chandrasek-harappa, P. Lockhart, A. Grimes, M. Bhave, D. Siemieniak et al., *Nat. Gen.*, **1993**, *3*, 20–25.

44. C. Vulpe, B. Levinson, S. Whitney, S. Packman, and J. Gitschier, *Nat. Gen.*, **1993**, *3*, 7–13.

45. R. E. Tanzi, K. Petrukhin, I. Chernov, J. L. Pellequer, W. Wasco, B. Ross, D. M. Romano, E. Parano, L. Pavone, L. M. Brzustowicz et al., *Nat. Gen.*, **1993**, *5*, 344–350.

46. Y. Yamaguchi, M. E. Heiny, and J. D. Gitlin, *Biochem. Biophys. Res. Commun.*, **1993**, *197*, 271–277.

47. P. C. Bull, G. R. Thomas, J. M. Rommens, J. R. Forbes, and D. W. Cox, *Nat. Gen.*, **1993**, *5*, 327–337.

48. S. G. Kaler, *Adv. Pediatr.*, **1994**, *41*, 263–304.

49. Z. Tumer and N. Horn, *J. Med. Gen.*, **1997**, *34*, 265–274.

50. D. M. Danks, B. J. Stevens, P. E. Campkell, E. C. Cartwright, J. M. Gillespie, R. R. Townley, J. Blomfield, B. B. Turner, V. Mayne, and J. A. Walker-Smith, *Birth Defects Orig. Artic Ser.*, **1974**, *10*, 132–137.

51. Y. Qian, E. Tiffany-Castiglioni, and E. D. Harris, *Am. J. Physiol.*, **1996**, *271*, 378–384.

52. M. J. Petris, D. Strausak, and J. F. Mercer, *Hum. Mol. Gen.*, **2000**, *9*, 2845–2851.

53. D. E. Hansel, V. May, B. A. Eipper, and G. V. Ronnett, *J. Neurosci.*, **2001**, *21*, 4625–4636.

54. T. C. Steveson, G. D. Ciccotosto, X. M. Ma, G. P. Mueller, R. E. Mains, and B. A. Eipper, *Endocrinology*, **2003**, *144*, 188–200.

55. R. El Meskini, V. C. Culotta, R. E. Mains, and B. A. Eipper, *J. Biol. Chem.*, **2003**, *278*, 12278–12284.

56. M. J. Petris, J. F. Mercer, J. G. Culvenor, P. Lockhart, P. A. Gleeson, and J. Camakaris, *Embo J.*, **1996**, *15*, 6084–6095.

57. K. Terada, T. Nakako, X. L. Yang, M. Iida, N. Aiba, Y. Minamiya, M. Nakai, T. Sakaki, N. Miura, and T. Sugiyama, *J. Biol. Chem.*, **1998**, *273*, 1815–1820.

58. A. S. Payne, E. J. Kelly, and J. D. Gitlin, *Proc. Natl. Acad. Sci. U. S.A.*, **1998**, *95*, 10854–10859.

59. J. F. Mercer, N. Barnes, J. Stevenson, D. Strausak, and R. M. Llanos, *Biometals*, **2003**, *16*, 175–184.

60. H. Roelofsen, H. Wolters, M. J. Luyn Van, N. Miura, F. Kuipers, and R. J. Vonk, *Gastroenterology*, **2000**, *119*, 782–793.

61. T. Saito, M. Okabe, T. Hosokawa, M. Kurasaki, A. Hata, F. Endo, K. Nagano, I. Matsuda, K. Urakami, and K. Saito, *Neurosci. Lett.*, **1999**, *266*, 13–16.

62. K. Petrukhin, S. Lutsenko, I. Chernov, B. M. Ross, J. H. Kaplan, and T. C. Gilliam, *Hum. Mol. Gen.*, **1994**, *3*, 1647–1656.

63. S. D. Moore and D. W. Cox, *Nephron*, **2002**, *92*, 629–634.

64. P. J. Lockhart, S. Fontaine La, S. D. Firth, M. Greenough, J. Camakaris, and J. F. Mercer, *Biochim. Biophys. Acta*, **2002**, *1588*, 189–194.

65. A. Grimes, C. J. Hearn, P. Lockhart, D. F. Newgreen, and J. F. Mercer, *Hum. Mol. Gen.*, **1997**, *6*, 1037–1042.

66. P. J. Lockhart and J. F. Mercer, *Eur. J. Cell. Biol.*, **2001**, *80*, 349–357.

67. P. J. Lockhart, S. A. Wilcox, H. M. Dahl, and J. F. Mercer, *Biochim. Biophys. Acta*, **2000**, *1491*, 229–239.

68. B. Levinson, C. Vulpe, B. Elder, C. Martin, F. Verley, S. Packman, and J. Gitschier, *Nat. Gen.*, **1994**, *6*, 369–373.

69. M. B. Theophilos, D. W. Cox, and J. F. Mercer, *Hum. Mol. Gen.*, **1996**, *5*, 1619–1624.

70. C. Cecchi and P. Avner, *Genomics*, **1996**, *37*, 96–104.

71. J. M. Howell and J. F. Mercer, *J. Comp. Pathol.*, **1994**, *110*, 37–47.

72. A. A. Michalczyk, J. Rieger, K. J. Allen, J. F. Mercer, and M. L. Ackland, *Biochem. J.*, **2000**, *352*, 565–571.

73. V. Coronado, M. Nanji, and D. W. Cox, *Mamm. Genome*, **2001**, *12*, 793–795.

74. J. Wu, J. R. Forbes, H. S. Chen, and D. W. Cox, *Nat. Gen.*, **1994**, *7*, 541–545.

75. K. Terada and T. Sugiyama, *Pediatr. Int.*, **1999**, *41*, 414–418.

76. K. T. Suzuki, *Res. Commun. Mol. Pathol. Pharmacol.*, **1995**, *89*, 221–240.

77. O. I. Buiakova, J. Xu, S. Lutsenko, S. Zeitlin, K. Das, S. Das, B. M. Ross, C. Mekios, I. H. Scheinberg, and T. C. Gilliam, *Hum. Mol. Gen.*, **1999**, *8*, 1665–1671.

78. P. Cunliffe, V. Reed, and Y. Boyd, *Genomics*, **2001**, *74*, 155–162.

79. V. Reed and Y. Boyd, *Hum. Mol. Gen.*, **1997**, *6*, 417–423.

80. S. Das, B. Levinson, C. Vulpe, S. Whitney, J. Gitschier, and S. Packman, *Am. J. Hum. Gen.*, **1995**, *56*, 570–576.

81. J. F. Mercer, A. Grimes, L. Ambrosini, P. Lockhart, J. A. Paynter, H. Dierick, and T. W. Glover, *Nat. Gen.*, **1994**, *6*, 374–378.

82. M. Mori and M. Nishimura, *Mamm. Genome*, **1997**, *8*, 407–410.

83. S. Lutsenko and J. H. Kaplan, *Biochemistry*, **1995**, *34*, 15607–15613.

84. M. Solioz and C. Vulpe, *Trends Biochem. Sci.*, **1996**, *21*, 237–241.

85. C. Rensing, B. Mitra, and B. P. Rosen, *Proc. Natl. Acad. Sci. U. S. A.*, **1997**, *94*, 14326–14331.

86. C. Rensing, Y. Sun, B. Mitra, and B. P. Rosen, *J. Biol. Chem.*, **1998**, *273*, 32614–32617.

87. Z. Hou and B. Mitra, *J. Biol. Chem.*, **2003**, *278*, 28455–28461.

88. J. Okkeri and T. Haltia, *Biochemistry*, **1999**, *38*, 14109–14116.

89. J. M. Arguello, *J. Membr. Biol.*, **2003**, *195*, 93–108.

90. C. Toyoshima, M. Nakasako, H. Nomura, and H. Ogawa, *Nature*, **2000**, *405*, 647–655.

91. C. Toyoshima and H. Nomura, *Nature*, **2002**, *418*, 605–611.

92. W. Kuhlbrandt, J. Zeelen, and J. Dietrich, *Science*, **2002**, *297*, 1692–1696.

93. W. J. Rice, H. S. Young, D. W. Martin, J. R. Sachs, and D. L. Stokes, *Biophys. J.*, **2001**, *80*, 2187–2197.

94. K. J. Sweadner and C. Donnet, *Biochem. J.*, **2001**, *356*, 685–704.

95. O. Radresa, K. Ogata, S. Wodak, J. M. Ruysschaert, and E. Goormaghtigh, *Eur. J. Biochem.*, **2002**, *269*, 5246–5258.

96. H. Ogawa and C. Toyoshima, *Proc. Natl. Acad. Sci. U. S. A.*, **2002**, *99*, 15977–15982.

97. D. L. Stokes, M. Auer, P. Zhang, and W. Kuhlbrandt, *Curr. Biol.*, **1999**, *9*, 672–679.

98. K. Melchers, T. Weitzenegger, A. Buhmann, W. Steinhilber, G. Sachs, and K. P. Schafer, *J. Biol. Chem.*, **1996**, *271*, 446–457.

99. K. Melchers, A. Schuhmacher, A. Buhmann, T. Weitzenegger, D. Belin, S. Grau, and M. Ehrmann, *Res. Microbiol.*, **1999**, *150*, 507–520.

100. K. J. Tsai, Y. F. Lin, M. D. Wong, H. H. Yang, H. L. Fu, and B. P. Rosen, *J. Bioenerg. Biomembr.*, **2002**, *34*, 147–156.

101. T. Yoshimizu, H. Omote, T. Wakabayashi, Y. Sambongi, and M. Futai, *Biosci. Biotechnol. Biochem.*, **1998**, *62*, 1258–1260.

102. K. D. Bissig, H. Wunderli-Ye, P. W. Duda, and M. Solioz, *Biochem. J.*, **2001**, *357*, 217–223.

103. M. Ralle, M. J. Cooper, S. Lutsenko, and N. J. Blackburn, *J. Am. Chem. Soc.*, **1998**, *120*, 13525–13526.

104. S. Lutsenko, K. Petrukhin, M. J. Cooper, C. T. Gilliam, and J. H. Kaplan, *J. Biol. Chem.*, **1997**, *272*, 18939–18944.

105. M. DiDonato, H. F. Hsu, S. Narindrasorasak, L. Que, Jr., and Sarkar, B, *Biochemistry*, **2000**, *39*, 1890–1896.

106. L. Banci, I. Bertini, S. Ciofi-Baffoni, D. L. Huffman, and T. V. O'Halloran, *J. Biol. Chem.*, **2001**, *276*, 8415–8426.

107. J. Gitschier, B. Moffat, D. Reilly, W. I. Wood, and W. J. Fairbrother, *Nat. Struct. Biol.*, **1998**, *5*, 47–54.

108. J. R. Forbes, G. Hsi, and D. W. Cox, *J. Biol .Chem.*, **1999**, *274*, 12408–12413.

109. D. Huster, M. Hoppert, S. Lutsenko, J. Zinke, C. Lehmann, J. Mossner, F. Berr, K. Caca. *Gastroenterology*, **2003**, *124* (2); 335–345.

110. I. Voskoboinik, D. Strausak, M. Greenough, H. Brooks, M. Petris, S. Smith, J. F. Mercer, and J. Camakaris, *J. Biol. Chem.*, **1999**, *274*, 22008–22012.

111. F. Arnesano, L. Banci, I. Bertini, S. Ciofi-Baffoni, E. Molteni, D. L. Huffman, and T. V. O'Halloran, *Genome Res.*, **2002**, *12*, 255–271.

112. L. Banci, I. Bertini, S. Ciofi-Baffoni, M. D'Onofrio, L. Gonnelli, F. C. Marhuenda-Egea, and F. J. Ruiz-Duenas, *J. Mol. Biol.*, **2002**, *317*, 415–429.

113. F. Arnesano, L. Banci, I. Bertini, F. Cantini, S. Ciofi-Baffoni, D. L. Huffman, and T. V. O'Halloran, *J. Biol. Chem.*, **2001**, *276*, 41365–41376.

114. F. Arnesano, L. Banci, I. Bertini, D. L. Huffman, and T. V. O'Halloran, *Biochemistry*, **2001**, *40*, 1528–1539.

115. A. K. Wernimont, D. L. Huffman, A. L. Lamb, T. V. O'Halloran, and A. C. Rosenzweig, *Nat. Struct. Biol.*, **2000**, *7*, 766–771.

116. P. Y. Jensen, N. Bonander, L. B. Moller, and O. Farver, *Biochim. Biophys. Acta*, **1999**, *1434*, 103–113.

117. S. Lutsenko, K. Petrukhin, T. C. Gilliam, and J. H. Kaplan, *Ann. New York Acad. Sci.*, **1997**, *834*, 155–157.

118. M. DiDonato, S. Narindrasorasak, J. R. Forbes, D. W. Cox, and B. Sarkar, *J. Biol. Chem.*, **1997**, *272*, 33279–33282.

119. M. DiDonato, J. Zhang, L. Que, Jr., B. Sarkar, *J. Biol. Chem.*, **2002**, *277*, 13409–13414.

120. P. L. Jorgensen, J. R. Jorgensen, and P. A. Pedersen, *J. Bioenerg. Biomembr*, **2001**, *33*, 367–377.

121. D. M. Clarke, T. W. Loo, and D. H. MacLennan, *J. Biol. Chem.*, **1990**, *265*, 22223–22227.

122. H. Ma, G. Inesi, and C. Toyoshima, *J. Biol. Chem.*, **2003**, *278*, 28938–28943.

123. I. Voskoboinik, J. Mar, and J. Camakaris, *Biochem. Biophys. Res. Commun.*, **2003**, *301*, 488–494.

124. H. P. Adamo, A. G. Filoteo, A. Enyedi, and J. T. Penniston, *J. Biol. Chem.*, **1995**, *270*, 30111–30114.

125. R. A. Farley, E. Elquza, J. Muller-Ehmsen, D. J. Kane, A. K. Nagy, V. N. Kasho, and L. D. Faller, *Biochemistry*, **2001**, *40*, 6361–6370.

126. F. Portillo, *Biochim. Biophys. Acta*, **2000**, *1468*, 99–106.

127. R. Tsivkovskii, B. C. MacArthur, and S. Lutsenko, *J. Biol. Chem.*, **2001**, *276*, 2234–2242.

128. P. L. Jorgensen, K. O. Hakansson, and S. J. Karlish, *Annu. Rev. Physiol.*, **2003**, *65*, 817–849.

129. S. Lutsenko, R. G. Efremov, R. Tsivkovskii, and J. M. Walker, *J. Bioenerg. Biomembr.*, **2002**, *34*, 351–362.

130. S. Danko, K. Yamasaki, T. Daiho, H. Suzuki, and C. Toyoshima, *FEBS Lett.*, **2001**, *505*, 129–135.

131. M. J. Petris, I. Voskoboinik, M. Cater, K. Smith, B. E. Kim, R. M. Llanos, D. Strausak, J. Camakaris, and J. F. Mercer, *J. Biol. Chem.*, **2002**, *277*, 46736–46742.

132. M. J. Petris, J. Camakaris, M. Greenough, S. LaFontaine, and J. F. Mercer, *Hum. Mol. Gen.*, **1998**, *7*, 2063–2071.

133. J. Borjigin, A. S. Payne, J. Deng, X. Li, M. M. Wang, B. Ovodenko, J. D. Gitlin, and S. H. Snyder, *J. Neurosci.*, **1999**, *19*, 1018–1026.

134. R. Tsivkovskii, J. F. Eisses, J. H. Kaplan, and S. Lutsenko, *J. Biol. Chem.*, **2002**, *277*, 976–983.

135. R. Tsivkovskii, R. G. Efremov, and S. Lutsenko, *J. Biol. Chem.*, **2003**, *278*, 13302–13308.

136. P. Wyler-Duda and M. Solioz, *FEBS Lett.*, **1996**, *399*, 143–146.

137. H. Wunderli-Ye and M. Solioz, *Biochem. Biophys. Res. Commun.*, **2001**, *280*, 713–719.

138. K. D. Bissig, S. Fontaine La, J. F. Mercer, and M. Solioz, *Biol. Chem.*, **2001**, *382*, 711–714.

139. N. Bal, C. C. Wu, P. Catty, F. Guillain, and E. Mintz, *Biochem. J.*, **2003**, *369*, 681–685.
140. R. Sharma, C. Rensing, B. P. Rosen, and B. Mitra, *J. Biol. Chem.*, **2000**, *275*, 3873–3878.
141. I. Voskoboinik, J. Mar, D. Strausak, and J. Camakaris, *J. Biol. Chem.*, **2001**, *276*.
142. J. E. Mahaney, J. P. Froechlich, and D. D. Thomas, *Biochemistry*, **1995**, *34*, 4864–4879.
143. K. D. Bissig, T. C. Voegelin, and M. Solioz, *FEBS Lett.*, **2001**, *507*, 367–370.
144. B. Mitra and R. Sharma, *Biochemistry*, **2001**, *40*, 7694–7699.
145. L. Segall, L. K. Lane, and R. Blostein, *J. Biol. Chem.*, **2002**, *277*, 35202–35209.
146. D. Curtis, M. Durkie, P. Balac, D. Sheard, A. Goodeve, I. Peake, O. Quarrell, and S. Tanner, *Hum. Mutat.*, **1999**, *14*, 304–311.
147. G. Loudianos, V. Dessi, M. Lovicu, A. Angius, B. Altuntas, R. Giacchino, M. Marazzi, M. Marcellini, M. R. Sartorelli, G. C. Sturniolo, N. Kocak, A. Yuce, N. Akar, M. Pirastu, and A. Cao, *J. Med. Gen.*, **1999**, *36*, 833–836.
148. G. R. Thomas, J. R. Forbes, E. A. Roberts, J. M. Walshe, and D. W. Cox, *Nat. Gen.*, **1995**, *9*, 210–217.
149. R. Serrano and F. Portillo, *Biochim. Biophys. Acta*, **1990**, *1018*, 195–199.
150. I. Hamza, M. Schaefer, L. W. Klomp, and J. D. Gitlin, *Proc. Natl. Acad. Sci. U. S.A.*, **1999**, *96*, 13363–13368.
151. T. D. Rae, P. J. Schmidt, R. A. Pufahl, V. C. Culotta, and T. V. O'Halloran, *Science*, **1999**, *284*, 805–808.
152. A. C. Rosenzweig, D. L. Huffman, M. Y. Hou, A. K. Wernimont, R. A. Pufahl, and T. V. O'Halloran, *Structure Fold Des.*, **1999**, *7*, 605–617.
153. T. V. O'Halloran and V. C. Culotta, *J. Biol. Chem.*, **2000**, *275*, 25057–25060.
154. M. Ralle, S. Lutsenko, and N. J. Blackburn, *J. Biol. Chem.*, **2003**, *278*, 23163–23170.
155. C. Srinivasan, M. C. Posewitz, G. N. George, and D. R. Winge, *Biochemistry*, **1998**, *37*, 7572–7577.
156. L. Banci, I. Bertini, R. Conte Del, S. Mangani, and W. Meyer-Klaucke, *Biochemistry*, **2003**, *42*, 2467–2474.
157. A. C. Rosenzweig and T. V. O'Halloran, *Curr. Opin. Chem. Biol.*, **2000**, *4*, 140–147.
158. M. D. Harrison, C. E. Jones, and C. T. Dameron, *J. Biol. Inorg. Chem.*, **1999**, *4*, 145–153.
159. S. J. Lin and V. C. Culotta, *Proc. Natl. Acad. Sci. U. S.A.*, **1995**, *92*, 3784–3788.
160. S. J. Lin, R. A. Pufahl, A. Dancis, T. V. O'Halloran, and V. C. Culotta, *J. Biol. Chem.*, **1997**, *272*, 9215–9220.
161. R. A. Pufahl, C. P. Singer, K. L. Peariso, S. J. Lin, P. J. Schmidt, C. J. Fahrni, V. C. Culotta, J. E. Penner-Hahn, and T. V. O'Halloran, *Science*, **1997**, *278*, 853–856.
162. M. E. Portnoy, A. C. Rosenzweig, T. Rae, D. L. Huffman, T. V. O'Halloran, and V. C. Culotta, *J. Biol. Chem.*, **1999**, *274*, 15041–15045.
163. D. L. Huffman and T. V. O'Halloran, *J. Biol. Chem.*, **2000**, *275*, 18611–18614.
164. I. H. Hung, R. L. Casareno, G. Labesse, F. S. Mathews, and J. D. Gitlin, *J. Biol. Chem.*, **1998**, *273*, 1749–1754.
165. I. Hamza, A. Faisst, J. Prohaska, J. Chen, P. Gruss, and J. D. Gitlin, *Proc. Natl. Acad. Sci. U. S.A.*, **2001**, *98*, 6848–6852.
166. D. Larin, C. Mekios, K. Das, B. Ross, A. S. Yang, and T. C. Gilliam, *J. Biol. Chem.*, **1999**, *274*, 28497–28504.
167. D. Strausak, M. K. Howie, S. D. Firth, A. Schlicksupp, R. Pipkorn, G. Multhaup, and J. F. Mercer, *J. Biol. Chem.*, **2003**, *278*, 20821–20827.
168. J. M. Walker, R. Tsivkovskii, and S. Lutsenko, *J. Biol. Chem.*, **2002**, *277*, 27953–27959.
169. D. L. Huffman and T. V. O'Halloran, *Annu. Rev. Biochem.*, **2001**, *70*, 677–701.
170. I. D. Goodyer, E. E. Jones, A. P. Monaco, and M. J. Francis, *Hum. Mol. Gen.*, **1999**, *8*, 1473–1478.
171. C. Cobbold, S. Ponnambalam, M. J. Francis, and A. P. Monaco, *Hum. Mol. Gen.*, **2002**, *11*, 2855–2866.
172. L. Ambrosini and J. F. Mercer, *Hum. Mol. Gen.*, **1999**, *8*, 1547–1555.
173. J. R. Forbes and D. W. Cox, *Hum. Mol. Gen.*, **2000**, *9*, 1927–1935.

174. M. J. Francis, E. E. Jones, E. R. Levy, R. L. Martin, S. Ponnambalam, and A. P. Monaco, *J. Cell. Sci.*, **1999**, *112*, 1721–1732.

175. M. J. Francis, E. E. Jones, E. R. Levy, S. Ponnambalam, J. Chelly, and A. P. Monaco, *Hum. Mol. Gen.*, **1998**, *7*, 1245–1252.

176. M. J. Petris and J. F. Mercer, *Hum. Mol. Gen.*, **1999**, *8*, 2107–2115.

177. D. Strausak, S. Fontaine La, J. Hill, S. D. Firth, P. J. Lockhart, and J. F. Mercer, *J. Biol. Chem.*, **1999**, *274*, 11170–11177.

178. B. E. Kim, K. Smith, C. K. Meagher, and M. J. Petris, *J. Biol. Chem.*, **2002**, *277*, 44079–44084.

179. J. P. Andersen, B. Vilsen, E. Leberer, and D. H. MacLennan, *J. Biol. Chem.*, **1989**, *264*, 21018–21023.

180. B. E. Kim, K. Smith, and M. J. Petris, *J. Med. Gen.*, **2003**, *40*, 290–295.

181. S. Fontaine La, M. B. Theophilos, S. D. Firth, R. Gould, R. G. Parton, and J. F. Mercer, *Hum. Mol. Gen.*, **2001**, *10*, 361–370.

182. I. Hamza, J. Prohaska, and J. D. Gitlin, *Proc. Natl. Acad. Sci. U. S.A.*, **2003**, *100*, 1215–1220.

183. S. M. Vanderwerf, M. J. Cooper, I. V. Stetsenko, and S. Lutsenko, *J. Biol. Chem.*, **2001**, *276*, 36289–36294.

184. I. Voskoboinik, R. Fernando, N. Veldhuis, K. M. Hannan, N. Marmy-Conus, R. B. Pearson, and J. Camakaris, *Biochem. Biophys. Res. Commun.*, **2003**, *303*, 337–342.

6

Bacterial Transport ATPases for Monovalent, Divalent and Trivalent Soft Metal Ions

Marco D. Wong, Bin Fan and *Barry P. Rosen*

6.1
Introduction

This chapter discusses bacterial transport ATPases that extrude ions of the heavy metals, transition metals and metalloids, including Cu(I), Ag(I), Zn(II), Cd(II), Pb(II), Co(II), As(III) and Sb(III). These ions of soft Lewis acids have high polarizing power (a large ratio of ionic charge to the radius of the ion). In contrast, the ions of hard Lewis acids are those in Group I such as Na^+, K^+, and Group II such as Mg^{2+} and Ca^{2+}. For simplicity, they are referred to here as *soft* and *hard* metal ions, respectively. Why is it important to distinguish between hard and soft metal ions? Biologically, hard metal ions serve as counterions for anions such as nucleotides or the carboxylates of aspartate and glutamate residues in proteins. While these ionic interactions are relatively weak, soft metal ions form strong bonds with functional groups in proteins such as the thiolates and imidazolium nitrogens. These strong interactions with cysteine and histidine residues in proteins account for some of the toxicity of soft metal ions. Other toxic effects result from the redox chemistry of transition metal ions such as copper.

All soft metal ions are toxic in excess, but some, such as Cu(I) and Zn(II), have important biological roles, and thus organisms from bacteria to humans require mechanisms to maintain intracellular concentrations at safe levels. Other ions, such as Cd(II), Pb(II) and As(III), appear to have no biological function, and require only detoxification. Before the atmosphere became oxidizing, the concentrations of dissolved metal ions in primordial oceans were probably high. It would have been critical to evolve homeostatic and detoxification mechanisms soon after the origin of life.

Resistance genes to inorganic salts of soft metals, including arsenic, antimony, lead, cadmium, copper, nickel, zinc, bismuth and mercury, are found both on extrachromosomal plasmids and in chromosomes of bacteria, archaea and eukaryotes. For metals such as copper and zinc, which are required in low concentrations but toxic in high concentrations, the efflux systems are components of the homeostatic mechanisms that maintain intracellular concentrations at optimal le-

Handbook of ATPases. Edited by M. Futai, Y. Wada, J. H. Kaplan
Copyright © 2004 WILEY-VCH Verlag GmbH & Co. KGaA, Weinheim
ISBN 3-527-30689-7

vels. Recent reviews on bacterial metal resistances allow this chapter to focus on three types of efflux pumps. Two are P-type ATPases that catalyze the efflux of monovalent [Cu(I)/Ag(I)] and divalent [Zn(II)/Cd(II)/Pb(II)/Co(II)] soft metal cations, respectively. The third is a novel anion-translocating ATPase that pumps arsenite or antimonite out of cells, hence providing resistance to these toxic metalloids.

P-type ATPases are so-called because they all form an acylphosphate intermediate with a conserved aspartate residue during the catalytic cycle. The superfamily is very large and includes five major evolutionary branches [1]. The most widely recognized are the hard metal cation pumps such as the Na^+,K^+-ATPase [2] (Figure 6.1). Some transport monovalent hard metal cations such as Na^+, K^+ or H^+ [3]. Others are divalent cation pumps for metals such as Ca^{2+} [4]. These hard metal ATPases are called Type II ATPases [1]. A related group of hard metal P-type ATPases, including the yeast proton [5] and bacterial Mg^{2+} pumps [6], form the branch of Type III ATPases. Soft metal P-type ATPases form a separate branch of the P-type ATPase superfamily [7], also called the Type I ATPases. As mentioned above, these also fall into two groups, monovalent and divalent pumps. Soft metal ion P-type ATPases are found in members of every kingdom, from bacteria to humans. In addition to the consensus regions common to all P-type ATPases [1], members of the soft metal ion translocating group have two regions characteristic

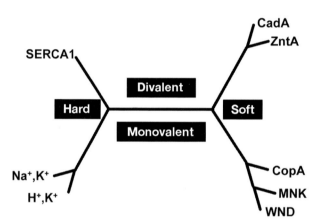

Figure 6.1 Hard- and soft-metal-ion-translocating P-type ATPases. Four sets of ion pumps are illustrated, with only a few representative members of the large and growing family of P-type ATPases shown. Accession numbers are given below in parentheses. Bottom left: monovalent hard metal ion pumps, including Na^+,K^+: 1023-residue human Na^+,K^+- ATPase, alpha 1 chain (A24414) and H^+,K^+: 1035-residue human H^+,K^+-ATPase (A35292). Top left: divalent hard metal ion pumps, including SERCA1: 1001-residue human sarcoplasmic reticulum calcium ATPase (NM 173201.1). Bottom right: monovalent soft metal ion pumps, including MNK: 1500-residue human Menkes disease-related protein (Q04656), WND: 1465-residue human Wilson disease-related protein (AAB52902) and CopA: 834-residue copper pump from E. coli (Q59385). Top right: divalent soft metal ion pumps, including CadA, 727-residue Zn(II)/Cd(II)/Pb(II) ATPase of S. aureus plasmid pI258 (AAB59154) and ZntA: 732-residue Zn(II)/Cd(II)/Pb(II) ATPase from E. coli (P37617).

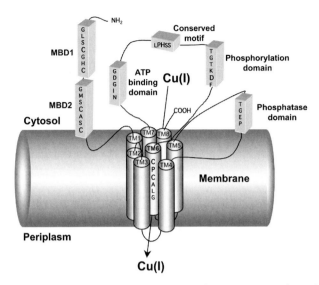

Figure 6.2 The CopA P-type ATPase. CopA has two N-terminal metal-binding domains (MBD1 and MBD2) and eight TMs [27]. The conserved phosphatase domain is in the loop connecting TM4 and TM5. TM6 has the conserved CysProCys sequence and is predicted to be part of the translocation domain. The phosphorylated aspartate residue, Asp-523, and the ATP binding domain are in the loop connecting TM6 and TM7, as is another conserved sequence characteristic of soft metal P-type ATPases.

of that branch of the family: one to six metal-binding domains (MBDs) with $Cys(X)_2Cys$ motifs, a CysProCys motif in the sixth transmembrane helix (TM6) and a conserved HisPro located 34 to 43 amino acids carboxyl-terminal to the CPC motif. In some members of the soft metal ion group the N-terminal MBDs are histidine-rich instead of cysteine-rich, and some have CysProHis or CysProSer in TM6. Hence members of this group are also called CPx-type ATPases [8]. While the hard metal ion translocating P-type ATPases have ten TMs, the soft metal ion pumps have eight TMs [9–11]. Six of the TMs form a common core in all P-type ATPases, the first six of the hard metal ion pumps, and the last six of the soft metal ion enzymes (Figure 6.2).

6.2
Monovalent Soft Metal Ion-translocating ATPases

6.2.1
Copper-translocating P-type ATPases

Copper is an essential trace element for both prokaryotes and eukaryotes. It serves as a cofactor for several enzymes involved in a diversity of biological processes. However, excess accumulation of copper ion is toxic. Copper can cycle between two oxidation states: Cu(I) and Cu(II). Copper(I) produces hydroxyl radicals, which are responsible for a variety of cell damage such as membrane lipid peroxidation, oxidation of proteins and cleavage of DNA and RNA [12]. Reactive oxygen species are thought to be a major contributing factor to the development of cancer, disease of the nervous system and aging. In addition, copper forms strong, nearly covalent bonds with functional groups of proteins such as the thiolates of cysteine residues or the imidazolium nitrogen of histidine residues. Furthermore, copper can compete with zinc or other enzyme cofactors and inhibit enzyme activity.

The importance of copper homeostatic mechanisms is evident in the human copper transport disorders Menkes (MNK) and Wilson (WND) diseases. Menkes disease is caused by mutations in the MNK ATPase, which is found in most tissues except the liver [13]. In Menkes disease, copper accumulates in gut epithelial cells and cannot be delivered to the blood. As a result, copper is deficient in most tissues of the body. Severe neurological impairment occurs soon after birth, and patients usually die at an early age. Wilson's disease is an autosomal recessive disorder characterized by excessive accumulation of copper in the liver with subsequent hepatic and neurological abnormalities. It is caused by mutations in the WND ATPase, which is found predominantly in the liver. WND is responsible for copper incorporation into ceruloplasmin and efflux into the biliary duct. Its malfunction results in copper accumulation in the body [14]. Both MNK and WND have six Cys(X)$_2$Cys metal-binding motifs and a CysProCys motif in putative TM6. Homologues of MNK and WND have been found in most sequenced genomes. To date over 150 bacterial homologues have been identified. The best characterized bacterial copper ATPases are those from *Enterococcus hirae* [15], *Archaeoglobus fulgidus* [16] and *Escherichia coli* [17].

The chromosomal *cop* operon of *E. hirae* consists of four genes: *copY, copZ, copA* and *copB*. The *copA* and *copB* genes encode two P-type ATPases. They exhibit 35 % sequence similarity and are homologues of human Menkes protein. The 727-residue CopA exhibits 35 % identity and 51 % overall sequence similarity with MNK, with a CysProCys motif in putative TM6 (Figure 6.2). CopA has only a single Cys(X)$_2$Cys metal-binding motif and forms an acylphosphate intermediate that is inhibited by orthovanadate and copper chelator [18]. Disruption of *copA* had no significant effect on copper tolerance, but the cells were considerably more tolerant to Ag(I), suggesting that *copA* is a monovalent soft metal ion importer [19]. The 745 residue CopB is more divergent from the mammalian copper pumps. It lacks the Cys(X)$_2$Cys metal-binding motif but may use a histidine-rich binding site, with 16 histidine residues in a stretch of 61 residues at the N-terminus. It also has a CysProHis motif in the putative TM6 rather than CysProCys. Disruption of *copB* renders cells copper sensitive [20]. ATP-driven copper and silver accumulation catalyzed by CopB was observed in native inside-out membrane vesicles of *E. hirae* [21]. These data suggest CopB catalyzes copper efflux in *E. hirae*.

The MNK homologue from the extremophile *Archaeoglobus fulgidus* is an 804-residue Cu(I)/Ag(I)-stimulated ATPase [16]. It exhibits 35 % identity and 50 % overall similarity with MNK. It has the typical membrane CysProCys motif but only a single N-terminal Cys(X)$_2$Cys metal-binding motif. While the purified enzyme exhibits ATPase activity that is stimulated by either copper or silver, the activity is rather low compared with the *E. coli* CopA or ZntA ATPases [22, 23].

Only two genes encoding soft-metal-ion-translocating P-type ATPases exist in the *E. coli* chromosome [24]. As described below, one is *zntA*, which codes for the Zn(II)-, Cd(II)-, and Pb(II)-translocating ATPase [25, 26]. The other, *copA*, encodes a protein of 835 amino acids [27]. CopA exhibits 29 % identity and 47 % overall sequence similarity with MNK. The similarity is greater within conserved domains. Based on the membrane topology of CopA from *H. pylori* [28], *E. coli* CopA can be predicted to have eight TMs, including $Cys_{479}ProCys$ in TM6. In the large loop between TM6 and TM7 are the conserved ATP binding domain, the conserved sequence HisPro, and the phosphorylation domain with the Asp_{523}. *E. coli* CopA has two N-terminal MBDs, $G_{11}LSCGHC$ and $G_{107}MSCASC$ [25].

An unanswered question is the determinants of metal ion specificity. *E. coli* CopA and ZntA are P-type ATPases that transport Cu(I) or Zn(II)/Cd(II)/Pb(II), respectively. CopA does not catalyze Zn(II) or Cd(II) extrusion, and ZntA does not pump copper ions out of cells [17]. Inspection of the sequence does not shed light on the basis for selectivity. Replacement of the N-terminus of the Zn(II) pump, ZntA, with the N-terminus of the WND copper pump (either all six MBDs or just MBD6) had no effect on Zn(II) resistance or Zn(II)-stimulated ATPase activity [29]. Moreover, the WND-ZntA chimera did not confer copper resistance. This suggests that it is the core of the protein that determines metal specificity, not the N-terminal MBDs. However, the residues or groups that form the metal-translocating channel have not been identified. The CysProCys motif that is distinctive in soft-metal-ion-translocating ATPases has been shown to be located in TM6 [9, 11, 28] and has been proposed to be part of the translocation domain [8]. This sequence corresponds to $Ile_{307}ProGlu$ in TM4 of the Ca^{2+}-ATPase of the sarcoplasmic reticulum [4]. The crystal structure of this hard-metal-ion-translocating P-type ATPase reveals that one of the two Ca^{2+} binding sites is formed almost on transmembrane helix 4 by the main-chain carbonyl oxygen atoms of Val-304, Ala-305 and Ile-307, and the side-chain oxygen atoms of Asn-796, Asp-800 and Glu-309 [4]. Results from mutational analysis suggest that Pro-308 and Glu-309 are involved in calcium translocation [30, 31]. Since cysteine thiolates and histidine imidazoles are better ligands for soft metal ions, the IleProGlu sequence of the hard metal ion pumps has been replaced by CysProCys/His in the soft metal ion pumps. To examine the requirement for the cysteines of CysProCys from *E. coli* CopA, the two cysteines were individually changed to alanine residues. Cells expressing either mutant became sensitive to copper, and neither mutant could form the acylphosphate intermediate [32]. A reasonable interpretation of this result is that the CysProCys motif of CopA is essential for Cu(I) binding, and that mutants with CysProAla or AlaProCys no longer bind metal. This is consistent with the known catalytic cycle for hard-metal-ion-translocating P-type ATPases, where binding of metal is required for acylphosphate formation at the conserved aspartate residue [33]. Thus, the CysProCys residues of CopA might be required for binding of Cu(I) in the ion translocation pathway and hence for Cu(I) transport.

As described above, a distinguishing feature of the soft metal P-type ATPases is the presence of one or more Cys(X)$_2$Cys metal-binding domains. Copper-translocating P-type ATPases identified in various species have different numbers of MBDs. For example, there are six in MNK and WND, five in Atp7b (the rat homologue of the human WND), three in Cua-1 of *C. elegans*, two in CopA of *E. coli*, and one in CopA of *E. hirae* [20]. As mentioned, the MBDs are not only present in copper pumps but also in other soft metal pumps, such as the single domain in *E. coli* ZntA and in the related *Staphylococcus aureus* CadA [26]. MBDs from different soft metal pumps are highly homologous, especially the Cys(X)$_2$Cys sequence. In addition, the MBDs are homologous with the metallochaperone MerP [34], which is involved in mercury transport, CopZ [35] and Atx1 [36], which are involved in copper transport.

There have been extensive studies on the function of the MBDs of copper pumps. N-Terminal polypeptides from MNK and WND bind Cu(I), and the bound metal ion protected the polypeptides against labeling with the cysteine-directed probe. This indicates that Cys residues in the MBDs are involved in coordination of copper [37]. The solution structure of the fourth MBD of MNK has been solved by NMR [38]. X-Ray absorption spectroscopy has revealed that each binding site of the WND N-terminus binds Cu(I) using two cysteine side-chains with distorted linear geometry. Copper binding triggers both secondary and tertiary structure changes [39]. In the absence of metal, the region containing the putative metal-binding sequence was disordered. Binding of Ag(I) (assumed to be similar to the copper-bound form) induced order to that domain.

As described above, the MBD does not confer metal ion specificity – so what is its biological role? In *E. coli* none of the four cysteine residues of the two N-terminal Cys(X)$_2$Cys motifs of CopA are required for copper resistance or transport, and the first MBD can be deleted without obvious effect [40]. A quadruple mutant lacking all four cysteine residues of the two N-terminal MBDs or CopA with a deletion of the first MBD was still phosphorylated [32]. These results are consistent with those of Voskoboinik and coworkers [41], who examined the effect of cysteine-to-serine mutations in all six MBDs of the human MNK. Wild-type and mutant MNK genes were heterologously expressed in CHO cells, and uptake of [64]Cu(I) was measured in Golgi-enriched membrane vesicles. Membranes from a mutant with cysteine-to-serine substitutions of all six MBDs retained 50–70% of wild-type transport activity. However, when the same mutant was heterologously expressed in *S. cerevisiae*, [64]Cu(I) transport was not observed, even though the MNK protein was still capable of forming a phosphoenzyme intermediate [42]. Overall, the results suggest that none of the six MBDs are required for human MNK catalysis.

In eukaryotes the MBD is required for trafficking of the pump to the appropriate membrane. The motifs do not appear to be essential for copper transport, but their removal appears to affect trafficking [43]. The MNK and WND are located in the trans-Golgi network under physiological conditions. When mammalian cells were exposed to high concentrations of copper, MNK and WND underwent a reversible copper-regulated trafficking event [44] that may be triggered by conformational changes induced by cooperative copper binding to the MBDs [45]. However, only the sixth MBD was necessary for the WND protein trafficking [46]. Mutation or deletion of the first four of the six MBDs of MNK had no apparent effect on copper-induced trafficking [47]. What are the functions of the other MBDs? Since most prokaryotes lack intracellular membranes or compartments, involvement in trafficking could not have been an ancestral function of the MBDs but must have evolved later. Thus it is likely that there are other functions of the MBDs that have not yet been identified.

Several hypotheses have been proposed for the role of copper binding to the MBDs in the overall functions of these P-type ATPases. First, the N-terminal MBDs may interact with metal chaperones. Yeast two-hybrid analysis demonstrated that Atx1, a copper chaperone protein containing one copper-binding motif, can directly interact with the putative copper-binding domain of the yeast MNK/WND orthologue, Ccc2. This interaction was dependent on copper ions and suggested that Atx1 could donate copper to Ccc2 by direct protein–protein interaction and copper exchange between homologous MBDs [48]. In addition, the copper exchange between the N-terminus of Ccc2 and Atx1 has been demonstrated *in vitro* [49]. The WND protein MBDs 1–4 can interact with HAH1p (human cytoplasmic copper chaperone) independently or in combination, but domains five and six do not [50]. In *Synechocystis* PCC 6803, a prokaryote which has thylakoids, interaction of a copper chaperone and copper-translocating P-type ATPase have been demonstrated [51]. In this organism, transfer of copper from the chaperone to PacS, the thylakoid copper pump, is involved in copper-dependent redox reactions required for photosynthesis [52]. In *B. subtilis* a copper chaperone, CopZ, has been shown to interact with the N-terminal MBD of the *B. subtilis* homologue of CopA [53]. In other prokaryotes such as *E. coli* no metal chaperones have been identified: there are no homologues of CopZ or the eukaryotic chaperones in the *E. coli* genome, although there could be other proteins that serve the same function.

Second, the copper-binding domain might play a role as a copper sensor. A phosphorylation assay on MNK with all six N-terminal Cys changed to Ser showed 2-fold less phosphoenzyme formation than wild-type protein at low copper concentration, suggesting that the mutant copper transporter had a lower affinity for copper [42]. In contrast, with *E. coli* CopA, a quadruple mutant lacking all four cysteine residues of the two N-terminal MBDs or CopA with a deletion of the first MBD showed a two- to three-fold decrease in the concentration of copper required for maximal phosphoenzyme formation [32]. However, cysteine mutations of *E. coli* CopA did not alter copper resistance [32], so it is questionable whether the *in vitro* results reflect a change in the efficiency of the pump *in vivo*. The contribution of the MBDs to CopA function may be apparent only under different physiological conditions.

It is presumed that copper pumps are specific for Cu(I) and not Cu(II). Since an N-terminal peptide containing the MBDs of the Wilson disease protein has been shown to bind Cu(II) and other divalent cations [45], it is possible that these proteins could pump both Cu(I) and Cu(II). Phosphorylation of WND and *E. hirae* CopA were not copper-dependent. This could have been due to copper contamination in the buffers, since copper chelators were found to inhibit [54, 55]. Phosphorylation of MNK and the *A. fulgidus* CopA showed copper dependency, but it was not clear whether the activity was stimulated by Cu(I), Cu(II) or both [16]. In both cases DTT was present in the phosphorylation assays, which would reduce some or all of the Cu(II) to Cu(I), but dithiothreitol (DTT) might also have been necessary to reduce cysteine thiolates in the proteins. The data reported by Fan and Rosen clearly show that Cu(I) but not Cu(II) stimulates phosphorylation of CopA [32]. The requirement for DTT in Cu(II) stimulated activity appears to be in reduction of copper rather than for maintenance of protein cysteine thiolates.

These Cu(I)-translocating P-type ATPases also pump Ag(I). Silver(I) was shown to be transported by CopA and CopB from *E. hirae* [15, 20, 21]. The purified CopA from *A. fulgidus* was activated by both Cu(I) and Ag(I). Silver(I)-stimulated ATPase activity was 4-fold more than Cu(I) [16]; Ag(I) could also stimulate *E. coli* CopA phosphoenzyme intermediate formation and ATPase activity [32]. These results suggest that these enzymes are actually pumps for monovalent soft metal ions rather than specific copper pumps.

6.2.2
Silver-translocating P-type ATPases

In contrast, some P-type ATPases appear to be specific Ag(I) pumps. Silver is a toxic and non-essential heavy metal. Silver products are widely used as antiseptics in clinical applications. Silver(I) ions strongly inhibit bacterial growth through inhibition of respiratory electron transport chains and components of DNA replication [56]. Although many Ag(I)-resistant bacteria have been isolated from clinical and environmental sources, the molecular basis of Ag(I) resistance (*sil*) from plasmid pMG101 has only been described relatively recently [57]. There are nine *sil* genes, *silESRCBAP*, and two un-named open reading frames, all contained in a 12.5 kbp of DNA. The product of *silP* is predicted to be a soft metal ion translocating P-type ATPase. SilP shares 39% sequence identity to *E. coli* CopA. However, SilP lacks cysteine residues in its amino terminus; instead it has a $(His)_5$-Asp-$(His)_2$ sequence. An internal deletion within *silP* resulted in reduced Ag(I) resistance [57].

6.3
Divalent Soft Metal Ion-Translocating ATPases

6.3.1
CadA, ZiaA and ZntA: Zn(II)/Cd(II)/Pb(II)-translocating P-type ATPases

CadA was the first soft-metal-ion-translocating ATPase to be identified. In 1968 Novick and Roth [58] characterized a plasmid, pI258, from *Staphylococcus aureus*, that conferred resistance to several metals, including cadmium and arsenic. Two decades later Nucifora and colleagues [59] sequenced the cadmium resistance genes from this plasmid. The two genes in the *cadCA* operon encode a repressor, CadC, and a P-type ATPase, CadA. The sequences for 178 bacterial, 8 archaeal and 7 plant homologues of CadA and its close homologue ZntA have been identified to date. Interestingly, no homologues of the divalent soft-metal-ion-translocating pumps have been recognized in animals, although it is admittedly difficult to distinguish between the monovalent and divalent soft metal pumps.

The plasmid pI258 CadA was expressed in *B. subtilis* and shown to catalyze ATP-dependent ^{109}Cd(II) transport into everted membrane vesicles, where accumulation in everted membrane vesicles is equivalent to efflux from intact cells [60]. CadA was also shown to form a Cd(II)-dependent phosphorylated enzyme intermediate during the catalytic cycle [61]. The membrane topology of pI258 CadA was determined by constructing fusions with the topological reporter genes *phoA* or *lacZ*. The results are consistent with the pI258 CadA ATPase having eight TMs [11]. The transmembrane topology of two P-type ATPases, a CopA homologue and a CadA homologue, from *Helicobacter pylori* has also been shown to have eight TMs [9]. Of these, the first two TMs are not found in the hard-metal-ion-translocating ATPases, while the other six TMs correspond to the central core of the hard-metal-ion-translocating ATPases. ZiaA is a Zn(II) ATPase efflux pump that mediates Zn(II) and Cd(II) resistance in *Synechocystis* sp. strain PCC 6803 [62].

CadA and ZiaA are negatively regulated by the CadC and ZiaR repressors, respectively. ZiaR and CadC are members of the ArsR family of metalloregulatory proteins [63, 64]. CadC is a Zn(II)/Cd(II)/Pb(II)-responsive repressor [25, 65]. It is a homodimer with two metal-binding sites, each composed of four cysteine residues from each monomer [66, 67]. Interestingly, each of the two metal-binding sites is composed of two cysteine residues from one subunit and two from the other [68].

CadA was identified as a plasmid-encoded cadmium resistance determinant. In contrast, the chromosomal *zntA* gene of *E. coli* was shown to encode a homologue involved both in Zn(II) homeostasis and resistance to Cd(II), Zn(II) [26, 69] and Pb(II) [25]. Did these pumps evolve to protect cells from toxic Pb(II) and Cd(II) and later acquire a function in Zn(II) homeostasis or vice versa? Zinc must be maintained within a narrow range since it is required in low concentrations but toxic in excess. The zinc quota is maintained by a balance between the rates of zinc uptake and efflux. Two ATP-dependent transporters, ZntA and ZnuABC, control Zn(II) homeostasis in *E. coli*. ZnuABC ATPase, an ATP binding cassette (ABC) transport ATPase, catalyzes Zn(II) uptake [70], while Zn(II) efflux is mediated by ZntA, the P-type ATPase.

Zn(II) homeostasis is controlled by two transcription factors. The ZnuABC ATPase is regulated by the *zur* gene product, Zur, a Fur homologue, (Zn(II) uptake regulator), which is a zinc-dependent dimeric repressor that senses the intracellular zinc concentration [70]. During conditions of zinc sufficiency, expression of the pump is repressed by Zur, which presumably binds to the bidirectional promoter region of *znuA* and *znuCB*. In low Zn(II) concentrations, the Zur repressor dissociates from the promoter, resulting in transcription of ZnuABC and zinc influx. Conversely, ZntA expression is regulated by ZntR, a member of the MerR family [71]. It represses transcription in the absence of Zn(II) by binding to the promoter. Upon metallation, ZntR becomes a transcriptional activator that distorts the promoter, facilitating binding of RNA polymerase to the promoter [72]. Thus low intracellular Zn(II) concentrations induce the uptake pump and repress the efflux pump, while high intracellular Zn(II) concentrations have the opposite effect. The interplay of these two Zn(II)-responsive regulatory elements control expression of the pumps and hence the level of intracellular Zn(II).

ZnuA is a 36 kDa periplasmic binding protein that donates Zn(II) to ZnuB, a 28 kDa integral membrane protein that catalyzes Zn(II) transport into the cytosol. ZnuC, the ABC subunit, is a 28 kDa ATPase. Genes for Znu homologues have been identified in 192 bacteria and seven archaea. A related sequence was also found in *Anopheles gambiae*, the African malaria mosquito, although the annotation suggests that this could be the result of bacterial contamination.

From its primary structure ZntA could clearly be identified as a member of the same group of P-type ATPases as CopA, but the phenotype of cells with a *zntA* deletion demonstrated that it was most likely a Zn(II)/Cd(II)/Pb(II) pump and not a Cu(I)/Ag(I) pump. This was demonstrated *in vitro* when ZntA was shown to catalyze ATP-coupled accumulation of ^{65}Zn(II) and ^{109}Cd(II) in everted membrane vesicles of *E. coli* [26]. Transport was sensitive to vanadate, an inhibitor of P-type ATPases. The unavailability of a Pb(II) radioisotope prevents direct demonstration that ZntA transports Pb(II). However, ATP-dependent Zn(II) transport by ZntA was inhibited to the same degree by either Pb(II) or Cd(II), and the latter is a demonstrated pump substrate [25].

ZntA with a C-terminal his$_6$ tag has been purified. ATPase assays with ZntA demonstrated metal-stimulated activity that was specific for Pb(II), Cd(II) or Zn(II), with Pb(II) being the best substrate. There was also some stimulation by Hg(II), although ZntA has not been shown to transport mercury. The highest activity was obtained when the metals were complexed with glutathione or cysteine compared with free metal cations [22]. Maximal ATPase activities of ZntA were over three times higher with the Pb(II)-thiolate complex compared with free Pb(II). Since GSH is the major intracellular thiol, these data suggest that the *in vivo* substrate of ZntA could be a metal-SG complex.

ZntA has a single N-terminal MBD. A truncated form of ZntA lacking the first 100 amino acids conferred resistance to Cd(II), Pb(II) and Zn(II) and had the same metal ion specificity as full-length ZntA [73]. Although there were some differences in the kinetics, the truncated form exhibited ATPase activity with metal thiolates of Cd(II), Pb(II) or Zn(II). These results suggest that the amino-terminal domain is not required for the function of ZntA or for conferring metal-binding specificity, and that the core of ZntA, not the N-terminus, is the determinate of transport and metal specificity. As mentioned above, replacing the N-terminus of ZntA with either the entire N-terminus of WND (with all six MBDs) or just MBD6 had no effect on metal ion specificity [29], supporting the concept that it is the core of the enzyme that determines its properties.

The N-terminal MBD of ZntA is homologous to the yeast Cu(I) metallochaperone Atx1. The solution structure of a peptide corresponding to ZntA residues

46–118 was determined by NMR in both the apo and Zn(II)-bound forms [74]. ZntA was found to form a novel zinc coordination environment composed of two cysteines, Cys59 and Cys62, and a carboxylate residue, Asp58. The Zn(II) metal-binding site had high solvent accessibility, which is a characteristic of domains of heavy metal ion transport proteins. Asp58 in the ZntA metal-ion-binding site may contribute to the specificity for the divalent Zn(II)/Pb(II)/Cd(II) over the monovalent soft metal ions Cu(I) and Ag(I).

6.3.2
CoaT, A Co(II)-translocating P-type ATPase

CoaT is a cobalt-transporting variant of P-type ATPases that confers Co(II) resistance *in Synechocystis* PCC 6803 [75]. CoaT lacks the signature N-terminal MBD of soft-metal-ion-translocating P-type ATPases and has a SerProCys sequence in putative TM6 rather than the more commonly found CysProCys motif.

6.3.3
Pb(II) Resistance System in *Ralstonia metallidurans* CH34

Although CadA and ZntA are Pb(II) pumps, they are, in fact, divalent soft metal ion transporters and not Pb(II)-specific. In contrast, the *pbr* operon of *Ralstonia metallidurans* (formerly *Alcaligenes eutrophus*) strain CH34 confers resistance specifically to Pb(II) [76]. Resistance to other divalent metal salts, including Zn(II), Cd(II), Co(II), Cu(II), Ni(II), and Hg(II), has not been observed, even after induction by Pb(II). The *pbr* operon contains a specific number of genes, including *pbrT, pbrR, pbrA, pbrB, pbrC, pbrD*, and is located on one of two endogenous megaplasmids, pMOL28 and pMOL30. PbrA is a soft metal ion-translocating P-type ATPase that is presumably a Pb(II) efflux pump. Although homologous to CadA and ZntA, PbrA has two putative MBDs with Cys-Pro-Thr-Glu-Glu motifs rather than Cys(X)$_2$Cys motifs. This difference in the MBDs suggests that Pb(II) may preferentially coordinate with oxygen ligands in PbrA rather than with sulfur ligands in CadA or ZntA. PbrT is a Pb(II) uptake protein functionally analogous to *merT*, which encodes the Hg(II) uptake protein of the *mer* operon. PbrB is a predicted integral membrane protein of unknown function, while *pbrC* encodes a predicted prolipoprotein signal peptidase. The *pbrD* gene encodes a Pb(II)-binding protein that is essential for lead sequestration. PbrD may function as a lead chaperone. PbrR, which belongs to the MerR family of metal-ion-responsive regulatory proteins, regulates Pb(II)-inducible transcription of the lead resistance operon [76].

6.4
A Trivalent Soft-metal-ion-translocating ATPase

Trivalent soft metals include As(III) and Sb(III); both metalloids are toxic, and no biological roles have been found for either [77]. Every bacterial genome thus far sequenced has a gene for a membrane protein that catalyzes extrusion of arsenite. There appear to be two unrelated proteins that perform this function. One is ArsB, a potential-driven arsenite uniporter [78, 79] that has been found in 110 bacterial species and 15 archaea. Although eukaryotic homologues have been identified, such as the human pink-eye dilution protein (P protein) [80], these are probably unrelated to arsenite detoxification. The other arsenite resistance membrane protein is the *yqcL* gene product from *B. subtilis* [81]. This is found in 66 bacterial species and nine archaea. A *Saccharomyces cerevisiae* homologue, Acr3p, is involved in arsenite resistance in yeast by catalyzing extrusion of arsenite [82, 83].

ArsB has a dual mode of energy coupling. Alone it is a uniporter that uses the positive exterior membrane potential, but in the presence of an ATPase, ArsA, it forms an ArsAB complex that is an As(III)/Sb(III)-translocating ATPase [84]. The 429-residue ArsB has 12 membrane-spanning segments, similar to many secondary carrier proteins [85]. Like many bacteria, *E. coli* has a chromosomal *arsRBC* operon that confers moderate resistance to arsenite [86]. ArsR is an As(III)/Sb(III)-responsive transcriptional repressor that is homologous to CadC and was the first identified member of the large and growing ArsR family of metal responsive repressors [63]. ArsC is a reductase that converts arsenate into arsenite [87].

In a few *ars* operons there are two additional genes, *arsD* and *arsA* [78]. ArsD is a second As(III)/Sb(III)-responsive transcriptional repressor [88]. ArsA is a 583-residue ATPase with two homologous halves, A1 and A2, each containing a nucleotide binding consensus sequence, connected by a short linker of approximately 25 residues [77]. There are homologues in most if not all organisms, from *E. coli* to humans, that have a 12-residue signature sequence DTAPTGHTIRLL [89]. Most (including all eukaryotic members) are half-sized proteins, equivalent to A1 or A2, and the physiological role of these ubiquitous enzymes is not known [89–92]. Only 12 full-length ArsAs have been identified, all in *ars* operons, most of which are plasmid-encoded. This suggests that ArsA arose rather recently by duplication and fusion of a smaller gene. We have postulated that arsenic resistance and transport function evolved relatively recently [93]. ArsA is a member of a family of ATPases with diverse functions that probably arose from GTPases. Most of the members of this family are half the size of ArsA [94]. Other members include NifH, the iron subunit of nitrogenase, and MinD, a bacterial cell division protein. However, it appears that only the 12 full-length ArsA ATPases are involved in arsenic detoxification.

While ArsA is normally membrane bound as part of the ArsAB complex, [95], overexpression results in cytosolic localization, allowing it to be characterized as a soluble protein. Both the A1 and A2 nucleotide binding sequences are required for arsenic resistance and transport *in vivo* and ATPase activity *in vitro*. Their role in catalysis has been the subject of considerable study [96]. The 2.3 Å crystal structure of the enzyme has provided considerable insight into the mechanism [97] (Figure 6.3). The structure was determined in the presence of MgATP and Sb(III). There are two MgADP bound in each nucleotide binding domain (NBD) and three Sb(III) bound in a single metal-binding domain (MBD) formed at the A1–A2 interface. That both NBDs are occupied with ADP means that ATP was hydrolyzed in those sites during crystallization. The two NBDs are located on the surface of the protein at opposite sides of the protein such that the two bound nucleotides are no closer than ca. 20 Å, and the two Mg^{2+} ions are 23 Å apart. Similarly, the MBD is distant from the two NBDs. The Sb(III) ions in the MBD are 20 Å from the Mg^{2+} ions in the NBDs and more than 30 Å from the adenine rings, yet binding of Sb(III) increases the affinity for nucleotide. This is accomplished by a physical connection by two seven residue sequences ($D_{142}TAPTGH_{148}$ and $D_{447}TAPTGH_{453}$) that form a 20 Å extended sequence connecting the single MBD to the two NBDs [98]. These sequences extend from the MBD to NBD1 and NBD2 and function as signal transduction domains to transmit binding occupancy of the MBD to the NBDs. We hypothesize that signal transduction is bi-directional, so that hydrolysis of ATP decreases affinity for As(III) or Sb(III), allowing those ions to dissociate from ArsA and flow through the transmembrane translocation pathway of ArsB. This implies that ArsA forms a tight seal over the entrance into ArsB.

From the crystal structure the residues that make up each of the domains are clearly visible [97] (Figure 6.4). Each NBD consists of residues from both A1 and A2, including the Walker A and Walker B sequences [99]. For example, ADP is bound in NBD1 by A1 P-loop residues Gly-18, Gly-20, Lys-21, Thr-22 and Ser-23.

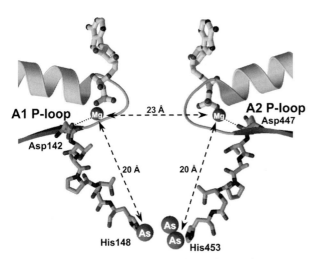

Figure 6.3 Domains in ArsA. The signal transduction domains are two 20 Å extended stretches of seven residues with the sequence $D_{142/447}TAPTGH_{148/453}$ that connect the two nucleotide binding domains, NBD1 and NBD2, to the single metal-binding domain (MBD). NBD1 and NBD2 are occupied with MgADP, and the MBD has three bound As(III). Strands, helices and the P-loops are drawn as ribbons.

Mutations in either A1 or A2 P-loop residues eliminate resistance and ATPase activity, demonstrating that both are involved in ArsA function [100, 101]. A2 residues Thr-501 and Thr-502 are also ligands to the nucleotide. In the crystal structure these residues can be seen to form a flap over the NBD, occluding it such that the ADP in NBD1 cannot be exchanged [102]. In NBD2 the corresponding residues are further from the nucleotide such that the ADP in NBD2 can be exchanged for ATP [102]. This visible asymmetry of NBD1 and NBD2 suggests an alternating catalytic sites mechanism similar to that of the F_0F_1 and P-glycoprotein ATPases [103, 104]. Both Mg^{2+} ions have octahedral coordination, including Asp-45 and Asp-364 in the Walker B sequences. Substitution of Asp-45 with a glutamate reduced the affinity for Mg^{2+} 10-fold [105]. A1 Asp-142 and A2 Asp-447 at the N-terminal ends of the signal transduction domains are also Mg^{2+} ligands, linking the NBDs to the MBD.

The MBD consists of six residues, three from A1 and three from A2. Each of the three Sb(III) in the MBD has two ArsA ligands, one from A1 and the other from A2. Each trivalent Sb(III) has a third non-protein ligand, perhaps a hydroxyl. For convenience the three ArsA-Sb(III) are termed site 1 (His-148 from A1 and Ser-420 from A2), site 2 (Cys-113 from A1 and Cys-422 from A2) and site 3 (Cys-172 from A1 and His-453 from A2). The three cysteine residues are each required for metal activation [106]. His-148 and His-453 form the C-terminal ends of the signal transduction domains, connecting the Sb(III) to the Mg^{2+}, and contribute to metal activation [107]. Our working hypothesis is that the interface of A1 and A2

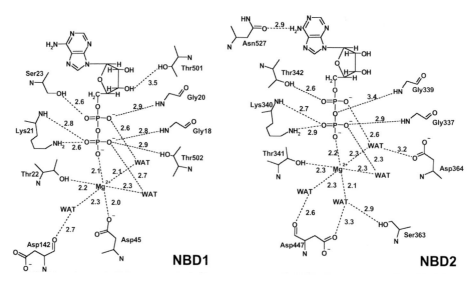

NBD1   **NBD2**

Figure 6.4 ArsA NBD1 and NBD2. Residues in NBD1 and NBD2 that are direct ligands to Mg^{2+} and ADP are shown. In both sites Mg^{2+} is octahedrally coordinated by several serine, threonine and aspartate residues, either directly or indirectly through water molecules. Signal transduction domains influence nucleotide binding through Asp-142 in NBD1 and Asp-447 in NBD2. NBD1 is occluded by short-range interactions with Thr-501 and Thr-502. Distances are in angstroms.

is a loose conformation in the absence of metalloid, so that the NBDs are not completed by residues from the opposite half (Figure 6.5). As(III) or Sb(III) binding at the MBD acts as a "molecular glue" that puts the A1 and A2 halves of ArsA into tight contact with each other, bringing residues from both A1 and A2 into the MBDs, activating catalysis.

While the crystal structure is enlightening, it does not define the catalytic mechanism. Spectroscopy has provided considerable information about conformational changes that occur during the catalytic cycle. ArsA has four tryptophan residues, of which only a single one, Trp-159, has fluorescent properties that change in response to nucleotide binding and hydrolysis [108]. By substitution of the tryptophans with tyrosines, ArsAs were constructed with tryptophan residues in strategic locations to serve as spectroscopic probes of nucleotide binding, hydrolysis and/or conformational changes [98, 105, 108, 109]. As mentioned above, in the crystal structure the two NBDs are confined within caverns formed at the A1–A2 interface [97]. The engineered tryptophans are thus "lamps" that light up these caverns, allowing us to view in real time what happens when nucleotides are bound or hydrolyzed [110, 111]. From these experiments a catalytic cycle has been proposed in which the rate-limiting step in the overall reaction in the absence of As(III)/Sb(III) is isomerization of a long-lived and highly fluorescent conformation of the enzyme; binding of As(III)/Sb(III) increases the rate of isomerization by two to three orders of magnitude, producing approximately a 10-fold increase in the steady state rate of hydrolysis. This is similar in outcome to the E1 to E2 transition of P-type ATPases.

As described above, the crystal structure shows that NBD1 and NBD2 are different; NBD1 is occluded while NBD2 is open, which suggests alternative site catalysis. Biochemical data support the idea that the two sites are different. First, NBD1 (but not NBD2) hydrolyzes ATP in the absence of As(III)/Sb(III), so basal hydrolysis has been termed unisite catalysis [112]. In the presence of As(III)/Sb(III), both sites hydrolyze ATP, so the activated rate is termed multisite catalysis. The fluorescence of two single tryptophan residues, Trp-141 and Trp-446, have been used to measure unisite and multisite hydrolysis [98, 109]. Trp-141 is next to the

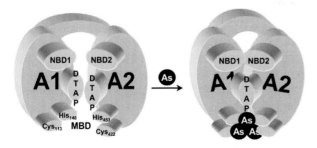

Figure 6.5 Activation of ArsA by As(III) or Sb(III). When the MBD is empty, NBD1 and NBD2 are not fully formed because of the loose interface of the A1 and A2 halves of ArsA. Under these conditions, NBD1 exhibits basal hydrolysis of ATP (unisite catalysis) [109,112]. As(III) or Sb(III) acts as "molecular glue" to make a tight interface between A1 and A2. Under these multisite conditions, the NBDs are complete and catalysis is accelerated [96, 98].

NBD1 Mg^{2+} ligand Asp-142, and Trp-446 is next to the equivalent position in NBD2, Asp-447. Results with Trp-141 demonstrate ATP hydrolysis in NBD1 under unisite conditions [absence of As(III)/Sb(III)] [98]. In contrast, results with measurements of Trp-447 fluorescence show that NBD2 is inactive under unisite conditions but hydrolyzes ATP under multisite conditions [109]. The use of more site-directed spectroscopic probes and the acquisition of additional crystal structures with and without ligands will greatly enhance our ability to understand the catalytic cycle of this trivalent soft-metal-ion-translocating ATPase.

Acknowledgments

Work from our laboratory described in this review was supported by grants GM55425, GM52216, AI45428 and ES10344 from the National Institutes of Health, a Wellcome Trust Biomedical Research Collaborative Grant and NIGMS and American Heart Association predoctoral fellowship # 0215161Z to BF.

References

1. K. B. Axelsen and M. G. Palmgren, *J. Mol. Evol.*, **1998**, *46*, 84–101.
2. P. L. Pedersen and E. Carafoli, *Trends Biochem. Sci.*, **1987**, *12*, 146–150.
3. J. B. Lingrel and T. Kuntzweiler, *J. Biol. Chem.*, **1994**, *269*, 19659–19662.
4. C. Toyoshima, H. Nakasako, and H. Nomura, H. Ogawa, *Nature*, **2000**, *405*, 647–655.
5. R. Serrano, M. C. Kielland-Brandt, and G. R. Fink, *Nature*, **1986**, *319*, 689–693.
6. M. E. Maguire, *J. Bioenerg. Biomembr.*, **1992**, *24*, 319–328.
7. C. Rensing, M. Ghosh, and B. P. Rosen, *J. Bacteriol.*, **1999**, *181*, 5891–5897.
8. M. Solioz, and C. Vulpe, *Trends Biochem. Sci.*, **1996**, *21*, 237–241.
9. K. Melchers, A. Schuhmacher, A. Buhmann, T. Weitzenegger, D. Belin, S. Grau, and M. Ehrmann, *Res. Microbiol.*, **1999**, *150*, 507–520.
10. K. Melchers, L. Herrmann, F. Mauch, D. Bayle, D. Heuermann, T. Weitzenegger, A. Schuhmacher, G. Sachs, R. Haas, G. Bode, K. Bensch, and K. P. Schafer, *Acta Physiol. Scand. Suppl.*, **1998**, *643*, 123–135.
11. K. J. Tsai, Y. F. Lin, M. D. Wong, H. H. Yang, H. L. Fu, and B. P. Rosen, *J. Bioenerg. Biomembr.*, **2002**, *34*, 147–156.
12. B. Halliwell and J. M. Gutteridge, *BioChem. J.*, **1984**, *219*, 1–14.
13. H. M. Darwish, J. E. Hoke and M. J. Ettinger, *J. Biol. Chem.*, **1983**, *258*, 13621–13626.
14. P. C. Bull, G. R. Thomas, J. M. Rommens, J. R. Forbes, and D. W. Cox, *Nat Genet.*, **1993**, *5*, 327–337.
15. A. Odermatt, R. Krapf, and M. Solioz, *Biochem. Biophys. Res. Commun.*, **1994**, *202*, 44–48.
16. A. K. Mandal, and W. D. Cheung, J. M. Arguello, *J. Biol. Chem.*, **2002**, *277*, 7201–7208.
17. C. Rensing, B. Fan, R. Sharma, B. Mitra, and B. P. Rosen, *Proc. Natl. Acad. Sci. U. S.A.*, **2000**, *97*, 652–656.
18. H. Wunderli-Ye and M. Solioz, *Biochem. Biophys. Res. Commun.*, **2001**, *280*, 713–719.
19. A. Odermatt, H. Suter, R. Krapf, and M. Solioz, *Ann. New York Acad. Sci.*, **1992**, *671*, 484–486.

20. A. Odermatt, H. Suter, R. Krapf, and M. Solioz, *J. Biol. Chem.*, **1993**, *268*, 12775–12779.

21. M. Solioz and A. Odermatt, *J. Biol. Chem.*, **1995**, *270*, 9217–9221.

22. R. Sharma, C. Rensing, B. P. Rosen, and B. Mitra, *J. Biol. Chem.*, **2000**, *275*, 3873–3878.

23. B. Fan and B. P. Rosen, *J. Biol. Chem.*, **2002**, *277*, 46987–46992.

24. F. R. Blattner, G. Plunkett III, C. A. Bloch, N. T. Perna, V. Burland, M. Riley, J. Collado-Vides, J. D. Glasner, C. K. Rode, G. F. Mayhew, J. Gregor, N. W. Davis, H. A. Kirkpatrick, M. A. Goeden, D. J. Rose, B. Mau, and Y. Shao, *Science*, **1997**, *277*, 1453–1474.

25. C. Rensing, Y. Sun, B. Mitra, and B. P. Rosen, *J. Biol. Chem.*, **1998**, *273*, 32614–32617.

26. C. Rensing, B. Mitra, and B. P. Rosen, *Proc. Natl. Acad. Sci. U. S. A.*, **1997**, *94*, 14326–14331.

27. C. Rensing, B. Fan, R. Sharma, B. Mitra, and B. P. Rosen, *Proc. Natl. Acad. Sci. U. S. A.*, **2000**, *97*, 652–656.

28. K. Melchers, T. Weitzenegger, A. Buhmann, W. Steinhilber, G. Sachs, and K. P. Schafer, *J. Biol. Chem.*, **1996**, *271*, 446–457.

29. Z. J. Hou, S. Narindrasorasak, B. Bhushan, B. Sarkar, and B. Mitra, *J. Biol. Chem.*, **2001**, *276*, 40858–40863.

30. B. Vilsen, J. P. Andersen, D. M. Clarke, and D. H. MacLennan, *J. Biol. Chem.*, **1989**, *264*, 21024–21030.

31. B. Vilsen and J. P. Andersen, *FEBS Lett.*, **1992**, *306*, 247–250.

32. B. Fan and B. P. Rosen, *J. Biol. Chem.*, **2002**, *277*, 46987–46992.

33. G. Inesi and M. R. Kirtley, *J. Bioenerg. Biomembr.*, **1992**, *24*, 271–283.

34. R. A. Steele and S. J. Opella, *Biochemistry*, **1997**, *36*, 6885–6895.

35. A. Odermatt and M. Solioz, *J. Biol. Chem.*, **1995**, *270*, 4349–4354.

36. S. J. Lin, R. A. Pufahl, A. Dancis, T. V. O'Halloran, and V. C. Culotta, *J. Biol. Chem.*, **1997**, *272*, 9215–9220.

37. S. Lutsenko, K. Petrukhin, M. J. Cooper, C. T. Gilliam, and J. H. Kaplan, *J. Biol. Chem.*, **1997**, *272*, 18939–18944.

38. J. Gitschier, B. Moffat, D. Reilly, W. I. Wood, and W. J. Fairbrother, *Nat. Struct. Biol.*, **1998**, *5*, 47–54.

39. M. DiDonato, H. F. Hsu, S. Narindrasorasak, L. Que, Jr., and B. Sarkar, *Biochemistry*, **2000**, *39*, 1890–1896.

40. B. Fan, G. Grass, C. Rensing, and B. P. Rosen, *Biochem. Biophys. Res. Commun.*, **2001**, *286*, 414–418.

41. I. Voskoboinik, H. Brooks, S. Smith, P. Shen, and J. Camakaris, *FEBS Lett.*, **1998**, *435*, 178–182.

42. I. Voskoboinik, J. Mar, D. Strausak, and J. Camakaris, *J. Biol. Chem.*, **2001**, *276*, 28620–28627.

43. J. R. Forbes, G. Hsi, and D. W. Cox, *J. Biol. Chem.*, **1999**, *274*, 12408–12413.

44. M. J. Petris, J. F. Mercer, J. G. Culvenor, P. Lockhart, P. A. Gleeson, and J. Camakaris, *EMBO J.*, **1996**, *15*, 6084–6095.

45. M. DiDonato, S. Narindrasorasak, J. R. Forbes, D. W. Cox, and B. Sarkar, *J. Biol. Chem.*, **1997**, *272*, 33279–33282.

46. M. Iida, K. Terada, Y. Sambongi, T. Wakabayashi, N. Miura, K. Koyama, M. Futai, and T. Sugiyama, *FEBS Lett.*, **1998**, *428*, 281–285.

47. D. Strausak, S. La Fontaine, J. Hill, S. D. Firth, P. J. Lockhart, and J. F. Mercer, *J. Biol. Chem.*, **1999**, *274*, 11170–11177.

48. R. A. Pufahl, C. P. Singer, K. L. Peariso, S. J. Lin, P. J. Schmidt, C. J. Fahrni, V. C. Culotta, J. E. Penner-Hahn, and T. V. O'Halloran, *Science*, **1997**, *278*, 853–856.

49. D. L. Huffman and T. V. O'Halloran, *J. Biol. Chem.*, **2000**, *275*, 18611–18614.

50. D. Larin, C. Mekios, K. Das, B. Ross, A. S. Yang, and T. C. Gilliam, *J. Biol. Chem.*, **1999**, *274*, 28497–28504.

51. S. Tottey, S. A. Rondet, G. P. Borrelly, P. J. Robinson, P. R. Rich, and N. J. Robinson, *J. Biol. Chem.*, **2002**, *277*, 5490–5497.

52. S. Tottey, P. R. Rich, S. A. Rondet, and N. J. Robinson, *J. Biol. Chem.*, **2001**, *276*, 19999–20004.

53. L. Banci, I. Bertini, S. Ciofi-Baffoni, R. Del Conte, and L. Gonnelli, *Biochemistry*, **2003**, *42*, 1939–1949.

54. P. Wyler-Duda and M. Solioz, *FEBS Lett.*, **1996**, *399*, 143–146.

55. R. Tsivkovskii, J. F. Eisses, J. H. Kaplan, and S. Lutsenko, *J. Biol. Chem.*, **2002**, *277*, 976–983.

56. S. M. Modak and C. L. Fox, Jr., *Biochem. Pharmacol.* **1973**, *22*, 2391–2404.

57. A. Gupta K. Matsui, J. F. Lo, and S. Silver, *Nat. Med.*, **1999**, *5*, 183–188.

58. R. P. Novick and C. Roth, *J. Bacteriol.*, **1968**, *95*, 1335–1342.

59. G. Nucifora, L. Chu, T. K. Misra, and S. Silver, *Proc. Natl. Acad. Sci. U. S.A.*, **1989**, *86*, 3544–3548.

60. K. J. Tsai, K. P. Yoon, and A. R. Lynn, *J. Bacteriol.*, **1992**, *174*, 116–121.

61. K. J. Tsai and A. L. Linet, *Arch. Biochem. Biophys.*, **1993**, *305*, 267–270.

62. C. Thelwell, N. J. Robinson, and J. S. Turner-Cavet, *Proc. Natl. Acad. Sci. U. S.A.*, **1998**, *95*, 10728–10733.

63. C. Xu and B. P. Rosen, Metalloregulation of soft metal resistance pumps. In: *Metals and Genetics*, B. Sarkar, ed., **1999**, Plenum Press, New York, p. 5–19.

64. W. Shi, J. Wu, and B. P. Rosen, *J. Biol. Chem.*, **1994**, *269*, 19826–19829.

65. G. Endo and S. Silver, *J. Bacteriol.*, **1995**, *177*, 4437–4441.

66. Y. Sun, M. D. Wong and B. P. Rosen, *J. Biol. Chem.*, **2001**, *276*, 14955–14960.

67. L. S. Busenlehner, N. J. Cosper, R. A. Scott, B. P. Rosen, M. D. Wong and D. P. Giedroc, *Biochemistry*, **2001**, *40*, 4426–4436.

68. M. D. Wong, Y. F. Lin, and B. P. Rosen, *J. Biol. Chem.*, **2002**, *277*, 40930–40936.

69. S. J. Beard, R. Hashim, J. Membrillo-Hernandez, M. N. Hughes, and R. K. Poole, *Mol. Microbiol.*, **1997**, *25*, 883–891.

70. S. I. Patzer and K. Hantke, *Mol. Microbiol.*, **1998**, *28*, 1199–1210.

71. K. R. Brocklehurst, J. L. Hobman, B. Lawley, L. Blank, S. J. Marshall, N. L. Brown, and A. P. Morby, *Mol. Microbiol.*, **1999**, *31*, 893–902.

72. C. E. Outten, F. W. Outten, and T. V. O'Halloran, *J. Biol. Chem.*, **1999**, *274*, 37517–37524.

73. B. Mitra and R. Sharma, *Biochemistry*, **2001**, *40*, 7694–7699.

74. L. Banci, I. Bertini, S. Ciofi-Baffoni, L. A. Finney, C. E. Outten, and T. V.

O'Halloran, *J. Mol. Biol.*, **2002**, *323*, 883–897.

75. J. C. Rutherford, J. S. Cavet, and N. J. Robinson, *J. Biol. Chem.*, **1999**, *274*, 25827–25832.

76. B. Borremans, J. L. Hobman, A. Provoost, N. L. Brown, and D. van Der Lelie, *J. Bacteriol.*, **2001**, *183*, 5651–5658.

77. B. P. Rosen, *FEBS Lett.*, **2002**, *529*, 86–92.

78. B. P. Rosen, *Trends Microbiol.*, **1999**, *7*, 207–212.

79. M. Kuroda, S. Dey, O. I. Sanders, and B. P. Rosen, *J. Biol. Chem.*, **1997**, *272*, 326–331.

80. S. J. Orlow and M. H. Brilliant, *Exp Eye Res.*, **1999**, *68*, 147–154.

81. T. Sato and Y. Kobayashi, *J. Bacteriol.*, **1998**, *180*, 1655–1661.

82. R. Wysocki, P. Bobrowicz and S. Ulaszewski, *J. Biol. Chem.*, **1997**, *272*, 30061–30066.

83. M. Ghosh, J. Shen, and B. P. Rosen, *Proc. Natl. Acad. Sci. U. S.A.*, **1999**, *96*, 5001–5006.

84. S. Dey and B. P. Rosen, *J. Bacteriol.* **1995**, 177, 385–389.

85. J. Wu, L. S. Tisa, and B. P. Rosen, *J. Biol. Chem.*, **1992**, *267*, 12570–12576.

86. A. Carlin, W. Shi, S. Dey, and B. P. Rosen, *J. Bacteriol.*, **1995**, *177*, 981–986.

87. R. Mukhopadhyay and B. P. Rosen, *Environ. Health Perspect* **2002**, *110*(Suppl 5), 745–748.

88. J. Wu and B. P. Rosen, *Mol. Microbiol.*, **1993**, *8*, 615–623.

89. H. Bhattacharjee, M. Ghosh, R. Mukhopadhyay, and B. P. Rosen, Arsenic transporters from *E. coli* to humans. In: *Transport Of Molecules Across Microbial Membranes*, J. K. Broome-Smith, et al., eds., **1999**, Society for General Micriobiology, Leeds, p. 58–79.

90. J. Shen, C. M. Hsu, B. K. Kang, B. P. Rosen, and H. Bhattacharjee, *Biometals*, **2003**, *16*, 369–378.

91. H. Bhattacharjee, Y. S. Ho, and B. P. Rosen, *Gene*, **2001**, *272*, 291–299.

92. B. Kurdi-Haidar, D. Heath, S. Aebi, and S. B. Howell, *J. Biol. Chem.*, **1998**, *273*, 22173–22176.

93. A. J. Driessen, B. P. Rosen, and W. N. Konings, *Trends Biochem. Sci.*, **2000**, *25*, 397–401.

94. D. D. Leipe, Y. I. Wolf, E. V. Koonin, and L. Aravind, *J. Mol. Biol.*, **2002**, *317*, 41–72.

95. S. Dey, D. Dou, L. S. Tisa, and B. P. Rosen, *Arch. Biochem. Biophys.*, **1994**, *311*, 418–424.

96. B. P. Rosen, H. Bhattacharjee, T. Zhou, and A. R. Walmsley, *Biochim. Biophys. Acta*, **1999**, *1461*, 207–215.

97. T. Zhou, S. Radaev, B. P. Rosen, and D. L. Gatti, *EMBO J.*, **2000**, *19*, 1–8.

98. T. Zhou and B. P. Rosen, *J. Biol. Chem.*, **1997**, *272*, 19731–19737.

99. J. E. Walker, M. Saraste, M. J. Runswick, and N. J. Gay, *EMBO J.*, **1982**, *1*, 945–951.

100. P. Kaur and B. P. Rosen, *J. Biol. Chem.*, **1992**, *267*, 19272–19277.

101. C. E. Karkaria, C. M. Chen, and B. P. Rosen, *J. Biol. Chem.*, **1990**, *265*, 7832–7836.

102. T. Zhou, S. Radaev, B. P. Rosen, and D. L. Gatti, *J. Biol. Chem.*, **2001**, *276*, 30414–30422.

103. A. E. Senior, *Acta Physiol. Scand. Suppl.*, **1998**, *643*, 213–218.

104. A. E. Senior, J. Weber, and M. K. al-Shawi, *Biochem. Soc. Trans.*, **1995**, *23*, 747–752.

105. T. Q. Zhou and B. P. Rosen, *J. Biol. Chem.*, **1999**, *274*, 13854–13858.

106. H. Bhattacharjee, J. Li, M. Y. Ksenzenko, and B. P. Rosen, *J. Biol. Chem.*, **1995**, *270*, 11245–11250.

107. H. Bhattacharjee and B. P. Rosen, *Biometals*, **2000**, *13*, 281–288.

108. T. Zhou, S. Liu, and B. P. Rosen, *Biochemistry*, **1995**, *34*, 13622–13626.

109. T. Zhou, J. Shen, Y. Liu, and B. P. Rosen, *J. Biol. Chem.*, **2002**, *8*, 23815–23820.

110. A. R. Walmsley, T. Zhou, M. I. Borges-Walmsley, and B. P. Rosen, *J. Biol. Chem.*, **1999**, *274*, 16153–16161.

111. A. R. Walmsley, T. Zhou, M. I. Borges-Walmsley, and B. P. Rosen, *J. Biol. Chem.*, **2001**, *276*, 6378–6391.

112. P. Kaur, *J. Biol. Chem.*, **1999**, *274*, 25849–25854.

7
Gastric H$^+$,K$^+$-ATPase

Jai Moo Shin, Olga Vagin, Keith Munson, and *George Sachs*

7.1
Gastric H$^+$,K$^+$-ATPase

The gastric H$^+$,K$^+$-ATPase exchanges cytoplasmic protons for extracytoplasmic potassium ions in an electroneutral manner and, by being coupled to a KCl pathway in the apical membrane of the parietal cell, is responsible for the elaboration of HCl by the parietal cell of the gastric mucosa. The transport of H$^+$ across the apical membrane is by means of conformational changes in the protein driven by cyclic phosphorylation and dephosphorylation of the catalytic subunit of the ATPase. This mechanism places the ATPase in the P$_2$ ATPase family.

The gastric H$^+$,K$^+$-ATPase is composed of two subunits, the α-subunit and the β-subunit. The α subunit with molecular mass of about 100 kDa has the catalytic site and the β subunit with peptide mass of 35 kDa is strongly but non-covalently associated with the α subunit. The β subunit has six or seven N-glycosylated sites exposed to the extra-cytoplasmic surface. There is about 60 % sequence homology between the Na$^+$,K$^+$-ATPase and the H$^+$,K$^+$-ATPase α subunit, while the Ca-ATPase of sarcoplasmic reticulum shows only about 15 % overall homology with the H$^+$,K$^+$-ATPase. The hydropathy profile is, however, very similar, as is the 3D structure based on site-directed mutagenesis. The β subunit has 35 % homology to the β$_2$ subunit of the Na$^+$,K$^+$-ATPase.

The gastric H$^+$,K$^+$-ATPase has been a therapeutic target in treatment of acid-related disease for the last 15 years. Substituted benzimidazoles that inhibit the H$^+$,K$^+$-ATPase are the newest and most effective class of anti-ulcer drugs used to treat various diseases of the upper GI tract.

7.1.1
α Subunit of Gastric H$^+$,K$^+$-ATPase

The primary sequences of the α subunits deduced from cDNA have been defined for pig [1], rat [2] and rabbit [3]. The hog gastric H$^+$,K$^+$-ATPase α subunit sequence deduced from its cDNA consists of 1034 amino acids and has an Mr of 114,285 [1].

Handbook of ATPases. Edited by M. Futai, Y. Wada, J. H. Kaplan
Copyright © 2004 WILEY-VCH Verlag GmbH & Co. KGaA, Weinheim
ISBN 3-527-30689-7

The sequence based on the known N-terminal amino acid sequence is one less than the cDNA derived sequence [4]. Rat gastric H^+,K^+-ATPase consists of 1033 amino acids and has a Mr of 114,012 [2], and rabbit gastric H^+,K^+-ATPase consists of 1035 amino acids, with $Mr = 114,201$ [3]. Notably, the initial publication of the rat sequence had an N-terminal fragment of retinol dehydrogenase that is used in some microarrays as a marker for the ATPase (Agilent). Homology of the α subunits among the animal species is extremely high (over 97 % identities). The gene sequence for human and the 5' part of the rat H^+,K^+-ATPase α subunits have been determined [5–7]. The human gastric H^+,K^+-ATPase gene has 22 exons and encodes a protein of 1035 residues, including the initiator methionine residue ($Mr = 114,047$). The N terminal amino acid is actually glycine. These H^+,K^+-ATPase α subunits show high homology (~60 % identity) with the Na^+,K^+-ATPase catalytic α subunits [5].

The gastric α subunit has conserved sequences along with the other P type ATPases, the sarcoplasmic reticulum Ca-ATPase and the Na^+,K^+-ATPase, for the ATP binding site, the phosphorylation site, the pyridoxal 5'-phosphate binding site and the fluorescein isothiocyanate binding site. These sites are thought to be within the ATP binding domain in the large cytoplasmic loop between membrane-spanning segments 4 and 5 (N domain).

Using the hog gastric vesicles containing the H^+,K^+-ATPase, it was shown that pyridoxal 5'-phosphate is bound at lys497 of the α subunit in the absence but not the presence of ATP [8], suggesting that lys497 is present in the ATP binding site or in its vicinity [9]. The phosphorylation site was observed to be at asp386 [10], which is well conserved in other P type ATPases. Fluorescein isothiocyanate (FITC) covalently labels the gastric H^+,K^+-ATPase in the absence of ATP [11]. The binding site of FITC was at lys518 [12]. Later, several additional lysines, such as those at 497 and 783, were shown to react with FITC during the inactivation of the Na^+,K^+-ATPase and to be protected from reaction with FITC when ATP was present in the incubation [13]. Based on these data, similar lysines of the H^+,K^+-ATPase could be near or in the ATP binding site.

The membrane topology of the α subunit has been extensively studied. The hydropathy plots is most commonly used to predict the location of membrane-spanning α helices [14]. These are based on determining a moving average of hydrophobic, neutral and hydrophilic amino acids using a variety of scales. Interpretation of the hydrophobicity plots of the α subunit of the H^+,K^+-ATPase based on the primary amino acid sequence suggested an 8 or 10 membrane-spanning segment model for the secondary structure. The first 4 membrane-spanning segments in the N terminal one-third of the α subunit are clearly defined; however, in the C terminal one-third of the protein a prediction from the hydropathy analysis is more difficult.

The C-terminal amino acids of the α subunit are tyr-tyr, which therefore can be iodinated with peroxidase–H_2O_2–^{125}I on the cytoplasmic side of intact hog gastric vesicles. Digestion with carboxypeptidase Y then released about 28 % of the counts incorporated into the α subunit, as would be predicted from a cytoplasmic location of the C terminal tyrosines [15]. These data show that there is an even number of transmembrane segments in the α subunit.

In either an 8- or 10-transmembrane segment model, each transmembrane pair connecting the luminal loop has at least one cysteine, which allows fluorescent labeling of the cysteines left after complete cleavage of the cytoplasmic domain by trypsin. The N-terminus of each fluorescent peptide fragment is then defined by N-terminal sequencing. The size of the fragment is determined from M_r measurement in the tricine gradient gels used for the separation. This then allows the C-termination of the peptide to be identified using the lys or arg cleavage sites at this end. Four transmembrane pairs connected by their luminal loop were detected in the hog gastric H⁺,K⁺-ATPase digest [16, 17]. A tryptic peptide fragment beginning at gln104 represents the H1/loop/H2 sector. The H3/loop/H4 sector was found at a single peptide beginning at thr291, and the H5/loop/H6 sector at a peptide beginning at leu776. The H7/loop/H8 region was found in a single peptide fragment of 11 kDa, beginning at leu853. These represent the first 4 transmembrane segment pairs of the proposed 10 transmembrane segment model [3]. From these studies, 4 membrane segment/loop/membrane segment sectors were identified, corresponding to H1/loop/H2 through H7/loop/H8. Although the hydropathy plot predicts two additional membrane-spanning segments (H9, H10) at the C-terminal region of the enzyme, no evidence was obtained for this pair biochemically, even though H9 is predicted to have 4 cysteine residues that should be able to bind fluorescein 5-maleimide (F-MI). The absence of F-MI labeling of these hydrophobic segments may suggest that the cysteines in H9 are post-translationally modified. Reduction, ester cleavage by hydroxylamine and chelating agents have not been successful in identifying this putative transmembrane pair after tryptic cleavage. The nature of the post-translational modification, if this is the reason for the lack of cysteine reactivity, is obscure.

Many monoclonal antibodies have been generated reacting with the H⁺,K⁺-ATPase. With the sequence of the α subunit deduced from cDNA, it is now possible to define the epitopes, and to determine the sidedness of these epitopes by staining intact or permeable cells with either fluorescent or immunogold techniques. Antibody 95 inhibits ATP hydrolysis in the intact vesicles, and appears to be K⁺ competitive [18]. Its epitope was identified by Western blotting of an *E. coli* expression library of fragments of the cDNA encoding the α subunit, as well as by Western analysis of tryptic fragments. The sequence recognized by this antibody was between the amino acid positions 529 and 561. Since it inhibits intact vesicles this epitope must be cytoplasmic. Its epitope is close to the region known to bind the cytoplasmic reagent FITC, namely in the loop between H4 and H5. The ability of this antibody to immunoprecipitate intact gastric vesicles showed that it was on the cytoplasmic surface of the pump. Antibody 1218 was shown by Western analysis of tryptic fragments and by recombinant methodology to have its major epitope between amino acid positions 665 and 689 [19], and the epitope was then more cleanly defined as seven consecutive amino acids, Asp682-Met-Asp-Pro-Ser-Glu-Leu688 [20]. This epitope is on the cytoplasmic surface of the enzyme also in the loop between H4 and H5. This antibody does not inhibit ATPase activity. A second epitope for mAb 1218 was also identified to locate between amino acid positions 853 and 946 according to Western blot analysis of tryptic fragments. A synthetic peptide (888–907) containing this second epitope displaced mAb 1218 from vesicles adsorbed to the surface of ELISA wells [18]. Monoclonal antibody 146 was generated against intact parietal cells and subsequently purified and shown to react with rat H⁺,K⁺-ATPase. In cells it reacts on the outside surface of the canaliculus, as shown by immunogold electron microscopy [21, 22]. Western analysis of rat, rabbit and hog enzyme showed, surprisingly, that it was present on the β subunit of rat and rabbit and absent from the β subunit of hog. Disulfide reduction eliminated reactivity of this antibody. Comparing sequence of the different β subunits, there is an arg, pro substitution in the hog for leu, val in the rat between the disulfide at position 161 and 178. This suggests that the β subunit epitope of mAb 146 is contained within this region of the subunit. Conversely, there was also an epitope on the α subunit of all three species recognized by mAb 146 (also using Western analysis). This epitope was defined, both by tryptic mapping and octamer walking, to be between positions 873 and 877 of the hog α subunit. This is on, or close to, the extra-cytoplasmic face of H7. The finding using Western blotting that there was an epitope both on the α and β subunits was confirmed by expressing the rabbit subunits individually in SF9 cells using baculovirus transfection. The β subunit reacted in the SF9 cells and on Western blots with mAb 146. The α subunit did not react in the cells but did on Western blots as if the epitope in the α subunit was difficult to access in the absence of a denaturing detergent such as SDS [19]. These data may indicate tight binding between the α and β subunits in this region of the enzyme.

A molecular biological method was developed to analyze not only for the presence of the membrane segments defined as above, but also to explore the nature of membrane insertion [23]. A cDNA encoding a fusion protein of the 102 N-terminal amino acids of the rabbit H⁺,K⁺-ATPase α subunit linked by a variable segment to the 177 most C-terminal amino acids of the rabbit H⁺,K⁺-ATPase β subunit was transcribed and translated in a rabbit reticulocyte lysate system using labeled methionine in the absence or presence of microsomes. The cDNA for the variable region containing one or more putative membrane-spanning segments is synthesized using selected primers in a PCR reaction and ligated into the cDNA construct. Since the β subunit region has 5 consensus N-glycosylation sites, translocation of the C-terminal β subunit part of the fusion protein into the interior of the microsomes can be determined by assessment of glycosylation. The presence of glycosylation is evidence for an odd number of transmembrane segments in the variable region preceding the β subunit part. The absence of glycosylation shows either the presence of an even number of membrane-spanning segments or absence of membrane insertion. From this approach, the first 4 membrane segments are present and are co-inserted with translation. The last 2 hydrophobic sequences, H9 and H10, also appear to be co-inserted. The information within the H5, H6 and H7 sequences is apparently insufficient for these to act on their own as either signal anchor or stop transfer sequences. The information within the H8 sequence is sufficient for this to act as a stop transfer sequence. However, a stop transfer sequence in the absence of a preceding signal anchor sequence may have no implications for membrane insertion of the stop transfer sequence [24]. That H5 through H7 are indeed membrane inserted in the mature enzyme is clear from the biochemical analyses. A reasonable hypothesis that may explain these translation data is that insertion of H5 through H7 is post-translational, depending on insertion of H9/10, and that H8 then acts as a stop transfer sequence. Alternatively, the presence of the β sequence may also be involved in tethering H8 and thence H5 to H7. The presence of the additional pair of membrane segments predicted by hydropathy at the C-terminus is strongly suggested by these translation data.

The combination of techniques described above, namely, sided trypsin proteolysis, epitope mapping, iodination, and *in vitro* translation, provides evidence for a 10 membrane segment model (Figure 7.1, see p. 184), with a large cytoplasmic loop between H4 and H5 and a large extra-cytoplasmic loop between H7 and H8. The 10 transmembrane segments determined and defined from the above evidence are named as TM1–10. These data are consistent with the 3D structure derived from the crystals of the SR Ca-ATPase. It is also worth noting that the reaction of the thiophilic luminal proton pump inhibitors provided conclusive evidence for the presence of TM5 and TM6, as discussed below.

7.1.2
β Subunit of Gastric H$^+$,K$^+$-ATPase

The β subunit was first identified in work analyzing the carbohydrates of gastric membranes [25]. Post-embedding electron-microscopic staining techniques showed that wheat germ agglutinin staining occurred on the inside face of the inside-out gastric vesicles. The same β subunit was found by determining the nature of the antigen in autoimmune gastritis [26]. Using a lectin-affinity chromatography, the H$^+$,K$^+$-ATPase α subunit was co-purified with the β subunit, showing that the α subunit interacts with the β subunit [27–29]. By cross-linking with low concentrations of glutaraldehyde, the α subunit was shown to be closely associated with the β subunit [30, 31].

The primary sequences of the β subunits have been reported for rabbit [32], hog [26], rat [33–36], mouse [37, 38], and human [39]. The hydropathy profile of the β subunit appears less ambiguous than the α subunit. There is one transmembrane-spanning region predicted by the hydropathy analysis, which is located between positions 38 and 63 near the N-terminus. Tryptic digestion of the intact gastric H$^+$,K$^+$-ATPase produces no visible cleavage of the β subunit on SDS gels. Wheat germ agglutinin (WGA) binding of the β subunit is retained. These data indicate that most of the β subunit is extra-cytoplasmic and glycosylated.

When lyophilized hog vesicles are cleaved by trypsin followed by reduction, a small, non-glycosylated peptide fragment is seen on SDS gels with the N terminal sequence AQPHYS, which represents C-terminal region beginning at position 236. This small fragment is not found either after trypsinolysis of intact vesicles or in the absence of reducing agents. A disulfide bridge must therefore connect this cleaved fragment to the β subunit containing the carbohydrates. The C-terminal end of the disulfide is at position 262. This leaves little room for an additional

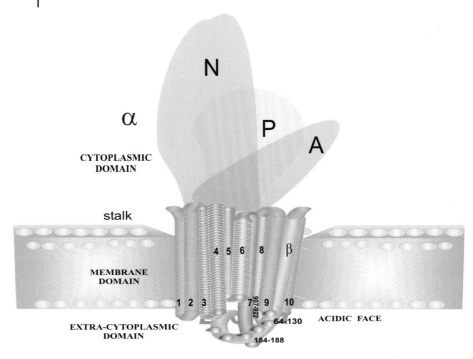

Figure 7.1 Proposed model of the gastric H^+,K^+-ATPase α subunit. There are three lobes in the cytoplasmic domain, N (ATP binding), P (phosphorylation), and A (activation) regions, and ten transmembrane segments. This model is based on the sr Ca-ATPase structures published by Toyoshima et al. [95]. The β subunit is shown with a single membrane-spanning domain and regions of interaction are numbered in both subunits.

membrane-spanning α helix. Hence it is likely that the β subunit has only one transmembrane-spanning segment. Reduction of the disulfides of the β subunit inhibits the activity of the H^+,K^+-ATPase [40]. With Na^+,K^+-ATPase, the effect of reducing agents on the ability of the enzyme to hydrolyze ATP and bind ouabain was quantitatively correlated with the reduction of disulfide bonds in the β subunit [41].

The β subunit of the Na^+,K^+-ATPase is necessary for targeting the complex from the endoplasmic reticulum to the plasma membrane [42, 43]. It also stabilizes a functional form of both the Na^+,K^+-ATPases [44, 45]. The H^+,K^+-ATPase β-subunit is also necessary to target the membrane after proper assembly and sorting [46–49]. When the α- and β-subunits of Na^+,K^+-ATPase and H^+,K^+-ATPase were expressed in Sf9 cells in different combinations, the hybrid ATPase with the Na^+,K^+-ATPase α-subunit and the H^+,K^+-ATPase β-subunit showed an ATPase activity, which was 12 % of the Na^+,K^+-ATPase activity, with decreased apparent K^+ affinity and about half the turnover number. Another hybrid ATPase with the H^+,K^+-ATPase α-subunit and the Na^+,K^+-ATPase β-subunit showed 9 % of the H^+,K^+-ATPase activity but increased the apparent K^+ affinity [50]. Another report has shown that both lipid and the β subunit affect the K$^+$ affinity of the gastric H$^+$,K$^+$-ATPase [51]. Since both enzymes are unique in being K$^+$ counter-transporters, perhaps the β is important to force counter-transport.

Seven putative N-glycosylated sites (AsnXaaSer and AsnXaaThr) are found in rabbit H$^+$,K$^+$-ATPase β-subunit [32], conserved in rat [33–36] and human [39], but only six putative N-glycosylation sites are present in the hog gastric β-subunit [26]. The structure of N-linked oligosaccharides of the β subunit of rabbit gastric H$^+$,K$^+$-ATPase was identified [52, 53]. All seven N-linked AsnX(Ser/Thr) sites at positions 99, 103, 130, 146, 161, 193, and 222 were fully glycosylated. Asn99 was modified exclusively with oligomannosidic-type structures, Man$_6$GlcNAc$_2$-Man$_8$GlcNAc$_2$, and Asn193 has Man$_5$GlcNAc$_2$-Man$_8$GlcNAc$_2$ and lactosamine-type structures. Asn 103, 146, 161, and 222 contain lactosamine-type structures. All the branches of the lactosamine-type structure were terminated with Galα-Galβ-GlcNAc extensions. Similarly, the structure of complex glycans of the β subunit of the Na$^+$,K$^+$-ATPase contain polylactosaminoglycans [54].

The role of the carbohydrate chains in membrane trafficking has been studied in HEK-293 cells and in polarized cells such as LLCPK and MDCK cells [55]. In HEK-293 cells, enzyme activity was not affected by removal of any single carbohydrate site of the β subunit but removal of all the glycosylation sites resulted in the complete loss of activity. The role of glycosylation in sorting and trafficking is discussed below.

The H$^+$,K$^+$-ATPase β subunit has the sequence Phe-Arg-His-Tyr in its cytoplasmic domain. This tyrosine is important in initiating the removal of the H$^+$,K$^+$-ATPase from the apical membrane of the parietal cells and in ensuring its return to the TVE compartment in order to terminate the process of acid secretion. The participation of a tyrosine-based signal in the retrieval of the H$^+$,K$^+$-ATPase suggests that this process involves interactions with adaptins and is mediated by clathrin-coated pit formation [56]. In mice deficient in the H$^+$,K$^+$-ATPase β-subunit, cells that express the H$^+$,K$^+$-ATPase α-subunit had abnormal canaliculi and were devoid of typical tubulovesicular membranes [57].

Chimeric β-subunits between the gastric H$^+$,K$^+$-ATPase and the Na$^+$,K$^+$-ATPase were constructed and co-transfected with the H$^+$,K$^+$-ATPase α-subunit cDNA in HEK-293 cells [58]. Whole cytoplasmic and transmembrane domains of H$^+$,K$^+$-ATPase β-subunit can be replaced by those of Na$^+$,K$^+$-ATPase β-subunit without losing the enzyme activity. Also, the extracellular segment between Cys152 and Cys178, which contains the second disulfide bond, was exchangeable between H$^+$,K$^+$-ATPase and Na$^+$, K$^+$-ATPase, preserving the ATPase activity intact. However, when four amino acids (^{76}QLKS^{79}) in the ectodomain of H$^+$,K$^+$-ATPase β-subunit were replaced by the corresponding amino acids (^{72}RVAP^{75}) of the Na$^+$,K$^+$-ATPase β-subunit, the ATPase activity was abolished. This region appears essential for effective interaction between the two subunits.

7.1.3
Regions of Association Between the α and β Subunits

For Na⁺,K⁺-ATPase, the last 161 amino acids of the α subunit are essential for effective association with the β subunit [59]. Furthermore, the last 4 or 5 C terminal hydrophobic amino acids of the Na⁺ pump β subunit are also essential for interaction with the α subunit whereas the last few hydrophilic amino acids are not [60]. Expression of sodium pump α subunit along with the β subunit of either sodium or proton pump in *Xenopus oocytes* has shown that the β subunit of the gastric proton pump can act as a surrogate for the β subunit of the sodium pump as far as membrane targeting and ⁸⁶Rb⁺ uptake, suggesting some homology in the associative domains of the β subunits of the two pumps [61].

To biochemically specify the region of the α subunit associated with the β subunit, the tryptic digest was solubilized using non-ionic detergents such as NP-40 or $C_{12}E_8$. These detergents allow the holoenzyme to retain ATPase activity. The soluble enzyme was then adsorbed onto a WGA affinity column. Following elution of the peptides not associated with the β subunit binding to the WGA column, elution of the β subunit by 0.1 N acetic acid also eluted almost quantitatively the TM7/loop/TM8 sector of the α subunit. These data show that this region of the α subunit is tightly associated with the β subunit such that non-ionic detergents are unable to dissociate it from the β subunit [62]. The antibody mAb 146-14 also recognizes the region of the α subunit at the extra-cytoplasmic face of the TM7 segment, and also recognizes the β subunit, a finding consistent with the association found by column chromatography. Using a yeast two-hybrid analysis, a fragment Leu855 to Arg922 of the α subunit was identified as interacting with the β subunit [63]. Also, two different domains in the β-subunit, Gln64 to Asn130 and Ala156 to Arg188, were identified as association domains with the α-subunit by this method.

7.1.4
Regions of Association of the α Subunits

There has been much suggestive evidence that the α-β heterodimeric H⁺,K⁺-ATPase exists as an oligomer. Such evidence includes target size [64, 65] and unit cell size of the enzyme in two-dimensional crystals [66]. It was shown that the enzyme did indeed exist as an $(\alpha\beta)_2$ heterodimeric dimmer, using blue native gel separation, and cross-linking with Cu^{2+}-phenanthroline. Membrane-bound H⁺,K⁺-ATPase reacted with Cu-phenanthroline to provide an α-α dimer. No evidence was obtained for β-β dimerization. ATP prevents this Cu^{2+}-phenanthroline-induced α-α dimerization. It was possible, by blocking free SH groups and subsequent reduction and labeling with fluorescein maleimide, to demonstrate that the site of Cu^{2+}-oxidative cross-linking was either at cys565 or cys616 [67]. Hence this region of the α subunit in the N domain is in close contact with its neighboring α subunit. Expression studies of the Na⁺,K⁺-ATPase in insect SF9 cells and immunoprecipitation reached the conclusion that a similar region of the α subunit of this enzyme was also in close contact [68].

7.2
Kinetics of the H^+,K^+-ATPase

H^+,K^+-ATPase exchanges intracellular hydrogen ions for extracellular potassium ions by consuming ATP. The H^+ for K^+ stoichiometry of the H^+,K^+-ATPase was reported to be one [69–71] or two [72, 73] per ATP hydrolyzed. The H^+/ATP ratio was independent of external KCl and ATP concentration [72]. When care is taken to use only tight vesicles at pH 6.1 the stoichiometry is $2H^+$ per ATP. Clearly, at a luminal pH of 0.8, the *in vitro* pH reached in the canaliculus, the stoichiometry must be $1H^+$ per ATP. This could be explained by retained protonation at one of the ion-binding site carboxylic acids at a pH well below the pK_a of the carboxylic amino acid.

The kinetics of the H^+,K^+-ATPase have defined many reaction steps, similar to those of the Na^+,K^+-ATPase [74]. The rate of formation of the phosphoenzyme (EP) and the K^+-dependent rate of phosphoenzyme breakdown are sufficiently fast to allow the phosphoenzyme to be an intermediate in the overall ATPase reaction. The initial step is the reversible binding of ATP to the enzyme in the absence of added K^+ ion, followed by a Mg^{2+} (and proton) dependent transfer of the terminal (γ) phosphate of ATP to asp386 of the catalytic subunit (E_1-P· H^+). The Mg^{2+} remains occluded in the P domain near asp730 [75] until dephosphorylation [76, 77]. The addition of K^+ to the enzyme-bound acyl phosphate results in a biphasic two-step dephosphorylation. The faster initial step depends on $[K^+]$, whereas the slower step is not affected by K^+ concentration. The second phase of EP breakdown is accelerated in the presence of K^+, but at $[K^+] > 500$ µM the rate becomes independent of K^+ concentration. This shows that two forms of EP exist. The first form, presumably E_1P, is K^+ insensitive and converts spontaneously in the rate-limiting step into E_2P, the K^+-sensitive form. ATP binding to the H^+,K^+-ATPase occurs in both the E_1 and the E_2 state, but with a lower affinity in E_2 state (2,000 times lower than for E_1) [78] (Figure 7.2).

H^+ or K^+ interacts competitively on the cytoplasmic surface of intact vesicles. The effects of H^+ and K^+ on formation and breakdown of phosphoenzyme were determined using transient kinetics [79]. Increasing hydrogen ion concentrations on the ATP-binding face of the vesicles accelerate phosphorylation, whereas increasing $[K^+]$ inhibit phosphorylation. Increasing $[H^+]$ reduces this K^+ inhibition of the phosphorylation rate. Decreasing $[H^+]$ accelerates dephosphorylation in the absence of K^+, and K^+ on the luminal surface accelerates dephosphorylation. Increasing $[K^+]$ at constant ATP decreases the rate of phosphorylation and increasing ATP concentrations at constant $[K+]$ accelerates ATPase activity and increases the steady-state phosphoenzyme level [80]. Therefore, inhibition by cations is due to cation stabilization of a dephospho form at a cytosolically accessible cation-binding site.

To determine the role of divalent cations in the reaction mechanism of the H^+,K^+-ATPase, calcium was substituted for magnesium, which is necessary for phosphorylation. Calcium ion inhibits K^+ stimulation of the H^+,K^+-ATPase by binding at a cytoplasmic divalent cation site. Ca·EP dephosphorylates 10–20

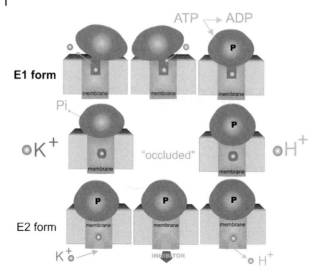

Figure 7.2 Pump cycle of the gastric H⁺,K⁺-ATPase. The initial step is proton binding to the enzyme (E$_1$·H⁺), followed the reversible binding of ATP to the enzyme in the absence of added K⁺ ion (E$_1$·ATP·H⁺). The phosphorylation step occurs by a Mg^{2+} (and proton) dependent transfer of the terminal (γ-) phosphate of ATP to the catalytic subunit (E$_1$-P·H+). After releasing the proton, the enzyme (E$_2$-P) binds K⁺ ion, forming E$_2$-P·K⁺. Dephosphorylation (E$_2$-P·K⁺ → E$_2$·K⁺) followed by release of K⁺ (E$_2$·K⁺ → E$_1$) are the last steps for exchange of H⁺ with K⁺.

times more slowly than Mg·EP in the presence of 10 mM KCl with either 8 mM CDTA or 1 mM ATP. The inability of the Ca·EP to dephosphorylate in the presence of K⁺, compared with Mg·EP, demonstrates that the type of divalent cation that occupies the catalytic divalent cation site required for phosphorylation is important for the conformational transition to a K⁺ sensitive phosphoenzyme. Calcium is tightly bound to the divalent cation site of the phosphoenzyme and the occupation of this site by calcium causes slower phosphoenzyme kinetics. Since the presence of CDTA or EGTA does not change the dephosphorylation kinetics of the EP·Ca form of the enzyme, it was concluded that the divalent cation remains occluded in the enzyme until dephosphorylation occurs [81].

7.3
Conformations of the H⁺,K⁺-ATPase

H⁺,K⁺-ATPase generates HCl in the stomach by the electroneutral exchange of H⁺ for K⁺, dependent on conformation changes in the protein. The alteration of enzyme conformation changes the affinity and sidedness of the ion-binding sites during the cycle of phosphorylation and dephosphorylation.

The ions transported from the cytoplasmic side are H⁺ at high pH. Since Na⁺ is transported as a surrogate for H⁺, it is possible that the hydronium ion, rather than the proton per se, is the species transported [82]. The ions transported inwards from the outside face of the pump are Tl^+, K^+, Rb^+ or NH_4^+ [74, 80, 83]. Presumably the change in conformation changes a relatively small ion-binding domain in the outward direction into a larger ion-binding domain in the inward direction.

The E_1 conformation of the H^+,K^+-ATPase binds the hydronium ion from the cytoplasmic side at high affinity. Following phosphorylation, the conformation changes from $E_1P \cdot H_3O^+$ to $E_2P \cdot H_3O^+$ form, which has high affinity for K^+ and low affinity for H_3O^+ allowing release of H_3O^+ and binding of K^+ from the extra-cytoplasmic surface of the enzyme. Breakdown of the E_2P form requires K^+ or its congeners on the outside face of the enzyme. With dephosphorylation, the $E_1 K^+$ conformation is produced with a low affinity for K^+, releasing K^+ to the cytoplasmic side, allowing rebinding of H_3O^+ [74]. The steps of phosphorylation of the H^+,K^+-ATPase were studied by measuring the inorganic phosphate, $P^{18}O_4$ and $P^{16}O_4$ distribution as a function of time at different H^+, K^+, and Pi concentrations [84]. The formation of E·Pi complex that exchanges ^{18}O with HOH was slower at pH 5.5 than at pH 8 and is not diffusion controlled, suggesting a unimolecular chemical transformation involving an additional intermediate in the phosphorylation mechanism such as, perhaps, a protein conformational change. From competitive binding of ATP and 2′,3′-O-(2,4,6-trinitrophenylcyclohexadienylidene) adenosine 5′-phosphate (TNP-ATP), two classes of nucleotide binding sites were suggested [85]. TNP-ATP is not a substrate for the H^+,K^+-ATPase. However, TNP-ATP prevents phosphorylation by ATP and inhibits the K^+-stimulated pNPPase and ATPase activities. The number of TNP-ATP binding sites was twice the stoichiometry of phosphoenzyme formation. Recently, two moles of phosphate were claimed to be liberated from one mole of phosphoenzyme of the gastric H^+,K^+-ATPase [86]. It was hypothesized that one mole of Pi is from a high-affinity ATP binding site and the other Pi is from enzyme-bound ATP at a low-affinity site during cross-talk between catalytic subunits. All-sites phosphorylation has been proposed by studies using Pi or acetyl phosphate [87, 88], which showed that the stoichiometry of the maximum amount of phosphoenzyme formed from ATP, that from acetyl phosphate, that from inorganic phosphate (Pi), and the maximum amount of ATP binding to the enzyme was close to 1:2:2:2. The phosphoenzyme formed was shown to be turning over. The addition of K^+ reduced the amount of phosphoenzyme from ATP to one-tenth but reduced those from acetyl phosphate or Pi to only a half.

Fluorescein isothiocyanate (FITC) binds to the H^+,K^+-ATPase at pH 9.0, inhibiting ATPase activity but not pNPPase activity [11]. Fluorescence of the FITC-labeled enzyme, representing the E_1 conformation, was quenched by K^+, Rb^+, and Tl^+ [11, 89]. The quenching of the fluorescence by KCl reflects the formation of E_2K^+. FITC binds at lys516 in the hog enzyme sequence [12]. This FITC binding site apparently becomes less hydrophobic when KCl binds to form the $E_2 \cdot K$ conformation.

The FITC labeled Na^+,K^+-ATPase has quite similar properties [90, 91]. Two K^+ ions are required to cause the conformational change from E1 to E2. The binding site of FITC was at lys501. However, several additional lysines at positions 480 and

766 were shown to react with FITC during inactivation of the Na^+,K^+-ATPase. These lysines were also protected from labeling in the presence of ATP [13].

The fluorescent 1-(2-methylphenyl)-4-methylamino-6-methyl-2,3-dihydropyrroloquinoline (MDPQ) was shown to inhibit the H^+,K^+-ATPase and the K^+ phosphatase competitively with K^+ [76]. MDPQ fluorescence is quenched by the imidazopyridine SCH 28080. The imidazopyridine Me-DAZIP binds to the TM1/loop/TM2 sector of the β subunit [92]. MDPQ binding to the extra-cytoplasmic surface of the pump enhances its fluorescence, suggesting that inhibitor binding occurs to a relatively hydrophobic region of the protein. The fluorescence was quenched by K^+, independently of Mg^{2+}. The binding of Mg-ATP increased the fluorescence due to the formation of an $E_2P\cdot[I]$ complex [76], suggesting that the binding pocket on the outside surface between TM5 and TM6, as discussed later for these compounds, changes conformation or position between the two major conformers of the enzyme.

The fluorescent changes with FITC and MDPQ may reflect relative motion of the cytoplasmic domain and outside surface of the enzyme. In the E_1 form, the FITC region is relatively closer to the membrane and the extra-cytoplasmic loop relatively hydrophilic. With the formation of the $E_2\cdot K^+$ form the FITC region is more distant from the membrane, whereas with formation of the E_2P form the MDPQ binding region moves towards the membrane. These postulated conformational changes are therefore reciprocal in the two major conformers of the enzyme.

The effect of trypsin on the gastric H^+,K^+-ATPase provides evidence for conformational changes as a function of ligand binding. Only K^+ of the ionic ligands provided significant protection against extensive tryptic hydrolysis [62]. Neither ATP nor ADP affected the tryptic pattern at high trypsin:protein ratios. (J. M. Shin, unpublished observations).

Two large fragments, 67 and 33 kDa, were found in the presence of ATP, Mg^{2+} and SCH 28080 as a K^+ surrogate whereas several fragments were produced in the absence of ligands. These data suggest that the E_2K^+, or more particularly the $E_2P\cdot[SCH]$ conformation of the H^+,K^+-ATPase, severely limits accessibility of trypsin to most of the lysines and arginines in the β subunit [17, 92].

Extensive tryptic digestion of the gastric H^+,K^+-ATPase in the presence of KCl provided a C-terminal peptide fragment of 20 kDa beginning at the TM7 transmembrane segment, a peptide of 9.4 kDa comprising the TM1/loop/TM2 sector beginning at asp84, and another peptide of 9.4 kDa containing the TM5/loop/M6 sector beginning at asn753 [62]. The C-terminal 20 kDa peptide fragment was suggested to be capable of Rb^+ occlusion [77]. When these digests in the absence and presence of KCl are compared, some regions near the membrane can be seen to be K^+ protected. The region between gly93 and glu104 near the TM1 segment, the region between asn753 and leu776 near the cytoplasmic side of the TM5 segment, and the region after the TM8 segment, especially the region between ile945 and ile963 containing 5 arginines and 1 lysine, are protected from the trypsin digestion in the $E_2\cdot K$ conformation [62]. Further, there must be protection prior to TM3 and for some distance subsequent to M4, since no fragment containing these segments was found at an M_r of less than 20 kDa.

When the gastric H^+,K^+-ATPase was cleaved by Fe^{2+}-catalyzed oxidation in the presence of various ligands, the cleavage patterns were different between the conformational changes. There are two Fe^{2+} cleavage sites. In Fe^{2+} site-1, the parallel appearance of the fragments at ^{230}ESE, near ^{624}MVTGD, and at ^{728}VNDS upon transition from E_1 to E_2(Rb) conformations were observed. Meanwhile, in Fe^{2+} site-2, the fragment near ^{299}HFIH was cleaved independently of conformational changes [75]. These cleavage patterns were identical to those of the Na^+,K^+-ATPase [93, 94]. The cleavage data showed that structural organization and changes in the cytoplasmic domains, association with E_1/E_2 transitions, are essentially the same for the H^+,K^+-ATPase, the Na^+,K^+-ATPase, and sr Ca-ATPase. Using the well-defined sr Ca-ATPase structure, the N-domain where ATP binds then inclines nearly 90° with respect to the membrane, and the A domain rotates by about 110° horizontally following E_1 to E_2 conformational changes [95].

7.4
Functional Residues of the H⁺,K⁺-ATPase

When the gastric H^+,K^+-ATPase was digested by trypsin in the presence of a high concentration of KCl, the tryptic membrane digest showed capability for Rb^+ occlusion [77], like the Na^+,K^+-ATPase [96]. As described above, some regions near the membrane were K^+ protected, such as the region between gly93 and glu104 near the TM1 segment, the region between asn753 and leu776 near the cytoplasmic side of the TM5 segment, and the region after the TM8 segment, especially the region between ile945 and ile963 containing 5 arginines and 1 lysine. Furthermore, when K^+ is removed from this membrane digest, the TM5–TM6 hairpin was released from the membrane, showing that this membrane hairpin is stabilized by K^+ ions and that this region is more flexible than the other membrane-spanning regions [97]. Proton pump inhibitors such as omeprazole, pantoprazole, lansoprazole, and rabeprazole bind to Cys813 of TM5–TM6, giving inhibition of activity [16, 98]. This biochemical study shows that TM5 and TM6 must be part of the ion pathway.

Using site-directed mutagenesis of the gastric H^+,K^+-ATPase transfected in HEK293 cells, the TM5–TM6 luminal loop was studied in terms of K^+ access to the ion binding domain [99]. Mutations of TM5, TM5–TM6, and M6 regions such as P798C, Y802L, P810A, C813A or S, F818C, T823V, and mutations of TM7–TM8 and TM8 such as E914Q, F917Y, G918E, T929L, F932L, reduced the affinity for SCH28080 up to 10-fold without affecting the affinity for the activating cation, NH_4^+. The L809F substitution in the loop between TM5 and M6 resulted in about a 100-fold decrease in inhibitor affinity. C813T mutant showed a 9-fold loss of SCH28080 affinity. All these data suggest that the binding domain for SCH28080 contains the surface between L^{809} and C^{813}, in the TM5–TM6 loop and the luminal end of TM6. Mutations of C^{813} and I^{816} in TM6 and M^{334} in TM4 also showed that the inhibitor binds close to the luminal surface of the enzyme [100–103].

When negatively charged amino acid residues in the α-subunit were mutated the results showed that carboxyl groups in the membrane-spanning domain are important for cation binding [104]. Mutation of E820Q showed less sensitivity to K^+ and the dephosphorylation was not stimulated by either K^+ or ADP, indicating that E^{820} might be involved in K^+ binding and transition to the E_2 form of the H^+,K^+-ATPase [105]. Mutation of E^{795} at the cytoplasmic side of TM5 showed a decrease of the phosphorylation rate and the apparent ATP affinity, indicating that E^{795} is involved in both K^+ and H^+ binding [106]. Mutation of E^{795} and E^{820} in TM5 and TM6 resulted in a K^+-independent, SCH28080-sensitive ATPase activity, due to a high spontaneous dephosphorylation rate [107, 108].

Similarly, the positively charged lysine in TM5/TM6 region was mutated. The mutants K800A and K800E of the Buffo bladder H^+,K^+-ATPase showed K^+-stimulated and ouabain-sensitive electrogenic transport. When the positive charge was conserved (K800R), no K^+-induced outward current could be measured, but rubidium transport was present. This shows that a single positive charged residue in TM5 can determine the electrogenicity [109]. The sixth transmembrane (TM6) segment of the catalytic subunit plays an important role in the ion recognition and transport in the P_2-type ATPase families. When all amino acid residues in the TM6 segment of gastric H^+,K^+-ATPase α-subunit were singly mutated with alanine, four mutants, L819A, D826A, I827A, and L833A, completely lost the K^+-ATPase activity. Mutant L819A was phosphorylated but hardly dephosphorylated in the presence of K^+, whereas mutants D826A, I827A, and L833A were not phosphorylated from ATP. Amino acids involved in the phosphorylation are located exclusively in the cytoplasmic half of the TM6 segment and those involved in the K^+-dependent dephosphorylation are in the luminal half. Several mutants, such as I821A, L823A, T825A, and P829A, partly retained the K^+-ATPase activity accompanying the decrease in the rate of phosphorylation [110].

The cation selectivity of the Na^+,K^+- and H^+,K^+-ATPase may be generated through a cooperative effort between residues of the transmembrane segments and the flanking loops that connect these transmembrane domains. Substituting three residues in the Na^+,K^+-ATPase sequence with their H^+,K^+-ATPase counterparts (L319F, N326Y, T340S) and replacing the TM3–TM4 ectodomain sequence with that of the H^+,K^+-ATPase result in a pump that gives 50 % of ATPase activity in the absence of Na^+ at pH 6. This effect was not seen when the ectodomain alone is replaced [111]. Many of these mutational results can be rationalized based on the crystal structure of the sr Ca-ATPase.

7.5
Structural Model of the H⁺,K⁺-ATPase

7.5.1
Crystal Structure of the H⁺,K⁺-ATPase

As discussed above, experimental evidence showed that the H^+,K^+-ATPase has 10 membrane spanning segments in the catalytic α-subunit and 1 membrane-spanning segment in the β-subunit.

The two-dimensional structure crystals of the H^+,K^+-ATPase formed in an imidazole buffer containing $VO3^-$ and Mg^{2+} ions was resolved at about 25 Å [64, 66]. The average cell edge of the H^+,K^+-ATPase was 115 Å, containing four asymmetric protein units of 50×30 Å. The structure of Na^+,K^+-ATPase was determined by electron crystallography at 9.5 Å from multiple small 2-D crystals induced in purified membranes [112]. The density map shows a protomer stabilized in the E2 conformation that extends approximately $65 \times 75 \times 150$ Å in the asymmetric unit of the P2 type unit cell. The unit cell dimension of the $Co(NH_3)_4ATP$-induced crystals of Na^+,K^+-ATPase was 141 Å [113]. Two-dimensional crystals of the Na^+,K^+-ATPase were reported to be best formed at pH 4.8 in sodium citrate buffer and to represent a unique lattice ($a = 108.7$, $b = 66.2$, $r = 104.2$) by electron cryomicroscopy. There are two high contrast parts in one unit cell [114].

The crystal structure of the sarcoplasmic reticulum Ca-ATPase was resolved at 2.6 Å resolution with two calcium ions bound in the transmembrane domain, which consists of ten α-helices [115]. Also, the calcium-free E_2 state of sr Ca-ATPase was compared with the calcium-bound E_1Ca state [95]. In the Ca-ATPase, the N-domain where ATP binds inclines nearly 90° with respect to the membrane and the A domain rotates by about 110° horizontally with a change from E_1 to E_2 conformation. Several attempts have been made to define the tertiary structure of this P_2 type-ATPase based on the structure of sr Ca-ATPase combined with site-directed mutagenesis, cleavage patterns of different conformations and molecular modeling [116–118].

7.5.2
Molecular Modeling of the Gastric H⁺,K⁺-ATPase

Gastric H^+,K^+-ATPase is a member of the P_2 type family of ATPases that transport ions against their concentration gradients across lipid bilayers at the expense of ATP hydrolysis. These enzymes, which differ in their ion and inhibitor specificities, nevertheless contain numerous stretches of identical amino acid sequence that not only place them in the family but imply a conserved three-dimensional structure. The implicit homology of mechanism has been substantiated during the past 30 years by an array of biochemical and spectroscopic results showing common conformational intermediates in all family members. These intermediates occur along a general reaction pathway which includes two fundamental conformations, E_1 in which the binding sites for the outwardly transported ion are accessible from the cytoplasm and E_2 where the binding sites for the inwardly transported ion are accessible from the side opposite the cytoplasm (the "outside").

Binding of the outwardly transported ion (H^+ or H_3O^+ for H^+,K^+-ATPase) favors ATP phosphorylation of a conserved aspartic acid side-chain giving E_1P. This phosphorylation shifts the conformational equilibrium in favor of E_2 and a separate conformer, E_2P, predominates. The fundamental reactive difference between these forms is that E_1P can reform ATP from ADP whereas E_2P cannot. Binding of the inwardly transported ion stimulates dephosphorylation and gives an occluded form $E_2[K^+]$ in the H^+,K^+-ATPase where the ion is inaccessible from either side of the membrane, similar to that observed in the Na^+,K^+-ATPase. Relaxation to the E_1 state results in release of the K^+ ion to the cytoplasm.

The ultimate goal of structural chemists in this area of research has been to define the mechanism of active ion transport in terms of the molecular structures of the reaction intermediates and, secondly, to understand the molecular basis of ion and inhibitor specificity. The first of these aims requires high-resolution crystal structures for each of the reaction intermediates of at least one member of the P_2-type ATPase family. A tremendous advance towards this requirement has been met recently with rabbit sr Ca-ATPase where high-resolution crystal structures have been reported for both $E_1 \cdot 2Ca^{2+}$ [115] and E_2[thapsigargin] [95] conformations. These structures have provided a molecular understanding of the general mechanism of ion transport [117].

The general shape of the sr Ca-ATPase is that of a "Y", where the lower half of the base of the Y (M domain) passes through the membrane and contains the ion transport pathway while the upper half extends out of the membrane to form the "stalk" and contains the site of phosphorylation (P domain). One arm of the "Y" makes up the N (nucleotide) domain which binds ATP and the other assists in the transfer of essential conformational effects from the active site to the M domain and is designated the A (for "activator") domain. In the $E_1 \cdot 2Ca^{2+}$ conformation the N and A domains are splayed open while the calcium ions are encaged side by side near the middle of the membrane in roughly octahedral complexes formed by side-chains' ligands from transmembrane (TM) helices 8, 5, and 6 (site-1) or 5, 6 and 4 (site-2). This splaying of the Y allows Ca^{2+} access to the ion-binding domain. The N domain in this form is too distant from P to allow for phosphorylation by bound ATP and it was surmised that N must be capable of rotation about a hinge to allow the γ-phosphate of ATP to reach the phosphorylated aspartate [119]. Subsequently, in the E_2 conformation which binds ATP with low affinity, the N domain was shown to be tilted over the P domain, which itself has tilted to a more vertical orientation with respect to the plane of the membrane, making the M, P and N domains nearly collinear. Further, ATP with a fully extended triphosphate can span the N and P domains in the E_2 state [120], proving the ability of N to achieve an orientation with P which is at least close to that in $E_1 \cdot ATP$ and, presumably, similar to E_1P (ADP still being bound and capable of reforming ATP). The position of the A domain in E_2 is raised and rotated upon conversion from E_1. Thus the A, P and N domains are substantially gathered together in E_2. These effects are transmitted to the membrane domain where the geometry of the ion ligands is altered, with TM4 rotating to move its carboxylate ligand away from ion site-2 and TM6 rotating with respect to TM4 and TM5. This rearrangement optimizes the transport site for release of outward ion and for entry of the counter-ion from the outside, while formation of the ion complex triggers dephosphorylation, occlusion, and return to E$_1$. Currently, the molecular structures of E$_1$P and E$_2$P, which are essential for fully understanding the conversion of energy into vectorial ion transport, remain unknown.

In contrast to the sr Ca-ATPase it has not been possible to crystallize either the Na$^+$,K$^+$- or H$^+$,K$^+$-ATPase in a form necessary for high-resolution structural determination. There have been two recent structures reported, however, for the Na$^+$,K$^+$- ATPase at 9.5 [112] and 11 Å [121] resolution determined by cryoelectron microscopy of the E$_2$·vanadate conformations. In each case the density envelope conformed fairly well to the high-resolution form of the sr Ca-ATPase E$_1$·2Ca^{2+} structure with the differences clearly explained by closer approximation to the E$_2$[thapsigargin] structure that has also been published [95]. The identity of the A, P, M, and N domains was clearly evident, demonstrating the expected homology in structure. The density contributed by the β subunit was tentatively identified in the transmembrane region of the higher-resolution structure [112]. Models for the N domain of the Na$^+$,K$^+$-ATPase have also been determined by NMR of this fragment expressed in *E. coli*. Changes in structure were shown in the presence of high concentrations of ATP and an ATP bound conformation was constructed [122]. This work is of particular benefit for homology modeling since the N domain sequences are considerably different in sr Ca-ATPase and Na$^+$,K$^+$-ATPase with several long deletions and insertions [116]. In contrast, the Na$^+$,K$^+$- and H$^+$,K$^+$-ATPases show high homology in the N domain, with no insertions or deletions in the sequence alignments.

For the closely related Na$^+$,K$^+$- and H$^+$,K$^+$-ATPases the more common strategy has been to use biochemical results to validate homology modeling based on the known sr Ca-ATPase structures [118]. In homology modeling, the peptide backbone of a known high-resolution structure is used to substitute the sequence of a homologous protein based on an amino acid sequence alignment. Steric clashes are removed and energy minimization then gives the final model. This method has been applied to the yeast H$^+$-ATPase [123], colonic H$^+$K$^+$-ATPase [124], and the Na$^+$,K$^+$- [125] and H$^+$,K$^+$-ATPases [126]. The predictions of a model can then tested experimentally.

For instance, mutational analyses have shown that most of the positions in the M domain which provide ion ligands are conserved, albeit with some substitutions to other oxygen-containing side-chains. It is assumed that these changes are at least partially responsible for the separate ion specificities, but mutating them to affect specificity changes is ineffective, implying the precise geometry is critically dependent on subtle differences in global conformation.

Similarly, site-specific mutagenesis has shown that many of the residues affecting specific inhibitor binding of ouabain to the Na$^+$,K$^+$-ATPase, and K$^+$ competitive imidazopyridine, SCH28080, to the H$^+$,K$^+$-ATPase ([103, 126, 127], are conserved in the two pumps. The binding sites are therefore formed from some of the same residues in the loops between TM3–TM4, TM5–TM6 and TM7–TM8 (Figure 7.3). Both of these inhibitors bind to E_2P conformations and prevent K^+ access to the ion binding site. Specificity presumably arises from a combination of a few side-chain changes near the site and global differences generated allosterically by distant substitutions whose importance for binding is nevertheless demonstrable by mutation. The presence of an inhibitor cavity or vestibule accessible from the outside near the TM5-M6 loop is further shown by covalent labeling of the H^+,K^+-ATPase at Cys813 at the beginning of TM6 by the covalent proton pump inhibitors (e. g. omeprazole). The homology model of the E_2 form of the H^+,K^+-ATPase with SCH28080 bound accounts accurately for the known active conformer of the inhibitor as well as for the effects given by specific amino acid mutations

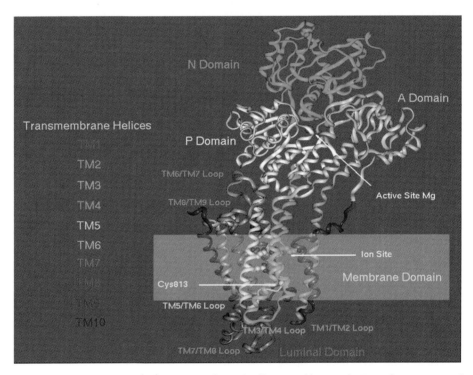

Figure 7.3 Gastric H^+,K^+-ATPase α subunit (backbone in ribbon) in the E_2 conformation viewed from the membrane (cytoplasmic side up). The model was constructed by substitution of the rabbit amino acid sequence on to the backbone of the known sr Ca-ATPase structure (pdb.1iwo, the E_2[thapsigargin] conformation) whose N domain backbone had been replaced with that of the Na^+,K^+-ATPase determined by NMR. Energy minimization was then applied to give the structure shown. The site of phosphorylation is adjacent to the active site magnesium (violet sphere). Ion transport sites are near the middle of the membrane. Proton pump inhibitors (imidazopyridines) block K^+ access and inactivate acid secretion by covalent modification of cys813 in a cavity formed by ala335 and phe332 from TM4 (green), tyr925 from TM8 (orange), pro808 and leu809 from loop TM5/TM6 (yellow), and TM1/TM2 (blue). The structure and location of the β subunit (not shown) is currently unknown although a major site of interaction is the M7/M8 luminal loop.

around the inhibitor site [126]. In addition, the model predicts disulfide bond linkages in the immediate vicinity of the bound inhibitor, which can be engineered into the protein via cysteine substitution. These results validate the accuracy of homology modeling in the membrane based on the sr Ca-ATPase structure. The fit of the model to the density envelopes reported for the homologous Na$^+$,K$^+$-ATPase mentioned above has led to predictions for the site of β subunit interaction and kinase phosphorylation, but these have yet to be tested experimentally.

7.6
Acid Secretion and the H$^+$,K$^+$-ATPase

The H$^+$,K$^+$-ATPase is present mainly in the gastric parietal cell. In the resting parietal cell it is present in smooth surfaced cytoplasmic membrane tubules. Upon stimulation of acid secretion, the pump is found on the microvilli of the secretory canaliculus of the parietal cell. This morphological change results in a several-fold expansion of the canaliculus [128, 129]. There are actin filaments within the microvilli and the subapical cytoplasm. In the cytoskeleton system, there is an abundance of microtubules among the tubulovesicles. Some microtubules appeared to be associating with tubulovesicles. Numerous electron-dense coated pits and vesicles were observed around the apical membrane vacuoles in cimetidine-treated resting parietal cells, consistent with an active membrane uptake in the resting state. The cultured parietal cells undergo morphological transformation under histamine stimulation, resulting in a great expansion of apical membrane vacuoles. Immunogold labeling of H$^+$,K$^+$-ATPase was present not only on the microvilli of expanded apical plasma membrane vacuoles but also in the electron-dense coated pits [130].

There is activation of a K$^+$ and Cl$^-$ conductance in the pump membrane which allows K$^+$ to access the extra-cytoplasmic face of the pump [131]. This allows H$^+$ for K$^+$ exchange to be catalyzed by the ATPase [132]. This conductance is probably due to the presence of individual proteins in the stimulated membrane.

Covalent inhibitors of the H$^+$,K$^+$-ATPase that have been developed to treat ulcer disease and esophagitis depend on the presence of acid secreted by the pump. They are also acid-activated prodrugs that accumulate in the acid space of the parietal cell. Hence their initial site of binding is only in the secretory canaliculus of the functioning parietal cell [133]. These data show also that the pump present in the cytoplasmic tubules does not generate HCl.

The upstream DNA sequence of the α subunit contains both Ca and cAMP responsive elements in rat H$^+$,K$^+$-ATPase [134]. There are gastric nuclear proteins present that bind selectively to a nucleotide sequence, GATACC, in this region of the gene [35]. These proteins have not been detected in other tissues. Another regulatory site has also been postulated [135].

Stimulation of acid secretion by histamine increases the level of mRNA for the α subunit of the pump [136]. Elevation of serum gastrin, which secondarily stimulates histamine release from the enterochromaffin-like cell in the vicinity of the parietal cell, also stimulates transiently the mRNA levels in the parietal cell [137]. H2 receptor antagonists block the effect of serum gastrin elevation on mRNA levels [138]. It seems, therefore, that activity of the H_2 receptor on the parietal cell determines in part gene expression of the ATPase. Chronic stimulation of this receptor might be expected therefore to upregulate pump levels whereas inhibition of the receptor would down-regulate levels of the ATPase.

However, chronic administration of these H2 receptor antagonists, such as famotidine, results in an increase in pump protein, whereas chronic administration of omeprazole (which must stimulate histamine release) reduces the level of pump protein in the rabbit [139, 140]. Regulation of pump protein turnover downstream of gene expression must account for these observations. Probably, the pump remains in the tubular state, preventing retrieval and degradation of part of the retrieved protein with receptor antagonist inhibition. This is consistent with the finding that the half-life of the protein is increased with H2 receptor inhibition [156].

7.7
Inhibitors of the H⁺,K⁺-ATPase

Two types of proton pump inhibitors have been developed to react exclusively with the extra-cytoplasmic surface of the gastric H⁺,K⁺-ATPase. One class of inhibitors now commercially available is substituted benzimidazoles such as omeprazole, lansoprazole, rabeprazole, and pantoprazole. The other class is a series of K⁺ competitive reagents, including imidazopyridines that are also known to react on the outside surface of the pump.

7.7.1
Substituted Benzimidazoles

The H⁺,K⁺-ATPase in the parietal cell secretes acid into the secretory canaliculus, generating a pH of < 1.0 in lumen of this structure. The acidity of this space is more than 1000-fold greater than anywhere else in the body, and allows accumulation of weak bases. Weak bases of a pK_a less than 4.0 would be accumulated only in this acidic space and no other acidic space in the body.

The first compound of this class with inhibitory activity on the enzyme and on acid secretion was the 2-(pyridylmethyl)sulfinylbenzimidazole, timoprazole [141]. Since a substituted benzimidazole was first reported to inhibit the H⁺,K⁺-ATPase, many inhibitors of the H⁺,K⁺-ATPase have been synthesized. The first pump inhibitor used clinically was 2-[(3,5-dimethyl-4-methoxypyridin-2-yl)methylsulfinyl]-5-methoxy-1H-benzimidazole, omeprazole [142]. The covalent inhibitors all belong to the substituted benzimidazole family [143–147] (Figure 7.4). These reagents are weak base, acid-activated compounds, which form cationic sulfenamides in acidic environments. The sulfenamides formed react with the SH group of cysteines in proteins to form relatively stable disulfides. Since the pump generates

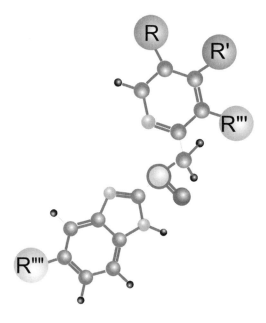

Figure 7.4 Proton pump inhibitors. These are substituted pyridylmethyl-sulfinyl benzimidazole. For omeprazole R and R''' are methyl groups and R' and R'''' are methoxy groups. For lansoprazole R''' and R'''' are H, R is a methyl group and R' is a trifluor-oethoxy group. Pantoprazole has R' and R''' as methoxy groups, R is H and R'''' is a difluoromethoxy group. For rabeprazole R''' and R'''' are H, R is a methyl group and R' is a methoxypropoxy group.

acid on its extra-cytoplasmic surface, only those cysteines available from that surface are accessible to these sulfenamides if labeling is carried out under acid transporting conditions.

Omeprazole is an acid-activated prodrug [148]. Omeprazole can be accumulated in the acidic space and easily converted into a reactive cationic sulfenamide species, which binds to SH group of cysteines in the H$^+$,K$^+$-ATPase [148–152]. Omeprazole has a stoichiometry of 2 mol inhibitor bound per mol phosphoenzyme under acid transporting conditions and is bound only to the α subunit even *in vivo* [151, 152].

Substituted benzimidazole inhibitors show slightly different effects depending on the inhibitor structure. Two irreversible inhibitors that form cysteine-reactive sulfenamides in the acid space generated by the pump, omeprazole and E3810, appear to inhibit the enzyme in different conformations [153]. Both omeprazole and E3810, 2-{[4-(3-methoxypropoxy)-3-methylpyridin-2-yl]methylsulfinyl}-1*H*-benzimidazole, are acid activated in luminal surface to form active sulfenamide derivatives, which can bind cysteines within the H$^+$,K$^+$-ATPase.

The omeprazole-bound enzyme has a lower FITC fluorescence, perhaps due to an E$_2$-like conformation. Both ATPase activity and steady-state phosphorylation were inhibited. The E3810 bound enzyme showed a high FITC fluorescence, more like a E$_1$ conformation. Fluorescence of the E3810 bound enzyme was quenched by K$^+$ in contrast to the omeprazole-derivatized FITC labeled enzyme. It is not known whether these effects are due to differences in structure of the inhibitors or to differences in location of binding site or both.

These proton pump inhibitors have different binding sites. Omeprazole (Figure 7.5) binds to cysteines in the extra-cytoplasmic regions of TM5/TM6 (cys813) and TM7/TM8 (cys892) [17]. Pantoprazole binds only to the cys813 and cys822 in M5/

REACTION PATHWAY of PROTON PUMP INHIBITORS

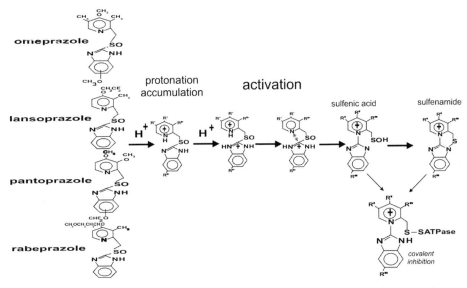

Figure 7.5 Proton pump inhibitor reaction pathway. Proton pump inhibitors have a pyridine moiety of pK_a ~ 4.0, which enables them to be selectively accumulated in the acidic space of the active parietal cell. A second protonation on the benzimidazole ring with a pK_a ~ 1.0 results in electron deficiency of C-2 of benzimidazole, where the unprotonated pyridine N can attack intra-molecularly. After this rearrangement, PPIs generate either a sulfenic acid or a cyclic sulfenamide, both of which are very reactive with thiol groups (cysteine in protein), giving a product that is a permanent cation, thereby restricting re-entry into the cytoplasm of the parietal cell.

M6 [16] and lansoprazole binds to cysteines in TM3/TM4(cys321), cys813 inTM5/ TM6, and cys892 in TM7/TM8 [154]. These data suggest that, of the 28 cysteines in the α subunit, only the cysteines present in the TM5/TM6 region are important for inhibition of acid secretion by the substituted benzimidazoles.

The proton pump inhibitors provided different acid recovery rates *in vivo*. In human, the half-life of the inhibitory effect on acid secretion is ~13 h for lansoprazole, ~28 h for omeprazole and ~46 h for pantoprazole [155]. The half-time of recovery of acid secretion and ATPase in rats following omeprazole treatment is ~ 15 h whereas the pump protein half-life is 54 h [156, 157]. Covalent inhibition of the ATPase results in inhibition of acid secretion extending for longer than the plasma half-life of the PPIs. Recovery from inhibition of acid secretion could occur in principle by either *de novo* synthesis of pump protein and/or reduction of the disulfide by an endogenous cellular reducing agent such as glutathione. The latter mechanism of reversal depends on whether glutathione or other reducing agents can gain access to a particular cysteine disulfide bond. Recovery of acid secretion following inhibition by all PPIs, other than pantoprazole, may depend on both protein turnover and reversal of the inhibitory disulfide bond. In contrast, recovery of acid secretion after pantoprazole may depend entirely on new protein synthesis [158].

7.7.2
Substituted Imidazo[1,2α]pyridines and other K⁺-competitive Antagonists

Reversible inhibitors contain protonatable nitrogens and have various structures. One type is represented by the imidazo-pyridine derivatives [159], others are piper-idinopyridines [160], substituted 4-phenylaminoquinolines [161], pyrrolo[3,2-c]qui-nolines [162], guanidino-thiazoles [163], substituted pyrimidines[164] and scopa-dulcic acid [165, 166]. Some natural products such as cassigarol A [167] and naphthoquinone [168] also showed inhibitory activity. Many of these reversible in-hibitors show K⁺ competitive characteristics, in contrast to the benzimidazole type.

Imidazo[1,2α]pyridine derivatives were shown to inhibit gastric secretion more rapidly than the benzimidazole type since the latter require acid secretion, accumu-lation and acid activation [169, 170]. SCH 28080, 3-cyanomethyl-2-methyl-8-(phe-nylmethoxy)imidazo[1,2α]pyridine, inhibited the H⁺,K⁺-ATPase competitively with K⁺ [171]. SCH 28080 binds to free enzyme extra-cytoplasmically in the ab-sence of substrate to form E2(SCH 28080) complexes. SCH 28080 inhibits ATPase activity with high affinity in the absence of K⁺ (Figure 7.6). SCH 28080 has no ef-fect on spontaneous dephosphorylation but inhibits K⁺-stimulated dephosphoryla-tion, presumably by forming a E_2-P·[I] complex. Hence SCH 28080 inhibits K⁺-sti-mulated ATPase activity by competing with K⁺ for binding to E_2P [172]. Steady-state phosphorylation is also reduced by SCH 28080, showing that this compound

SCH 28080   **Me-DAZIP⁺**

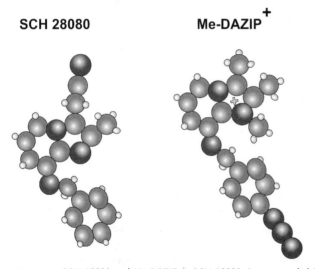

Figure 7.6   SCH 28080 and Me-DAZIP⁺. SCH 28080, 3-cyanomethyl-2-methyl-8-(phenyl-methoxy)imidazo[1,2α]pyridine, inhibited H⁺,K⁺-ATPase competitively with K⁺. SCH 28080 binds to free enzyme extra-cytoplasmically in the absence of substrate to form E_2(SCH 28080) com-plexes. SCH 28080 inhibits ATPase activity by forming an E_2-P·[I] complex. Another imidazo-pyridine derivative, 8-[(4-azidophenyl)methoxy]-1-tritiomethyl-2,3-dimethylimidazo-[1,2α]pyridinium iodide (Me-DAZIP⁺), inhibits enzyme activity as SCH28080 does.

also binds to the free enzyme. Using a photoaffinity reagent, 8-[(4-azidophenyl)-methoxy]-1-tritiomethyl-2,3-dimethylimidazo[1,2α]pyridinium iodide (Me-DAZIP+), part of the binding site of this class of K+ competitive inhibitor was identified to be in or close to the loop between the TM1 and TM2 segments [92]. Site-directed mutagenesis shows that the binding region is also in the vestibule bounded by the loop between TM5 and TM6 [103, 126]. This binding region, as discussed earlier, was identified by site-directed mutagenesis.

Another type of fluorescent K+ competitive arylquinoline, 1-(2-methylphenyl)-4-methylamino-6-methyl-2,3-dihydropyrrolo[3,2-c]quinoline (MDPQ), shows enhanced hydrophobicity of its environment with formation of the E_2-P·[I] conformer of the enzyme, as if the H1/loop/H2 segment moves further into the membrane in the E2 conformation [76].

7.8
Trafficking of the H+,K+-ATPase

Early electron microscopy suggested stimulation of acid secretion was accompanied by a morphological change in the parietal cell whereby membrane vesicles changed locale from the cytoplasm to microvilli of the secretory canaliculus [129, 173]. Disappearance of membrane vesicles from the cytoplasm that is observed by immuno-electron microscopy in association with acid secretion [174] was also consistent with the pump moving into microvilli. Fluorescence microscopy then demonstrated acid secretion takes place into the microvillus-lined secretory canaliculus [175]. The association of a cytoskeletal component, ezrin, with stimulated but not with resting membranes [176] and the presence of other SNARE proteins such as Rab 11 [177] in the parietal cell was taken as further support of the vesicle fusion hypothesis.

Freeze–fracture images of rapidly frozen fixed tissue indicated the membrane vesicles were in fact tubules [178, 179], or even a network of tubules [180]. These data imply the microvilli result from tubule followed by tubule fusion-eversion, rather than vesicle fusion to form the microvillus. Stimulation of secretion with a tubular network structure presumably results in eversion of this network to form the microvillar network of the stimulated secretory canaliculus without the need even for individual tubule fusion events [181, 182].

Thus, regulation of acid secretion in gastric parietal cells reflects the appearance of the pump in the microvilli of the secretory canaliculus as compared with a cytoplasmic location. [181, 182]. As mentioned above, a tyrosine-based motif in a cytoplasmic tail of the β-subunit appears to be responsible for the H+,K+-ATPase relocation to the cytoplasmic compartment of the cell [56]. When this tyrosine was mutated to alanine and mutant β-subunit was expressed in mice, transgenic animals continuously secreted acid in a stimulus-independent manner. It was suggested that the same motif might be recognized as a basolateral sorting signal when the H+,K+-ATPase β-subunit was expressed in MDCK cells [183] but the experimental data did not support this suggestion [184].

Parietal cells, similar to other polarized cells, sort proteins into two delivery routes, either to apical or basolateral domains of the plasma membrane. Newly synthesized proteins are sorted to the trans-Golgi network and recycling proteins are sorted in endosomes. Sorting is based on the presence of intrinsic sorting signals that are recognized by specific sorting machinery [185–187]. Information on the sorting signals within the H^+,K^+-ATPase subunits has been inferred from expression studies.

When expressed in LLC-PK1 cells, which do not contain tubulovesicular elements, the gastric H^+,K^+-ATPase is located exclusively on the apical membrane [188]. Studies in which the H^+,K^+-ATPase β-subunit or chimeric complexes of the H^+,K^+-ATPase α- or β-subunit with the appropriate subunit of the Na^+,K^+-ATPase were expressed in LLC-PK1 cells have shown that both α- and β-subunits of the H^+,K^+-ATPase contain apical trafficking signals [183, 188].

The apical sorting signal in the α-subunit appears to reside in the fourth transmembrane domain and to act through long-range interactions with its flanking cytoplasmic loop domains [189]. However, the nature of the apical sorting signal(s) within the β-subunit remains unknown. N-Glycosylation sites might be considered as potential candidates for the apical signals since the gastric H^+,K^+-ATPase β-subunit is heavily glycosylated and N-glycosylation sites act as apical signals in several secreted and membrane proteins [190–192]. If expressed in non-polarized HEK-293 cells, the H^+,K^+-ATPase is localized on the plasma membrane [102, 193]. Mutational studies in HEK-293 cells have shown that six of the seven glycosylation sites in the gastric H^+,K^+-ATPase β-subunit are essential for trans-Golgi network

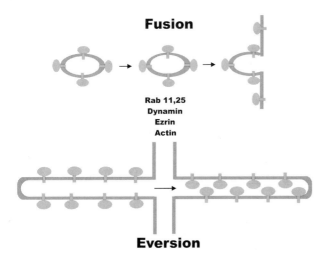

Figure 7.7 Hypothetical models of stimulation of acid secretion. The vesicle fusion hypothesis suggests that upon stimulation the H^+,K^+-ATPase containing vesicles relocate from the cytoplasm to the microvilli and fuse with the apical membrane, with a contribution of cytoskeletal components and SNARE proteins. The tubule-eversion model suggests that stimulation of acid secretion results in fusion and eversion of the H^+,K^+-ATPase containing tubules rather than individual vesicle fusion events.

to plasma membrane trafficking of the H⁺,K⁺-ATPase β-subunit in HEK-293 cells [194]. The second glycosylation site (Asn103), which is not conserved among the β-subunits from different species, is not critical for the plasma delivery of the protein. The trafficking step that is affected by the removal of N-glycosylation sites is the route from the Golgi to the plasma membrane and not from ER to Golgi. It is possible that, similar to polarized cells, HEK-293 cells have machinery which recognizes sorting signal(s) and places the β-subunit into specific cargo vesicles in trans-Golgi network that deliver the protein to the plasma membrane.

There is increasing evidence that apical and basolateral sorting and trafficking pathways exist not only in polarized but also in non-polarized cells [195]. When apical and basolateral proteins are expressed in non-polarized cells, in trans-Golgi network they are sorted into different containers that travel separately until they fuse with plasma membrane [196–199] (Figure 7.7). This implies that trans-Golgi network to plasma membrane delivery in HEK-293 cells depends on the presence or absence of apical sorting information encoded by particular glycosylation sites within the extracellular domain of the β-subunit [200]. The decrease in apical surface expression of the H⁺,K⁺-ATPase β-subunit in polarized LLC-PK1 due to the mutation of N-glycosylation sites support this conclusion [201].

References

1. M. Maeda, J. Ishizaki, and M. Futai, *Biochem. Biophys. Res. Commun.*, **1988**, *157*, 203–209.
2. G. E. Shull and J. B. Lingrel, *J. Biol. Chem.*, **1986**, *261*, 16788–16791.
3. K. Bamberg, F. Mercier, M. A. Reuben, Y. Kobayashi, K. B. Munson, and G. Sachs, *Biochim. Biophys. Acta*, **1992**, *1131*, 69–77.
4. L. K. Lane, T. L. Kirley, and W. J. Ball, Jr., *Biochem. Biophys. Res. Commun.*, **1986**, *138*, 185–192.
5. M. Maeda, K. Oshiman, S. Tamura, and M. Futai, *J. Biol. Chem.*, **1990**, *265*, 9027–9032.
6. P. R. Newman, J. Greeb, T. P. Keeton, A. A. Reyes, and G. E. Shull, *DNA Cell. Biol.*, **1990**, *9*, 749–762.
7. K. Oshiman, K. Motojima, S. Mahmood, A. Shimada, and S. Tamura, et al., *FEBS Lett.*, **1991**, *281*, 250–254.
8. S. Tamura, M. Tagaya, M. Maeda, and M. Futai, *J. Biol. Chem.*, **1989**, *264*, 8580–8584.
9. M. Maeda, M. Tagaya, and M. Futai, *J. Biol. Chem.*, **1988**, *263*, 3652–3656.
10. M. O. Walderhaug, R. L. Post, G. Saccomani, R. T. Leonard, and D. P. Briskin, *J. Biol. Chem.*, **1985**, *260*, 3852–3859.
11. R. J. Jackson, J. Mendlein, and G. Sachs, *Biochim. Biophys. Acta*, **1983**, *731*, 9–15.
12. R. A. Farley and L. D. Faller, *Biochemistry*, **1989**, *28*, 5764–5772.
14. J. Kyte and R. F. Doolittle, *J. Mol. Biol.*, **1982**, *157*, 105–132.
15. D. R. Scott, K. Munson, N. Modyanov, and G. Sachs, *Biochim. Biophys. Acta*, **1992**, *1112*, 246–250.
16. J. M. Shin, M. Besancon, A. Simon, and G. Sachs, *Biochim. Biophys. Acta*, **1993**, *1148*, 223–233.
17. M. Besancon, J. M. Shin, F. Mercier, K. Munson, and M. Miller, et al., *Biochemistry*, **1993**, *32*, 2345–2355.
18. D. Bayle, J. C. Robert, K. Bamberg, F. Benkouka, and A. M. Cheret, et al., *J. Biol. Chem.*, **1992**, *267*, 19060–19065.

19. F. Mercier, D. Bayle, M. Besancon, T. Joys, and J. M. Shin, et al., *Biochim. Biophys. Acta*, **1993**, *1149*, 151–165.

20. A. J. Smolka, K. A. Larsen, and C. E. Hammond, *Biochem. Biophys. Res. Commun.*, **2000**, *273*, 942–947.

21. F. Mercier, H. Reggio, G. Devilliers, D. Bataille, and P. Mangeat, *J. Cell. Biol.*, **1989**, *108*, 441–453.

22. F. Mercier, H. Reggio, G. Devilliers, D. Bataille, and P. Mangeat, *Biol. Cell.*, **1989**, *65*, 7–20.

23. K. Bamberg and G. Sachs, *J. Biol. Chem.*, **1994**, *269*, 16909–16919.

24. D. Bayle, D. Weeks, and G. Sachs, *J. Biol. Chem.*, **1995**, *270*, 25678–25684.

25. K. Hall, G. Perez, D. Anderson, C. Gutierrez, and K. Munson, et al., *Biochemistry*, **1990**, *29*, 701–706.

26. B. H. Toh, P. A. Gleeson, R. J. Simpson, R. L. Moritz, and J. M. Callaghan, et al., *Proc. Natl. Acad. Sci. U. S. A.*, **1990**, *87*, 6418–6422.

27. C. T. Okamoto, J. M. Karpilow, A. Smolka, and J. G. Forte, *Biochim. Biophys. Acta*, **1990**, *1037*, 360–372.

28. J. M. Callaghan, B. H. Toh, J. M. Pettitt, D. C. Humphris, and P. A. Gleeson, *J. Cell. Sci.*, **1990**, *95*, 563–576.

29. J. M. Callaghan, B. H. Toh, R. J. Simpson, G. S. Baldwin, and P. A. Gleeson, *Biochem. J.*, **1992**, *283*, 63–68.

30. K. Hall, G. Perez, G. Sachs, and E. Rabon, *Biochim. Biophys. Acta*, **1991**, *1077*, 173–179.

31. E. C. Rabon and M. A. Reuben, *Annu. Rev. Physiol.*, **1990**, *52*, 321–344.

32. M. A. Reuben, L. S. Lasater, and G. Sachs, *Proc. Natl. Acad. Sci. U. S. A.*, **1990**, *87*, 6767–6771.

33. V. A. Canfield, C. T. Okamoto, D. Chow, J. Dorfman, and P. Gros, et al., *J. Biol. Chem.*, **1990**, *265*, 19878–19884.

34. G. E. Shull, *J. Biol. Chem.*, **1990**, *265*, 12123–12126.

35. M. Maeda, K. Oshiman, S. Tamura, S. Kaya, and S. Mahmood, et al., *J. Biol. Chem.*, **1991**, *266*, 21584–21588.

36. P. R. Newman and G. E. Shull, *Genomics*, **1991**, *11*, 252–262.

37. V. A. Canfield and R. Levenson, *Proc. Natl. Acad. Sci. U. S. A.*, **1991**, *88*, 8247–8251.

38. G. P. Morley, J. M. Callaghan, J. B. Rose, B. H. Toh, P. A. Gleeson, and I. R. Driel van, *J. Biol. Chem.*, **1992**, *267*, 1165–1174.

39. J. Y. Ma, Y. H. Song, S. E. Sjostrand, L. Rask, and S. Mardh, *Biochem. Biophys. Res. Commun.*, **1991**, *180*, 39–45.

40. D. C. Chow, C. M. Browning, and J. G. Forte, *Am. J. Physiol.*, **1992**, *263*, 39–46.

41. T. L. Kirley, *J. Biol. Chem.*, **1990**, *265*, 4227–4232.

42. K. J. Renaud, E. M. Inman, and D. M. Fambrough, *J. Biol. Chem.*, **1991**, *266*, 20491–20497.

43. P. Jaunin, J. D. Horisberger, K. Richter, P. J. Good, B. C. Rossier, and K. Geering, *J. Biol. Chem.*, **1992**, *267*, 577–585.

44. U. Ackermann and K. Geering, *FEBS Lett.*, **1990**, *269*, 105–108.

45. K. Geering, *Soc. Gen. Physiol. Ser.*, **1991**, *46*, 31–43.

46. C. J. Gottardi and M. J. Caplan, *J. Biol. Chem.*, **1993**, *268*, 14342–14347.

47. D. L. Roush, C. J. Gottardi, and M. J. Caplan, *Ann. New York Acad. Sci.*, **1994**, *733*, 212–222.

48. M. J. Caplan, *Curr. Opin. Cell. Biol.*, **1998**, *10*, 468–473.

49. L. A. Dunbar and M. J. Caplan, *Eur. J. Cell. Biol.*, **2000**, *79*, 557–563.

50. J. B. Koenderink, H. G. Swarts, H. P. Hermsen, and J. J. Pont De, *J. Biol. Chem.*, **1999**, *274*, 11604–11610.

51. H. P. Hermsen, H. G. Swarts, L. Wassink, F. J. Dijk, and M. T. Raijmakers, et al., *Biochim. Biophys. Acta*, **2000**, *1480*, 182–190.

52. K. Tyagarajan, R. R. Townsend, and J. G. Forte, *Biochemistry*, **1996**, *35*, 3238–3246.

53. K. Tyagarajan, P. H. Lipniunas, R. R. Townsend, and J. G. Forte, *Biochemistry*, **1997**, *36*, 10200–10212.

54. M. J. Treuheit, C. E. Costello, and T. L. Kirley, *J. Biol. Chem.*, **1993**, *268*, 13914–13919.

55. S. Asano, K. Kawada, T. Kimura, A. V. Grishin, M. J. Caplan, and N. Takeguchi, *J. Biol. Chem.*, **2000**, *275*, 8324–8330.

56. N. Courtois-Coutry, D. Roush, V. Rajendran, J. B. McCarthy, and J. Geibel, et al., *Cell*, **1997**, *90*, 501–510.

57. K. L. Scarff, L. M. Judd, B. H. Toh, P. A. Gleeson, and I. R. Driel Van, *Gastroenterology*, **1999**, *117*, 605–618.

58. S. Asano, T. Kimura, S. Ueno, M. Kawamura, and N. Takeguchi, *J. Biol. Chem.*, **1999**, *274*, 22257–22265.

59. M. V. Lemas, K. Takeyasu, and D. M. Fambrough, *J. Biol. Chem.*, **1992**, *267*, 20987–20991.

60. A. T. Beggah, P. Beguin, P. Jaunin, M. C. Peitsch, and K. Geering, *Biochemistry*, **1993**, *32*, 14117–14124.

61. J. D. Horisberger, P. Jaunin, M. A. Reuben, L. S. Lasater, and D. C. Chow, et al., *J. Biol. Chem.*, **1991**, *266*, 19131–19134.

62. J. M. Shin and G. Sachs, *J. Biol. Chem.*, **1994**, *269*, 8642–8646.

63. D. Melle-Milovanovic, M. Milovanovic, S. Nagpal, G. Sachs, and J. M. Shin, *J. Biol. Chem.*, **1998**, *273*, 11075–11081.

64. E. Rabon, M. Wilke, G. Sachs, and G. Zampighi, *J. Biol. Chem.*, **1986**, *261*, 1434–1439.

65. E. C. Rabon, R. D. Gunther, A. Bassilian, and E. S. Kempner, *J. Biol. Chem.*, **1988**, *263*, 16189–16194.

66. H. Hebert, Y. Xian, I. Hacksell, and S. Mardh, *FEBS Lett.*, **1992**, *299*, 159–162.

67. J. M. Shin and G. Sachs, *J. Biol. Chem.*, **1996**, *271*, 1904–1908.

68. J. C. Koster, G. Blanco, and R. W. Mercer, *J. Biol. Chem.*, **1995**, *270*, 14332–14339.

69. W. W. Reenstra and J. G. Forte, *J. Membr. Biol.*, **1981**, *61*, 55–60.

70. G. S. Smith and P. B. Scholes, *Biochim. Biophys. Acta*, **1982**, *688*, 803–807.

71. S. Mardh and L. Norberg, *Acta Physiol. Scand. Suppl.*, **1992**, *607*, 259–263.

72. E. C. Rabon, T. L. McFall, and G. Sachs, *J. Biol. Chem.*, **1982**, *257*, 6296–6299.

73. A. T. Skrabanja, H. T. Hijden van der, and J. J. Pont De, *Biochim. Biophys. Acta*, **1987**, *903*, 434–440.

74. B. Wallmark, H. B. Stewart, E. Rabon, G. Saccomani, and G. Sachs, *J. Biol. Chem.*, **1980**, *255*, 5313–5319.

75. J. M. Shin, R. Goldshleger, K. B. Munson, G. Sachs, and S. J. Karlish, *J. Biol. Chem.*, **2001**, *276*, 48440–48450.

76. E. Rabon, G. Sachs, S. Bassilian, C. Leach, and D. Keeling, *J. Biol. Chem.*, **1991**, *266*, 12395–12401.

77. E. C. Rabon, K. Smillie, V. Seru, and R. Rabon, *J. Biol. Chem.*, **1993**, *268*, 8012–8018.

78. P. Brzezinski, B. G. Malmstrom, P. Lorentzon, and B. Wallmark, *Biochim. Biophys. Acta*, **1988**, *942*, 215–219.

79. B. Stewart, B. Wallmark, and G. Sachs, *J. Biol. Chem.*, **1981**, *256*, 2682–2690.

80. P. Lorentzon, G. Sachs, and B. Wallmark, *J. Biol. Chem.*, **1988**, *263*, 10705–10710.

81. J. Mendlein and G. Sachs, *J. Biol. Chem.*, **1989**, *264*, 18512–18519.

82. C. Polvani, G. Sachs, and R. Blostein, *J. Biol. Chem.*, **1989**, *264*, 17854–17859.

83. P. Lorentzon, D. Scott, S. Hersey, B. Wallmark, E. Rabon, and G. Sachs, *Prog. Clin. Biol. Res.*, **1988**, *273*, 247–254.

84. L. D. Faller and R. A. Diaz, *Biochemistry*, **1989**, *28*, 6908–6914.

85. L. D. Faller, *Biochemistry*, **1989**, *28*, 6771–6778.

86. K. Abe, S. Kaya, T. Imagawa, and K. Taniguchi, *Biochemistry*, **2002**, *41*, 2438–2445.

87. D. W. Martin and J. R. Sachs, *Biochemistry*, **1999**, *38*, 7485–7497.

88. H. Eguchi, S. Kaya, and K. Taniguchi, *Biochem. Biophys. Res. Commun.*, **1993**, *196*, 294–300.

89. S. Markus, Z. Priel, and D. M. Chipman, *Biochemistry*, **1989**, *28*, 793–799.

90. S. J. Karlish, *J. Bioenerg. Biomembr.*, **1980**, *12*, 111–136.

91. I. N. Smirnova and L. D. Faller, *J. Biol. Chem.*, **1993**, *268*, 16120–16123.

92. K. B. Munson, C. Gutierrez, V. N. Balaji, K. Ramnarayan, and G. Sachs, *J. Biol. Chem.*, **1991**, *266*, 18976–18988.

93. R. Goldshleger and S. J. Karlish, *J. Biol. Chem.*, **1999**, *274*, 16213–16221.

94. R. Goldshleger and S. J. Karlish, *Proc. Natl. Acad. Sci. U. S. A.*, **1997**, *94*, 9596–9601.

95. C. Toyoshima and H. Nomura, *Nature*, **2002**, *418*, 605–611.

96. S. J. Karlish, R. Goldshleger, and W. D. Stein, *Proc. Natl. Acad. Sci. U. S. A.*, **1990**, *87*, 4566–4570.

97. C. Gatto, S. Lutsenko, J. M. Shin, G. Sachs, and J. H. Kaplan, *J. Biol. Chem.*, **1999**, *274*, 13737–13740.

98. M. Besancon, A. Simon, G. Sachs, and J. M. Shin, *J. Biol. Chem.*, **1997**, *272*, 22438–22446.

99. O. Vagin, S. Denevich, K. Munson, and G. Sachs, *Biochemistry*, **2002**, *41*, 12755–12762.

100. S. Asano, S. Matsuda, S. Hoshina, S. Sakamoto, and N. Takeguchi, *J. Biol. Chem.*, **1999**, *274*, 6848–6854.

101. K. B. Munson, N. Lambrecht, and G. Sachs, *Biochemistry*, **2000**, *39*, 2997–3004.

102. N. Lambrecht, K. Munson, O. Vagin, and G. Sachs, *J. Biol. Chem.*, **2000**, *275*, 4041–4048.

103. O. Vagin, K. Munson, N. Lambrecht, S. J. Karlish, and G. Sachs, *Biochemistry*, **2001**, *40*, 7480–7490.

104. E. C. Rabon, M. Hoggatt, and K. Smillie, *J. Biol. Chem.*, **1996**, *271*, 32137–32146.

105. H. G. Swarts, C. H. Klaassen, M. Boer de, J. A. Fransen, and J. J. Pont De, *J. Biol. Chem.*, **1996**, *271*, 29764–29772.

106. H. P. Hermsen, J. B. Koenderink, H. G. Swarts, and J. J. Pont De, *Biochemistry*, **2000**, *39*, 1330–1337.

107. H. P. Hermsen, H. G. Swarts, L. Wassink, J. B. Koenderink, P. H. Willems, and J. J. Pont De, *Biochemistry*, **2001**, *40*, 6527–6533.

108. H. G. Swarts, J. B. Koenderink, H. P. Hermsen, P. H. Willems, and J. J. Pont De, K*J. Biol. Chem.*, **2001**, *276*, 36909–36916.

109. M. Burnay, G. Crambert, S. Kharoubi-Hess, K. Geering, and J. D. Horisberger, *J. Biol. Chem.*, **2003**, *278*, 19237–19244.

110. S. Asano, T. Io, T. Kimura, S. Sakamoto, and N. Takeguchi, *J. Biol. Chem.*, **2001**, *276*, 31265–31273.

111. M. Mense, V. Rajendran, R. Blostein, and M. J. Caplan, *Biochemistry*, **2002**, *41*, 9803–9812.

112. H. Hebert, P. Purhonen, H. Vorum, K. Thomsen, and A. B. Maunsbach, *J. Mol. Biol.*, **2001**, *314*, 479–494.

113. H. Hebert, E. Skriver, M. Soderholm, and A. B. Maunsbach, *J. Ultrastruct. Mol. Struct. Res.*, **1988**, *100*, 86–93.

114. Y. Tahara, S. Ohnishi, Y. Fujiyoshi, Y. Kimura, and Y. Hayashi, *FEBS Lett.*, **1993**, *320*, 17–22.

115. C. Toyoshima, M. Nakasako, H. Nomura, and H. Ogawa, *Nature*, **2000**, *405*, 647–655.

116. K. J. Sweadner and C. Donnet, *Biochem. J.*, **2001**, *356*, 685–704.

117. A. G. Lee, *Biochim. Biophys. Acta*, **2002**, *1565*, 246–266.

118. P. L. Jorgensen, K. O. Hakansson, and S. J. Karlish, *Annu. Rev. Physiol.*, **2003**, *65*, 817–849.

119. C. Xu, W. J. Rice, W. He, and D. L. Stokes, *J. Mol. Biol.*, **2002**, *316*, 201–211.

120. H. Ma, G. Inesi, and C. Toyoshima, *J. Biol. Chem.*, **2003**, *278*, 28938–28943.

121. W. J. Rice, H. S. Young, D. W. Martin, J. R. Sachs, and D. L. Stokes, *Biophys. J.*, **2001**, *80*, 2187–2197.

122. M. Hilge, G. Siegal, G. W. Vuister, P. Guntert, S. M. Gloor, and J. P. Abrahams, *Nat. Struct. Biol.*, **2003**, *10*, 468–474.

123. W. Kuhlbrandt, J. Zeelen, and J. Dietrich, *Science*, **2002**, *297*, 1692–1696.

124. M. L. Gumz, D. Duda, R. McKenna, C. S. Wingo, and B. D. Cain, Molecular modeling of the rabbit colonic (HKalpha2a) H(+), K(+) ATPase. *J. Mol. Model. (Online)*, **2003**, *9*, 283–289.

125. H. Ogawa and C. Toyoshima, *Proc. Natl. Acad. Sci. U. S. A.*, **2002**, *99*, 15977–15982.

126. K. Munson, O. Vagin, G. Sachs, and S. Karlish, *Ann. New York Acad. Sci.*, **2003**, *986*, 106–110.

127. E. L. Burns, R. A. Nicholas, and E. M. Price, *J. Biol. Chem.*, **1996**, *271*, 15879–15883.

128. H. F. Helander and B. I. Hirschowitz, *Gastroenterology*, **1972**, *63*, 951–961.

129. H. F. Helander and B. I. Hirschowitz, *Gastroenterology*, **1974**, *67*, 447–452.

130. A. Sawaguchi, K. L. McDonald, S. Karvar, and J. G. Forte, *J. Microsc.*, **2002**, *208*, 158–166.

131. J. M. Wolosin and J. G. Forte, *J. Membr. Biol.*, **1983**, *76*, 261–268.

132. G. Sachs, H. H. Chang, E. Rabon, R. Schackman, M. Lewin, and G. Saccomani, *J. Biol. Chem.*, **1976**, *251*, 7690–7698.

133. D. R. Scott, H. F. Helander, S. J. Hersey, and G. Sachs, *Biochim. Biophys. Acta*, **1993**, *1146*, 73–80.

134. S. Tamura, K. Oshiman, T. Nishi, M. Mori, M. Maeda, and M. Futai, *FEBS Lett.*, **1992**, *298*, 137–141.

135. M. Kaise, A. Muraoka, J. Yamada, and T. Yamada, *J. Biol. Chem.*, **1995**, *270*, 18637–18642.

136. A. Tari, V. Wu, M. Sumii, G. Sachs, and J. H. Walsh, *Biochim. Biophys. Acta*, **1991**, *1129*, 49–56.

137. A. Tari, G. Yamamoto, K. Sumii, M. Sumii, and Y. Takehara, et al., *Am. J. Physiol.*, **1993**, *265*, 752–758.

138. A. Tari, G. Yamamoto, Y. Yonei, M. Sumii, and K. Sumii, et al., *Am. J. Physiol.*, **1994**, *266*, 444–450.

139. D. R. Scott, M. Besancon, G. Sachs, and H. Helander, *Dig. Dis. Sci.*, **1994**, *39*, 2118–2126.

140. J. M. Crothers, J r., D. C. Chow and J. G. Forte, *Am. J. Physiol.*, **1993**, *265*, 231–241.

141. E. Fellenius, T. Berglindh, G. Sachs, L. Olbe, and B. Elander, et al., *Nature*, **1981**, *290*, 159–161.

142. B. Wallmark, A. Brandstrom, and H. Larsson, *Biochim. Biophys. Acta*, **1984**, *778*, 549–558.

143. H. Nagaya, H. Satoh, K. Kubo, and Y. Maki, *J. Pharmacol. Exp. Ther.*, **1989**, *248*, 799–805.

144. H. Fujisaki, H. Shibata, K. Oketani, M. Murakami, and M. Fujimoto, et al., *Biochem. Pharmacol.*, **1991**, *42*, 321–328.

145. J. C. Sih, W. B. Im, A. Robert, D. R. Graber, and D. P. Blakeman, *J. Med. Chem.*, **1991**, *34*, 1049–1062.

146. W. A. Simon, C. Budingen, S. Fahr, B. Kinder, and M. Koske, *Biochem. Pharmacol.*, **1991**, *42*, 347–355.

147. T. Arakawa, T. Fukuda, K. Higuchi, K. Koike, H. Satoh, and K. Kobayashi, *Jpn. J. Pharmacol.*, **1993**, *61*, 299–302.

148. P. Lindberg, P. Nordberg, T. Alminger, A. Brandstrom, and B. Wallmark, *J. Med. Chem.*, **1986**, *29*, 1327–1329.

149. D. J. Keeling, C. Fallowfield, K. J. Milliner, S. K. Tingley, R. J. Ife, and A. H. Underwood, *Biochem. Pharmacol.*, **1985**, *34*, 2967–2973.

150. W. B. Im, J. C. Sih, D. P. Blakeman, and J. P. McGrath, *J. Biol. Chem.*, **1985**, *260*, 4591–4597.

151. P. Lorentzon, R. Jackson, B. Wallmark, and G. Sachs, *Biochim. Biophys. Acta*, **1987**, *897*, 41–51.

152. D. J. Keeling, C. Fallowfield, and A. H. Underwood, *Biochem. Pharmacol.*, **1987**, *36*, 339–344.

153. M. Morii and N. Takeguchi, *J. Biol. Chem.*, **1993**, *268*, 21553–21559.

154. G. Sachs, J. M. Shin, M. Besancon, and C. Prinz, *Aliment. Pharmacol. Ther. 7*, **1993**, *29*, 4–12.

155. M. Katashima, K. Yamamoto, Y. Tokuma, T. Hata, Y. Sawada, and T. Iga, *Eur. J. Drug. Metab. Pharm.*, **1998**, *23*, 19–26.

156. K. Gedda, D. Scott, M. Besancon, P. Lorentzon, and G. Sachs, *Gastroenterology*, **1995**, *109*, 1134–1141.

157. W. B. Im, D. P. Blakeman, and G. Sachs, *Biochim. Biophys. Acta*, **1985**, *845*, 54–59.

158. J. M. Shin and G. Sachs, *Gastroenterology*, **2002**, *123*, 1588–1597.

159. J. J. Kaminski and A. M. Doweyko, *J. Med. Chem.*, **1997**, *40*, 427–436.

160. Y. Hioki, J. Takada, Y. Hidaka, H. Takeshita, M. Hosoi, and M. Yano, *Arch. Int. Pharmacol. Ther.*, **1990**, *305*, 32–44.

161. R. J. Ife, T. H. Brown, D. J. Keeling, C. A. Leach, and M. L. Meeson, et al., *J. Med. Chem.*, **1992**, *35*, 3413–3422.

162. C. A. Leach, T. H. Brown, R. J. Ife, D. J. Keeling, and S. M. Laing, et al., *J. Med. Chem.*, **1992**, *35*, 1845–1852.

163. J. L. LaMattina, P. A. McCarthy, L. A. Reiter, W. F. Holt, and L. A. Yeh, *J. Med. Chem.*, **1990**, *33*, 543–552.

164. K. S. Han, Y. G. Kim, J. K. Yoo, J. W. Lee, and M. G. Lee, *Biopharm. Drug. Dispos.*, **1998**, *19*, 493–500.

165. T. Hayashi, S. Asano, M. Mizutani, N. Takeguchi, and T. Kojima, et al., *J. Nat. Prod.*, **1991**, *54*, 802–809.

166. S. Asano, M. Mizutani, T. Hayashi, N. Morita, and N. Takeguchi, *J. Biol. Chem.*, **1990**, *265*, 22167–22173.

167. S. Murakami, I. Arai, M. Muramatsu, S. Otomo, and K. Baba, et al., *Biochem. Pharmacol.*, **1992**, *44*, 33–37.

168. A. H. Dantzig, P. L. Minor, J. L. Garrigus, D. S. Fukuda, and J. S. Mynderse, *Biochem. Pharmacol.*, **1991**, *42*, 2019–2026.

169. D. J. Keeling, A. G. Taylor, and C. Schudt, *J. Biol. Chem.*, **1989**, *264*, 5545–5551.

170. D. J. Keeling, C. Fallowfield, K. M. Lawrie, D. Saunders, S. Richardson, and R. J. Ife, *J. Biol. Chem.*, **1989**, *264*, 5552–5558.

171. B. Wallmark, C. Briving, J. Fryklund, K. Munson, and R. Jackson, et al., *J. Biol. Chem.*, **1987**, *262*, 2077–2084.

172. J. Mendlein and G. Sachs, *J. Biol. Chem.*, **1990**, *265*, 5030–5036.

173. T. Forte and J. G. Forte, *J. Ultrastruct. Res.*, **1971**, *37*, 322–334.

174. A. Smolka, H. F. Helander, and G. Sachs, *Am. J. Physiol.*, **1983**, *245*, 589–596.

175. P. Mangeat, T. Gusdinar, A. Sahuquet, D. K. Hanzel, J. G. Forte, and R. Magous, *Biol. Cell.*, **1990**, *69*, 223–231.

176. D. Hanzel, H. Reggio, A. Bretscher, J. G. Forte, and P. Mangeat, *Embo J.*, **1991**, *10*, 2363–2373.

177. X. Yao and J. G. Forte, *Annu. Rev. Physiol.*, **2003**, *65*, 103–131.

178. G. Peranzi, D. Bayle, M. J. Lewin, and A. Soumarmon, *Biol. Cell.*, **1991**, *73*, 163–171.

179. T. Jons, S. Lehnardt, H. Bigalke, H. K. Heim, and G. Ahnert-Hilger, *Eur. J. Cell. Biol.*, **1999**, *78*, 779–786.

180. T. Namikawa, K. Araki, and T. Ogata, *Arch. Histol. Cytol.*, **1998**, *61*, 47–56.

181. T. Ogata and Y. Yamasaki, *Microsc. Res. Tech.*, **2000**, *48*, 282–292.

182. T. Ogata and Y. Yamasaki, *Anat. Rec.*, **2000**, *258*, 15–24.

183. D. L. Roush, C. J. Gottardi, H. Y. Naim, M. G. Roth, and M. J. Caplan, *J. Biol. Chem.*, **1998**, *273*, 26862–26869.

184. A. S. Duffield, A. N. Brown, H. Folsch, and I. C. Mellman. Presented at the *42nd American Society for Cell Biology Annual Meeting*, San Francisco, **2002**.

185. L. M. Traub and S. Kornfeld, *Curr. Opin. Cell. Biol.*, **1997**, *9*, 527–533.

186. C. Yeaman, K. K. Grindstaff, and W. J. Nelson, *Physiol. Rev.*, **1999**, *79*, 73–98.

187. E. Ikonen and K. Simons, *Semin. Cell. Dev. Biol.*, **1998**, *9*, 503–509.

188. C. J. Gottardi and M. J. Caplan, *J. Cell. Biol.*, **1993**, *121*, 283–293.

189. L. A. Dunbar, P. Aronson, and M. J. Caplan, *J. Cell. Biol.*, **2000**, *148*, 769–778.

190. K. Matter and I. Mellman, *Curr. Opin. Cell. Biol.*, **1994**, *6*, 545–554.

191. P. Scheiffele, J. Peranen, and K. Simons, *Nature*, **1995**, *378*, 96–98.

192. A. Gut, F. Kappeler, N. Hyka, M. S. Balda, H. P. Hauri, and K. Matter, *EMBO J.*, **1998**, *17*, 1919–1929.

193. T. Kimura, Y. Tabuchi, N. Takeguchi, and S. Asano, *J. Biol. Chem.*, **2002**, *277*, 20671–20677.

194. O. Vagin, S. Denevich, and G. Sachs, *Am. J. Physiol. Cell. Physiol.*, **2003**, *285*, 968–976.

195. P. Keller and K. Simons, *J. Cell. Sci.*, **1997**, *110*, 3001–3009.

196. P. Keller, D. Toomre, E. Diaz, J. White, and K. Simons, *Nat. Cell. Biol.*, **2001**, *3*, 140–149.

197. A. Musch, H. Xu, D. Shields, and E. Rodriguez-Boulan, *J. Cell. Biol.*, **1996**, *133*, 543–558.

198. A. Rustom, M. Bajohrs, C. Kaether, P. Keller, and D. Toomre, et al., *Traffic*, **2002**, *3*, 279–288.

199. T. Yoshimori, P. Keller, M. G. Roth, and K. Simons, *J. Cell. Biol.*, **1996**, *133*, 247–256.

200. F. T. Wieland, M. L. Gleason, T. A. Serafini, and J. E. Rothman, *Cell*, **1987**, *50*, 289–300.

201. O. Vagin, S. Turdikulova, I. Yakubov, and G. Sachs. *Molecular Biology of the Cell*, **2003**, *14*, p. 80a. (Abstracts of the 43rd Annual Meeting of the American Society for Cell Biology, Dec 13–17, 2003, San Francisco.)

8
Plasma Membrane Calcium Pumps

Ernesto Carafoli, Luisa Coletto, and *Marisa Brini*

8.1
Introduction

The transport of Ca^{2+} out of the cytosol of eukaryotic cells, which is essential to the maintenance of cellular Ca^{2+} homeostasis, is accomplished by two systems: a low-affinity, high-capacity Na^+/Ca^{2+} exchanger, which is particularly active in excitable tissues, and a high-affinity, low-capacity Ca^{2+}-ATPase (the plasma membrane Ca^{2+} pump, PMCA), which is active in all eukaryotic cells. The high affinity of the ATPase enables it to interact with Ca^{2+} with adequate efficiency even when its concentration is at the very low level prevailing in the cytosol of cells at rest (100–200 nM). Thus, the PMCA pump is the fine tuner of cell Ca^{2+}: It counteracts the action of the plasma membrane channels, across which a limited and carefully controlled amount of Ca^{2+} penetrates into the cytosol [1]. The ATPase was discovered in erythrocytes in 1966 [2] and was later characterized as a P-type pump [3]. In the nearly 40 years that have elapsed since its discovery, the work has developed as with other transport ATPases, gradually evolving from an initial phase focused on the properties of the transport process and on the reaction mechanism to a later phase in which the enzyme was dissected molecularly and characterized genetically. Knowledge on the PMCA pump has progressed rapidly, particularly in recent years, establishing the enzyme as a central actor in the precise control of Ca^{2+} homeostasis in the cells, and thus in their proper functioning. Several reviews, comprehensively covering earlier research on the pump or focusing on particular aspects, have appeared over the years [4–13]. They should be consulted to complement the information contained in the present chapter, which will deal in a comparatively rapid way with the earlier work on the pump, to concentrate on more recent research achievements.

Handbook of ATPases. Edited by M. Futai, Y. Wada, J. H. Kaplan
Copyright © 2004 WILEY-VCH Verlag GmbH & Co. KGaA, Weinheim
ISBN 3-527-30689-7

8.2
Pre-cloning Period

8.2.1
Early History

The pump was discovered in 1966 by Schatzmann [2]. In a now classical experiment, he found that Ca^{2+} "emerged" from erythrocyte ghosts loaded with Ca^{2+} and ATP at a rate faster than from ghosts not loaded with ATP. The requirement for an energy source, and the fact that Ca^{2+} was exported to a medium in which its concentration was higher than in the ghosts made it clear that a process of active transport, i.e., a pump, was involved. Subsequently, this initial observation was extended to numerous other cell types, and the reaction mechanism of the pump was studied in detail. It was soon established that the system operated according to the principles of P-type ATPases, i.e., it formed an aspartyl-phosphate intermediate during the reaction cycle (Figure 8.1). Since the intermediate is sufficiently stable it can be detected and characterized in phosphate polyacrylamide gels, allowing a simple means to assess the activity of the pump. At variance with the Ca^{2+} pump of the sarcoplasmic reticulum, which transports two Ca^{2+} ions per molecule of ATP hydrolyzed, the Ca^{2+}/ATP stoichiometry of the PMCA pump is 1.0. The phosphorylated pump exists in two conformational states, termed E_1-P and E_2-P, Ca^{2+} being most likely translocated across the protein in the transition from E_1-P to E_2-P. Basically, in the E_1 state the Ca^{2+}-binding site has higher Ca^{2+} affinity, whereas in the E_2 state its Ca^{2+} affinity drops. The type of detailed structural information now available on the SERCA pump and on the location and dynamics of its Ca^{2+} binding sites is not yet available for the PMCA enzyme. Thus, it can only be suggested that the single Ca^{2+} binding site is located on the cytosolic site of the membrane prior to the translocation step, and faces the exterior of the cell at its end. After re-

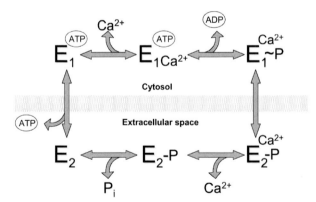

Figure 8.1 Catalytic cycle of the PMCA pump. E_1 and the E_2 notations correspond to the two major conformational states of the pump. The symbols "~" and "–" in the phosphorylated intermediates indicate, respectively, the high and low energy content of the complex between the pump and the phosphate.

lease of Ca^{2+} to the exterior of the cell, the pump becomes dephosphorylated and returns to the E_1 conformation. All arrows in the catalytic cycle of Figure 8.1 are doubly barbed to indicate that the cycle is reversible, i. e., the phosphorylated intermediate can be formed, under appropriate conditions, from inorganic phosphate as well [14]. As expected, the two general inhibitors of P-type ATPases orthovanadate ($[VO_3(OH)]^{2-}$) and La^{3+} inhibit the pump. While the action of orthovanadate, which is a stereo-analogue of phosphate, appears to be canonical, i. e., it interacts with the unphosphorylated E_2 form of the pump blocking the final $E_2 \rightarrow E_1$ transition, that of La^{3+} is peculiar, and deserves a special comment. At variance with all other P-type pumps, in which La^{3+} decreases the steady-state concentration of the aspartyl-phosphate intermediate, with the PMCA pump La^{3+} instead increases it. The effect could be conveniently attributed to the inhibition of the hydrolysis of the intermediate by La^{3+}. In addition to being mechanistically interesting, the effect is of practical value in laboratory work, since it allows the detection of the PMCA intermediate in the phosphate gels of membrane protein preparations that contain, as is customary, much larger amounts of other P-type pumps.

8.2.2
Regulation of the Pump

One important aspect of the PMCA pump which distinguishes it from the SERCA pump, and from all other P-type transport ATPases, is the multiplicity of regulatory mechanisms that act on it. The pump regulator par excellence is generally considered to be calmodulin, whose stimulatory effect was discovered about 25 years ago by Gopinath and Vicenzi [15] and Jarrett and Penniston [16]. Basically, calmodulin decreases the apparent K_m, of the pump for Ca^{2+} very significantly, bringing it down from values in excess of 10 μM to 0.4–0.5 μM. However, in addition to calmodulin, in the cell environment other factors could also contribute to the activation process. Of particular interest are acidic phospholipids (and long-chain polyunsaturated fatty acids), whose stimulatory effect on the pump in the red cell membrane was discovered by Ronner et al. [17] and defined in detail by Niggli et al. [18] in one of the first studies on the purified pump (see below). Niggli et al. reconstituted the purified enzyme in liposomes containing various proportions of acidic and zwitterionic phospholipids, and found that the pump was activated to about 50 % of the value attained with calmodulin when the liposomal membrane contained between 20–40 % acidic phospholipids. When transferring this information to the original erythrocyte membrane from which the enzyme had been purified, it was calculated that the native phospholipid ambient would permanently activate the pump to about 50 % of maximum. Kinase-mediated phosphorylations have also been claimed to activate the pump. The effect was first found in the heart sarcolemma membrane [19] and attributed to the cAMP dependent protein kinase, since it was removed by PKA inhibitors. The phosphorylation step increased both the V_{max} and the Ca^{2+} affinity of the pump, decreasing the K_m (Ca^{2+}) to about 1 μM in the absence of calmodulin. Later work has identified a consensus site for protein kinase A (KRNSS) downstream of the calmodulin-binding domain in the C-terminal region of one of the isoforms of the pump (PMCA1) [20]. S1178 in this site was phosphorylated by PKA: although potential PKA consensus sites (DKAS in iso-forms 2 and 4) are present in the C-terminal region of other isoforms it is not known whether they are phosphorylated. The matter of the activation by protein kinase C, which was originally described by Smallwood et al. [21], is more complex, since the magnitude of the activation reported in different studies has varied very significantly. One of the consensus sites for protein kinase C was identified around a threonine within the C-terminal calmodulin-binding domain [22]. Phosphoryla-tion of this site may be important in preventing the autoinhibitory association of the calmodulin-binding domain with the region of the pump that contains the active site [23] (see below). The other two activating mechanisms that have been described are limited proteolysis and oligomerization. Several proteases, e. g., tryp-sin, catalyze the controlled degradation of the pump, inducing its irreversible acti-vation. The activation of trypsin, however interesting as an experimental tool, has no physiological meaning. Activation of the pump by the controlled degradation induced by the intracellular Ca^{2+}-dependent protease calpain may instead have physiological significance. The problem with calpain is its dependence on Ca^{2+} concentrations that must be at least in the µM range, and are thus presumably never experienced by cells under normal conditions. Possibly, therefore, calpain de-gradation only occurs under conditions of pathological cytosolic Ca^{2+} overload, as a means to provide suffering cells with a more efficient means to extrude Ca^{2+}. Unfortunately, the relief is only temporary, since the truncated pump product initially generated by calpain is further degraded by other proteases to smaller, in-active fragments.

Activation by oligomerization is related to the suggestion that the PMCA pump (as other P-type pumps) may function in the membrane as a dimer [24]. Self-asso-ciation (oligomerization) has been shown to occur through the calmodulin-binding domain [25] and to make the pump calmodulin insensitive. Since self-association only occurs at concentrations of the pump much higher than those that can be expected in the native membrane environment [26] its physiological meaning appears doubtful. However, the pump has been shown to be concentrated in the caveolae [27, 28], where conditions favouring the self-association process could pos-sibly prevail.

8.2.3
Isolated Pump

The purification of the pump from erythrocyte ghosts in 1979 was a turning point for research in the area [29]. The purification procedure exploited the activation of the pump by calmodulin, isolating it from calmodulin columns as an active protein of molecular mass about 140 KDa, on which a number of important properties, in-cluding some mentioned above, were established. The ATPase was soon isolated from other cell types as well, and successfully reconstituted as an active pump in li-posomes. An important development of the work on liposomes was the detailed analysis of the role of acidic phospholipids as activators of the pump [18]. In the

first purification experiments the pump was maximally active in the absence of calmodulin, a puzzling observation which was rapidly shown to be due to the presence
of acidic phospholipids in the purification column. Their replacement with phos-
phatidyl-choline soon yield purified pump preparations that were fully responsive
to calmodulin. The early work on the purified enzyme also established that the doubly phosphorylated derivative of phosphatidyl-inositol (PIP_2) was the most effective
among activatory phospholipids [6]. Another important result of the early work on
the purified enzyme was the precise definition of the pattern of activatory proteolysis
by trypsin [30], and, especially, by calpain (see above). The latter protease removes in
a first cut about half of the calmodulin-binding domain and the portion of the pump
downstream of it. A second cut then truncates the pump at the beginning of the calmodulin-binding domain [31]. Later on, in combination with data on the primary
structure of the pump (see below), the resulting 124 kDa fragment, which is fully
active in the absence of calmodulin and which can be easily purified since it is
not retained by calmodulin columns, has been used to demonstrate that the calmodulin-binding sequence acts as an autoinhibitory domain that interacts with two "receptors" in the cytosolic portion of the pump close to the active site, maintaining the
pump inactive [32, 33] (Figure 8.2). One "receptor" domain is located between the
catalytic D residue and the K that is part of the ATP binding site in the main cytosolic
unit of the pump. The other is located in the cytosolic unit protruding from the second transmembrane domain of the enzyme. Calmodulin activates the pump by interacting with its binding domain, removing it from the cytosolic "receptor" site.
Later studies then clarified the molecular details of the interaction of calmodulin
with its binding domain [34].

An important result of the work on the purified pump performed prior to the
solution of its primary structure in 1988 was the identification and characterization
of three crucial functional domains: The fluorescent isothiocyanate-binding domain, which is part of the ATP binding site [35], the phosphorylation domain,
which has the canonical sequence CSDKTG [36], and, especially, the calmodulin-
binding domain [37]. To identify the latter domain use was made a bifunctional,
photoactivatable, radioactive, cleavable cross-linker [38, 39], which was conjugated
to K residues in calmodulin – through an oxysuccinimide moiety located at one of
its ends it was covalently attached to the purified erythrocyte pump through photo-
activation of the aryl-azide at its opposite end. The cross-linker was then cleaved
with dithionite at the azo linkage in the middle, after which calmodulin was removed by dialysis, leaving the pump radioactively labeled in the calmodulin-binding domain. Cleavage with CNBr generated a labeled 33-residue peptide which had
the predominance of basic residues alternating with hydrophobic residues, which
had been found in the few – still putative – calmodulin-binding domains known at
that time. It also had the frequently conserved W residue in its N-terminal portion,
and showed a strong propensity to form an amphiphilic helix. The location of the
newly identified calmodulin-binding domain within the pump molecule had to
wait for the cloning of the enzyme, which was successfully accomplished a few
months later, but indications from chymotrypisin cleavage work on the purified
pump strongly suggested a peripheral location [37].

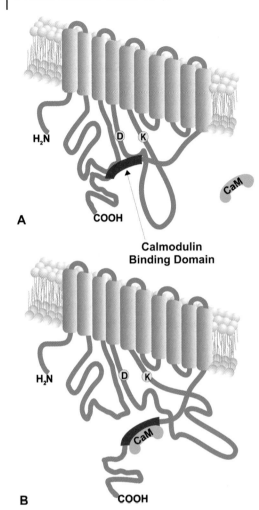

**Calmodulin
Binding Domain**

Figure 8.2 Models of the interaction between the PMCA pump and calmodulin. The 10 transmembrane domains of the pump (blue) are connected on the external side by loops represented by the blue ribbon. Catalytic D465, and K591, which are part of the ATP binding site, are enclosed in yellow circles. (A) Pump is represented in the autoinhibited form, in which the calmodulin-binding domain (purple box) binds to two "receptor" sites in the first and second major cytosolic units, protruding into the cytosol. (B) Calcium-calmodulin complex (green) binds to the calmodulin-binding domain and removes it from the pumps, freeing access to the active site and liberating the pumps from autoinhibition.

8.3
Cloning and Beyond

The primary structure of the pump was eventually deduced independently by Shull and Greeb [40] and Verma et al. [41] in 1988, using rat brain and human teratoma cDNA libraries, respectively. The original cloning effort immediately revealed the existence of isoforms: The rat brain library yielded two isoforms, containing 1176 (129.5 KDa) and 1198 (132.6 KDa) amino acids, respectively. The human library yielded a sequence of 1220 residues (134.7 KDa), which was 99 % identical to the first rat sequence in the first 1117 residues, differing significantly from it only in the C-terminal portion (later work showed the difference to be due to differential RNA splicing involving a 154 bp exon at the end of the pump gene). Both the human and the rat pumps were predicted to be organized with 10 transmembrane domains, with about 80 % of the pump mass protruding into the cytosol with two main domains and a long C-terminal chain downstream of the 10th transmembrane domain (Figure 8.3). The first cytosolic unit protrudes between transmembrane domains two and three: It contains a basic stretch of amino acids which is the main site of interaction with acidic phospholipids [42]. It also contains one of the two "receptor" sites for the autoinhibitory calmodulin-binding domain. The second (and largest) cytosolic unit connects transmembrane domains four and five, and contains the active site of the pump and the other "receptor" site for the C-terminal calmodulin-binding domain. The cytosolic C-terminal tail contains the most important regulatory sites of the pump, including the calmodulin-binding domain and consensus sites for protein kinase A and C phosphorylation. The calmodulin-binding domain, owing to its basic character, also binds acidic phospholipids [43] and could thus also mediate the response of the pump to them. It also contains Ca^{2+}-binding sites different from the catalytic site, which are likely to have an allosteric role [44].

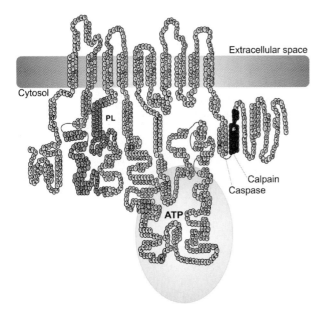

Figure 8.3 Membrane topography and important domains of the isoforms of the PMCA pump. The pump is organized in the membrane with 10 transmembrane domains connected on the external side by short loops. The cytosolic portion of the pump contains the catalytic centre and other functionally important domains, which have been color-coded in the sequence of PMCA4, the isoform taken here as reference. The calmodulin-binding domain and the phospholipid-binding domain (PL) are shown in red and in cyan, respectively. D465, the residue phosphorylated during the catalytic cycle and K591, the residue labeled by fluorescein isothiocyanate (FITC) in the ATP-binding domain, are enclosed in green circles. T1102, in the calmodulin-binding domain is the residue phosphorylated by protein kinase C, and is enclosed in pink. The ATP-binding domain is indicated by a gray shadowed ellipse. The amino acids that form the receptor site for the calmodulin-binding domain are in gray and those forming the caspase cutting site are in violet.

8.3.1
Isoforms of the PMCA Pump

Soon after the appearance of the first cloning contributions it became clear that the pump was the product of a multigene family: Four distinct members were recognized in animals. In humans PMCA 1 and 4 genes are located on chromosomes 12 (q21-q23) and 1 (q25-q37), respectively [45], PMCA2 gene is located on chromosome 3 (3p26-p25) [46–48] and PMCA3 gene on chromosome X (Xq28) [48]. Two of the four gene products (PMCAs 1 and 4) were detected in all tissues and are thus now considered as housekeeping pumps (however, PMCA4 appears to be absent in rat liver [49]). By contrast, PMCA2 was found primarily in brain and heart, and PMCA3 in brain and skeletal muscles. Significant sequence differences (homology 70–80%) was detected among the products of the four genes (Figure 8.4), and additional isoform variability was produced by alternative mRNA splicing at two locations, named splice sites A and C. An additional alternative splicing site has been proposed (site B) in the C-terminal portion of the pump, leading to the exclusion of a 108 nt exon, which would cause the loss of the 10th transmembrane domain of the protein. This probably represents an aberrant splicing site: Even if

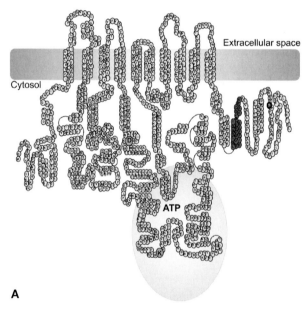

A

Figure 8.4 Sequence and membrane topography of isoforms 1–3 of the PMCA pump. Isoform 1 (A), 2 (B) and 3 (C). The catalytic aspartic acid and the ATP-binding lysine are enlarged. The ATP-binding domain is marked by the gray ellipse. S1178 is enclosed in a black circle in a PKA consensus site (KRNSS). S1178 has been shown to be phosphorylated by PKA in isoform 1. Isoforms 2 and 4 contain potential PKA phosphorylation sites (DKSS and DKAS, respectively) but their phosphorylation has not been documented. Isoform 3 does not contain serines that could be phosphorylated by PKA in the C-terminal region.

mRNAs lacking the 108 nt exon have been detected in intestine and liver, no protein lacking the 10th transmembrane domain has so far been identified [50, 51].

Alternative splicing at site A generates insertions in the mid-portion of the cytosolic loop between transmembrane domains 2 and 3, and is thus close to the phospholipid-binding domain. The splicing involves an exon (or portion thereof) that can be optionally inserted or excluded in the mature transcripts of all pumps

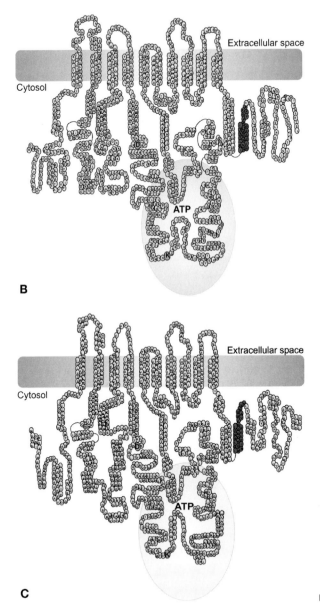

Figure 8.4

with the exception of PMCA1, in which a 39 nt exon is always included. Given the location of alternative splicing at site A, differences in the response of the pump to acidic phospholipids could have been expected. However, a study in which three variants of PMCA2 with site A inserts of 0, 31 and 45 amino acids at K302 were overexpressed in Sf9 insect cells with the help of the baculoviruses failed to show differences in general activity and in response to phospholipids [52]. Splicing at site C occurs within the calmodulin-binding domain in the C-terminal portion of the pump, and has been studied more frequently. It involves the inclusion of a 154 nt exon (or of portions thereof) in isoforms 1 and 3, and of 227 nt or 178nt exons in isoforms 2 and 4, respectively. The insertion of the full exon causes the loss of the reading frame and the truncation of the pump at a stop codon located about 50 residues upstream of the C-terminus. Splicing at site C was found to confer pH dependence to the regulation of the activity of the PMCA1 pump by calmodulin [53]. Expressed C-terminal portions of PMCA1 that contained the regular calmodulin-binding domain bound calmodulin with the same affinity at pH 7.2 and 5.9, whereas expressed peptides containing the insert showed decreased calmodulin affinity at acidic pH. It was thus suggested that these variants could be preferentially activated in a slightly acidic cell environment, such as could prevail in tissues like skeletal muscle. A variant termed 3f (3CVI), in which a short insert after the 15th residue of the calmodulin-binding domain led to the loss of most of the C-terminal portion down stream of the calmodulin-binding domain and to the formation of an isoform which had the shortest C-terminal cytosolic unit described in PMCA pumps, has been described as the major pump variant in skeletal muscles and was only negligibly stimulated by calmodulin in membrane preparations of over-expressing COS cells, even if it still bound to calmodulin columns [54, 55]. The effects of alternative splicing at site C have also been extensively studied on expressed PMCA4 variants, and data on PMCA3 have recently appeared as well [56] (Table 8.1). The insertion of the entire 178 bp exon in the PMCA4 gene generated a truncated isoform with reduced calmodulin affinity (K_m, 600–700 nM as compared with 30–40 nM for the corresponding full-length variant) and reduced

Table 8.1 Properties of PMCA Isoforms.

Parameter	PMCA1CI (b)	PMCA2CI (b)	PMCA3CI (b)	PMCA4CI (b)
Tissue distribution	Ubiquitous	Restricted (brain)	Restricted (brain)	Ubiquitous
K_d CaM (nM)	40–50	2–4	8	30–40
K_d ATP (nM)	100	200–300	ND	700
Calpain sensitivity	High	Low	Low	Low
Parameter	PMCA1CII (a)	PMCA2CII (a)	PMCA3CII (a)	PMCA4CII (a)
Tissue distribution	Ubiquitous	Restricted (brain)	Restricted (brain)	Ubiquitous
K_d CaM (nM)	ND	ND	5–10	600–700

Ca^{2+} affinity [57, 58]. This was expected, as the insert occurred in the middle of the calmodulin-binding domain, separating its C- and N-terminal halves. Surprisingly, however, a similarly truncated variant of PMCA3, which could have been expected to behave like the truncated version of PMCA4, revealed instead a calmodulin affinity that was even higher (K_m, 5–10 nM) than that of the full-length variant of PMCA4, and in the same range as that of PMCA2, which has the highest calmodulin affinity of all PMCAs.

8.3.2
Isoform Nomenclature

After the discovery of the complex pattern of alternative splicing at site C described above [13], a nomenclature was adopted in which the full-length variant of the pump without insertions was designated with the letter "b", whereas the truncated variant, resulting from the insertion of the entire novel exon and the formation of a premature stop codon, was designated as "a". The two intermediate variants with insertion of portions of the novel exon were termed "c" and "d". Later on, when splicing at site A became known, the nomenclature was extended to "wxyz". Later discoveries of additional variants of C-site type splicing added the designations "e" and "f". The shortcomings of this nomenclature were not evident at the beginning of the cloning period, basically because it was not suspected that so many splicing processes occurring at different sites would generate such bewildering isoform variability. However, the shortcoming soon became obvious, the most important being perhaps the lack of indication of the splice site. An improved nomenclature was thus proposed [59] in which the splice site was designated with a capital letter (A or B) and the type of insertion with progressive roman numerals. For example, variant 4b became 4CI variant 4a, 4CII, etc. As of today, however, even if the advantages of the new nomenclature are obvious, most workers in the field still use the old system. Therefore, in the discussion to follow, the PMCA isoforms will be designated with the notations of both nomenclatures.

8.3.3
Tissue Distribution of the Isoforms

The discovery of an ever-increasing number of spliced isoforms has complicated the initially simple subdivision of PMCAs in housekeeping and tissue (neurons) specific isoforms. Some information gradually became available on functional differences among pump variants. Even if only generated on purified pumps or on membrane preparations of overexpressing cells, it nevertheless suggested that the peculiar tissue distribution of pump variants reflected particular Ca^{2+} homeostasis demands of different cell types. Future work on the PMCA pump will have to privilege the search for (subtle) functional differences among isoforms, to be related to tissue peculiarities in the control of cell Ca^{2+}. A tissue map of PMCA isoforms based on transcript and/or protein studies is a necessary step in this direction, and is thus be discussed below.

Information on the preference of four of the basic gene products (PMCA2 and PMCA3) for neurons can now be extended and completed by adding that, within the brain, PMCA2 is expressed at particularly high levels in the cerebellum whereas PMCA3 is particularly abundant in the choroids plexus. However, pancreatic islet β cells have also been found to express substantial amounts of PMCA2 and PMCA3 variants [60].

As for the two ubiquitous basic gene products (PMCA1 and PMCA4) their relative expression is not constant in tissues, although PMCA1 is generally more abundant [61].

Examination of the tissue-specific distribution of alternatively spliced pump products (mRNA and/or proteins) is also unexpectedly complex, most splice variants being tissue-restricted. PMCA1CII (1a) was detected in spinal cord, brain, skeletal muscles and heart, whereas the full-length counterpart PMCA1CI (1b) was detected in a wide spectrum of tissues (spleen, thymus, pancreas, skeletal muscle, kidney, liver, lung, heart, spinal cord, brain and adrenal glands). PMCA2 mRNA has also been detected in brain, liver, spinal cord and adrenal gland. PMCA3 was found in spinal cord and brain. PMCA4CI mRNA was present in all tissues and the truncated variant PMCA4CII (4a) has been detected primarily in skeletal muscles, small intestine, heart, spinal cord and brain [62]. The brain deserves an especial mention in discussing spliced pump variants, since it contains more and larger amounts of them than other tissues. Essentially, the four PMCA genes are expressed in all sub-regions of the human brain; however, consistent differences were found in their distribution and especially in that of the alternatively spliced options. The cortical regions and the hippocampus are characterized by the presence of isoforms of the CII (a) type, whereas the inferior olive and the olfactory bulb show a relative preponderance of the CI (b) type [63]. In situ hybridization work on sections of human hippocampus has permitted a more detailed definition of the cellular distribution of the pump variants in that brain nucleus. PMCA1 and PMCA3 mRNAs were low throughout the entire hippocampus, whereas the transcripts of PMCA2 and PMCA4 had a more abundant and similar pattern of distribution in the distinct hippocampus regions differences: They were much more abundant in the granule cell layer of the dentate gyrus, and in the CA2 region as compared with the CA1 and CA3 regions of Ammon's Horn [64]. The finding that the two isoforms with the highest calmodulin affinity (PMCA2 and PMCA4) were abundant in an area known to be resistant to Ca^{2+}-linked excitotoxicity [65] has suggested that differences in the amount/or isoform pattern of PMCAs may influence the susceptibility of groups of neurons to Ca^{2+}-mediated excitotoxicity.

The PMCA distribution in the auditory system deserves a special mention. Ca^{2+} is a key modulator of mechanoelectrical transduction by hair cells, the sensory cells of the inner ear. The sensory organelle of the hair cell, the hair bundle, which is exposed to an extracellular fluid that is high in K^+ and low in Na^+, relies on mobile Ca^{2+} buffers and on the plasma membrane Ca^{2+}-ATPase to regulate Ca^{2+} levels in stereocilia [66, 67]: Most Ca^{2+} that enters through the transduction channels is removed by the PMCA pump. Indeed, hair cells contain a high amount of the PMCA pump, particularly in the hair bundle. Interestingly, PMCA1CI (1b) is the major enzyme of hair cells and of the supporting cell basolateral membranes, whereas PMCA2CII (2a) is the only isoform present in the hair bundles [68]. This site C-spliced variant must evidently be able to remove Ca^{2+} from the intracellular environment very efficiently.

One important point that has emerged from the work in different laboratories is the demonstration that the process of alternative splicing is related to the embryonic development and to the differentiation process. The first descriptions of differences in the PMCA expression in the developing rat brain revealed that PMCA1CI (1b) was present from embryonic day 10, while the alternative spliced forms 1CII and 1CIII (1a, 1c) and the other PMCA gene transcripts only appeared at the end of the gestation [69]. Further work on the developmental pattern of expression of the PMCA isoforms has recently applied in situ hybridization techniques on tissue sections of mouse embryos [70]. The study has revealed that PMCA1 appeared earliest during embryo development, together with its ubiquitous expression this confirmed that this isoform may be considered as "house-keeping". All other isoforms only became detectable at day 12.5. As expected, PMCA2 expression was confined to the nervous system. Surprisingly PMCA3, at variance with the adult, was widely expressed in the mouse embryo, pointing to its possible importance during the development of different organs. The other ubiquitous pump, PMCA4, was expressed in the embryo at much lower levels than the other isoforms. Again this was at variance with human tissues where PMCA4 is frequently as abundant, and normally as ubiquitous, as PMCA1. A similar study on developing rat brain found that PMCA1 and PMCA3 showed a similar pattern of expression, both being detected in the early phases of development. PMCA2 appeared instead at day 18, at a time when PMCA1CI (1b) and PMCA3CI (3b) became replaced by the truncated variants, suggesting that the CII (a) forms could be the brain-specific variants. PMCA4 mRNA was absent in the developing rat brain, confirming the very low level of expression of PMCA4 found by Keeton and Shull [49].

One interesting aspect of the pattern of expression of PMCA pump isoforms that has emerged is the regulation of their expression by Ca^{2+} itself. Work has shown that in developing neurons the mRNAs transcription and the transduction of the corresponding proteins PMCAs are affected by Ca^{2+} [71]. Work on cerebellar granule cells in culture has shown that the expression of PMCA2 and PMCA3 increased markedly during the maturation process. The increase is brought about by a modest change in cellular Ca^{2+} produced by partially depolarizing the plasma membrane of the cultured cells obtained by increasing the concentration of K^+ in the medium. These changes were accompanied by the up-regulation of the expression of the alternatively spliced variant CII (a) of PMCA1. The changes became noticeable after 2–3 days of culture and were fully evident after 4–5 days. In sharp contrast with these three isoforms, PMCA4 became rapidly downregulated under these conditions to disappear completely in 2–3 h [72] in a process that was mediated by calcineurin. At the end of the maturation process, despite the down regulation of isoform 4, however, the total amount of PMCA protein was significantly increased. It may be reasonably speculated that the changes in the PMCA isoform expression pattern reflect the changing demands of cells in terms of Ca^{2+} homeostasis and thus of Ca^{2+} pumping during the maturation process. The modest increase of intracellular Ca^{2+} linked to increased penetration during maturation (which is also essential to the long-term survival of the neurons in culture) suggests that granular cells re-adjust their (PMCA) pumping capacity to finely tune the Ca^{2+} signaling they now demand. What still needs to be understood is how the changed pattern of expression of the different isoforms is eventually translated into the goal of letting cell Ca^{2+} increase about 3-fold. It won't be an easy task: On one hand, neurons overexpress very efficient Ca^{2+} export systems (i. e., PMCA2); on the other hand, they downregulate another efficient system (PMCA4) while up-regulating one with much lower Ca^{2+} affinity (PMCA1CII).

Differentiation can also induce changes in the splicing of PMCA mRNA in muscle and in the pheochromocytoma cell line PC12 [73], and in the human neuroblastoma cell line IMR-32 [74]. In the latter, the PMCA-mediated Ca^{2+} efflux increased more than three-fold during 12–16 days of differentiation. The increase was suggested to be mediated by a pronounced up-regulation of PMCA2, 3 and 4 isoforms that appeared to be independent of voltage-gated Ca^{2+} influx.

8.3.4
How do PMCA Pump Isoforms Differ Functionally?

According to the prevailing view, the four basic PMCA isoforms should be considered necessary for the fine tuning of basal Ca^{2+} levels in the cell at large. However, emerging evidence now indicates that they may play important roles in modulating specific cellular functions. The existence of numerous PMCA isoforms and of their splice variants suggests specific roles in the shaping of Ca^{2+} signals, not merely in different cell types and in their development, but in the specialised domain of the cells as well.

Assessment of the physiological role of each basic PMCA isoform (and splice variants) in the context of intact cells is a very difficult problem, chiefly because most cells express more than one PMCA isoform and due to the absence of specific inhibitors. Some information has nevertheless been collected using different experimental approaches. Even if fragmentary and altogether scarce, some of the data deserve to be mentioned.

An interesting model is represented by polarized cells, e. g., those of pancreatic acini in which Ca^{2+} mobilizing agonists have been shown to directly activate the plasma membrane pump. [75, 76]. Several reports had shown that the PMCA pump was mainly confined to the basolateral membrane of the pancreatic acinar cell, but more recent work has shown that Ca^{2+} efflux in stimulated acinar cell occurred instead predominantly from the luminal pole [77], contributing to maintaining the high Ca^{2+} concentration in the acinar lumen necessary to endocytosis. This may have been due to a relatively high density of calcium pumps in the secretory granule region, but it cannot be excluded that the calcium pumps in that area would have properties different from those of the pumps in the basolateral membrane [78]. Unfortunately, the identity of the PMCA isoforms expressed in the lu-

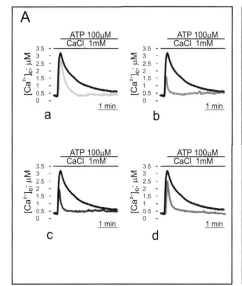

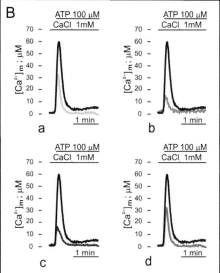

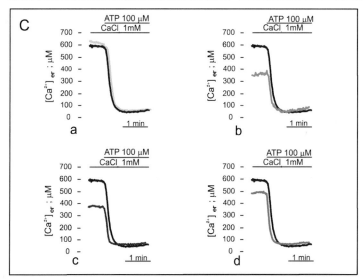

Figure 8.5 Monitoring Ca^{2+} in CHO cells overexpressing the isoforms of the PMCA pump. Cells were transfected with mammalian expression plasmids for the PMCA isoforms and for the Ca^{2+}-sensitive photoprotein aequorin, which monitor the Ca^{2+} concentration in the cytosol (panel A), for aequorin targeted to mitochondrial matrix (panel B) and for aequorin targeted to the lumen of the endoplasmic reticulum (panel C). The prosthetic group coelenterazine was added to the incubation medium to reconstitute active aequorin before carrying out Ca^{2+} measurements. Where indicated ATP was added as an agonist linked to generating $InsP_3$, liberating Ca^{2+} from the endoplasmic reticulum and causing a cytosolic Ca^{2+} transient and rapid accumulation of Ca^{2+} by mitochondria. Detailed technical information on the experiment is given in Ref. 56, from which the figures have been modified. Symbols: er, endoplasmic reticulum; c, cytosol; m, mitochondria. Color-code of the traces: blue, transfection with the various aequorins without the cDNAs of the pumps. Yellow, cotransfection of the aequorins with the cDNA of PMCA1. Cyan, cotransfection of the aequorins with the PMCA2 cDNA. Red, cotransfection of the aequorins with the PMCA3 cDNA. Green, cotransfection of the aequorins with the PMCA4 cDNA.

minal and in the basolateral membrane is still unknown. A similar study was carried out on another type of epithelial polarized cells: The Madin–Darby canine kidney tubular (MDCK) cells. The PMCA pump in the kidneys of several species has been suggested by immunohistochemical analysis to be concentrated in the basolateral membrane of distal kidney tubular cells [79]. Recent work has indeed detected abundant amounts of PMCA1CI (1b) and PMCA4CI (4b) proteins in the basolateral membrane of MDCK cells. Experiments with antisense probe against isoform 4CI (4b) showed that Ca^{2+} efflux from the cells was reduced by about 50 % [80].

The differential role of PMCA isoforms in the control of Ca^{2+} homeostasis has been repeatedly studied by overexpressing them in model cell system. One of the first studies used L6 myogenic cell lines stably overexpressing human PMCA isoform 4CI (4b) [81]. The study found that the resting cytosolic calcium was significantly reduced (20–30 %) with respect to control cells. It also found that the differentiation process was remarkably accelerated, suggesting that the control of Ca^{2+} homeostasis by the PMCA pump played an active role in the control of myogenic differentiation. At variance with these results, the overexpression of rat isoform 1CII (1a) in rat aortic endothelial cells [82] failed to produce significant differences in the Ca^{2+} concentration of cells at rest. However, the overexpression of the isoform decreased the height of the Ca^{2+} transient induced by cell stimulation. A similar study in CHO cells compared the effects of transfecting human PMCA4CI (4b) and the endoplasmic reticulum Ca^{2+} pump (SERCA): The reduction of the Ca^{2+} transient elicited by cell stimulation was much faster in cells overexpressing the PMCA pump than in those overexpressing the SERCA counterpart [83]. The most important point that emerges from these studies is that the PMCA pump indeed plays a significant role in maintaining the resting cytosolic Ca^{2+} levels in the nanomolar range. They also suggest that it contributes to the clearance of cytosolic Ca^{2+} transients generated by cell stimulation, perhaps even to a greater extent than the SERCA pump.

The studies above, even performed in the native cell environment and thus advantageous with respect to the work on preparations of purified pump or on membrane fractions, offer no clues as to the differences in the physiological functions of the different basic isoforms. A recent study in which the four PMCA basic isoforms and two of their truncated variants [PMCA3CII (3a) and 4CII (4a)] were expressed in CHO cells together with the Ca^{2+} indicator probe engineered to direct it to the various cell compartments [56] found that the two ubiquitous isoforms PMCA1 and PMCA4 were less effective in controlling the homeostasis of Ca^{2+} than the two tissue (neuron)-specific isoforms PMCA2 and PMCA3 (Figure 8.5, see p. 225). Surprisingly, it also showed that the truncated version CII (a) of two of the pumps (one ubiquitous, one tissue-specific) were just as effective in controlling cellular Ca^{2+} as the corresponding full-length variants. Since at least one of these truncated variants has mediocre calmodulin affinity, the results suggest that in the cellular ambient the availability of calmodulin is not critical to the function of the PMCA pumps. This may be relevant to the discussion in one of the preceding sections of the mechanisms for the activation of the pump, which still is an open question, when one considers the *in vivo* situation. Perhaps, in the native cell environment, mechanisms alternative to calmodulin (acidic phospholipids) may take primacy as activators of the pump. A recent study in rat sensory neurons in which Ca^{2+} efflux was accelerated by the stimulation of PMCA isoform 4 by neuromodulators via a PKC-dependent pathway could be usefully quoted in this context [84].

One plausible way to rationalize the existence of so many PMCA isoforms is to suggest that they play differential roles in the control of Ca^{2+} signaling in selected Ca^{2+} microdomains, i.e., different PMCA isoforms could be recruited to specific regions of the plasma membrane by forming differential complexes with partner proteins.

Recent evidence indicates that this may indeed occur. PMCA isoforms 2CI (2b) and 4CI (4b) interact promiscuously and selectively with members of the Membrane-Associated Guanylate Kinase family (MAGUK or SAP) of PDZ (PSD95) domain-containing proteins which are known to cluster ion channels with other signaling proteins [85, 86]. The interaction has been demonstrated by two-hybrid yeast screening using the C-terminal tail of PMCA2CI (2b) and 4CI (4b), and in co-immunoprecipitation experiments. The work suggests that PMCA-MAGUK(SAP) interactions mediate the recruitment and maintenance of the PMCA at specific membrane domains involved in local Ca^{2+} regulation. It was also reported that PMCA2CI (2b) interacts preferentially with the regulatory factor 2 of the Na^{+}/H^{+} exchanger (NHERF2), another PDZ domain-containing protein involved in the targeting, retention and regulation of membrane protein like the β2-adrenergic receptor, the cystic fibrosis transmembrane conductance regulator and trp4 Ca^{2+} channels [87]. Interestingly, GFP-tagged PMCA2CI (2b) overexpressed in MDCK cells became sorted to the apical membrane of the cells where it co-localized with transfected NHERF2. Again, this suggests a possible mechanism for the apical sorting of PMCAs in epithelial cells. These initial results are interesting, but should still be taken with caution, since apical targeting of the pump also occurred in the absence of exogenously expressed NHERF2. Interaction between the PMCA pump and the inducible nitric oxide synthase I (NOS-I) was recently described [88]. PMCA isoform 4CI (4b), which has been claimed to be concentrated in the caveolae [27, 28], which contain several effectors of cellular signal transduction, interacts through its C-terminal amino acids with the PDZ domain of NOS-I. The interaction appears to be specific for NOS-I, as it failed to occur with the endothelial NOS. The authors have proposed a model in which a PMCA molecule interacting with NOS-I might reduce the local (caveolar) calcium concentration, leading to reduced NOS-I activity. The proposal was supported by experiments in which increasing PMCA4CI (4b) expression reduced the NOS-I-dependent cGMP-cyclase activity, strengthening the proposal that PMCA4CI (4b), by virtue of its interaction with nitric oxide synthase I, decreased the concentration of Ca^{2+} in its immediate vicinity, thus down-regulating its activity. Finally, another member of the MAGUKs family, CASK (calcium/calmodulin-dependent serine protein kinases), was recently found to interact with PMCA4CI (4b). CASK is of particular interest since in adult brain it is located at neuronal synapses, where the presence of a Ca^{2+} extrusion system in its immediate proximity could contribute to the regulation of its activity [89].

8.3.5
Connection of PMCA Pumps to Apoptosis

An emerging role for PMCA pumps is the prevention of cell death by acting as safeguards against inappropriate Ca^{2+} rises in the cytosol. In many cell types, cell death induced by various insults is accompanied by excess influx of Ca^{2+} and by an increase in its intracellular concentration. The plasma membrane Ca^{2+}-ATPase may be instrumental in reducing Ca^{2+}-mediated cytotoxicity since PMCA4CI (4b) overexpressing model cell clones were less vulnerable to Ca^{2+}-mediated cell death induced by Ca^{2+} ionophores, whereas "antisense" clones, in which the endogenous expression of the pump was suppressed, were significantly more susceptible [90]. The ability of cells to regulate the expression of Ca^{2+} homeostatic mechanisms in response to situations of Ca^{2+} overloads appears to determine their destiny when faced with different pathological stimuli. The finding quoted above that Ca^{2+} influx in cerebellar granule cells regulated the expression of PMCA's genes to prevent their premature demise [71, 72] validates the hypothesis. Convincing support is also provided by the demonstration in two recent studies that PMCA pumps are targets of activated caspases [91, 92]. One study claimed that the cleavage activated the pump, since it removed the entire calmodulin-binding domain, freeing the pump from autoinhibition [91]. Ca^{2+} accumulation would thus be prevented in apoptotic cells. The other study [92] showed that PMCA cleavage induced instead inactivation of the PMCA pump (isoforms 2 and 4) leading to Ca^{2+} overload and eventual cell death. The discrepancy in the effects of caspases on the activity of the pump may be linked to differences in the experimental protocols in the two studies. It emerges, however, that apoptotic cells act on their PMCA pumps in ways that will eventually lead to their demise.

8.3.6
Pathology

Increases in intracellular Ca^{2+} up to overload have often been associated with cell dysfunction, injury and death, especially in the central nervous system (CNS) [93, 94]. It is reasonable to expect that malfunction of Ca^{2+}-exporting systems would generate pathological phenotypes. Such phenotypes have been described for the malfunction of the SERCA pump: Brody's [95] and Darier's [96] diseases are linked to severe malfunctions of the SERCA pump. Alterations in the pumping function of the PMCA pumps have been described in several genetic conditions; however, it is difficult to establish whether the disease phenotypes were due to direct defects in pumping activity, or were secondary to the impairment of mechanisms that would control pump function. Considering that the PMCA pump is modulated by numerous factors, the latter possibility seems realistic. No human disease has so far been linked to defects in any of the four PMCA genes, but mouse models have been described. Two mouse hearing defects (Wriggle Mouse Sagami, *wri* and Deaf Waddler Mouse, *dfw*) have been linked to mutations in PMCA2, which is abundantly expressed in the outer hair cells of the organ of Corti (see above). In one mutant a

G, which is embedded in a sequence highly conserved among all PMCA isoforms, is replaced by an S in the first cytosolic loop between transmembrane domains 2 and 3 (G283) [97]. In the other mutant, an E → K replacement occurs in transmembrane domain 4 (E412) [98]. Homozygous *wri* mice suffer from a neurological disorder characterized by motor imbalance and progressive hearing loss. A third mutation has been described: a 2 bp deletion causes a loss of the reading frame that leads to a truncated pump (*dfw*^{2j}) and to its disappearance from the stereocilia and the basolateral membrane of cochlea haircells [97, 99].

The PMCA2 gene has been knocked out [100], producing mice with equilibrium defects and deafness. Histology has shown an increased number of Purkinje neurons in the cerebellum, and severe abnormalities in the vestibular system and in the organ of Corti. Finally, a recent report has shown that the amounts of PMCA2 transcript and protein were dramatically decreased in neuronal cells of the lumbar spinal cord of rats with symptoms of experimental autoimmune encephalomyelitis (a model for multiple sclerosis). Glutamate suppressed neuronal PMCA2 expression, leading to the alteration of Ca^{2+} homeostasis at the initial clinical phase [101].

8.4
Final Remarks

The PMCA pump, long a relatively neglected topic as compared with the vastly more popular SERCA pump, has now come of age. The past few years have seen impressive advances in knowledge on many aspects of its functional properties, its biochemical regulation, and the role of functional domains. An aspect of the pump that is attracting increasing interest is that of the isoforms and of their tissue-specific role. The rationale for the isoform multiplicity is still obscure, but light on this problem will hopefully be shed soon. It has now emerged that the PMCA pumps are not only critical to the maintenance of the cytosolic Ca^{2+} at the appropriate level over the long term, but also in shaping Ca^{2+} signals in a spatially and temporally defined manner. Finally, it is becoming increasingly evident that the PMCA pumps are necessary for the proper development and survival of neuronal systems.

References

1. E. Carafoli, *Annu. Rev. Biochem.*, **1987**, *56*, 395–433.
2. H. J. Schatzmann, *Experientia Basel*, **1966**, *22*, 364–368.
3. P. L. Pedersen and E. Carafoli, *Trends Biochem. Sci.*, **1987**, *12*, 146–150.
4. A. F. R. Rega and P. J. Garrahan, *The Ca²⁺ Pump Of Plasma Membrane*, CRC, Boca Raton, **1985**.
5. H. J. Schatzmann, *Membrane Transport of Calcium*. E. Carafoli, ed., Academic Press, London **1982**, pp. 41–108.
6. E. Carafoli and M. Zurini, *Biochim. Biophys. Acta*, **1982**, *683*, 279–301.
7. J. T. Penniston, *Calcium and Cell Function*. W. Y. Cheung, ed., Academic Press, New York **1983**, pp. Vol IV, 99–147.
8. G. R. Monteith and B. D. Roufgalis, *Cell. Calcium*, **1995**, *18*, 459–470.
9. E. Carafoli, *Physiol. Rev.*, **1991**, *71*, 129–153.
10. E. Carafoli, *J. Biol. Chem.*, **1992**, *267*, 2115–2118.
11. E. Carafoli, *FASEB J.*, **1994**, *8*, 993–1002.
12. E. Carafoli and M. Brini, *Curr. Opin. Chem. Biol.*, **2000**, *4*, 152–161.
13. E. E. Strehler and D. A. Zacharias, *Physiol. Rev.*, **2001**, *81*, 21–50.
14. M. Chiesi, M. Zurini, and E. Carafoli, *Biochemistry*, **1984**, *23*, 2595–2600.
15. R. M. Gopinath and F. F. Vicenzi, *Am. J. Haematol.*, **1981**, *7*, 303–312.
16. H. R. Jarret and J. T. Penniston, *Biochem. Biophys. Res. Commun.*, **1977**, *77*, 1210–1216.
17. P. Ronner, P. Gazzotti, and E. Carafoli, *Arch. Biochem. Biophys.*, **1977**, *179*, 578–583.
18. V. Niggli, E. S. Adunyah, J. T. Penniston, and E. Carafoli, *J. Biol. Chem.*, **1981**, *256*, 395–401.
19. P. Caroni and E. Carafoli, *J. Biol. Chem.*, **1981**, *256*, 9371–9373.
20. P. H. James, M. Pruschy, T. Vorherr, J. T. Penniston, and E. Carafoli, *Biochemistry*, **1989**, *28*, 4253–4258.
21. J. I. Smallwood, B. Gügi, and H. Rasmussen, *J. Biol. Chem.*, **1988**, *263*, 2195–2202.
22. K. Wang, L. C. Wright, C. L. Madian, B. G. Allen, A. D. Conigrave, and B. D. Roufogalis, *J. Biol. Chem.*, **1991**, *266*, 9078–9085.
23. F. Hofmann, J. Anagli, E. Carafoli, and T. Vorherr, *J. Biol. Chem.*, **1994**, *269*, 24298–24303.
24. J. D. Caviers, *Biochem. Biophys. Acta*, **1984**, *771*, 241–244.
25. T. Vorherr, T. Kessler, F. Hofmann, and E. Carafoli, *J. Biol. Chem.*, **1991**, *266*, 22–27.
26. D. Kosk-Kosicka and T. Bzdega, *J. Biol. Chem.*, **1988**, *263*, 18184–18189.
27. T. Fujimoto, *J. Cell. Biol.*, **1993**, *120*, 1147–1157.
28. A. Hammes, S. Oberdorf-Maass, T. Rother, K. Nething, F. Gollnick, K. W. Linz, R. Meyer, K. Hu, H. Han, and P. Gaudron, et al., *Circ. Res.*, **1998**, *83*, 877–888.
29. V. Niggli, J. T. Penniston, and E. Carafoli, *J. Biol. Chem.*, **1979**, *254*, 9955–9958.
30. M. Zurini, J. Krebs, J. T. Penniston, and E. Carafoli, *J. Biol. Chem*, **1984**, *259*, 618–627.
31. P. James, T. Vorherr, J. Krebs, A. Morelli, G. Castello, D. J. McCormick, J. T. Penniston, A. De Flora and E. Carafoli, *J. Biol. Chem.*, **1989**, *264*, 8289–8296.
32. R. Falchetto, T. Vorherr, J. Brunner, and E. Carafoli, *J. Biol. Chem.*, **1991**, *266*, 2930–2936.
33. R. Falchetto, T. Vorherr, and E. Carafoli, *Protein Sci.*, **1992**, *1*, 1613–1621.
34. B. Elshorst, M. Hennig, H. Försterling, A. Diener, M. Maurer, P. Schulte, H. Schwalbe, C. Griesinger, J. Krebs, H. Schmid, T. Vorherr, and E. Carafoli, *Biochemistry*, **1999**, *38*, 12320–12332.
35. A. G. Filoteo, J. P. Gorski, and J. T. Penniston, *J. Biol. Chem.*, **1987**, *262*, 6526–6530.

36. P. James, E. Zvaritch, M. S. Hakhparanov, J. T. Penniston, and E. Carafoli, *Biochem. Biophys. Res. Commun.*, **1987**, *149*, 7–12.

37. P. James, M. Maeda, R. Fischer, A. K. Verma, J. Krebs, J. T. Penniston, and E. Carafoli, *J. Biol. Chem.*, **1988**, *263*, 2905–2910.

38. C. L. Jaffe, H. Lis, and N. Sharon, *Biochemistry*, **1980**, *19*, 4423–4429.

39. J. B. Denny and G. Blobel, *Proc. Natl. Acad. Sci. U. S. A.*, **1984**, *81*, 5286–5290.

40. G. E. Shull and J. Greeb, *J. Biol. Chem.*, **1988**, *263*, 8646–8657.

41. A. K. Verma, A. G. Filoteo, D. R. Stanford, E. D. Wieben, J. T. Penniston, E. E. Strehler, R. Fisher, R. Heim, G. Vogel, S. Mathews, M. A. Strehler-Page, P. James, T. Vorherr, J. Krebs, and E. Carafoli, *J. Biol. Chem.*, **1988**, *263*, 14152–14159.

42. E. Zvaritch, P. James, T. Vorherr, R. Falchetto, N. Modyanov, and E. Carafoli, *Biochemistry*, **1990**, *29*, 8070–8076.

43. P. Brodin, R. Falchetto, T. Vorherr, and E. Carafoli, *Eur. J. Biochem.*, **1992**, *204*, 939–946.

44. F. Hofmann, P. James, T. Vorherr, and E. Carafoli, *J. Biol. Chem.*, **1993**, *268*, 10252–10259.

45. S. Olson, M. G. Wang, E. Carfoli, E. E. Strehler and O. W. Mcbride, *Genomics*, **1991**, *9*, 629–641.

46. P. Brandt, E. Ibrahim, G. A.P. Bruns, and R. L. Neve, *Genomics*, **1992**, *14*, 484–487.

47. F. Latif, F. M. Duh, J. Gnarra, K. Tory, I. Kuzmin, M. Yao, T. Stackhouse, W. Modi, L. Geil, L. Schmidt, H. Li, M. L. Orcutt, E. Maher, F. Richards, M. Phipps, M. Ferguson-Smith, D. Le Paslier, W. M. Linehan, B. Zbar, and M. I. Lerman, *Cancer Res.*, **1993**, *53*, 861–867.

48. M. G. Wang, H. Yi, H. Hilfiker, E. Carafoli, E. E. Strehler, and O. W. McBride, *Cytogenet. Cell Gen.*, **1994**, *67*, 41–45.

49. T. P. Keeton and G. E. Shull, *Biochem. J.*, **1995**, *306*, 779–785.

50. A. Howard, S. Legon, and J. R.F. Walters, *Am. J. Physiol. Gastrointest. Liver Physiol.*, **1993**, *265*, 917–925.

51. A. Howard, N. F. Barley, F. Legon, and J. R.F. Walters, *Biochem. J.*, **1994**, *303*, 275–279.

52. H. Hilfiker, D. Guerini and E. Carafoli, *J. Biol. Chem.*, **1994**, *269*, 26178–26183.

53. F. Kessler, R. Falchetto, R. Heim, R. Meili, T. Vorherr, E. E. Strehler, and E. Carafoli, *Biochemistry*, **1992**, *31*, 11785–11792.

54. A. G. Filoteo, N. L. Elwess, A. Enyedi, A. Caride, H. H. Aung, and J. T. Penniston, *J. Biol. Chem.*, **1997**, *272*, 23741–23747.

55. A. G. Filoteo, A. Enyedi, A. K. Verma, N. L. Elves, and J. T. Penniston, *J. Biol. Chem.*, **2000**, *275*, 4323–4328.

56. M. Brini, L. Coletto, N. Pierobon, N. Kraev, D. Guerini, and E. Carafoli, *J. Biol. Chem.*, **2003**, *278*, 24500–24508.

57. A. Enyedi, T. Vorherr, P. James, D. J. McCormick, A. G. Filoteo, E. Carafoli, and J. T. Penniston, *J. Biol. Chem.*, **1989**, *264*, 12313–12321.

58. B. S. Seiz-Preiano, D. Guerini, and E. Carafoli, *Biochemistry*, **1996**, *35*, 7946–7953.

59. E. Carafoli and T. Stauffer, *J. Neurobiol.*, **1994**, *25*, 312–324.

60. A. Kamagate, A. Herchuelz, A. Bollen van, and F. Eylen, *Cell Calcium*, **2000**, *27*, 231–246.

61. T. P. Stauffer, D. Guerini, and E. Carafoli, *J. Biol. Chem.*, **1995**, *270*, 12184–12190.

62. P. Brandt, R. L. Neve, A. Kammesheidt, R. E. Rhoadas, and T. C. Vanaman, *J. Biol. Chem.*, **1992**, *267*, 4376–4385.

63. D. A. Zacharias, S. J. Dalrymple, and E. E. Strehler, *Brain Res. Mol. Brain*, **1995**, *28*, 263–272.

64. D. A. Zacharias, S. J. DeMarco, and E. E. Strehler, *Mol. Brain Res.*, **1997**, *45*, 173–176.

65. M. P. Mattson, P. B. Gutbrie, and S. B. Kater, *Prog. Clin. Biol. Res.*, **1989**, *317*, 333–351.

66. E. A. Lumpikin and A. J. Hudspeth, *J. Neurosci.*, **1998**, *18*, 6300–6318.

67. A. J. Ricci, Y. C. Wu, and R. J. Fettiplace, *J. Neurosci.*, **1998**, *18*, 8261–8277.

68. R. A. Dumont, U. Lins, A. G. Filoteo, J. T. Penniston, B. Kachar, and

P. G. Gillespie, *J. Neurosci.*, **2001**, *21*, 5066–5078.

69. P. Brandt and R. L. Neve, *J. Neurochem.*, **1992**, *59*, 1566–1569.

70. D. A. Zacharias and C. Kappen, *Biochim. Biophys. Acta,* **1999**, *1428*, 397–405.

71. D. Guerini, E. Garcìa-Martin, A. Gerber, C. Volbracht, M. Leist, C. Gutierrez Merino, and E. Carafoli, *J. Biol. Chem.*, **1999**, *274*, 1667–1676.

72. D. Guerini, X. Wang, L. Li, A. Genazzani, and E. Carafoli, *J. Biol. Chem.*, **2000**, *275*, 3706–3712.

73. A. Hammes, S. Oberdorf, E. E. Strehler, T. Stauffer, E. Carafoli, H. Vetter, and L. Neyses, *FASEB J.*, **1994**, *8*, 428–435.

74. Y. M. Usachev, S. L. Toutenhoofd, G. M. Goellner, E. E. Strehler, and S. A. Thayer, *J. Neurochem.*, **2001**, *76*, 1756–1765.

75. B. X. Zhang, H. Zhao, P. Loessberg, and S. Muallem, *J. Biol. Chem.*, **1992**, *267*, 15419–15425.

76. A. V. Tepikin, S. G. Voronina, D. V. Gallacher, and O. H. Petersen, *J. Biol. Chem.*, **1992**, *267*, 3569–3572.

77. P. V. Belan, O. V. Geransimenko, A. V. Tepikin, and O. H. Petersen, *J. Biol. Chem.*, **1996**, *271*, 7615–7619.

78. P. V. Belan, O. V. Geransimenko, O. H. Petersen, and A. V. Tepikin, *Cell Calcium*, **1997**, *22*, 5–10.

79. J. L. Borke, A. Caride, A. K. Verma, J. T. Penniston, and R. Kumer, *Am. J. Physiol. Renal Fluid Electrolyte Physiol.*, **1989**, *257*, F842–F849.

80. S. N. Kip and E. E. Strehler, *Am. J. Physiol. Renal Physiol.*, **2003**, *284*, 122–132.

81. A. Hammes, S. Oberdorf-Maass, S. Jenatschke, T. Pelzer, A. Maass, F. Gollnick, R. Meyer, J. Afflerbach, and L. Neyses, *J. Biol. Chem.*, **1996**, *48*, 30816–30822.

82. B. F. Liu, X. Xu, R. Fridman, S. Muallem, and T. H. Kuo, *J. Biol. Chem.*, **1996**, *271*, 5536–5544.

83. M. Brini, D. Bano, S. Manni, R. Rizzuto, and E. Carafoli, *EMBO J.*, **2000**, *19*, 4926–4935.

84. Y. M. Usachev, S. J. DeMarco, C. Campbell, E. E. Strehler, and S. A. Thayer, *Neuron*, **2002**, *33*, 113–122.

85. E. Kim, S. J. DeMarco, S. M. Marfatia, A. H. Chishti, M. Sheng, and E. E. Strehler, *J. Biol. Chem.*, **1998**, *273*, 1591–1595.

86. S. J. DeMarco and E. E. Strehler, *J. Biol. Chem.*, **2001**, *276*, 21594–21600.

87. S. J. DeMarco, M. C. Chicka, and E. E. Strehler, *J. Biol. Chem.*, **2002**, *277*, 10506–10511.

88. K. Schuh, S. Uldrijan, M. Telkamp, N. Rothlein, and L. Neyses, *Cell Biol.*, **2001**, *155*, 201–205.

89. K. Schuh, S. Uldrijan, S. Gambaryan, N. Roethlein, and L. Neyses, *J. Biol. Chem.*, **2003**, *278*, 9778–9783.

90. M. L. Garcia, Y. M. Usachev, S. A. Thayer, E. E. Strehler, and A. J. Windebank, *J. Neurosci. Res.*, **2001**, *64*, 661–669.

91. K. Paszty, A. K. Verma, R. Padani, A. G. Filoteo, J. T. Penniston, and A. Enyedy, *J. Biol. Chem.*, **2002**, *277*, 6822–6829.

92. B. L. Schwab, D. Guerini, C. Didszun, D. Bano, E. Ferrando-May, E. Fava, J. Tam, D. Xu, S. Xanthoudakis, D. W. Nicholson, E. Carafoli, and P. Nicotera, *Cell Death Differ.*, **2002**, *9*, 818–831.

93. W. Paschen, *Cell Calcium*, **2001**, *29*, 1–11.

94. A. O'Neill, P. V. Nicot, Y. Ratnakar, C. C. Ron, S. Chen, and S. Elkabes, *Brain*, **2003**, *126*, 398–412.

95. A. Odermatt, P. E. M. Taschner, V. K. Khanna, H. F. M. Busch, G. Karpati, C. K. Jablecki, M. H. Breuning, and D. H. MacLennan, *Nat. Genet.*, **1996**, *14*, 191–194.

96. A. Sakuntabhai, V. Ruiz-Perez, S. Crater, N. Jacobsen, S. Burge, S. Monk, M. Smith, C. S. Munro, M. O'Donovan, N. Craddock, R. Kucherlapati, J. L. Rees, M. Owen, G. M. Lathrop, A. P. Monaco, T. Strachan, and A. Hovnanian, *Nat. Genet.*, **1999**, *21*, 271–277.

97. V. A. Street, J. W. McKee Johnson, B. Fonseca, L. Tempel, and K. Noben-Trauth, *Nat. Genet.*, **1998**, *19*, 390–394.

98. K. Takahashi and K. Kitamura, *Biochem. Biophys. Res. Commun.*, **1999**, *261*, 773–778.

99. K. Noben-Trauth, Q. Y. Zheng, K. R. Johnson, and P. M. Nishina, *Genomics*, **1997**, *44*, 266–272.

100. P. J. Kozel, R. A. Friedman, L. C. Erway, En. Yamoah, L. H. Liu, T. Riddle, J. J. Duffy, T. Doetschman, M. L. Miller, E. L. Cardell, and G. E. Schull, *J. Biol. Chem.*, **1998**, *273*, 18693–18696.

101. A. Nicot, P. V. Ratnakar, Y. Ron, C. C. Chen, and S. Elkabes, *Brain*, **2003**, *126*, 398–412.

Part II
F-type ATPases

Handbook of ATPases. Edited by M. Futai, Y. Wada, J. H. Kaplan
Copyright © 2004 WILEY-VCH Verlag GmbH & Co. KGaA, Weinheim
ISBN 3-527-30689-7

9

Proton Translocating ATPases: Introducing Unique Enzymes Coupling Catalysis and Proton Translocation through Mechanical Rotation

Masamitsu Futai, Ge-Hong Sun-Wada, and *Yoh Wada*

Abbreviations

DCCD, dicyclohexylcarbodiimide; Pi, inorganic phosphate.

9.1
Introduction

The mechanism of ATP synthesis has been a focus of biochemists for more than four decades. ATP synthase was first identified in the mitochondrial inner membrane, and then found successively in chloroplasts and bacterial membranes (for reviews, see Refs. 1–8). This enzyme synthesizes ATP from ADP and phosphate (Pi) coupled with an electrochemical proton gradient, and is also called proton-translocating ATPase or F-ATPase (F-Type ATPase) because of the reversible proton pumping upon ATP hydrolysis. The name F-ATPase originated from the coupling factor of oxidative phosphorylation sensitive to oligomycin [1]. *Escherichia coli* F-ATPase is composed of a membrane extrinsic F_1 sector and a transmembrane F_o, formed from five (α_3 β_3 γ δ ϵ) and three (a b_2 c_{10-14}) subunit assemblies with different stoichiometries, respectively (Figure 9.1). It can also be divided into a catalytic α_3 β_3 hexamer, stalks ($\gamma\epsilon ab_2$), and membrane (a b_2 c_{10-14}) domains. Two stalks have been observed by electron microscopy [9]. The mitochondrial enzyme has additional subunits, possibly with regulatory functions.

The higher-ordered structure of the bovine α_3 β_3 γ complex has been solved by X-ray crystallography [10]. Following this breakthrough, the structures of crystals of bovine F_1 inhibited by efrapeptin, aurovertin, NBD-Cl (7-chloro-4-nitrobenzo-2-oxa-1,3-diazole), and DCCD (dicyclohexylcarbodiimide) were solved by the same group [11–14]. Bovine F_1 containing 1 mol MgADP-trifluoroaluminate [15] and two moles MgADP trifluoroaluminate [16] has also been crystallized, and the structures determined. These studies, together with biochemical analysis, have contributed greatly to an understanding of the catalytic site and mechanism. The structures of F_1 sectors of other origins have been reported [17–20]. The catalytic site

Handbook of ATPases. Edited by M. Futai, Y. Wada, J. H. Kaplan
Copyright © 2004 WILEY-VCH Verlag GmbH & Co. KGaA, Weinheim
ISBN 3-527-30689-7

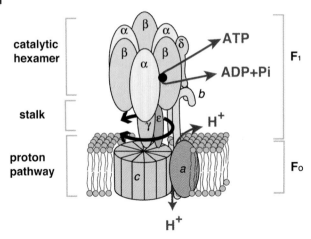

Figure 9.1 Schematic model of F-ATPase, showing the subunit structures of the catalytic hexamer ($\alpha_3\beta_3$), stalk and membrane domain. The membrane extrinsic F_1 and transmembrane F_o sector are shown together with energy coupling between the ATP synthesis–hydrolysis and subunit rotation.

is mainly located in the β subunit of the $\alpha_3\beta_3$ hexamer, and the three sites (one in each β) show strong cooperativity. The amino and carboxyl terminal helices of the γ subunit are located in the center of the $\alpha_3\beta_3$ hexamer, and form a central stalk with the ε subunit. The second or peripheral stalk is formed from the *a* and *b* subunits, and the δ subunit is located near the top of the α subunit [8]. The proton pathway is located at the interface between the *a* and *c* subunits, and the hairpin structure of the purified *c* subunit has been solved by NMR [21]. A ring structure formed from multiple *c* subunits was suggested by early studies involving electron [22] and atomic force microscopy [23, 24], and was extensively analyzed recently through NMR structure and genetic approaches [25, 26]. The X-ray structure of yeast F_1 with a *c* subunit ring has also been reported [27].

The mechanism of coupling of proton transport and ATP synthesis or hydrolysis has been a major question in research on this complicated enzyme [6–8]. The binding change mechanism proposes rotation of the γ subunit relative to the $\alpha_3\beta_3$ hexamer coupled with the chemistry at the catalytic sites [7]. Rotation of the $\gamma\varepsilon c_{10-14}$ complex relative to $\alpha_3\beta_3\delta ab_2$ upon ATP addition was shown recently, indicating that continuous rotation of the assembly of F_1 and F_o subunits is involved in the coupling between ATP hydrolysis and proton transport [28–31].

Vacuolar-type ATPase (V-ATPase) with significant similarity to F-ATPase was introduced later to a family of proton-pumping ATPases (for reviews, see Refs. 32–35). V-ATPase is apparently different from F-ATPase in its physiological roles, and forms an acidic luminal pH in endomembrane organelles including lysosomes and endosomes, and in extracellular compartments such as resorption lacuna formed between osteoclasts and the bone surface. It has extrinsic membrane V_1 and V_o transmembrane domains formed from the *A*, *B*, *C*, *D*, *E*, *F*, *G* and *H* subunits, and *a*, *c*, *c'*, *c''* and *d*, respectively. F- and V-ATPase share significant homology, especially in their catalytic subunits and proton pathways, but also some unique differences, including their membrane sector and stalk region subunit compositions. Thus, comparative studies of the two proton pumps are pertinent for understanding the molecular mechanisms of both.

In this chapter we discuss recent progress in the understanding of F-ATPase, focusing mainly on the energy coupling between the chemistry and proton transport through subunit rotation, and briefly on comparison with V-ATPase. It should be interesting for readers to follow a series of biochemical studies to show rotational catalysis of the F-ATPase holoenzyme leading to that of V-ATPase. We also hope this chapter will serve as a general introduction for the F- and V-ATPase sections of this book.

9.2
Catalytic Mechanism of F-ATPase

As expected from its complicated structure, F-ATPase is not a simple Michaelis–Menten type enzyme. Furthermore, the overall mechanism includes catalysis (chemistry), subunit rotation and proton translocation. X-Ray structure and kinetic studies, especially substrate binding analysis with an intrinsic tryptophan probe, revealed that the three catalytic sites are asymmetric. Senior and coworkers have recently extensively summarized and discussed the molecular catalytic mechanism of F-ATPase [36]. Thus, we discuss the catalytic mechanism and active site only

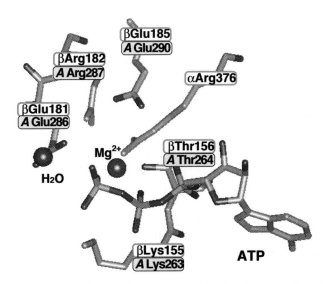

Figure 9.2 Catalytic sites of F-ATPase and V-ATPase. The catalytic residues of *E. coli* F-ATPase are shown together with bound ATP. Their positions are cited according to the bovine crystal structure [10]. Corresponding residues of yeast V-ATPase are also shown [110, 111].

briefly here, and leave the details to the excellent review and to related chapters of this book.

ATP synthesis and hydrolysis could be carried out kinetically at a single site (uni-site catalysis), two sites operating together (bi-site catalysis), or all three sites working together (tri-site catalysis). Uni-site catalysis has only been demonstrated for ATP hydrolysis, and can be measured experimentally with an ATP:F_1 ratio of less than 1:3 [3]. This rate is 10^5–10^6-fold lower than that of steady-state (multisite) catalysis. The enzyme cross-linked chemically, and thus could not rotate, but could still carry out uni-site catalysis [37]. This catalysis is not a part of the steady state, which includes subunit rotation.

Catalytic residues have been identified by analyzing uni-site catalysis of the purified *E. coli* mutant F_1 sector (Figure 9.2). They were discussed in detail previously for mechanistic implication [36, 38–41]. Briefly, βLys155 of the β subunit is required for binding of the γ phosphate moiety, as shown by studies involving affinity labeling [42] and mutant enzymes such as βLys155Ala (βLys155 → Ala) or βLys155Ser [43–46]. The βArg182 residue is also involved in the binding. Enzymes with substitutions of βThr156 showed similar properties to those of βLys155 [43]. The hydroxyl moiety of βThr156 is essential, possibly for Mg^{2+} binding, since it can only be replaced by a serine residue [43, 47]. βGlu181 is a critical catalytic residue [44]. Its side-chain forms a hydrogen bond with a water molecule located near the γ phosphate of ATP [10]. However, it was shown later that this residue was not involved in nucleotide binding or Mg^{2+} coordination [45, 48]. βTyr331, which is stacked close to the adenine ring [10], is required for the binding of ADP or ATP [40]. Results of analysis of the tryptophan fluorescence of the βTyr331Trp mutant F_1 are consistent with the role of the βTyr331 residue [6, 7]. The bovine residues corresponding to those discussed above are located close to the phosphate moiety or the adenine ring of bound-ATP or ADP in the X-ray structure [10] (Figure 9.2, see p. 239).

Recent results [16, 18, 40, 49–52], including the F_1 structure of all three sites filled with nucleotides [16, 18], support tri-site catalysis, in which the three sites are working together during the steady state, and the notion that bi-site catalysis does not occur [36]. We leave convincing discussion of these points to the article of Senior and coworkers [36].

It can be assumed that mutant enzymes defective in steady state catalysis should show impaired uni-site catalysis. In this regard, αArg376 mutant enzymes are of interest [53]. The αArg376 residue of the α subunit is located close to the β- or γ-phosphate of ATP or ADP and Mg at the catalytic site [10] (Figure 9.2). However, αAg376 does not directly participate in the chemistry of ATP hydrolysis or synthesis, as shown by the uni-site catalysis of mutant F_1 sectors [53]. The αArg376Lys r αArg376Ala mutant enzyme showed 2×10^3-fold lower steady-state ATP hydrolysis than the wild type. However, the mutant enzymes showed essentially the same kinetics for uni-site catalysis as the wild type, suggesting that they can pass through the transition state. These results indicate that αArg376 is essential for promotion of catalysis to the steady-state turnover. This notion is different from the previously suggested roles of αArg376 deduced from the structural model [10] and fluoroaluminate binding, an indicator of the formation of the pentacovalent transition state [54]. The βGlu185 residue, located close to the γ-phosphate and Mg at the catalytic site, may have a similar role to αArg376 because the mutant enzymes maintain uni-site catalysis but are defective in multisite catalysis [55]. However, detailed analysis could not be carried out because the mutant F_1 was unstable after solubilization from membranes.

The binding change mechanism of Boyer [7] proposes that the three sites are involved sequentially in ATP synthesis or hydrolysis: at a specific time point during the steady state, the chemical reaction occurs reversibly at one site, and ATP release and/or binding of ATP + Pi at two other sites with the expenditure of energy. Evidence supporting catalytic cooperativity and the binding change mechanism has accumulated, as reviewed previously [7]. It includes early kinetic evidence of uni-site and steady-state catalysis showing three K_m values and three interacting nucleotide binding sites, ^{18}O isotope exchange reactions, inhibitor studies, etc. These studies suggested that the chemistry, "ADP + Pi $\leftrightarrow$ ATP + H_2O", at the catalytic site is reversible, and provides essentially no energy change. The mechanism is strongly supported by the X-ray structure showing asymmetric catalytic sites and continuous γ subunit rotation [10]. However, it can not be concluded that the binding change mechanism was proven at the molecular level [36], although the mechanism is conceptually accepted and has been extremely useful for understanding the enzyme.

9.3
Roles of the γ subunit: energy coupling by mechanical rotation

9.3.1
Roles of the γ Subunit in Energy Coupling

The essential role of the γ subunit in catalysis and assembly was shown by early reconstitution experiments: a catalytic core complex exhibiting ATPase activity could be reconstituted from the purified E. coli α, β and γ subunits, but not without the γ subunit [56, 57]. Consistently, assembly of the F_1 sector is strongly affected by γ subunit mutations, especially those of residues interacting with the β subunit [58]. The isolated $\alpha_3\beta_3\gamma$ complex could functionally bind to the F_o sector only after its assembly with the δ and ε subunits [57]. These results suggest that the two minor subunits are required for the functional binding of $\alpha_3\beta_3\gamma$.

Early genetic approaches focused on the γ subunit carboxyl terminal region [2, 59, 60]. The amino acid sequences of the γ subunits are weakly conserved among different species [4, 59, 60], although their X-ray structures are similar [10, 18–20]. When the known γ sequences were aligned, only 28 of the 286 residues of the E. coli subunit are conserved, and mostly in the carboxyl and amino terminal helices located at the center of the $\alpha_3\beta_3$ hexamer [60]. Seventeen residues are conserved between residues 242 and 286 of the γ subunit. An enzyme with a nonsense mutation lacking 10 residues at the carboxyl terminus is still capable of in vivo ATP synthesis [59], indicating that the three conserved residues in this region are not required. Structural flexibility of the carboxyl terminus was also demonstrated by a frameshift mutation: the enzyme was still active with the γ subunit having seven additional residues at its carboxyl terminus together with nine altered residues downstream of γThr277 [61].

The enzyme with the nonsense mutation (γGln269End) was inactive, and substitution of a conserved residue (γGln269, γThr273, or γGlu275) between γGln269 and γLeu276 gave enzymes with reduced ATPase activity and energy coupling [59]. We noticed that mutations resulted in different ratios of ATPase catalysis and proton transport; three mutants (βThr277End, βGln269Leu, and βGlu275Lys) exhibited about 15 % of the wild-type ATPase activity, but showed various degrees of ATP-dependent H^+ transport and *in vivo* ATP synthesis. These results suggest active role(s) of the γ subunit in ATPase activity and energy coupling.

In addition to the carboxyl terminus, the only other conserved region is near the amino terminus [60]. The importance of this region was first suggested by the mutant lacking residues between γLys21 and γAla27, which resulted in failure of assembly of the F_1 complex [62]. We introduced amino acid substitutions systematically into the amino terminal region of the γ subunit [63]. Most of the changes between γIle19 and γLys33, γAsp83 and γCys87, or at γAsp65 had no effect, even with a drastic replacement such as hydrophobic to hydrophilic or acidic to basic. Interesting exceptions were the γMet23Arg and γMet23Lys substitutions. These mutants grew only slowly on succinate through oxidative phosphorylation, indicating that they are impaired in ATP synthesis. However, the membranes prepared from the γArg23 and γLys23 strains showed 100 and 65 % of the wild-type ATPase activity, but formed only 32 and 17 % of the electrochemical proton gradient, respectively, indicating that these mutants are defective in energy coupling between ATP hydrolysis and proton transport. In the X-ray structure, the γMet23 residue is located close to the DELSEED (βAsp380–βAsp386) loop of the β subunit [10]. Thermodynamic and kinetic analyses of the purified γMet23Lys enzyme suggested that the introduced γLys23 residue forms an ionized hydrogen bond with βGlu381 in the loop [64]. Consistent with this interpretation, the phenotype of γMet23Lys was restored by the second mutation, βGlu381Gln [65]. The βGlu381Lys mutation also caused deficient energy coupling. These results suggest that the interaction between the regions around γMet23 and βGlu381, and thus the γ subunit residue and β subunit DELSEED loop, are involved in energy coupling.

The γMet23Lys mutation was suppressed by substitution of carboxyl terminal residues, including γArg242, γGlu269, γAla270, γIle272, γThr273 γGlu278, γIle279k, and γVal280 [66]. From these results, we initially suggested that γMet23, γArg242, and the region between γGlu269 and γVal280 are three interacting domains that are required for efficient energy coupling. The X-ray structure shows that γMet23 located in the amino terminal helix is near γArg242, but γGly269 and γVal280 in the carboxyl terminal helix are near the top of the $\alpha_3\beta_3$ hexamer [10]. We further isolated second site mutations that suppressed the primary mutations, γGln269Glu and γThr273Val [67]. These mutations were mapped to the amino (residues 18, 34, and 35) and carboxyl (residues 236, 238, 242, and 262) termini. The higher-ordered structure clearly shows that γGlu269 or γThr273 does not interact directly with the residues of the second mutations. The occurrence of suppression at a distance may suggest that the two α helices of the γ subunit located at the center of the $\alpha_3\beta_3$ complex undergo long-range conformational changes during catalysis [68]. As expected, the relative orientations of the γ subunit to the three β subunits (β_E, β_{DP}, and β_{TP}) are different in the crystal structure, strongly supporting γ subunit rotation during ATP hydrolysis or synthesis [10].

9.3.2
γ Subunit Rotation

The higher-ordered structure of $\alpha_3\beta_3\gamma$ indicated that α and β are arranged alternately around the amino- and carboxyl-terminal α helices of the γ subunit [8, 10]. The binding change mechanism predicts that the catalytic sites in the three β subunits participate sequentially in ATP synthesis or hydrolysis via conformation transmission through the γ subunit rotation [7]. This rotation was suggested by experiments on chemical cross-linking between γCys87 and βCys380 (originally βAsp380) in the DELSEED loop [69, 70], and analysis of polarized absorption recovery after photobleaching of a probe linked to the carboxyl terminus of the chloroplast γ subunit [71, 72].

The continuous unidirectional γ subunit rotation in F_1 was recorded directly by Yoshida and Noji and their colleagues [73] (Figure 9.3). They immobilized the *Bacillus* $\alpha_3\beta_3\gamma$ complex on a glass surface through a histidine-tag introduced into the β subunit. The fluorescent actin filament connected to the γ subunit rotated continuously in an anticlockwise direction during ATP hydrolysis. The rotation became slower with an increase in the filament length, and generated a frictional torque of ~40 pN nm. The ε subunit rotation was also shown using the same approach [74], consistent with the tight association of γ and ε [27, 75]. A 120° step rotation

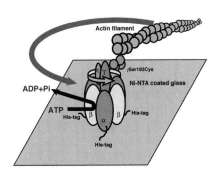
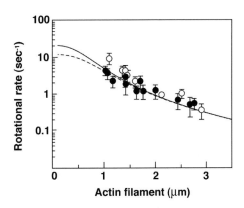

Figure 9.3 Rotation of the γ subunit of F-ATPase. F_1 was immobilized through a histidine tag introduced to the α or β subunit, and an actin filament was connected to the carboxyl terminus of the γ subunit. ATP-dependent rotation of the wild-type (open circles) and γ Met23Lys mutant (filled circles) F_1 sector are shown. Taken from Omote et al. [78].

was shown in the presence of a dilute ATP concentration, indicating that the γ subunit rotates, interacting with the three β subunits successively [76]. Furthermore, a refined measurement system involving gold beads revealed that the 120° step could be divided into 90° and 30° steps [77]. These two steps were proposed to correspond to ATP binding and ADP release, respectively, although this model is difficult to prove unambiguously. Interestingly, the rotation rates observed were much higher than those expected from the V_{max} of steady-state catalysis.

We could observe the rotation of an actin filament connected to the γ subunit of *E. coli* F_1 [78] using a similar system to that described for the *Bacillus* $α_3β_3γ$ complex [73]. The rotation was anticlockwise, inhibited by azide (F-ATPase inhibitor), and generated a frictional torque of ~40 pN·nm, as reported for the *Bacillus* complex. The F_1 sector, proven to be a chemically driven motor, should be connected functionally to the F_o sector to complete ATP-driven proton transport. Conversely, proton transport through F_o should be coupled to the γ subunit rotation and chemistry at the catalytic sites. Questions on the energy coupling between chemistry, rotation and proton transport could be answered with the *E. coli* enzyme by taking advantage of the accumulated genetic and biochemical information.

9.3.3
Mutational Analysis of the γ Subunit Rotation

We were interested in characterizing the γ rotation of a series of mutant enzymes defective in energy coupling and catalytic cooperativity. As described above, the γMet23Lys mutant is defective in energy coupling between catalysis and proton transport [63]. We thought that the γ rotation in the mutant F_1 may be defective, since the γ residue was replaced. Similar to the original mutant, the γMe23Lys enzyme engineered for rotation observation exhibited essentially the same ATPase activity as the engineered wild type. However, the enzyme could not form an electrochemical proton gradient in membrane vesicles, or carry out *in vivo* oxidative phosphorylation [78]. Unexpectedly, an actin filament connected to the γ subunit of γMet23Lys rotated, and generated essentially the same torque as that of the wild type (Figure 9.3). These results suggest that the γMet23Lys mutant F-ATPase could couple between chemistry and rotation, but was defective in transforming mechanical work into proton translocation or vice versa. In this regard, Al-Shawi et al. showed that the γMet23Lys mutant is defective in communication between F_1 and F_o [64]. These results also suggested that analysis of F_1 sector rotation is not enough to understand F-ATPase. It became imperative to determine which subunit complex is rotating in the F-ATPase holo enzyme purified or embedded in the membrane.

Mutant F_1 sectors with substitution of the βSer174 residue and their suppressors have been useful for understanding the rotation mechanism (Figure 9.4). Replacing βSer-174 with other residues altered the ATPase activity to between 150 and 10 % of the wild-type level, and the larger the side-chain of the residue introduced the lower the ATPase activity observed [79]. Both the βSer174Leu and βSer174Phe enzymes retained about 10 % of the wild-type ATPase activity. However, the two mutants showed a difference in energy coupling: the βSer174Leu mutant could still grow on succinate by oxidative phosphorylation and transport protons into the isolated membrane vesicles, whereas the βSer174Phe mutant could not grow and showed no proton transport. Consistent with these observations, their F_1 sectors differed in γ subunit rotation [80]. The F_1 sector with βSer174Phe showed apparently slower rotation than the wild type, and generated significantly lower frictional torque (~17 pN nm), whereas the F_1 with βSer174Leu was similar to the wild type. These results suggest that the rotation or torque generation is closely related to the energy coupling with proton transport.

Biochemical defects of the βSer174Phe mutation were suppressed by a second-site mutation, βGly149 to Ser, Cys, or Ala [81]. Double mutants such as βSer174Phe/βGly149Ser showed essentially the same ATPase activity and proton transport as the wild type. As expected from these results, the double mutant F_1 generated the wild-type torque (~40 pN nm) [80]. The high-resolution structure of the bovine F_1 predicts that βSer174 is located on the β subunit surface within the loop between an α helix (helix B) and a β sheet (β sheet 4) (Figure 9.4). βGly149, the first residue of the phosphate binding P loop connected to helix B, is close to the catalytic site. The structure of the βGly149–βSer174 region is significantly different between the nucleotide-bound and empty β subunits. Thus, it can be assumed that the conformational change of the catalytic site, followed by that of

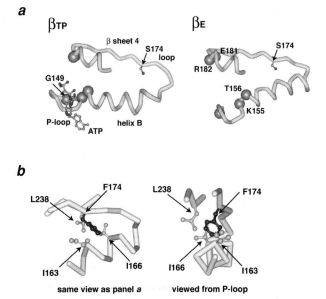

Figure 9.4 Models of the *E. coli* β subunit domain including βGly149 and βSer174. (a) Domain structure for the ATP-bound ($β_{TP}$) and empty ($β_E$) β subunit. The bovine structure [10] was used to model *E. coli* domains. Positions of ATP and residues discussed in the text are indicated. Nomenclatures for the α helix and β sheet are those of previous workers [10]. (b) Models of the βSer174Phe mutant domain structure. The positions of the residues discussed in the text are shown. Taken from Iko et al. [80].

the βGly149–βSer174 domain, leads to the γ subunit rotation for the energy coupling to proton transport. The βSer174Phe mutation was also suppressed by αArg296Cys of the α subunit [79]. It was of interest to analyze the rotation of the double mutant and related strains. However, the F_1 sector of βser174Phe/αArg296Cys was difficult to purify.

Energy minimization with a simple potential function of the modeled βSer174Phe mutant F_1 predicts that the side-chain of βPhe174 interacts with that of βIle163 or βIle166 of the β subunit with no nucleotide bound [80]. We substituted βIle163 or βIle166 with a less bulky Ala residue in the βSer174Phe mutant. As expected, the F_1 with the βIle163Ala/βSer174Phe or βIle166Ala/βSer174Phe double mutant could rotate and generate almost similar torque to the wild type. These results suggest that the βGly149–βSer174 domain plays important roles in the rotation and torque generation essential for energy coupling.

The role of the DELSEED loop has been focused on because the conformations of the three β subunits ($β_E$ and $β_T$ or $β_D$) are significantly different [10], which is consistent with the roles of γMet23 [63] and βGlu381 (the second residue of the DELSEED loop) [64] in energy coupling. However, mutagenesis studies involving replacement of residues in the DELSEED loop of thermophilic *Bacillus* $α_3β_3γ$ did not reveal significant effects on the torque generation [82]. The negative results may suggest that the loop is not related to the conformation change driving the β subunit rotation. However, this is difficult to conclude from the directed mutations without structural analysis. Furthermore, the experimental system with actin filaments gives rotation rates and torque values with high deviations. Extensive substitution of related residues and their second mutations should be analyzed for the final conclusion, as discussed above for a region around βSer174. Furthermore, analysis of rotation in the F-ATPase holoenzyme may be necessary, as pointed out for the γMet23Lys mutation [78].

9.4
Rotational Catalysis of the F-ATPase Holoenzyme

9.4.1
Structure of F_o Sector and Proton Transport Pathway

As discussed above, the F_o membrane sector is composed of *a*, *b* and *c* subunits, whose stoichiometry (1:2:10 ± 1, for *a:b:c*) was first determined from the stained bands on polyacrylamide-gel electrophoresis [83]. Based on structural prediction, Cox et al. proposed a model of F_o in which the *a* and *b* subunit helices are surrounded by a ring of *c* subunits [84]. However, this model was not consistent with the electron [22] and atomic force [23, 24] microscopic images. In the model derived from the images, the *a* and *b* subunits are attached to one side of the symmetric ring structure formed by the *c* subunits. An atomic force microscope image indicated <12 *c* subunits in the ring. The ab_2 complex was purified recently after solubilization of F_o from membranes utilizing a histidine tag introduced at the *a* subunit amino terminus [85]. The proton pathway (F_o) was reconstituted from the ab_2 complex and *c* subunit. These results confirmed that ab_2 and the *c* ring are two functional subcomplexes of the F_o sector.

Subunit *a* spans the membrane with five helices, and its fourth helix could be cross-linked to the carboxyl terminal helix of a *c* subunit when Cys residues were introduced at appropriate positions [86, 87]. *a*Arg210 in the fourth helix is involved in proton transport, and all the substitutions, even *a*Arg210 to Lys, gave a F_o sector with no ability to transport protons [25, 26, 88]. The structure of the amino terminus (residues 1–34) of the *b* subunit, including the trans-membrane helix, was solved by NMR in a chloroform–methanol–water mixture [89], and extramembrane helices interact with the δ and α subunits at the top of the F_1 sector [90, 91], which is consistent with the model proposing that the *b* subunit closely interacts with an $\alpha_3\beta_3$ hexamer [26]. The *b* subunit dimer interaction with the F_1 sector is necessary for a second stalk [89]. The carboxyl terminal helix of a *c* subunit can be cross-linked to the membrane helix of the *b* subunit [92]. Fillingame and coworkers solved the structure of the *E. coli c* subunit by NMR [21]: monomeric *c* (in a mixture of chloroform, methanol and water, pH 5) gave a hairpin-like structure formed from the two helices connected by a polar region that interacts with the γ and ε subunits. Structural models of F_o and the *c* ring were reviewed recently by Fillingame and coworkers [25, 26]. Thus, we discuss here only the pertinent points of the rotational catalysis.

The front face of one *c* subunit packs with the back face of a second *c* subunit to form a dimer, consistent with the results of mutant studies [25]. Fillingame and coworkers proposed that dimers form a ring of 12 monomers, with the amino and carboxyl α helices of each *c* subunit in the interior and at the periphery, respectively, and the ring was modeled from the results of molecular dynamic calculations [93]. The model was supported by cross-linking analysis [94]. However, a ring containing more than 10 monomers was not found in purified F_oF_1 [95], and recent recalculation gave essentially the same but a slightly smaller ring formed from 10 copies of the *c* subunit with the two helices in similar orientations [26, 96]. In an alternative model, the carboxyl and amino terminal α helices form inner and outer rings, respectively [97]. However, this model is not supported by cross-linking experiments [25]. The X ray structure indicated a yeast F_1 tightly bound with a *c* ring formed from ten monomers [27].

These results established that the *c* subunits form a ring of ten monomers. However, the number of monomers forming the ring is still controversial [27, 28]: atomic force microscopy of chloroplast [98] and bacterial [99] F_o demonstrated rings of 14 and 11 *c* subunits, respectively. The difference in the copy number may be due to the species difference, loss of a part of the monomers, or reorganization of the ring during purification. Structural studies on F-ATPase or the F_o sector will eventually explain the discrepancy.

*c*Asp61, in the middle of the second transmembrane α helix, is responsible for proton transport, and close to *c*Ala24 and *c*Ile28, of which substitutions by other residues reduced the DCCD reactivity of *c*Asp61 [24]. The stoichiometry of the *a* and *c* subunits (1:10) indicates that one *a*Arg210 and multiple *c*Asp61 are required for proton translocation in F-ATPase. As the pKa of the cAsp61 carboxyl moiety is 7.1, the NMR c subunit structure at pH 5 is at the fully protonated stage. The interaction between cAsp61 and aArg210 may lower the pKa to facilitate proton release into the proton pathway [25, 26]. Restogi and Girvin solved the c subunit structure at pH 8, with cAsp61 being in an almost completely deprotonated form [100]. The difference between the c structures at pH 5 and 8 is that the carboxyl terminal α helix was rotated by 140° with respect to the amino terminal helix. From the two structures and the results of cross-linking experiments, Fillingame et al. suggested that the carboxyl terminal α helix rotates during proton transport, interacts with aArg210, and deprotonates cAsp61 [25]. This c subunit structural change possibly drives stepwise rotation of the c ring.

9.4.2
Rotational Catalysis of the F-ATPase Holoenzyme

To complete ATP hydrolysis-dependent proton transport, the γ rotation should be transmitted to the F_o membrane sector. Conversely, for ATP synthesis, proton transport through F_o should generate torque to drive the γ subunit rotation coupled to chemistry. The γ rotation should be coupled to protonation–deprotonation of cAsp61 in either direction. Depending on the modeling of F_o (a and b subunit inside the c ring), rotation of a, b, γ, δ and ε relative to the complex of α, β and the c ring once has been suggested [84]. This speculation led to experimental tests, although an alternative model in which the c subunit ring attached by the a and b subunit helices is supported experimentally, as discussed above.

Different mechanisms could be proposed for the coupling between γ rotation within the $α_3β_3$ hexamer and proton transport through the F_o sector: (a) the γ subunit rotates on the surface of the c ring; and (b) the γ subunit and the c ring rotate as a single complex. Models should be consistent with the proton transport continuously utilizing one aArg210 and 10–14 cAsp61 residues of the a subunit and the c ring, respectively. Consistent with mechanism (b), energy transduction of F-ATPase with a counter-rotating rotor (γε c ring) and stator has been proposed [72, 101]. Convincing evidence supporting this mechanism includes the results of chemical cross-linking. Cross-linking between the γ and c subunits did not affect F-ATPase activity, indicating that sequential interaction of the rotating γ subunits with multiple c subunits is not necessary during ATP synthesis–hydrolysis [102]. Similarly, cross-linking between the γ and ε subunits did not affect ATP hydrolysis [103], supporting the rotation of an actin filament connected to either subunit. However, α–γ, α–ε, β–γ and β–ε cross-linking resulted in loss of the activity [104, 26], confirming that ε γ rotates against $α_3β_3$. These results suggest that a complex formed from γ, ε and c subunits is a mechanical unit for relative rotation to the $α_3β_3$ hexamer.

We obtained direct evidence of continuous rotation of the *E. coli* $εγc_{10–14}$ complex during ATP hydrolysis [28] (Figure 9.5). F-ATPase engineered for rotation was solubilized from membranes, purified, and immobilized through histidine tags introduced into the α subunit. Upon the addition of ATP, the filament connected to the

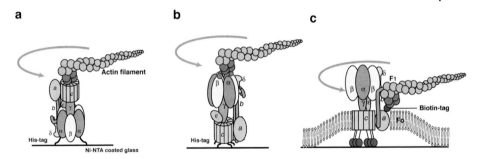

Figure 9.5 Rotational catalysis of F-ATPase. F_oF_1 was immobilized through a histidine-tag attached to the α subunit (a) or c ring (b, c), and a fluorescent actin filament was connected as a probe to the c (a), α (b), or a (c) subunit to observe rotational catalysis. Continuous rotation of the probe connected to the purified F_oF_1 (a or b) or membrane-embedded F_oF_1 (c) was observed upon addition of ATP.

c ring rotated anticlockwise continuously, and generated similar torque to that observed for the γ rotation in the F_1 sector. The rotation was inhibited by venturicidin, a specific inhibitor for F-ATPase but not for ATPase activity of the F_1 sector. The rotation and ATPase activity became less sensitive to the antibiotic when the cI-le38Thr mutation was introduced into the c subunit [30], indicating that antibiotic binding to the c ring inhibited rotation [28].

Pänke et al. also observed c subunit rotation during ATP hydrolysis [29]. They immobilized F_oF_1 through a histidine tag introduced into the β subunit, and an actin filament was connected to the c subunit through a streptag and streptavidin. The characteristics of the rotation observed were the same as those with the above systems. These results are consistent with the recent experiments showing that complete cross-linking of the γ, ε and c subunit had no effect on ATP hydrolysis, proton translocation, or ATP synthesis [104].

The important conclusion drawn from these experiments is that the c subunit ring rotates when F-ATPase is immobilized through the α or β subunit. As discussed by Pänke et al. [29] and Wada et al. [105], it is not easy to prove that all the subunits were integrated into F-ATPase rotating under the microscope. However, indirect evidence supports the intactness of the enzyme used for rotation: reconstitution studies indicated that the a, b and c subunit are required to form F_o capable of F_1 binding [106], and all F_1 subunits are required for binding to the F_o sector [57]. Early genetic and biochemical studies led to the conclusion that all three F_o subunits are required for F_1 binding (see Ref. 2 for a review). The intactness of F-ATPase in these observations was also supported by the results for the membrane-bound enzyme [31], which rotated in essentially a similar manner to the purified F-ATPase. As discussed previously [30, 105], criticism [107] regarding the initial observation [28] was based mainly on the negative observation under different experimental conditions.

We further addressed the basic question of whether the rotor and stator are interchangeable in F-ATPase [30] (Figure 9.5b). Thus, F-ATPase was immobilized on a glass surface through a histidine tag introduced into the c subunit, and an actin filament was connected to the β subunit through the biotin-binding domain of transcarboxylase (biotin-tag). We could observe ATP-dependent filament rotation generating the same frictional torque as observed above, similar to the case of F-ATPase immobilized through the α or β subunit and with an actin filament connected to the c subunit [28, 29]. Thus, either of the two subcomplexes ($\varepsilon\gamma c_{10-14}$ or $\alpha_3\beta_3\delta ab_2$) could be a rotor or a stator. Their actual roles *in vivo* will depend on the viscous drag due to the cytosol and membranes.

9.4.3
Rotational Catalysis of F-ATPase in Membranes

Although rotation of the $\gamma\varepsilon c_{10-14}$ complex has been shown using purified F-ATPase [28–30], a more favorable experimental system is the isolated membrane because the original integrity of the enzyme is maintained. The membrane could be used to answer one of the obvious questions of whether the c ring rotates relative to the a subunit. Such rotation would support the model of proton transport through the interface of the two subunits (Figure 9.5c). It may be difficult to test the rotation of a probe connected to the c ring embedded in membranes by immobilizing F-ATPase through the α or β subunit. As described above, purified F-ATPase can be immobilized through a histidine-tag connected to the c ring, and an actin filament can be connected to the α or β subunit through a biotin-tag [41]. This experimental system was modified to test the rotation of F-ATPase in the membrane. As histidine and biotin-tags face the periplasm and cytoplasm of the intact *E. coli* cell, respectively, membrane preparation for rotation should be carefully considered. Right-side out or everted membrane vesicles can not be used for testing rotation, because only one of the two tags in these vesicles is accessible from the medium: the histidine-tag faces the medium for right-side out vesicles, but the biotin-tag is inside the same vesicles. Similarly, the histidine-tag is inside everted vesicles, and thus cannot be used for immobilizing F-ATPase in these vesicles. Therefore, F-ATPase in planar membranes should be used to test for the rotation.

We prepared membrane fragments by passing *E. coli* cells through a French press by a slight modification of the procedure used to prepare everted vesicles [31]. A labeling experiment with streptavidine-conjugated gold particles indicated that the preparation contained a significant number of planar membranes. Using this preparation, we observed ATP-dependent rotation of an actin filament connected to the β subunit. As expected from the previous studies, the rotation was counter-clockwise and sensitive to DCCD.

To complete a model of proton translocation through the rotation in F-ATPase, it is essential to show rotation of the c ring relative to the a subunit (Figure 9.5c). The c subunit (cMet65Cys) cross-linked with the a subunit (αAsn214Cys) was highly protected from labeling with ^{14}C-DCCD [108]. However, when F-ATPase was labeled with ^{14}C-DCCD and subjected to the ATPase reaction before cross-linking there was significantly increased labeling of the a–c dimer. This experiment supports the rotational catalysis in F_o, although no information on the mechanism could be obtained. We could show rotation of the actin filament connected to the a subunit relative to the c ring using membrane fragments [31] (Figure 9.6).

These results and those obtained with purified F_oF_1 or the F_1 sector indicate that $\gamma\varepsilon c_{10-14}$ and $\alpha_3\beta_3\delta ab_2$ are mechanical units, i. e., an interchangeable rotor and stator, respectively. Notably, the results obtained with membrane-embedded F-ATPase and those with the purified one are similar. Furthermore, the experimental system will

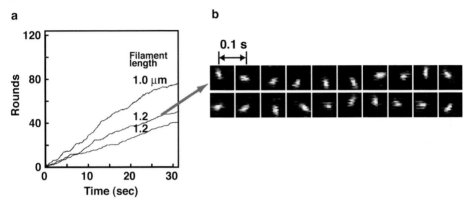

Figure 9.6 Relative rotation of the a subunit in membrane-embedded F-ATPase. Time courses of rotating filaments connected to the a subunit of membrane F-ATPase immobilized through the c subunit are shown, along with time courses of rotating filaments of varying length (a) and typical sequential video images (video interval, 100 ms) (b). Taken from Nishio et al. [31].

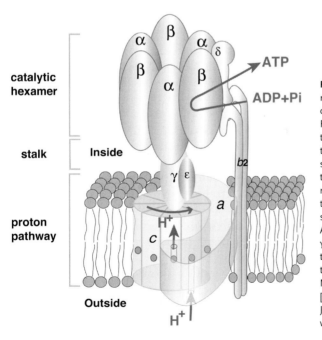

Figure 9.7 Schematic mechanism of rotational catalysis by F-ATPase. For ATP synthesis, electrochemical proton transport changes the c subunit conformation that drives the rotational movement of the c ring together with the γ and ε subunits. Conversely, ATP hydrolysis drives the γ and ε subunit rotation together with the c ring to transport protons. Modified from Oster [101], Fillingame [25, 26], Junge [72] and their co-workers.

be extremely valuable for studying the mechanism of coupling of rotation and proton transport, although further modification(s) may be necessary.

These studies established rotational catalysis by F-ATPase (Figure 9.7). For ATP hydrolysis, the F-ATPase is a chemically driven motor rotating $\gamma\varepsilon c_{10-14}$ to drive proton transport. Studies on the mechanism of the rotation in the F_o sector were started only recently. The rotation is inhibited by venturicidin or DCCD, suggesting that the tight or covalent binding of these bulky chemicals to the c subunit inhibits mechanical rotation [20]. Junge and coworkers showed that the c subunit ring with the cAsp61Asn mutation can still rotate during ATP hydrolysis, indicating that the proton transport is not obligatory for the chemically driven rotation (ATP hydrolysis-dependent c ring rotation) [109]. For ATP synthesis, the same enzyme is a potential-driven motor rotating $\gamma\varepsilon c_{10-14}$ through an electrochemical proton gradient. This membrane system will be useful for studying the rotation in this direction.

9.5
Rotational catalysis of V-ATPase

9.5.1
Catalytic Site and Proton Pathway

V-ATPase (vacuolar-type ATPase) acidifies the lumens of endomembrane organelles such as lysosomes, endosomes and synaptic vesicles (for reviews, see Refs. 32–35). The same enzyme in the plasma membranes of specialized cells, including osteoclasts and renal intercalated cells, pumps protons into extra cellular compartments such as resorption lacunae and collecting ducts, respectively. Despite significant physiological differences, V-ATPase exhibits similarities with F-ATPase. The catalytic A subunit of V-ATPase is homologous to F-ATPase β. The homology is striking in the phosphate binding P-loop and other sequences, including the F-ATPase catalytic residues discussed above [33]. *E. coli* F-ATPase residues at the catalytic site, βLys155, βThr156, βGlu181, βArg182 and βGlu185, correspond to ALys263, AThr264, AGlu266, AArg267 and AGlu290 of the yeast V-ATPase A subunit, respectively (Figure 9.8). Mutational studies of yeast V-ATPase supported the notion that they are catalytic residues [110, 111]. The kinetics indicate that V-ATPase has three catalytic sites exhibiting cooperativity [112]. The cysteine residue in the V-ATPase P-loop may be involved in regulation [113]. F-ATPase with Cys introduced at the corresponding position became sensitive to sulfhydryl reagents, similar to V-ATPase [114]. &&

The membrane V_o sector of V-ATPase has a more complicated subunit composition: yeast V_o is formed from c, c', c'', a and d subunits [32]. Of the three proteolipid subunits, the c and c' having four transmembrane helices are a duplicated form of the F-ATPase c subunit, and show 56% identity in amino acid residues with each other. The carboxyl moieties (cGlu137 and c'Glu145) for proton transport are located on the fourth helix of the c and c' subunits, respectively [115, 116]. The c'' subunit, exhibiting some homology with c and c', is also required in proton

```
Yeast G    1 ------------------------------------------------------------MSQKN     5
E.coli b   1 VNLNATILGQAIAFVLFVLFCMKYVWPPLMAAIEKRQKEIADGLASAERAHKDLDLAKAS            60


           6 GIATLLQAEKEAHEIVSKARKYRQDKLKQAKTDAAKEIDSYKIQKDKELKEFEQKNAGGV           65
          61 ATDQLKKAKAEAQVIIEQANKRRSQILDEAKAEAEQERTKIVAQAQAEIEA-ERKRARE-           118


          66 GELEKKAEAG-VQGELAEIKKIAEKKKD-DVVKILIETVIKPSAEVHINAL                    114
         119 -ELRKQVAILAVAGAEKIIERSVDEAANSDIVDKLVAEL------------                    156



Yeast G    1 ----------------------------------MSQKNGIATLLQAEKEAHEIVSK             23
E.coli δ   1 MSEFITVARPYAKAAFDFAVEHQSVERWQDMLAFAAEVTKNEQMAELLSGALAPETLAES            60

          24 ARKYRQDKLKQAKTDAAKEIDSYK-IQKDKELK-EFEQKNAGGVG--ELEKKAEAGVQGE           79
          61 FIAVCGEGLDENGQNLIRVMAENGRLNALPDVLEQFIHLRAVSEATAEVDVISAAALSEQ          120

          80 -LAEIKKIAEKKKDDVVKIL--IETVIKPSAEVHINAL------------------             114
         121 QLAKISAAMEKRLSRKVKLNCKLDKSVMAGVIIRAGDMVIDGSVRGRLERLADVLQS            177
```

Figure 9.8 Similarities of the V-ATPase G subunit with the F-ATPase b or δ subunit. The subunits are aligned to obtain maximal homology. Identical amino acid residues are boxed.

transport: c''Glu188 in the middle of the third α helix is also implicated for proton transport [115]. The c'' subunit is conserved in mammalians [117, 118], whereas the c' subunit is only found in yeast. The stoichiometry of the c:c':c'' subunits in yeast is n:1:1 [119]. The three proteolipids may form a ring structure similar to the F_o c ring. Assuming that 4, 1, and 1 molecules of the c, c' and c'' subunits [115], respectively, form a ring, V_o has ~6 proton translocating residues altogether. This is in contrast with 10–14 residues for F_o. Mutation of V-ATPase aArg735 abolished proton translocation similar to that of F-ATPase aArg210 [120], although the subunits a of V_o and F_o exhibit little homology. Thus, V_o has multiple carboxyl moieties and one arginine essential for proton translocation.

The similarities between the two ATPases raised the interesting question of whether V-ATPase can synthesize ATP. Unlike F-ATPase localized with the respiratory chain in the mitochondrial membrane, mammalian or yeast membranes containing V-ATPase do not have a system for generating an electrochemical proton gradient. Does V-ATPase synthesize ATP when a membrane potential or proton-gradient is generated? To answer this question, we expressed a plant proton translocating pyrophosphatase in yeast vacuoles [121]. Upon hydrolysis of pyrophosphate, the vacuolar membrane vesicles could form an electrochemical proton gradient, which was lowered by the addition of ADP + Pi. As expected from this result, we observed ATP synthesis sensitive to the V-ATPase specific inhibitor bafilomycin. The results further support the similarities between the two ATPases. Plant vacuoles having both V-ATPase and pyrophosphatese may synthesize ATP similar to the hybrid system studied in yeast.

9.5.2
Subunit Rotation of V-ATPase During Catalysis

Despite the similarities of V- and F-ATPase discussed above, they are also significantly different. Typically, F-ATPase of *E. coli* has eight subunits, whereas yeast V-ATPase has 13. Most of the subunits unique to V-ATPase are located in the stalk region. Furthermore, the presence of isoforms of subunits B, E, G, d, and a has been found for V-ATPase, whereas the information on isoforms is limited for F-ATPase. As pointed out above, V_o, possibly its ring structure, may be different from F_o. These differences and similarities led us to examine the rotational catalysis in V-ATPase.

As described above, rotation of an actin filament connected to the a, α, or β subunit was observed when F-ATPase was immobilized through the c subunit ring. The correspondence of minor subunits in the stalk region between the two ATPases was difficult to determine (Figure 9.8). The G subunit was shown to exist in the stalk domain of a recent model, and to be accessible from the cytosol [34]. The G subunit and F-ATPase b exhibit $\sim 24\%$ homology, although the G subunit lacks a transmembrane region. However, recent cross-linking studies suggested that the G subunit is located near the top of V_1 [122, 123], similar to the F-ATPase δ subunit [90, 91]. Thus, the G subunit may correspond to the b or δ subunit of F_1, and could be a candidate for connecting a probe to observe rotation, although the homology is quite limited.

Based on these considerations, we engineered the yeast chromosome (*VMA3* and *VMA10* for the c and G subunit, respectively), solubilized V-ATPase from vacuolar membranes, and tested rotational catalysis. An actin filament connected to the G subunit rotated upon ATP hydrolysis [124] (Figure 9.9). The rotation was inhibited by nitrate and concanamycin, similar to their effects on ATPase activity. Concanamycin inhibition was striking: the rotation terminated within a few seconds after the additionFigure 9., which is consistent with its tight affinity [112]. This antibiotic is similar to bafilomycin, which binds to the V_o sector, possibly to its c subunit [125], indicating that concanamycin, possibly bound to the rotor–stator interface, terminated the rotation. Torque generated by the rotation was similar to that with F-ATPase. We observed rotation in a sonicated vacuolar membrane preparation; however, we could not obtain enough data for analysis (unpublished observation).

These results suggest that the A_3B_3 hexamer rotates together with the G subunit, when the c subunit ring is immobilized. An interesting question is "Which part of the V-ATPase rotates *in vivo*?" In this regard, Holiday and coworkers reported that the B subunit interacts with the cytoskeleton, possibly actin [126]. Thus, the c subunit ring together with the subunit corresponding to the γ (possibly D) subunit may be rotating *in vivo*, if the V-ATPase holo enzyme is immobilized with the cytoskeleton. However, it is also possible that the rotor–stator is determined only by the slight difference in viscous drag applied to each sub-complex.

Rotational catalysis of the peripheral membrane sector of *T. thermophilus* ATPase was shown recently [127]. This *Archaean* ATPase is more similar to F-ATPase than

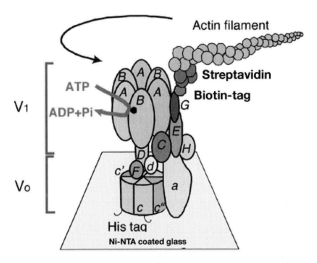

Figure 9.9 Rotation of an actin filament connected to the g subunit of V-ATPase. Solubilized V-ATPase was immobilized on a glass surface through the c subunit, and an actin filament was connected to the G subunit. Rotation was observed upon the addition of ATP, and inhibited by nitrate and concanamycin. Taken from Hirata et al. [124].

V-ATPase, as pointed out by the authors. In this regard, the ATPase of *Archae* (*Archaean bacterial*) plasma membranes has been classified as an A-type ATPase [128]. Thus, three distantly related proton translocating ATPases carry out common rotational catalysis.

9.6
Epilogue

F-ATPase has been a fascinating membrane enzyme for generations of biochemists. As Paul Boyer called it "A splendid molecular machine" [7], it is more than an ordinary enzyme. F-ATPase synthesizes or hydrolyzes ATP coupling between proton translocation and chemistry, and has become a real molecular machine in this sense, since mechanical rotation of its subunit complex was included recently in its mechanism. For the engineering aspect of the mechanism, the F_1 sector immobilized on a metal surface can rotate a metal plate [129].

As described above, experimental systems have been established for further studies. Details of the rotation mechanism will be revealed by studies involving *E. coli* F-ATPase together with the progress regarding physical methods and the higher-ordered X-ray structure determinations of the F_0 sector.

It was interesting to learn that rotational catalysis has been expanded to V-ATPase and also distantly related Archaean ATPase. The basic mechanism of V-ATPase rotation may be similar to that of F-ATPase, as expected from the similarities between the two enzymes. Thus, the rotational mechanism of *E. coli*

F-ATPase should be studied extensively, using its genetic and biochemical advantages. V-ATPase may have a more fascinating regulatory mechanism that is supported by a unique V_o structure and a series of isoforms in the stalk region. The regulatory role of stalk subunit(s) including subunit *E* in energy coupling has already been discussed [130].

Finally, owing to limited space, the present chapter is not a comprehensive survey, and so the reader is directed to the reviews listed above for references not cited here. We hope that readers will learn more interesting aspects of F- and V-ATPase in other Chapters of this book.

Acknowledgments

We wish to thank our coworkers, whose names appear in the references, from our laboratory. We are grateful to Ms S. Shimamura and M. Nakashima for preparation of the manuscript. The research in our own laboratories was supported by Grants-in-Aid from the Ministry of Education, Science, and Culture of Japan, the Takeda Foundation, and the Japan Science and Technology Corporation.

References

1. E. Racker, *A New Look at Mechanisms in Bioenergetics*, Academic Press, New York, **1976**.
2. M. Futai, T. Noumi, and M. Maeda, *Annu. Rev. Biochem.*, **1989**, *58*, 111–136.
3. H. S. Panefsky and R. L. Cross, *Adv. Enzymol. Relat. Areas Mol. Biol.*, **1991**, *64*, 173–214.
4. M. Futai and H. Omote, *Handbook of Biological Physics*. W. N. Konings, H. R. Kaback, and J. S. Lolkema, eds., Elsevier Science, Amsterdam **1996**, Vol. 2, pp 47–74.
5. R. H. Fillingame, *Curr. Opin. Struct. Biol.*, **1996**, *6*, 491–498.
6. J. Weber and A. E. Senior, *Biochim. Biophys. Acta*, **1997**, *1319*, 19–58.
7. P. D. Boyer, *Annu. Rev. Biochem.*, **1997**, *66*, 717–749.
8. D. Stock, C. Gibbons, I. Arechaga, A. G.W. Leslie, and J. E. Walker, *Curr. Opin. Struct. Biol.*, **2000**, *10*, 672–679.
9. S. Wikens and R. A. Capaldi, *Nature*, **1998**, *393*, 29–29.
10. J. P. Abrahams, A. G.W. Leslie, R. Lutter, and J. E. Walker, *Nature*, **1994**, *370*, 621–628.
11. J. P. Abrahams, S. K. Buchanan, M. J. Raaji Van, I. M. Fearnley, A. G.W. Leslie, and J. E. Walker, *Proc. Natl. Acad. Sci. U.S.A.*, **1996**, *93*, 9420–9424.
12. M. J. Raaji Van, J. P. Abrahams, A. G.W. Leslie, and J. E. Walker, *Proc. Natl. Acad. Sci. U.S.A.*, **1996**, *93*, 6913–6917.
13. G. L. Orriss, A. G.W. Leslie, K. Braig, and J. E. Walker, *Structure*, **1996**, *6*, 831–837.
14. C. Gibbons, M. G. Montgomery, A. G.W. Leslie, and J. E. Walker, *Nat. Struct. Biol.*, **2000**, *7*, 1055–1061.
15. K. Braig, R. I. Menz, M. G. Montgomery, A. G.W. Leslie, and J. E. Walker, *Structure*, **2000**, *8*, 567–573.
16. R. I. Menz, J. E. Walker, and A. G.W. Leslie, *Cell*, **2001**, *106*, 331–341.
17. Y. Shirakihara, A. G.W. Leslie, J. P. Abrahams, J. E. Walker, T. Ueda, Y.

Sekimoto, M. Kambara, K. Saika, Y. Kagawa, and M. Yoshida, *Structure*, **1997**, *5*, 825–836.

18. M. A. Bianchet, J. Hullihen, P. L. Pederson, and L. M. Amzel, *Proc. Natl. Acad. Sci. U. S.A.*, **1998**, *95*, 11065–11070.

19. A. C. Hausrath, G. Gruber, B. W. Matthews, and R. A. Capaldi, *Proc. Natl. Acad. Sci. U. S.A.*, **1999**, *96*, 13697–13702.

20. G. Groth and E. Pohl, *J. Biol. Chem.*, **2001**, *276*, 1345–1352.

21. M. E. Girvin, V. K. Rastogi, F. Abildgaard, J. L. Markley, and R. H. Fillingame, *Biochemistry*, **1998**, *37*, 8817–8824.

22. R. Birkenhäger, M. Hoppert, G. Deckers-Hebestreit, F. Mayer, and K. Altendorf, *Eur. J. Biochem.*, **1995**, *230*, 58–67.

23. K. Takeyasu, H. Omote, S. Nettikadan, F. Tokumasu, A. Iwamoto-Kihara, and M. Futai, *FEBS Lett.*, **1996**, *392*, 110–113.

24. S. Sign, P. Turina, C. J. Bustamante, D. J. Keller, and R. A. Capaldi, *FEBS Lett.*, **1996**, *397*, 30–34.

25. R. H. Fillingame, C. M. Angevine, and O. Y. Dmitriev, *Biochim. Biophys. Acta*, **2002**, *1555*, 29–36.

26. R. H. Fillingame and O. Y. Dmitriev, *Biochim. Biophys. Acta*, **2002**, *1565*, 232–245.

27. D. Stock, A. Leslie and J. E. Walker, *Science*, **1999**, *286*, 1700–1705.

28. Y. Sambongi, Y. Iko, M. Tanabe, H. Omote, A. Iwamoto-Kihara, I. Ueda, T. Yanagida, Y. Wada, and M. Futai, *Science*, **1999**, *286*, 1722–1724.

29. O. Pänke, K. Gumbiowski, W. Junge, and S. Engelbrecht, *FEBS Lett.*, **2000**, *472*, 34–38.

30. M. Tanabe, K. Nishio, Y. Iko, Y. Sambongi, A. Iwamoto-Kihara, Y. Wada, and M. Futai, *J. Biol. Chem.*, **2001**, *276*, 15269–15274.

31. K. Nishio, A. Iwamoto-Kihara, A. Yamamoto, Y. Wada, and M. Futai, *Proc. Natl. Acad. Sci. U. S.A.*, **2002**, *99*, 13448–13452.

32. N. Nelson and W. R. Harvey, *Physiol. Rev.*, **1999**, *79*, 361–385.

33. M. Futai, T. Oka, G.-H. Sun-Wada, Y. Moriyama, H. Kanazawa, and Y. Wada, *J. Exp. Biol.*, **2000**, *203*, 107–116.

34. T. Nishi and M. Forgac, *Nat. Rev. Mol. Cell. Biol.*, **2002**, *3*, 94–103.

35. G.-H. Sun-Wada, Y. Wada, and M. Futai, *J. Bioenerg. Biomembr.*, **2003**, *35*, 347–358.

36. A. E. Senior, S. Nadanaciva, and J. Weber, *Biochim. Biophys. Acta*, **2002**, *1553*, 188–211.

37. J. J. Garcia and R. A. Capaldi, *J. Biol. Chem.*, **1998**, *273*, 15940–15945.

38. H. Omote and M. Futai, *Acta Physiol. Scand.*, **1998**, *163 (Suppl)*, 177–183.

39. M. Futai, H. Omote, Y. Sambongi, and Y. Wada, *Biochim. Biophys. Acta*, **2000**, *1458*, 276–288.

40. H. Ren and W. S. Allison, *Biochim. Biophys. Acta*, **2000**, *1458*, 221–233.

41. A. E. Senior, S. Nadanaciva, and J. Weber, *J. Exp. Biol.*, **2000**, *203*, 35–40.

42. K. Ida, T. Noumi, M. Maeda, T. Fukui, and M. Futai, *J. Biol. Chem.*, **1991**, *266*, 5424–5429.

43. H. Omote, M. Maeda, and M. Futai, *J. Biol. Chem.*, **1992**, *267*, 20571–20576.

44. M. Y. Park, H. Omote, M. Maeda, and M. Futai, *J. Biochem.*, **1994**, *116*, 1139–1145.

45. S. Löbau, J. Weber, S. Wilke-Mounts, and A. E. Senior, *J. Biol. Chem.*, **1997**, *272*, 3648–3456.

46. S. Nadanaciva, J. Weber, and A. E. Senior, *Biochemistry*, **1999**, *38*, 7670–7677.

47. A. E. Senior and M. K. Al-Shawi, *J. Biol. Chem.*, **1992**, *267*, 21471–21478.

48. J. Weber, S. T. Hammond, S. Wike-Mounts, and A. E. Senior, *Biochemistry*, **1998**, *37*, 608–614.

49. J. Weber, R. S.V. Lee, E. Grell, J. G. Wise, and A. E. Senior, *J. Biol. Chem.*, **1992**, *267*, 1712–1718.

50. R. K. Nakamoto, C. J. Ketchum, and M. K. Al-Shawi, *Annu. Rev. Biophys. Biomol. Struct.*, **1999**, *28*, 205–234.

51. J. Weber and A. E. Senior, *Biochem. Biophys. Acta*, **2000**, *1458*, 300–309.

52. J. Weber and A. E. Senior, *J. Biol. Chem.*, **2001**, *276*, 35422–35428.

53. N. P. Le, H. Omote, Y. Wada, M. K. Al-Shawi, R. K. Nakamoto, and M. Futai, *Biochemistry*, **2000**, *39*, 2778–2783.

54. S. Nadanaciva, J. Weber, S. Wilke-Mounts, and A. E. Senior, *Biochemistry*, **1999**, *38*, 15493–15499.

55. H. Omote, N. P. Le, M.-Y. Park, M. Maeda, and M. Futai, *J. Biol. Chem.*, **1995**, *270*, 25656–25660.

56. M. Futai, *Biochem. Biophys. Res. Commun.*, **1977**, *79*, 1231–1237.

57. S. D. Dunn and M. Futai, *J. Biol. Chem.*, **1980**, *265*, 113–118.

58. H. Omote, K. Tainaka, A. Iwamoto-Kihara, Y. Wada, and M. Futai, *Arch. Biochem. Biophys.*, **1998**, *308*, 277–282.

59. A. Iwamoto, J. Miki, M. Maeda, and M. Futai, *J. Biol. Chem.*, **1990**, *265*, 5043–5048.

60. R. K. Nakamoto, K. Shin, A. Iwamoto, H. Omote, M. Maeda, and M. Futai, *Ann. New York Acad. Sci.*, **1992**, *671*, 335–344.

61. C. J. Beukelaer De, H. Omote, A. Iwamoto-Kihara, M. Maeda, and M. Futai, *J. Biol. Chem.*, **1995**, *270*, 22850–22854.

62. H. Kanazawa, H. Hama, B. P. Rosen, and M. Futai, *Arch. Biochem. Biophys.*, **1985**, *241*, 364–370.

63. K. Shin, R. K. Nakamoto, M. Maeda, and M. Futai, *J. Biol. Chem.*, **1992**, *267*, 20835–20839.

64. M. K. Al-Shawi, C. J. Ketchum, and R. K. Nakamoto, *J. Biol. Chem.*, **1997**, *272*, 2300–2306.

65. C. J. Ketchum, M. K. Al-Shawi, and K. K. Nakamoto, *Biochem. J.*, **1998**, *330*, 707–712.

66. R. K. Nakamoto, M. Maeda, and M. Futai, *J. Biol. Chem.*, **1993**, *268*, 867–872.

67. R. K. Nakamoto, M. K. Al-Shawi, and M. Futai, *J. Biol. Chem.*, **1995**, *270*, 14042–14046.

68. M. Futai and H. Omote, *J. Bioenerg. Biomembr.*, **1996**, *28*, 409–414.

69. R. Aggeler, M. A. Haughton, and R. A. Capaldi, *J. Biol. Chem.*, **1995**, *270*, 9185–9191.

70. T. M. Dancan, V. V. Bulygin, Y. Zou, M. L. Hutcheon, and R. L. Cross, *Proc. Natl. Acad. Sci. U. S. A.*, **1995**, *92*, 10964–10968.

71. D. Subbert, S. Engelbrecht, and W. Junge, *Nature*, **1996**, *381*, 623–625.

72. W. Junge, H. Lill, and S. Engelbrecht, *Trends Biochem. Sci.*, **1997**, *22*, 420–423.

73. H. Noji, R. Yasuda, H. Yoshida, and K. Kinoshita, Jr., *Nature*, **1997**, *386*, 299–302.

74. K. Kato-Yamada, H. Noji, R. Yasuda, K. Kinoshita, Jr., and M. Yoshida, *J. Biol. Chem.*, **1998**, *273*, 19375–19377.

75. S. D. Dunn, *J. Biol. Chem.*, **1982**, *267*, 7354–7359.

76. R. Yasuda, H. Noji, and K. Kinoshita, Jr. and M. Yoshida, *Cell*, **1998**, *93*, 1117–1124.

77. R. Yasuda, H. Noji, H. Yoshida, and K. Kinoshita, Jr., *Nature*, **2001**, *410*, 898–904.

78. H. Omote, N. Sambonmatsu, K. Saito, Y. Sambongi, A. Iwamoto-Kihara, T. Yanagida, Y. Wada, and M. Futai, *Proc. Natl. Acad. Sci. U. S. A.*, **1999**, *96*, 7780–7784.

79. H. Omote, M.-Y. Park, M. Maeda, and M. Futai, *J. Biol. Chem.*, **1994**, *269*, 10265–10269.

80. Y. Iko, Y. Sambongi, M. Tanabe, A. Iwamoto-Kihara, K. Saito, I. Ueda, Y. Wada, and M. Futai, *J. Biol. Chem.*, **2001**, *276*, 47508–47511.

81. A. Iwamoto, H. Omote, H. Hanada, N. Tomioka, A. Itai, M. Maeda, and M. Futai, *J. Biol. Chem.*, **1991**, *266*, 16350–16355.

82. K. Y. Hara, H. Noji, D. Bald, R. Yasuda, K. Kinoshita, Jr., and M. Yoshida, *J. Biol. Chem.*, **2000**, *275*, 14260–14263.

83. D. L. Foster and R. H. Fillingame, *J. Biol. Chem.*, **1982**, *257*, 2009–2015.

84. G. B. Cox, A. L. Fimmel, F. Gibson, and L. Hatch, *Biochim. Biophys. Acta*, **1986**, *849*, 62–69.

85. W.-D. Stalz, J.-C. Greie, G. Deckers-Hebestreit, and K. Altendorf, *J. Biol. Chem.*, **2003**, *278*, 27068–27071.

86. J. C. Long, S. Wang, and S. B. Vik, *J. Biol. Chem.*, **1998**, *273*, 16235–16240.

87. F. I. Valiyaveetil and R. H. Fillingame, *J. Biol. Chem.*, **1998**, *273*, 16241–16247.

88. S. Eya, M. Maeda, and M. Futai, *Arch. Biochem. Biophys.*, **1991**, *284*, 71–77.

89. O. Y. Dmitriev, P. C. Jones, W. P. Jiang, and R. H. Fillingame, *J. Biol. Chem.*, **1999**, *274*, 15598–15604.

90. A. J.W. Rodgers and R. A. Capaldi, *J. Biol. Chem.*, **1998**, *273*, 29406–29410.

91. D. T. Mc Lachlin and S. D. Dunn, *Biochemistry*, **2000**, *39*, 3486–3490.

92. P. C. Jones, J. Hermolin, W. P. Jiang, and R. H. Fillingame, *J. Biol. Chem.*, **2000**, *275*, 31340–31346.

93. O. Y. Dmitriev, P. C. Jones, and R. H. Fillingame, *Proc. Natl. Acad. Sci. U. S.A.*, **1999**, *96*, 7785–7790.

94. P. C. Jones and R. H. Fillingame, *J. Biol. Chem.*, **1998**, *273*, 29701–29705.

95. W. P. Jiang, J. Hermolin, and R. H. Fillingame, *Proc. Natl. Acad. Sci. U. S.A.*, **2001**, *98*, 4966–4971.

96. R. H. Fillingame, W. P. Jiang, and O. Y. Dimitriev, *J. Bioenerg. Biomembr.*, **2000**, *32*, 433–439.

97. G. Groth and J. E. Walker, *FEBS Lett.*, **1997**, *410*, 117–123.

98. H. Seelert, A. Poetsch, N. A. Dencher, A. Engel, H. Stahlberg, and D. J. Müller, *Nature*, **2000**, *405*, 418–419.

99. H. Stahlberg, D. J. Müller, K. Suda, D. Fotiadis, A. Engel, T. Meier, U. Matthey, and P. Dimroth, *EMBO Rep.*, **2001**, *2*, 229–233.

100. V. K. Restogi and M. E. Girvin, *Nature*, **1999**, *402*, 263–268.

101. T. Elston, H. Wang, and G. Oster, *Nature*, **1998**, *391*, 510–513.

102. S. D. Watts and R. A. Capaldi, *J. Biol. Chem.*, **1997**, *272*, 15065–15068.

103. C. L. Tang and R. A. Capaldi, *J. Biol. Chem.*, **1996**, *271*, 3018–3024.

104. S. P. Tsunoda, R. Aggeler, M. Yoshida, and R. A. Capaldi, *Proc. Natl. Acd Sci. U. S.A.*, **2001**, *98*, 898–902.

105. Y. Wada, Y. Sambongi, and M. Futai, *Biochim. Biophys. Acta*, **2000**, *1549*, 499–505.

106. E. Schneider and K. Altendorf, *EMBO J.*, **1985**, *4*, 515–518.

107. S. P. Tsunoda, R. Aggeler, H. Noji, K. Kinoshita, Jr, M. Yoshida, and R. A. Capaldi, *FEBS Lett.*, **2000**, *470*, 244–248.

108. M. L. Hutcheon, T. M. Duncan, H. Ngai, and R. L. Cross, *Proc. Natl. Acad. Sci. U. S.A.*, **2001**, *98*, 8519–8524.

109. K. Gumbiowski, O. Pänke, W. Junge, and S. Engelbrecht, *J. Biol. Chem.*, **2002**, *277*, 31287–31290.

110. Q. Liu, P. M. Kane, P. R. Newman, and M. Forgac, *J. Biol. Chem.*, **1996**, *271*, 2018–2022.

111. K. J. MacLeod, E. Vasilyeva, J. D. Baleja, and M. Forgac, *J. Biol. Chem.*, **1998**, *273*, 150–156.

112. H. Hanada, Y. Moriyama, M. Maeda, and M. Futai, *Biochem. Biophys. Res. Commun.*, **1990**, *170*, 873–878.

113. Q. Liu, X. H. Leng, P. R. Newman, E. Vasilyeva, P. M. Kane, and M. Forgac, *J. Biol. Chem.*, **1997**, *272*, 11750–11756.

114. A. Iwamoto, Y. Orita, M. Maeda, and M. Futai, *FEBS Lett.*, **1994**, *352*, 243–246.

115. R. Hirata, L. A. Graham, A. Takatsuki, T. H. Stevens, and Y. Anraku, *J. Biol. Chem.*, **1997**, *272*, 4975–4803.

116. N. Umemoto, Y. Ohya, and Y. Anraku, *J. Biol. Chem.*, **1999**, *266*, 24526–24532.

117. G.-H. Sun-Wada, H. Murakami, H. Nakai, Y. Wada, and M. Futai, *Gene*, **2001**, *274*, 93–99.

118. T. Nishi, S. Kawasaki-Nishi, and M. Forgac, *J. Biol. Chem.*, **2001**, *276*, 34122–34130.

119. B. Powell, L. A. Graham, and T. H. Stevens, *J. Biol. Chem.*, **2000**, *275*, 26354–23660.

120. S. Kawasaki-Nishi, T. Nishi, and M. Forgac, *Proc. Natl. Acd Sci. U. S.A.*, **2001**, *98*, 12397–12402.

121. H. Hirata, N. Nakamura, H. Omote, Y. Wada, and M. Futai, *J. Biol. Chem.*, **2000**, *275*, 386–389.

122. Y. Arata, J. D. Baleja, and M. Forgac, *J. Biol. Chem.*, **2002**, *277*, 3357–3363.

123. Y. Arata, J. D. Baleja, and M. Forgac, *Biochemistry*, **2002**, *41*, 11301–11307.

124. T. Hirata, A. Iwamoto-Kihara, G.-H. Sun-Wada, T. Okajima, Y. Wada, and M. Futai, *J. Biol. Chem.*, **2003**, *278*, 23741–23719.

125. M. Huss, G. Ingenhorst, S. Konig, M. Gassel, S. Dorse, A. Zeeck, K. Altendorf, and H. Wieczorek, *J. Biol. Chem.*, **2002**, *277*, 40544–40548.

126. L. S. Holiday, M. Lu, B. S. Lee, R. D. Nelson, S. Solivan, L. Zhang, and S. L. Gluck, *J. Biol. Chem.*, **2000**, *275*, 32331–32337.

127. H. Imamura, M. Nakano, H. Noji, E. Mureyuki, S. Ohkuma, M. Yoshida, and K. Yokoyama, *Proc. Natl. Acad. Sci. U. S.A.*, **2003**, *100*, 2312–2315.

128. K. Ihara, T. Abe, K. I. Sugimura, and Y. Mukohata, *J. Exp. Biol.*, **1992**, *172*, 475–485.

129. R. K. Soong, G. D. Bachand, H. P. Neves, A. G. Olkhovets, H. G. Craighead, and C. D. Montemagno, *Science*, **2000**, *290*, 1555–1558.

130. G.-H. Sun-Wada, Y. Imai-Senga, A. Yamamoto, Y. Murata, T. Hirata, Y. Wada, and M. Futai, *J. Biol. Chem.*, **2002**, *277*, 18098–18105.

10
Rotation of F_1-ATPase

Eiro Muneyuki and *Masasuke Yoshida*

Abbreviations

TF_1, F_1-ATPase from thermophilic *Bacuilus* PS3; EF_1, F_1-ATPase from *Escherichia coli*; MF_1, F_1-ATPase from mitochondria; CF_1, F_1-ATPase from chloroplasts; rps, round per second; fps, frame per second.

10.1
Introduction

F_oF_1-ATP synthase plays a central role in ATP synthesis during oxidative or photo-phosphorylation. This enzyme is special in that it utilizes the protonic electroche-mical potential created by an electron-transfer reaction by light or respiration to drive thermodynamically unfavorable ATP synthesis, as predicted by Mitchell [1, 2]. It can also drive uphill proton transport by hydrolyzing ATP. These properties are clearly in contrast with those of usual enzymes which promote only thermody-namically favorable chemical reactions by lowering the activation energy. In addi-tion, this enzyme has an unusual mechanistic property in that the rotation of a part of the enzyme complex relative to the rest plays an essential role in its function, as predicted by Boyer [3–6].

ATP synthase is a large protein complex ($\sim$500 kDa) is composed of a mem-brane-embedded portion, F_o (pronounced as 'ef oh'), and a peripheral component F_1 (Figure 10.1). F_o acts as a proton channel by itself, and isolated F_1 – often called the F_1-ATPase – catalyzes ATP hydrolysis, (which is considered a reverse reaction of ATP synthesis. (Strictly speaking, the ATP hydrolysis by F_1-ATPase cannot be the exact reverse reaction of ATP synthesis by F_oF_1-ATPase. Depending on the environ-mental conditions, F_1-ATPase or F_oF_1-ATP synthase may switch the form of the en-zyme and it is safe to regard the reactions under different conditions as different reactions are carried out by different forms of enzyme.). The subunit structures of ATP synthases have been well conserved during evolution but there are some var-iations among sources. The simplest version is bacterial ATP synthases in which F_o

Handbook of ATPases. Edited by M. Futai, Y. Wada, J. H. Kaplan
Copyright © 2004 WILEY-VCH Verlag GmbH & Co. KGaA, Weinheim
ISBN 3-527-30689-7

contains three kinds of transmembrane subunit with a stoichiometry $a_1b_2c_{10-14}$ (some bacteria contain 11 copies of c subunits), and F_1 contains five kinds of subunit with a stoichiometry $\alpha_3\beta_3\gamma_1\delta_1\varepsilon_1$. The catalytic sites for ATP hydrolysis are found mainly on the β subunits of F_1, but residues of the α subunits also contribute. The α subunits contain a non-catalytic nucleotide-binding site, the function of which is yet poorly understood. The central stalk is made of the γ and ε subunits, and the side stalk is made of the $F_1\delta$ and F_ob_2 subunits (Figure 10.1).

In this chapter we first describe the phenomenological, biochemical aspect of the kinetics of this enzyme, which eventually lead to Boyer's binding change mechanism and rotary catalysis hypothesis. Secondly, we briefly describe structural studies, mainly carried out by Walker's group. Then we describe the results and implications drawn by direct observation of the rotation by this enzyme. Finally, we attempt to present possible reaction schemes of this enzyme and define remaining problems. Throughout this manuscript, we regard the $\alpha_3\beta_3\gamma$ subcomplex of F_1-ATPase as essentially the same as F_1-ATPase and call them simply F_1-ATPase. The amino acid residue numbers are those of TF_1 (F_1-ATPase from thermophilic *Bacillus* PS3) unless otherwise stated.

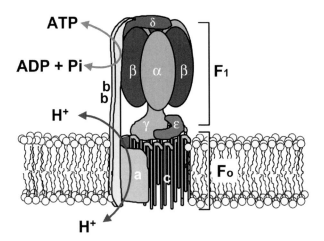

Figure 10.1 Schematic illustration of F_oF_1-ATP synthase. A bacterial F_oF_1-ATP synthase is illustrated as the simplest version of ATP synthases. It is composed of a water-soluble protein complex of ~380 kDa, F_1, and a hydrophobic transmembrane portion, F_o. F_o acts as a proton channel by itself, and isolated F_1 – often called the F_1-ATPase – catalyzes ATP hydrolysis. Subunit structures of ATP synthases have been well conserved during evolution, but there are some variations among sources. The simplest is bacterial ATP synthase, in which F_1 contains five kinds of subunit with a stoichiometry $\alpha_3\beta_3\gamma_1\delta_1\varepsilon_1$, and F_o contains three kinds of transmembrane subunit with a stoichiometry $a_1b_2c_{10-14}$. Chloroplast ATP synthase has the same subunit composition except that two kinds of F_o b homologue exist. Mitochondrial ATP synthase has only one copy of F_o b subunit and at least six kinds of additional accessory subunit. Confusing subunit nomenclature remains for historical reasons (for example, mitochondrial δ corresponds to bacterial ε, and the mitochondrial subunit named OSCP corresponds to bacterial δ); in this review we use the names of subunits according to the bacterial enzyme.

10.2
Biochemical Characterization

In the very early studies of this enzyme, it was already noted that the ATP concentration dependency of the reaction rate of this enzyme did not show a simple Michaelis–Menten type behavior. Rather, it exhibited an apparent negative cooperativity (Figure 10.2), which was regarded as an indication of some interaction between multiple catalytic sites [7]. Concurrently, it was found many anions affect the ATPase activity and apparent negative cooperativity. Although there were some discrepancies in the quantitative aspects, these findings were repeatedly reported by many researchers. But the explanations were not necessarily the same. Explanations of these features involved cooperativity between catalytic nucleotide binding sites, some control function by noncatalytic nucleotide binding sites, or the effect of some anion-specific binding sites [8–13].

As early as the finding of these characteristics, it was also noted that Mg^{2+} ion, which is required for ATP hydrolysis and synthesis, had a strong inhibitory effect on ATPase activity [14]. ADP, a product of ATP hydrolysis, was also found to have a strong inhibitory effect. As the inhibition by ADP required Mg^{2+}, the Mg^{2+} inhibi-

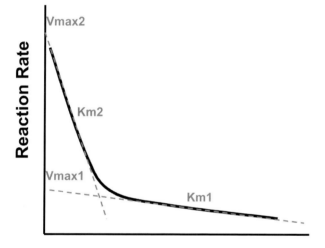

Reaction rate/[ATP]

Figure 10.2 Schematic illustration of a kinetic plot of F_1-ATPase activity. The apparent negative cooperativity of F_1-ATPase is most clearly recognized when the reaction rate is plotted against (reaction rate)/[ATP] (an Eadie–Hofstee plot). When the reaction rate is expressed by a Michaelis–Menten type scheme, the plot is straight and the slope and ordinate intercept give $-K_m$ and V_{max}, respectively. For F_1-ATPase, however, the plot is often concave, showing apparent negative cooperativity. By extrapolating the linear part of the plot, or using a nonlinear regression method, two or three sets of K_m and V_{max} are often deduced [13, 31, 76, 77]. However, the apparent cooperativity may not necessarily reflect interactions between catalytic sites. Rather, it seems to reflect the complex effects of ADPMg inhibition and activation by ATP binding to noncatalytic site(s). See text for details.

tion and ADP inhibition were supposed to share a common mechanism in which bound ADPMg exerts an inhibitory effect [15–17]. When Mg^{2+} alone, without externally added ADP, inhibits the enzyme, it is most likely that ADP endogenously bound to the purified enzyme forms a complex with added Mg^{2+} and causes the inhibition. Even in the presence of an ATP regenerating system, which keeps ADP concentration low, the inhibition proceeds slowly, over a time range of 10 s. Therefore, when the time course of the ATPase reaction was followed by the ATP regenerating system, it shows a rapid hydrolysis phase during the first several seconds, followed by a slow inactivated hydrolysis phase. Furthermore, another slow reactivation phase following the inactivated hydrolysis phase at several ten μM ATP concentration range is frequently observed (Figure 10.3) [18]. Extensive analyses of the kinetics of ADPMg inhibition concluded that, with ADPMg formed during the ATP hydrolysis or bound from external medium, the enzyme fell into an inhibited state that hardly releases the bound ADPMg [15, 16, 19]. NaN_3, a potent F₁-ATPase inhibitor, acts by stabilizing the ADPMg inhibited state. Only one ADPMg at a catalytic site is enough to induce the inhibition [19]. But the rate of the progress of ADPMg inhibition or NaN_3 inhibition shows strong ATP concentration dependency [20]. The ATP concentration which gives half-maximal rate of inhibition was 3–10 μM. From this concentration range it was previously concluded that when two catalytic sites were occupied by nucleotides the rate of inhibition was promoted [21]. However, compared with the nucleotide binding profile measured fluorescently with genetically introduced Trp near catalytic sites [22–24], this concentration range corresponds to filling of the third catalytic site rather than the second catalytic site. The slow reactivation was attributed to ATP binding to noncatalytic nucleotide binding sites [18]. After the crystallographic structure of F₁-ATPase was reported [25], a noncatalytic site-deficient mutant was produced and that ATP binding to noncatalytic site recovers ADPMg inhibition was firmly established [26]. Activation anions are assumed to reduce the ADPMg inhibition – such an effect of phosphate has been confirmed. By monitoring nucleotide binding to catalytic sites using genetically introduced Trp, it was concluded that phosphate binding to the γ phosphate position of ADP bound catalytic site shifts the equilibrium from ADPMg inhibited state to an active state without releasing the ADPMg [27]. Sulfate may also activate the ADPMg inhibited enzyme by binding to the catalytic site, as seen in the crystal structure [28]. The mechanisms of the action of the other anions are not very clear. So far, no anion-specific binding site that may explain the anion effects has been identified in the crystal structure.

Despite the long history of research, the origin of the apparent negative cooperativity (Figure 10.2) is not fully understood and is still a matter of debate. At present, it seems most plausible that the previously observed high K_m of several hundred μM or higher, which was sometimes regarded as ATP binding to the third catalytic site, reflects ATP binding to a noncatalytic site(s) that reactivates the ADPMg inhibited form [18, 24, 29]. The K_m of several μM to several tens μM, which was previously assumed to reflect ATP binding to the second catalytic site [30], seems to correspond to ATP binding to the third catalytic site, judging from the fluorescent titration of nucleotide binding to catalytic sites using genetically introduced Trp

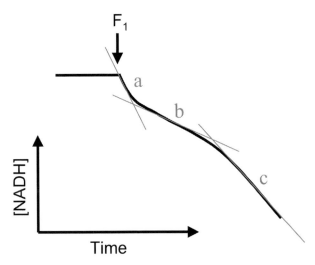

Figure 10.3 Schematic illustration of the time course of ATP hydrolysis by F_1-ATPase. When measured with an ATP regenerating system that couples ATP hydrolysis with NADH oxidation, the time course of ATP hydrolysis is continuously followed as a decrease in NADH concentration. This figure is a schematic illustration of such a measurement, emphasizing the time-dependent change in ATPase activity. The slope of the trace is proportional to the ATPase activity. (Part a) Just after the start of the reaction by adding F_1-ATPase to the assay mixture (indicated by an arrow), most of the F_1-ATPase is active and an initial burst is observed. (b) Gradually, ADPMg inhibition proceeds and the activity is decelerated. (c) Activity increases again, due to a slow binding of ATP to noncatalytic sites which reduces ADPMg inhibition [18].

[22–24]. However, the relationship of other apparent K_ms in the sub-µM to µM range, which were sometimes reported, to ATP binding to catalytic or noncatalytic sites is not clear. It may actually be only an apparent K_m caused by the progress of ADPMg inhibition and reactivation during ATP hydrolysis. In fact, depending on when the rate of hydrolysis is defined, apparent kinetic parameters significantly change [31]. If we can eliminate the effects of ADPMg inhibition and reactivation by ATP binding to noncatalytic site(s), the ATP hydrolysis activity of this enzyme may turn out to follow a simple Michaelis–Menten type kinetics. Some careful studies seem to point to this possibility [29, 70].

10.2.1
Oxygen Exchange Reaction and Boyer's Binding Change Mechanism

Among many biochemical studies, Boyer tried to elucidate the mechanism of this enzyme by analyzing oxygen exchange reactions. When ATP is synthesized from ADP and Pi by F_oF_1-ATPsynthase, one molecule of water is released whose oxygen atom was originally contained in Pi. Likewise, when ATP is hydrolyzed by this enzyme, the resultant Pi contains an oxygen atom from the water medium. During hydrolysis of one ATP, if one or more re-synthesis and hydrolysis occur on the enzyme before product release, the resultant Pi may contain more than one oxygen atom from water medium [32–34]. Using ^{18}O isotope, Boyer carefully examined the number of oxygen atoms exchanged during hydrolysis and synthesis of ATP, and found that reversible hydrolysis and synthesis of ATP occur while bound to the catalytic site of this enzyme. In particular, he examined the substrate concentration dependency and the effect of uncoupler on the exchange reaction and reached the following important conclusions. (1) The equilibrium constant of the step of ATP hydrolysis (chemical bond cleavage) on the enzyme is close to unity rather than the order of 10^5 M in the aqueous medium. (2) Proton translocation in the F_o portion is energetically coupled with the substrate binding and product release steps rather than the chemical reaction on the enzyme. (3) Multiple catalytic sites interact with each other such that they alternate their role in catalysis. Furthermore, Boyer unified these conclusions as a form of Binding Change Mechanism [3–5] (Figure 10.4). According to this mechanism, all catalytic sites have different affinity and catalytic properties at any time. When only one ATP binds to the enzyme in a hydrolytic reaction, it binds to the high-affinity catalytic site where reversible hydrolysis and synthesis occur. The product release from this tight binding site is slow, but binding of the second ATP to another low-affinity site induces conformational change of the first catalytic site to promote the hydrolysis and product release. In this binding change step, the second binding site becomes the high-affinity site and the previous high-affinity site becomes the low-affinity site. In such a way, multiple catalytic sites alternately change their binding affinity for the substrate and product during steady-state catalysis. The "binding change" step (affinity change for substrate and product), rather than the chemical bond cleavage, was concluded to be linked to the proton translocation in the F_o portion. As there are three catalytic sites on the F_1-ATPase, each of these sites was assumed to be in an "open", "loose", or "tight" conformation at a time. The "open" site has a very low affinity for substrates, the "loose" site can bind substrates reversibly and the "tight" site has a very high affinity such that ATP can form spontaneously from ADP and inorganic phosphate (Figure 10.4). The basic concept of the binding change mechan-

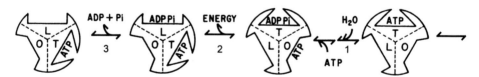

Figure 10.4 Binding change mechanism. The bi-site version of the binding change mechanism in which, maximally, two catalytic sites are occupied with AT(D)P is shown. Three catalytic sites take one of the following three conformations: L, loose binding for ligands and catalytically inactive; T, tight binding for ligands and catalytically active; O, open site with very low affinity for ligands and catalytically inactive. When only ATP binds to a catalytic site, it binds to a tight form catalytic site and in a near 1:1 equilibrium with ADP and Pi, allowing oxygen exchange reactions (right half of step 1). When a second ATP binds to an open site (left half of step 1), the catalytic sites of F_1 are interconverted between open, tight, and loose conformations, producing energy (step 2). Finally, ADP and Pi are released from the loose site (step 3). This Scheme shows one-third of the enzyme cycle. Figure is modified from the 1981 view of Cross [35]. Further modifications have been suggested by many researchers.

ism can be accommodated in a kinetic scheme that assumes nucleotide binding to maximally two [4, 35] or three catalytic sites [3]. Actually, whether the simultaneous nucleotide binding to the three catalytic sites is essential for the rapid and energy-linked catalysis, or simultaneous occupation of two catalytic sites by nucleotide is essential for the function, is still a matter of debate [36–38].

The binding change mechanism was supported by the so-called uni-site catalysis experiments first carried out by Grubmeyer, Cross and Penefsky [39]. Using mitochondrial F_1-ATPase, they found that when the molar ratio of ATP to F_1-ATPase was less than 1, the added ATP bound to F_1 very tightly and was hydrolyzed at $10–20 \text{ s}^{-1}$, more than ten times slower than the V_{max} rate. The free energy change of ATP binding was similar to that of ATP hydrolysis under physiological conditions. The release of the hydrolyzed product was very slow and a near 1:1 pseudo-equilibrium between bound ATP and ADP + Pi was established on the high-affinity site. Furthermore, the hydrolysis of bound ATP and release of the products were enormously accelerated by chase addition of excess ATP. Although these typical uni-site features were not necessarily observed for F_1-ATPase of other sources [40, 41], they coincided with what the binding change mechanism had predicted, thereby strongly supporting Boyer's hypothesis. The finding by Feldman and Sigman [42] that tightly bound ATP is synthesized on CF_1-ATPase incubated with high concentration ADP and Pi gave further support in that, once ADP and Pi are tightly bound to the catalytic site, tightly bound ATP is spontaneously formed without external energy.

10.2.2
Rotary Catalysis Hypothesis by Boyer

According to the binding change mechanism, three catalytic sites on the β subunits are in different states at any instance, but they change their role alternately and sequentially so that they behave in the same way with different phases. In order that the three β subunits behave exactly in the same way, Boyer postulated that the smaller one-copy subunits, γ, δ and ε, together, physically rotate in the ring of $\alpha_3\beta_3$ so that they always face a β subunit of the same state (Figure 10.5). About the same time, Oosawa in Japan independently predicted the rotation in ATP synthase [43] based on a theoretical consideration.

Several lines of experimental results indirectly supported the rotary catalysis hypothesis [3, 4]. The first was the single hit inactivation. If the three catalytic sites alternately and sequentially change their role during catalysis, inactivation of one of the three β subunits should stop the alternation and result in abolishment of total activity. Such an inhibitor or chemical modification was indeed found and their effects were explained as the rotating subunit(s) was stopped at the inactivated β subunit. Such an inhibition or modification study was often hampered by its incomplete reaction yield; however, Amano et al. succeeded in preparing an $\alpha_3\beta_3\gamma$ subcomplex of TF_1-ATPase in which exactly one, two or three β subunits were inactivated by mutation [44], thereby sweeping away the ambiguity of previous results.

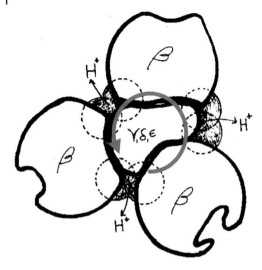

Figure 10.5 Rotation hypothesis. According to the binding change mechanism, all three catalytic sites are in different conformations at any time, but all pass sequentially through the same conformations. If this asymmetry between catalytic sites is caused by the interactions between the $\alpha_3\beta_3$ portion and single copy subunits, the sequential conformational change of the catalytic sites must accompany the rotational movement of the single copy subunits relative to the $\alpha_3\beta_3$ portion. Modified from Ref. 3.

The second line of results was the reversible crosslinking study carried out by Cross's group [45]. In their experiment, they first crosslinked a labeled β subunit with a γ subunit and cleaved the crosslinking, and then regenerated the crosslinking again. They found that if they added ATP between the cleavage and re-crosslinking, the γ subunit was crosslinked to a β subunit different from the originally labeled one. The result indicated some re-orientation of the γ subunit during hydrolysis of ATP. Conversely, Junge's group applied a different method [46]. They labeled the γ subunit with a fluorescent dye and quenched the fluorescence in a specific orientation using a polarized laser. From the recovery of the fluorescence, they estimated the change of the γ subunit was greater than $200°$ during ATP hydrolysis.

Together with the high-resolution crystal structure of F_1-ATPase reported in 1994 [25], these experiments strongly supported the rotation of the central γ subunit in F_1-ATPase, but none of them directly indicated the continuous rotation of the central γ subunit in the F_1-ATPase at work.

10.3
Structural Study

10.3.1
"Native" Structure Revealed

After a long effort, in 1994, Walker's group first reported the high-resolution crystal structure of bovine heart mitochondrial F_1-ATPase [25]. The revealed structure had an enormous impact in the field of ATP synthase.

In the initial crystal structure of the $\alpha_3\beta_3\gamma$ portion of bovine mitochondrial F_1 (termed the 'native' structure), three α subunits and three β subunits are arranged alternately, forming a cylinder of $\alpha_3\beta_3$ around the coiled-coil structure of the γ sub-unit (Figure 10.6). The α and β subunits have a similar fold, as would be expected from their sequence similarity. All of the α subunits are bound to the ATP analogue AMP–PNP, and the three subunits adopt very similar conformations, except that the noncatalytic nucleotide binding site of an α subunit (termed α_{TP}, see Figure 10.6) is more open than the others. The three β subunits, however, are in three nu-cleotide-bound states: the first, termed β_{TP}, has AMP–PNP in the catalytic site, the second (β_{DP}) has ADP, and the third (β_E) has no bound nucleotide (Figure 10.6). The γ subunit seems to prevent β_E from adopting a nucleotide-binding conforma-tion by obstructing the rotation of its lower half towards the central axis. In conse-

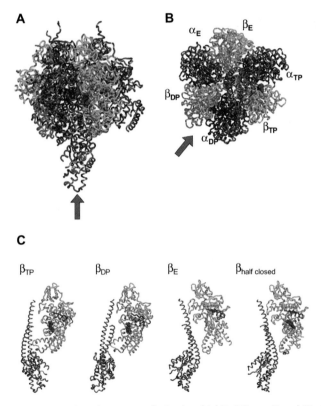

Figure 10.6 Crystal structure of mitochondrial F_1-ATPase. (A and B) side and bottom view of bovine mitochondrial F_1-ATPase. The view points for (A) and (B) are shown by red arrows. Structural data are of DCCD bound F_1 (PDB ID = 1E79) [52]. Nucleotides bound to catalytic and noncatalytic sites are shown by red and pink spheres, respectively. (C) Different forms of β subunit. Structural data of β_{TP}, β_{DP} and β_E are of DCCD bound F_1 (PDB ID = 1E79) as in (A) and (B). Data of $\beta_{half-closed}$ are of (ADP·AlF$_4^-$)$_2$F$_1$ (PDB ID = 1H8E) [28]. β_{TP}, β_{DP} represent the closed conformation and β_E and $\beta_{half-closed}$ represent open and half-closed conformation, respectively. The structure around the putative hinge region is colored as in Figure 10.11. The DELSEED region of the β subunit (residues 394–400) is in purple. Molecular graphics generated using PyMOL (Warren L. DeLano, DeLano Scientific, San Carlos, CA. http://www.pymol.org).

quence, the lower halves of β_E and α_{TP} do not interact, and the middle and carboxyl terminal domains of α_{TP} have moved away from the central axis. Thus, the β_E adopts the clear 'open' form, whereas β_{TP} and β_{DP} have the 'closed' form. The different nucleotide binding states of three β subunits accord with what Boyer's model had predicted. The three β subunits, β_E, β_{TP}, and β_{DP} were regarded to correspond to "open", "loose" and "tight" catalytic sites, respectively, as postulated in the binding change mechanism. Together with the three characteristic structures of the β subunit, the coiled-coil structure of the γ subunit penetrating the cylinder of $\alpha_3\beta_3$ fascinatingly urged people to believe that the central subunit rotates during catalysis. The structure of the protruding part of γ subunit was, however, disordered. Several crystals of bovine F₁-ATPase were reported such as those inhibited by efrapeptin [47], aurovertin [48], 4-chloro-7-nitrobenzofurazan chloride [49] and Mg-ADP fluoroaluminate [50], but the structure of the protruding part of γ subunit remained unclear. After the relatively modest resolution crystal structure of F_oF_1-ATP synthase from *Saccharomyces cerevisiae*, which revealed the architecture of the central stalk linked to a ring of 10 c subunits in the F_o domain [51], a high-resolution structure of the central stalk was obtained with a crystal of F₁-ATPase inhibited with DCCD [52]. The nucleotide binding state and overall conformation of α and β subunit were similar to the previously obtained structures, but the coiled-coil region of the γ subunit was slightly twisted (about 10°) compared with the previous structure. This structural change seems in part due to the closer packing of the F₁ complexes in the crystal lattice, involving more extensive interaction between the central stalk and the head of an adjacent F₁ complex.

10.3.2
(ADP·AlF₄⁻)₂F₁ Structure Revealed

A novel crystal structure of bovine F₁, with all three catalytic sites occupied by nucleotides, was subsequently reported [28] in 2001. Crystals were grown in Mg-ADP and aluminium fluoride (AlF₄⁻; a dummy phosphate inhibitor that is expected to stabilize the conformations of a catalytic transition state). In this structure –termed (ADP·AlF₄⁻)₂F₁ – the two βs, which are identified as being equivalent to β_{TP} and β_{DP} in the native structure from their relative positions to the asymmetric γ subunit, hold Mg-ADP·AlF₄⁻ at their catalytic sites. Their structures are very similar to each other and to their equivalents in the native structure. However, the β subunit equivalent to β_E takes a 'half-closed' conformation (Figure 10.6C), and retains Mg-ADP and sulfate (a mimic of phosphate) at the catalytic site. In the half-closed β form, the lower part of the nucleotide binding domain and carboxyl terminal domain rotated about 20° and the main-chain torsion angle of β_E188 (MF₁ βE188 corresponds to TF₁ βE190) changed largely. In addition, α_ER373 (MF₁ αR373 corresponds to TF₁ αR365) intrudes between ADP and sulfate and the distance between the β-phosphate of ADP and the sulfate would be too long to re-synthesize ATP, even if sulfate were replaced by phosphate. As it is not possible to accommodate an ATP molecule in the half-closed form, this form may selectively bind ADP and Pi in the large excess of ATP during ATP synthesis. Conversely, the β_{TP}

form may preferentially bind ATP during ATP hydrolysis, which is in accord with Boyer's bi-site activation model [37]. Another prominent difference between the native and $(ADP \cdot AlF_4^-)_2F_1$ structure is seen in the γ subunit. While the C terminal residues (γ 259–272 [MF_1 γ 259–272 corresponds to TF_1 γ 269–282]) superimpose well on the native γ structure, the residues γ 234–244 (MF_1 γ 234–244 corresponds to TF_1 γ 244–254) which form the coiled coil with residues γ 20–10 (MF_1 γ 20–10 corresponds to TF_1 γ 21–11) of the $(ADP \cdot AlF_4^-)_2F_1$ twist −20° compared with the native structure. (We define the positive angle as that of the rotation of the γ subunit by ATP hydrolysis described later.) The structural change of the γ subunit seems closely linked to the formation of the half closed β subunit, and, if so, may imply an essential role in catalysis. [The interpretation of the twist of the γ subunit is actually not very easy because, except for the lower middle part, there is not a significant difference in the γ between the $(ADP \cdot AlF^-)_2F_1$ structure and native structure.]

10.4
Rotation

10.4.1
Rotation of γ Subunit Observed at Video Rate

Although Boyer's binding change mechanism was widely accepted and rotary catalysis was expected to occur, concrete and direct evidence for the unidirectional rotation of the central γ subunit could not be obtained. In 1997, Noji, Yasuda, and others succeeded in direct observation of the rotation of the γ subunit in the $α_3β_3$ cylinder of F_1 using the $α_3β_3γ$ subcomplex from a thermophilic bacterium [53]. To visualize the rotation of the tiny γ subunit under an optical microscope, they attached a probe of fluorescently labeled actin filament to the γ subunit via biotin–streptavidin interaction and fixed the His tagged-$α_3β_3$ cylinder to the Ni-NTA modified surface of a glass plate (Figure 10.7).

In spite of the large viscous load of the actin filament, the rotation was strikingly clearly observed under an epifluorescence microscope. Viewed from the F_o side, the γ subunit rotated anti-clockwise. With reference to the crystal structure, the direction was such that one β subunit undergoes a transition in the order $β_{TP}$, $β_{DP}$ and $β_E$, consistent with ATP-hydrolysis-driven rotation expected from the crystal structure [25]. The rotation was not observed in the absence of ATP and inhibited by F_1-ATPase specific inhibitor NaN_3. Thus, this direct observation gave decisive experimental evidence for Boyer's rotary catalysis hypothesis and swept away all skepticism (but Ref. 54). Long actin filaments rotated slowly and short ones rotated more rapidly, and the torque of rotary motion was calculated from the rotational velocity of an actin filament and the frictional resistance of water (viscous load). When the rotational velocity was plotted against the actin length, the highest rates which were most reliable at each actin length were on the iso-torque line corresponding to 40 pN nm (Figure 10.8). If this torque is produced at the β–γ interface at the radius of ~1 nm from the center of $\alpha_3\beta_3$ cylinder, the force that makes the γ subunit slide past the β subunit would be as high as 40 pN. Thus the $\alpha_3\beta_3\gamma$ subcomplex produced a constant force irrespective of the load under the experimental conditions. When compared with individual linear motors producing a sliding force of 3–6 pN (myosin [55–57]), 5 pN (kinesin [58]), or 14 pN (RNA polymer-

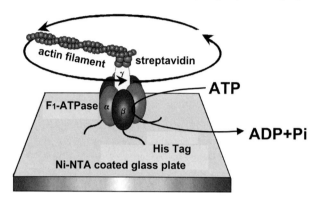

Figure 10.7 System used to observe the rotation of the γ subunit. For this assay system, mutations were introduced into the $\alpha_3\beta_3\gamma$ complex. A new cysteine was introduced into the γ subunit, to which biotin, streptavidin and biotinylated actin were attached in order. A sequence of 10 histidines at the N terminus of the β subunit (His-tag) immobilized the $\alpha_3\beta_3$ cylinder on a Ni^{2+}-nitrilotriacetic acid (Ni-NTA)-coated glass plate.

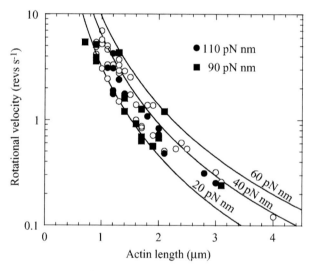

Figure 10.8 Load-dependence of the rotational rate of the F₁-ATPase from thermophilic *Bacillus* PS3. (Open circles) Measurements at [ATP] = 2 mM (because of the ATP-regenerating system, ΔG cannot be defined). (Filled circles) Measurements at [ATP] = 2 mM, [ADP] = 10 μM and [Pi] = 0.1 mM (ΔG = 110 pN·nm). (Filled squares) Measurements at [ATP] = 2 mM, [ADP] = 10 μM and [Pi] = 10 mM (ΔG = 90 pN nm). The mean work done by a one-third rotation is 80 pN nm. Theoretical lines give torques of 40, 20 and 60 pN nm. 1 pN nm = 10^{-21} J. Modified from Ref. 66.

ase [59]), it can be said that the F_1 motor exerts by far the strongest force. That the $\alpha_3\beta_3\gamma$ subcomplex produced a constant force irrespective of the viscous load is in contrast with the case of kinesin, whose force decreases as viscous load increases [60]. For the $\alpha_3\beta_3\gamma$ subcomplex-actin filament system, the torque may be first stored in a softest elastic element in the system and then released to induce the rotational motion of the actin filament. The energy required to produce this magnitude of torque is $\sim 8 \times 10^{-20}$ J per 120° rotation; which is similar to the free energy change of the hydrolysis of one ATP molecule under physiological conditions. Strictly speaking, most of the rotation experiments were carried out without ADP or Pi and ΔG_{ATP} could not be defined. In some cases, low concentrations of ADP and Pi were added to ATP to control the free energy of ATP hydrolysis to 110, 90 [61] (Figure 10.8) or 80 pN nm [62], but the average rotational rate was not affected significantly compared with that in the presence of an ATP regenerating system, where ΔG_{ATP} should be a large negative value. Under these conditions, something other than the input free energy, most likely the frictional load, limits the average rotational speed.

10.4.2
Effect of Nucleotide Binding on Torque Generation

Purine nucleotides, such as ATP, GTP and ITP, support rotation of the γ subunit whereas pyrimidine nucleotides such as UTP and CTP do not, although UTP was hydrolyzed at an appreciable rate [62]. Although the binding rate constant and K_m of ATP, GTP and ITP were significantly different, they exerted similar average rotational torque at saturating nucleotide concentrations. The effect of nucleotide binding to the rotation was further examined by mutating the nucleotide binding site. βPhe414 and βPhe420, which surround the adenine moiety of bound ATP, were mutated to Glu to reduce affinity for ATP [63]. In spite of the significant decrease in affinity for ATP, the average rotational torque of the mutant was almost the same as the wild-type enzyme again. The insensitivity of the rotational torque to the difference in nucleotide binding affinity seems contradictory to the binding change mechanism, which insists nucleotide binding and release are the energy conversion steps. However, this apparent discrepancy may be solved if we divide the binding step into two sub-processes: docking of ATP and zipping of hydrogen bonds between phosphate moiety of ATP and residues of the catalytic site. The torque force may not be generated during the docking process but during the zipping [64]. Both of the introduced two mutations (βF414E and βF420E) are located at the adenine-binding pocket rather than phosphate binding region. It is plausible that these mutations might reduce the chance of docking of ATP, but once the ATP-binding mode can transit into the zipping process by chance, the same torque force is exerted. The finding that ATP, GTP, and ITP all exert similar torque may be understood in a similar way. The difference between purine and pyrimidine nucleotides may be that the former can trigger and let the zipping process proceed while the latter cannot. When the βF414E and βF420E mutations were introduced to only one or two of the three β subunits, only the ATP binding to the mutated β subunit was retarded, and binding to the normal β subunit was not affected by the adjacent mutant subunit [63].

10.4.3
Resolution of the Continuous Rotation into 120° Steps and into Substeps

The rotation rate of attached actin filaments (~4 rps) at high ATP concentration (>2 mM) was 20 times lower than that expected from the $\frac{1}{3}$ of the rate of ATP hydrolysis (250 s^{-1}) in solution without viscous load. It was mostly because the hydrodynamic friction against a rotating actin filament limited the rotational rate. Actually, when ATP concentration was lowered and ATP binding, rather than the frictional load, was the limiting factor, the rotational rate coincided with (or a little higher than) the rate of ATP hydrolysis/3 [61]. Under these conditions, clear 120° steps corresponding to three β subunits were observed at video rate (30 fps). The torque for the individual 120° steps was calculated and the data indicated that the steps were made at an average torque of some 40 pN nm irrespective of the frictional load, as was also the case for the average velocity at high ATP concentrations. The dwell time histogram of successive steps at 20 or 60 nM ATP indicated that one ATP binding triggers one 120° step, with a binding rate constant of 2–3 × 10^7 M^{-1} s^{-1}. When a colloidal gold bead of 40 nm diameter, for which the viscous friction is 10^{-3} to 10^{-4} times that for actin filament, was used as a probe of rotation the intrinsic ATPase cycle, rather than the frictional load of the probe, was the limiting factor of the rotation even at 2 mM ATP. The high-speed rotation (up to 130 rps) of the gold bead was observed with a laser dark field microscopy and recorded at 8000 fps. The rotational rate was slightly higher than that of the rate of ATP hydrolysis/3 at all ATP concentrations examined. The slight difference between the rotational rate and ATPase rate was explained as that the latter is an average rate of active enzyme fraction and ADPMg inhibited fraction. The ATP concentration dependency of the rotational rate followed a simple Michaelis–Menten kinetics with a K_m of 15 µM and V_{max} of 130 rps. The value of $3V_{max}/K_m$, which is roughly k_{on} for ATP when k_{off} for ATP is negligible, was 2.6 ± 0.5 × 10^7 M^{-1} s^{-1}, which agreed with the results obtained with actin filament at low ATP concentrations. Although subtle deviation from Michaelis–Menten kinetics cannot be excluded, the results suggest that a single mechanism operates from 20 nM to 2000 µM ATP. Another important finding revealed by high-speed imaging was the existence of a substep. At 2 mM ATP, only 120° steps were seen, but at 20 µM or 2 µM ATP, the 120° step was further split into roughly 90° and 30° substeps (Figure 10.9). ATP concentration dependency of the substeps indicated that during the interval between the 30° and 90° substep ("0° dwell") (Figure 10.9B), F₁ waits for the binding of one ATP. Then the ATP binding triggers the 90° substep. The following interval between 90° and 30° substep ("90° dwell") includes a process or processes independent of ATP concentration. The reason that only 120° steps were observed at 2 mM ATP was that ATP binding occurs very fast under the conditions (2.6 × 10^7 M^{-1} s^{-1} × 2 mM ≈ 5 × 10^4 s^{-1}) and duration of the 0° dwell was below the detection limit. In other words, the absence of a 0° dwell at 2 mM ATP means that the 90° substep occurs instantaneously upon ATP binding (within 0.1 ms). Further statistical analyses revealed that the 90° dwell contains (at least) two processes, both of which take roughly 1 ms on average, and the latter is terminated by the 30° substep. ATP hydrolysis, ADP release or Pi release may happen during the two 1 ms processes, but a subtle conformational change without distinct chemical reaction cannot be excluded as the candidate for the event.

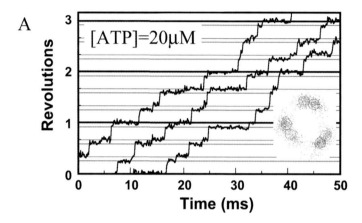

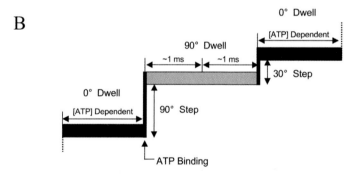

Figure 10.9 Time courses of stepping rotation. (A) Time courses of stepping rotation of 40 nm beads at 20 μM ATP. All curves in a panel are continuous; later curves are shifted, to save space. Grey horizontal lines are placed 30° below black lines. Insets: positions of a bead within 0.25 ± 0.5 ms before (green) and after (red) the main (90° or 120°) steps; runs lasting 2 s were analyzed. Circles indicate projection of 0° and 90° dwell points on an obliquely situated circular trajectory that best fit the data. (B) Schematic illustration of the events during stepping rotation. We call the interval between a 30° substep and a 90° substep a '0° dwell' and the interval between 90° and 30° substeps a '90° dwell'. F_1 waits for the arrival of ATP during the 0° dwell, which is terminated by a 90° substep induced by ATP binding. Therefore, the lifetime of the 0° dwell depends on ATP concentration. The lifetime of the 90° dwell is independent of ATP concentration. Statistical analyses revealed that during the 90° dwell two (at least) events of 1 ms lifetime occur. Modified from Ref. 29.

10.4.4
Direct Observation of ATPMg Inhibition at the Single Molecule Level

When the rotation was observed for long periods, occasional pauses of rotation that last for several seconds are recognized. Such pauses are often caused by obstructions on the glass plate, such as tiny dusts or denatured proteins. In this case, the pauses occur at a peculiar point which is independent of the three-fold symmetry of the $\alpha_3\beta_3$ structure of F₁-ATPase. But when rotating actin filaments or beads that make long pauses at every 120° were carefully selected and analyzed [65], it was revealed that the duration of the pauses did not depend on the frictional load. The duration of the pause longer than 1 s was analyzed and two kinds of pauses, a "short pause" with a lifetime of 1.7 s and a "long pause" with a lifetime of 22 s, were revealed. For the short pause the incidence was not statistically random and its origin remains unknown. However, statistical analyses of the long pause revealed that the lifetime of the paused state and actively rotating state roughly coincided with that of ADPMg inhibition deduced from a kinetic study. More importantly, the position of the long pause coincided with the position of 90° dwell (Figure 10.10). (Actually, the position of the long pause was near 80°, rather than 90° reported by Yasuda et al. [29]. But the difference between 80°and 90° may be within experimental error and we regard them as essentially the same position and call them 90° here.) Thus, the well-known phenomenon of ADPMg inhibition was first recognized at the single molecule level.

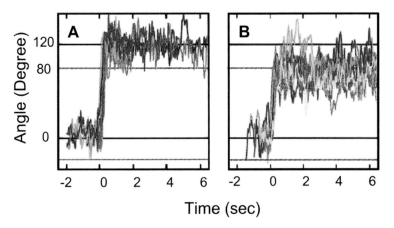

Figure 10.10 Motions of rotation probe stepping from the ATP-waiting position (A) to the next ATP waiting positions or (B) to the ADP-Mg inhibition. ATP concentration was 200 nM. (A) Stepping motions (of which subsequent pauses were shorter than 5 s) are collected and overlaid ($n = 13$). (B) Stepping motions (of which subsequent pauses were longer than 2 min) are collected and overlaid ($n = 13$). Whereas steps between two adjacent ATP-waiting positions were 120°, the long pauses always started after ~90° rotation. Modified from Ref. 65.

10.4.5
Effects of Amino Acid Substitutions on the Rotation

Trials to understand the mechanism of rotation in relation to the role of amino acid residues were carried out by the combination of single molecule rotational assay and site-directed mutagenesis. Masaike et al. [66] compared the dihedral angle of peptide bonds of the open form and closed form of the β subunit and found that His179, Gly180 and Gly181 [His179, Gly180 and Gly181 of TF$_1$ β correspond to His177, Gly178 and Gly179 of MF$_1$ β (Figure 10.11)] form a hinge region between β sheet 4 and α helix b for the open-close transition of the β subunit (Figure 10.11). They mutated these residues to Ala; however, there was not a significant change in the rotational torque calculated from the average rate. Rather, these mutations affected the sensitivity to the ADPMg inhibition. It was concluded that the closed form of the β subunit binds inhibitory ADP tightly so that mutation which stabilizes the closed β conformation enhances ADPMg inhibition. Likewise, a mutation that stabilizes the open conformation reduces ADPMg inhibition. Iko et al.

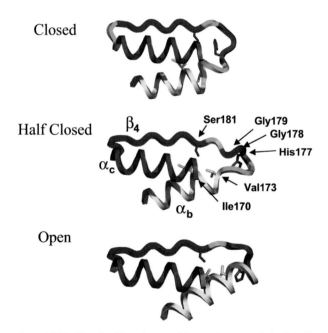

Figure 10.11 Putative hinge loop and its motion; α$_b$, α helix B (yellow); α$_c$, α helix C (red); β$_4$, β sheet 4 (blue); the hinge loop is in purple. Nomenclature is taken from the crystal structure (see [25]). Structural data of closed, half closed, and open form are taken from chain D of 1BMF (PDB ID), chain E of 1H8E and chain E of 1BMF, respectively. Assignment of the secondary structure was according to the structural data base. Amino acid residues and their number shown are of bovine MF$_1$. Ile170, Val173, His177, Gly178, Gly179, and Ser181 of MF$_1$ correspond to Ile172, Ile175, His179, Gly180, Gly181, and Ser183 of TF$_1$ and Ile163, Ile166, His170, Ser171, Gly172, and Ser174 of EF$_1$, respectively. Molecular graphics generated using PyMOL (Warren L. DeLano, DeLano Scientific, San Carlos, CA. http://www.pymol.org).

mutated βSer174 [Ser174 of EF₁ β corresponds to Ser183 of TF₁ and Ser 181 of MF₁ β (Figure 10.11)] of EF₁ β subunit which locates on β sheet 4 and protrudes to α helix b [67]. When β Ser174 was mutated to a bulky residue of Phe, the torque calculated from averaged rotational rate decreased to 40 % of the wild-type F₁. The decrease was not simply due to the decrease in the ATPase activity of the mutant (10 % of the wild-type) because another mutant with similarly low ATPase activity (βS174L) exerted essentially the same average rotational torque as the wild-type F₁. The decrease in the torque of βS174F mutant seems to be the result of steric hindrance to the open–close transition of the hinge region. Interestingly, βG149A, βI163A or βI166A mutations suppressed the effect of βS174L in EF₁ [Gly149, Ile163 and Ile166 of EF₁ β correspond to Gly158, Ile172 and Ile175 of TF₁ β, and Gly156, Ile170 and Val173 of MF₁ β, respectively (Figure 10.11)].

The crystal structure of F₁-ATPase revealed that β subunit contacts the γ subunit mainly at the "switch" region and helix-turn-helix containing "DELSEED" region [68] (Figure 10.6C). Especially, the latter region is rich in negative charge and undergoes large displacement upon open–close transition of the β subunit. Extensive Ala scanning was carried out in this region [69]. Some Ala replacement reduced the ATPase activity in solution, but the average rotational torque was hardly affected by any Ala mutation. This result may imply that the side-chains of this region do not contribute to the rotation mechanism.

10.5
Mechanistic Implication and Possible Mechanism

Despite the vast knowledge piled up by extensive biochemical, structural and single molecule studies, there are still many unanswered questions. Therefore, it is still premature to present an unambiguous picture about the mechanism of F₁-ATPase. Here we summarize and examine the implications of the experimental results described in the previous sections and try to illustrate a possible mechanism of F₁-ATPase. The following schemes should be taken as working hypotheses that we hope will facilitates further consideration and experimental study.

10.5.1
Problems in ATPase Catalytic Scheme

To understand the mechanism of any enzyme it is absolutely necessary to clarify the kinetics of the enzyme and construct a kinetic scheme. For F₁-ATPase, however, the kinetic mechanism has not been fully understood. As has been described, there are three β subunits and each has a catalytic site on it. Therefore, we can formally imagine three kinds of catalytic pathways in which at most one, two or three catalytic sites are filled during catalysis (They are called uni-site catalysis, bi-site catalysis, and tri-site catalysis, respectively [30].) Boyer's binding change mechanism [3, 4] proposed that, when only one catalytic site is filled with a substrate, the substrate binds very tightly and hydrolytic reaction proceeds slowly, establishing a pseudo 1:1 equilibrium between bound substrate and products. The hydrolysis and product release are greatly enhanced by another subsequently bound substrate; this was indeed demonstrated by uni-site experiment for mitochondrial F_1-ATPase [39]. However, whether the reaction rate is fully accelerated when only two catalytic sites are filled with substrate (bi-site catalysis) or filling of three catalytic sites is absolutely required for the significant rate of hydrolysis (tri-site catalysis) is still a matter of debate. Weber and Senior measured the nucleotide binding using the quenching of Trp introduced near the catalytic site of EF_1 [22]. The technique is excellent in that it can selectively measure the nucleotide binding to catalytic sites without disturbing the binding equilibrium. Using this technique, they concluded that F_1-ATPase has significant catalytic activity only in the tri-site catalysis mode. They further insist that the bi-site catalysis is not coupled with the rotation of the γ subunit, if it existed [38]. Conversely, Boyer insists that filling of two catalytic sites is enough for both high ATPase activity and ATP synthesis activity based on the kinetic study [70] and direct quantification of bound nucleotides [71]. Boyer pointed out several difficulties in the Trp fluorescence measurements [36, 37, 72]. Among them, the influence of ADPMg inhibition seems to be the most serious. During the fluorescence titration, the accumulated ADPMg most likely induces ADPMg inhibition and the sticky-bound inhibitory ADP may cause overestimation of the amount of bound nucleotides and underestimation of the contribution of bi-site catalysis. Weber and Senior insist that there is little ADPMg inhibition for EF_1, but even if so, it seems logically difficult to exclude the possibility that the bi-site catalysis is coupled with rotation of the γ subunit. Boyer's proposal of bi-site activation with three-site filling insists that the filling of three catalytic sites by nucleotide occurs due to retarded product release and the essential mechanism of the enzyme is the bi-site mechanism where binding of two nucleotides is enough for the rapid, rotation-coupled catalysis. As continuous rotation was observed even at 20 nM ATP [61] it is tempting to believe that bi-site catalysis is coupled with rotation. However, the reported K_m for rotation (some 15 μM) was in the same range as the tri-site K_m deduced by fluorescence titration reported by Allison's group for TF_1 $\alpha_3\beta_3\gamma$ complex [23, 24]. If a single mechanism operates from 20 nM to 2000 μM ATP and the K_m for rotation corresponds to the ATP binding to the third catalytic site (tri-site catalysis), it follows that two catalytic sites should become filled with nucleotide with less than 20 nM ATP.

After all, there is no direct evidence as to whether bi-site catalysis is coupled with rotation or not and it is hard to reach an unambiguous conclusion at present. Therefore, we tentatively assume that the bi-site catalysis has a potential to drive the rotation of γ subunit while the maximum rate of ATP hydrolysis is achieved in the tri-site catalysis. This point, however, should be replaced by a model which is based on solid and decisive experimental results in the near future. As for the relationship between the ATP concentration dependency of the hydrolytic reaction, we assume that the frequently observed K_m above 100 μM represents ATP binding to noncatalytic site and K_m of 10–30 μM represents ATP binding to the third catalytic site. The K_m which corresponds to the ATP binding to the second catalytic site seems obscure for possible ADPMg inhibition. It seems at most 1 μM and might be much smaller. The K_m corresponding to the ATP binding to the first catalytic site should be very low, at most 0.01 µM. These assignments are for bacterial ATPase, and mitochondrial or chloroplast enzyme may have significantly different parameters [70].

10.5.2
Correlation Between Crystal Structures and Rotation Intermediates

The second problem is the assignment of the known crystal structures to the possible intermediates during catalysis. This is of course based on the assumption that the crystal structures are similar, if not identical, to the structures of catalytic intermediates. Briefly, there are two major crystal structures, the "native" structure reported in 1994 [25] and the "(ADP·AlF₄⁻)₂F₁" structure reported in 2001 [28]. The major differences between the two structures are (1) the native structure has two closed form β subunits with bound nucleotide (β_{DP} and β_{TP}) and one open form without bound nucleotide (β_E). In the (ADP·AlF₄⁻)₂F₁ structure, the two closed β subunits are similar to the native structure (β_{DP} and β_{TP}) but the third β subunit binds ADP and sulfate (an analogue of phosphate) and takes a half-closed conformation. (2) The lower middle part of the γ subunit of (ADP·AlF₄⁻)₂F₁ structure is about −20° rotated compared with the native structure. (The positive angle is defined according to the direction of the rotation by ATP hydrolysis.) Together with the fact that there is a substep at the 90° position during rotation of the γ subunit, the (ADP·AlF₄⁻)₂F₁ structure, whose γ is at 100° (=120° −20°) may correspond to the 90° dwell intermediate (However, see the note in parentheses at the end of Section 10.3.2). Upon ATP binding, the γ subunit rotates to the 90° position. Therefore, the state with three catalytic sites filled with nucleotides is consistent with the 90° dwell intermediate. However, as the C terminus part (upper part in Figure 10.6) precedes the lower middle part of the γ subunit in this structure, it seems as if the upper part is rotated first and the lower middle part follows it, which is somewhat contradictory to the idea that the motion of the helix turn helix region containing DELSEED sequence in the β subunit pushes the bottom part of the γ subunit to rotate it. It may correspond to a transient intermediate in the direction of ATP synthesis in which the lower middle part of the γ subunit is rotated clockwise and the upper part is following it. The above argument suggests that the native structure corresponds to an intermediate during 0° dwell. Conversely, a single molecule study revealed that the γ subunit in ADPMg inhibited state is also at the 90° position [65]. As the native structure was obtained in the presence of NaN₃, which stabilizes ADPMg inhibited state, it may correspond to ADPMg inhibited state and therefore a structure similar to the 90° dwell intermediate. If so, the upper part in the (ADP·AlF₄⁻)₂F₁ structure represents the 90° position and the lower part represents a transient intermediate that has not reached the 90° position or driven to the direction of ATP synthesis. Alternatively, the 20° difference of the lower middle part of γ subunit between the native and (ADP·AlF₄⁻)₂F₁ structure may arise from the different packing of the F₁ complexes in the crystal lattice. Actually, in the DCCD structure, the γ subunit is twisted about −10° compared with the native structure [52]. Therefore, it seems that the $(ADP \cdot AlF_4^-)_2 F_1$ structure corresponds to the 90° dwell intermediate, but we cannot be 100% sure on this point and cannot confidently assign the native structure to a specific rotation intermediate.

10.5.3
Correlation Between ATPase Catalytic Intermediates and Rotation Intermediates

The third problem is the assignment of the ATPase catalytic events to 0° dwell and 90° dwell. One thing seems clear: during the 0° dwell, the enzyme is waiting for an ATP molecule. Therefore, in the bi-site catalysis mode, the enzyme should have only one ATP at a catalytic site which is hydrolyzed slowly and in pseudo-equilibrium with bound ADP and Pi during the 0° dwell. In this state, there are two empty catalytic sites. It seems reasonable that one of them has a high affinity for ATP and the other a high affinity for ADP, as Boyer proposed [37]. Then, ATP binding to the second site instantly drives the 90° rotation. During the subsequent 90° dwell, at least two events of 1 ms occur. ATP hydrolysis or the shift of the equilibrium to the hydrolyzed product at the first catalytic site is one candidate. The release of ADP and Pi must be another candidate; however, there is no reason to exclude a contribution of a conformational change at the third, empty catalytic site. If we accept the tri-site catalysis scenario, during the 0° dwell, two catalytic sites are occupied by AT(D)P and waiting for ATP binding to the third catalytic site. Upon ATP binding, the γ subunit rotates 90°. During the 90° dwell, catalytic events similar to the bi-site scenario may occur. The $(ADP \cdot AlF_4^-)_2 F_1$ structure may correspond to an intermediate during the 90° dwell in which hydrolysis has finished and ADP and Pi are to be released from the half-closed β subunit, as stated above. In this case, the half-closed β form was generated from the closed β_{DP} form rather than from the open β_E form observed in the native structure.

10.5.4
Possible Working Hypotheses

The totality of the results and discussion up to this point indicates that there are still so many ambiguities that it is still too premature to make out a detailed scheme. Therefore, we present two tentative schemes here, hoping that they might serve as useful working hypotheses.

10.5.5
Model Assuming the Native Structure is in the 0° Dwell

The first scheme assumes that the native crystal structure corresponds to an intermediate during 0° dwell. In principle, it is reasonable to assume that each β subunit undergoes transformation from the (empty – open) form to (ATP bound – open), (ATP bound – closed), (ATP⇔ADP·Pi – closed), (ADP + Pi – half closed), (empty –half closed), returning to the (empty – open) form (Figure 10.12A). [The present assumption that there is a form of β subunit on which ATP and ADP +

Pi are in equilibrium means that interconversion between ATP and ADP + Pi on this β subunit does not necessarily affect the conformations of the adjacent two β subunits. The possibility that the conformational change of the adjacent β subunits is tightly coupled with the interconversion between ATP and ADP + Pi on the central β subunit merits consideration although it is not included here.] Here it is assumed that the (empty – open) form favors ATP binding over ADP + Pi binding and the (empty – half closed) form favors ADP + Pi binding over ATP binding. The way of favoring ATP binding or ADP + Pi binding may be thermodynamic or kinetic. Some of the forms may exist only transiently and have not been observed crystallographically, but are likely to exist. Three β subunits carry out the same transformations but with different phases. Therefore, the problem is how to adjust the phase shift of the transformation of the three β subunits in accordance with the experimental data or to make a plausible scheme. Figure 10.12B is one such example. Intermediate "a" is the starting point for bi-site catalysis cycle. In "a", reversible transformation between ATP and ADP plus Pi occurs in the (ATP⇔ADP·Pi – closed) form β subunit, allowing multiple oxygen exchange reaction. Upon ATP binding to an (empty – open) form β subunit (intermediate "b"), conformational change of the β subunit to (ATP bound – closed) form occurs, which drives 90° rotation of the γ subunit (intermediate "c"). Further transformation of the (empty – half closed) β subunit to the (empty – open) state (intermediate "d", similar to the native structure) drives the 30° substep. Along the bi-site catalytic pathway, (ATP⇔ADP·Pi – closed) β transforms into the (ADP + Pi – half closed) state (intermediate "e"), extinguishing the oxygen exchange reaction at the catalytic site. Instead, (ATP bound – closed) β subunit transforms to (ATP⇔ADP·Pi – closed) form, where multiple oxygen exchange reaction occurs. Subsequently, ADP and Pi release from the (ADP + Pi – half closed) β to regenerate the starting intermediate with the 120° rotation (intermediate "a + 120°"). Compared with the experimental data, the "c" to "d" transition should include (at least) two 1 ms events,

Figure 10.12 Schemes for ATPase and rotation. (A) Assumed forms of β subunit during catalysis. ▶ 1, (empty – open) form, which is similar to the $β_E$ observed in the MF₁ native structure and open conformation in the binding change mechanism (see Figure 10.4). 2, (ATP bound – open) form. This is not observed in the crystal structure. This form may appear only transiently. 3, (ATP bound – closed) form. This corresponds to $β_{TP}$ observed in the MF₁ native structure and the loose conformation in the binding change mechanism. 4, (ATP⇔ADP·Pi – closed) form. This corresponds to $β_{DP}$ observed in the MF₁ native structure and the tight conformation in the binding change mechanism. 5. (ADP + Pi – half closed) form, corresponding to the half closed β observed in the MF₁ (ADP·AlF₄⁻)₂ structure. 6. (empty – half closed) form. The conformation of the β subunit resembles that of the half closed β observed in the MF₁ (ADP·AlF₄⁻)₂ structure, but without bound nucleotide. This is not observed in the crystal structure and may appear only transiently. (B) A scheme which assumes that the native structure resembles an intermediate in the 0° dwell and that the (ADP·AlF₄⁻)₂ structure resembles an intermediate in the 90° dwell. The direction of the γ subunit is expressed with a green triangle. In the (ADP·AlF₄⁻)₂ structure, there is a twist in the γ subunit. To express this, the upper part of the γ is expressed with a light green and the lower part of the γ is expressed with dark green. (C) A scheme which assumes both the native structure and (ADP·AlF₄⁻)₂ structure resemble intermediates in the 90° dwell. See text for details.

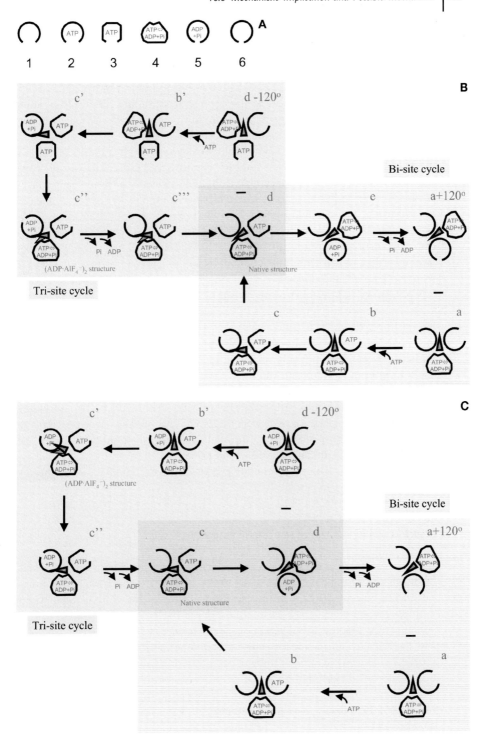

but only one conformational change, from (empty – half closed) β to (empty – open) β, is illustrated. Obviously some other change should be included if this scheme is basically correct. Product release was put after 30° step, because if it is put before this step, the native like structure (intermediate "d") cannot be an intermediate in the 0° dwell in the bi-site cycle. Taking into account that, after 30° substep, the 90° substep occurs within 0.1 ms at saturating ATP concentration we must admit that these are weak points in our bi-site scheme. When the transformation from "d" into "e" is slower than ATP binding to the (empty – open) β in intermediate "d", a tri-site cycle may start. In Figure 10.12B, the tri-site cycle is illustrated stating from "d-120°" for easy comparison between the bi-site cycle and tri-site cycle. In the tri-site cycle, ATP binding transforms "d-120°" intermediate to a "b'" intermediate. Conformational change from the (ATP bound – open) β to (ATP bound – closed) β drives a 90° rotation of the γ subunit (intermediate "c'"), as is in the "b" to "c" transition in the bi-site cycle. Then a "c'" to "c''" transition with a conformational change of (ATP bound – closed) to (ATP ⇔ ADP·Pi – closed) form drives the upper part of the γ subunit by 30°. Rotation of the lower part of γ is retarded. The "c''" intermediate corresponds to the (ADP·AlF₄⁻)₂F₁ structure. After product release and conformational change of the (empty – half closed) β to (empty – open) β, the starting intermediate "d" is regenerated. Here, we assumed that the 90° rotation is driven by a conformational change of a β subunit in the (ATP bound – open) form to (ATP bound – closed) form and the 30° rotation is driven by a conformational change of a β subunit in the (empty – half closed) form to (empty – open) form in both the bi-site and tri-site catalytic cycles. Similar tri-site schemes have been proposed by Senior [73], Allison [74], Walker [28], Noji [75] and us [6]. They are all similar and, especially, the tri-site scheme illustrated in Figure 10.12B is essentially the same as both Noji's and our previous scheme. Actually, when we fix the native structure as the starting intermediate in the 0° dwell and the (ADP·AlF₄⁻)₂F₁ structure as a 90° dwell intermediate, there is only a small degree of freedom to change the scheme, except for the combinations of bound nucleotide species.

10.5.6
Model Assuming the Native Structure is in the 90° Dwell

If we assume that the native structure resembles an intermediate in the 90° dwell, other schemes may be possible. Figure 10.12C is one of such examples. As is Figure 10.12B, intermediate "a" is the starting point for bi-site catalysis cycle, where multiple oxygen exchange reaction occurs in the (ATP⇔ADP·Pi – closed) form β subunit. ATP binding to an (empty – open) form β subunit (intermediate "b") and successive conformational change of the β subunit to the (ATP bound – closed) form drives the 90° rotation of the γ subunit (intermediate "c"). We regard this intermediate "c" as resembling the native structure. Further transformation of the (ATP bound – closed) β subunit into the (ATP⇔ADP·Pi – closed) form and the (ATP⇔ADP·Pi – closed) β subunit into (ADP + Pi – half closed) form drives the 30° substep (intermediate "d"). The transformation of the (ATP⇔ADP·Pi – closed) β into the (ADP + Pi – half closed) state extinguishes the oxygen exchange reaction at the catalytic site. Instead, oxygen exchange reaction occurs in the newly formed (ATP⇔ADP·Pi – closed) β subunit in intermediate "d". Subsequently, ADP and Pi release from the (ADP + Pi – half closed) β to regenerate the starting intermediate with the 120° rotation (intermediate "a + 120°"). Here, the "c" to "d" transition includes conformational changes of two subunits and they may be the origin of experimentally observed two 1 ms events. As in Figure 10.12B, product release was placed after the 30° step in the bi-site cycle. This is necessary because the tri-site cycle, which branches from the bi-site cycle, should start from an intermediate with two catalytic sites filled and the γ subunit at 0° position. This again seems to contradict the experimental results that the 90° step occurs within 0.1 ms after the 30° step at saturating ATP concentration. It may be possible that the experiments at saturating ATP concentration were observing the characteristics of tri-site cycle and not the bi-site cycle. When the transformation from "d" into "a + 120°" is slower than ATP binding to the (empty – open) β in intermediate "d", a tri-site cycle may start, as in Figure 10.12B. In Figure 10.12C, the tri-site cycle is illustrated stating from "d-120°", again for easy comparison between the bi-site cycle and tri-site cycle. In the tri-site cycle, ATP binding transforms the "d-120°" intermediate into the "b′″" intermediate. Conformational change from the (ATP bound – open) β in "b′″" into (ATP bound – closed) β in "c′″" drives the 90° rotation of the γ subunit, as is in the "b" to "c" transition in the bi-site cycle, but the rotation of the lower middle part of the γ subunit is some 20° retarded. We regard this intermediate "c′″" as resembling the $(ADP·AlF_4^-)_2F_1$ structure. The "c′″" intermediate spontaneously relaxes to the "c′″″" intermediate in which the whole part of the γ subunit is at the 90° position. Then, product release occurs from the (ADP + Pi – half closed) β to regenerate "c" intermediate, which is similar to the native structure.

10.6
Epilogue

The schemes described in the previous section are only two examples of many plausible alternatives. There may be better schemes than those presented here. Yet it is obvious that without the establishment of concrete relationships among ATPase catalytic states, known crystal structures, and rotation intermediates, any scheme cannot be convincing.

After the direct observation of the rotation of this enzyme, the era of research on this enzyme, which had started since the establishment of the chemiosmotic theory, has finished and we have surely entered a new era. Some may think that the elucidation of the mechanism of this enzyme has finished. But we think that the most outstanding feature of this enzyme, which other enzymes never have, has not been fully appreciated. Further elucidation of the mechanism of this enzyme will have a great impact not only in the field of biological science, but also in chemistry and physics and will enrich our understanding of nature.

Acknowledgments

E. M. especially thanks Drs P. D. Boyer, K. Kinosita, Jr., and H. Noji for continuous encouragement, critical and stimulating discussion.

References

1. P. Mitchell, *Nature*, **1961**, *191*, 144–148.
2. Y. Kagawa, et al., *J. Bioenerg. Biomembr.*, **1979**, *11*, 39–78.
3. P. D. Boyer, *FASEB J.*, **1989**, *3*, 2164–2178.
4. P. D. Boyer, *Biochim. Biophys. Acta*, **1993**, *1140*, 215–250.
5. P. D. Boyer, *Annu. Rev. Biochem.*, **1997**, *66*, 717–749.
6. M. Yoshida, E. Muneyuki, and T. Hisabori, *Nat. Rev. Mol. Cell. Biol.*, **2001**, *2*, 669–677.
7. R. E. Ebel and H. A. Lardy, *J. Biol. Chem.*, **1975**, *250*, 191–196.
8. E. M. Larson and A. T. Jagendorf, *Biochim. Biophys. Acta*, **1989**, *2*, 973.
9. E. M. Larson, A. Umbach, and A. T. Jagendorf, *Biochim. Biophys. Acta*, **1989**, *3*, 973.
10. A. F. Lodeyro, N. B. Calcaterra, and O. A. Roveri, *Biochim. Biophys. Acta*, **2001**, *1506*, 236–243.
11. D. Recktenwald and B. Hess, *FEBS Lett.*, **1977**, *76*, 25–28.
12. O. A. Roveri and N. B. Calcaterra, *FEBS Lett.*, **1985**, *192*, 123–127.
13. W. Sin-Yin, M.-Y. Akemi, and H. Youssef, *Biochemistry*, **1984**, *23*, 5004–5009.
14. J. Moyle and P. Mitchell, *FEBS Lett.*, **1975**, *56*, 55–61.
15. E. A. Vasilyeva, et al., *Biochem. J.*, **1982**, *202*, 15–23.
16. E. A. Vasilyeva, et al., *Biochem. J.*, **1982**, *202*, 9–14.
17. J. M. Jault, et al., *J. Biol. Chem.*, **1996**, *271*, 28818–28824.
18. J. M. Jault and W. S. Allison, *J. Biol. Chem.*, **1993**, *268*, 1558–1566.
19. Y. M. Milgrom and P. D. Boyer, *Biochim. Biophys. Acta*, **1990**, *1020*, 43–48.
20. D. A. Harris, *Biochim. Biophys. Acta*, **1989**, *974*, 156–162.
21. E. Muneyuki, et al., *Biochim. Biophys. Acta*, **1993**, *1144*, 62–68.
22. J. Weber, et al., *J. Biol. Chem.*, **1993**, *268*, 20126–20133.
23. C. Dou, P. A. G. Fortes, and W. S. Allison, *Biochemistry*, **1998**, *37*, 16757–16764.
24. H. Ren and W. S. Allison, *J. Biol. Chem.*, **2000**, *275*, 10057–10063.
25. J. P. Abrahams, et al., *Nature*, **1994**, *370*, 621–628.
26. T. Matsui, et al., *J. Biol. Chem.*, **1997**, *272*, 8215–8221.
27. N. Mitome, et al., *Eur. J. Biochem.*, **2002**, *269*, 53–60.
28. R. I. Menz, J. E. Walker, and A. G. W. Leslie, *Cell*, **2001**, *106*, 331–341.
29. R. Yasuda, et al., *Nature*, **2001**, *410*, 898–904.
30. R. L. Cross, C. Grubmeyer, and H. S. Penefsky, *J. Biol. Chem.*, **1982**, *257*, 12101–12105.
31. Y. Kato, et al., *Biochim. Biophys. Acta*, **1995**, *1231*, 275–281.
32. D. D. Hackney and P. D. Boyer, *Proc. Natl. Acad. Sci. U. S.A.*, **1978**, *75*, 3133–3137.
33. D. D. Hackney and S. Kerstin, *Methods Enzymol.*, **1980**, *64*, 60–83.
34. P. Boyer, et al., *Ann. New York Acad. Sci*, **1982**, *402*, 65–83.
35. R. L. Cross, *Annu. Rev. Biochem.*, **1981**, *50*, 681–714.
36. P. D. Boyer, *Biokhimiya*, **2001**, *66*, 1058–1066.
37. P. D. Boyer, *FEBS Lett.*, **2002**, *512*, 29–32.
38. J. Weber and A. E. Senior, *J. Biol. Chem.*, **2001**, *276*, 35422–35428.

39. C. Grubmeyer, R. L. Cross, and H. S. Penefsky, *J. Biol. Chem.*, **1982**, *257*, 12092–12100.

40. M. Yohda and M. Yoshida, *J. Biochem.*, **1987**, *102*, 875–883.

41. E. Muneyuki, et al., *Biochim. Biophys. Acta*, **1991**, *1058*, 304–311.

42. R. I. Feldman and D. S. Sigman, *J. Biol. Chem.*, **1982**, *257*, 1676–1683.

43. F. Oosawa and S. Hayashi, *Adv. Biophys.*, **1986**, *22*, 151–183.

44. T. Amano, et al., *J. Biol. Chem.*, **1996**, *271*, 18128–18133.

45. T. M. Duncan, et al., *Proc. Natl. Acad. Sci. U.S.A.*, **1995**, *92*, 10964–10968.

46. D. Sabbert, S. Engelbrecht, and W. Junge, *Nature*, **1996**, *381*, 623–625.

47. J. P. Abrahams, et al., *Proc. Natl. Acad. Sci. U.S.A.*, **1996**, *93*, 9420–9424.

48. M. J. Raaij van, et al., *Proc. Natl. Acad. Sci. U.S.A.*, **1996**, *93*, 6913–6917.

49. G. L. Orriss, et al., *Structure*, **1998**, *6*, 831–837.

50. K. Braig, et al., *Structure Fold Des.*, **2000**, *8*, 567–573.

51. D. Stock, A. G.W. Leslie, and J. E. Walker, *Science*, **1999**, *286*, 1700–1705.

52. C. Gibbons, et al., *Nat. Struct. Biol.*, **2000**, *7*, 1055–1061.

53. H. Noji, et al., *Nature*, **1997**, *386*, 299–302.

54. A. F. Hartog and J. A. Berden, *Biochim. Biophys. Acta*, **1999**, *1412*, 79–93.

55. J. T. Finer, R. M. Simmons, and J. A. Spudich, *Nature*, **1994**, *368*, 113–119.

56. H. Miyata, et al., *Biophys. J.*, **1995**, *68*, 286s–290s.

57. A. Ishiima, et al., *Biochem. Biophys. Res. Commun.*, **1994**, *199*, 1057–1063.

58. K. Svoboda, et al., *Nature*, **1993**, *365*, 721–727.

59. H. Yin, et al., *Science*, **1995**, *270*, 1653–1657.

60. A. J. Hunt, F. Gittes, and J. Howard, *Biophys. J.*, **1994**, *67*, 766–781.

61. R. Yasuda, et al., *Cell*, **1998**, *93*, 1117–1124.

62. H. Noji, et al., *J. Biol. Chem.*, **2001**, *276*, 25480–25486.

63. T. Ariga, et al., *J. Biol. Chem.*, **2002**, *277*, 24870–24874.

64. G. Oster and H. Wang, *Biochim. Biophys. Acta*, **2000**, *1458*, 482–510.

65. Y. Hirono-Hara, et al., *Proc. Natl. Acad. Sci. U.S.A.*, **2001**, *98*, 13649–13654.

66. T. Masaike, et al., *J. Exp. Biol.*, **2000**, *203*, 1–8.

67. Y. Iko, et al., *J. Biol. Chem.*, **2001**, *276*, 47508–47511.

68. H. Noji, T. Amano, and M. Yoshida, *J. Bioenerg. Biomembr.*, **1996**, *28*, 451–457.

69. K. Y. Hara, et al., *J. Biol. Chem.*, **2000**, *275*, 14260–14263.

70. Y. M. Milgrom, M. B. Murataliev, and P. D. Boyer, *Biochem. J.*, **1998**, *330*, 1037–1043.

71. J.-M. Zhou and P. D. Boyer, *J. Biol. Chem.*, **1993**, *268*, 1531–1538.

72. P. D. Boyer, *Biochim. Biophys. Acta*, **2000**, *1458*, 252–262.

73. J. Weber and A. E. Senior, *Biochim. Biophys. Acta*, **2000**, *1458*, 300–309.

74. H. Ren and W. S. Allison, *Biochim. Biophys. Acta*, **2000**, *1458*, 221–233.

75. H. Noji, *Molecular Motors*. M. Schliwa, ed., Wiley-VCH Verlag Gmbh & Co, Weinheim **2003**, pp. 141–151.

76. M. J. Gresser, J. A. Myers, and P. D. Boyer, *J. Biol. Chem.*, **1982**, *257*, 12030–12038.

77. E. Muneyuki and H. Hirata, *FEBS Lett.*, **1988**, *234(2)*, 455–458.

11

Coordinating Catalysis and Rotation in the F_1-ATPase

Robert K. Nakamoto, Nga Phi Le, and *Marwan K. Al-Shawi*

Abbreviations

AMPPNP, 5'adenylylimidodiphosphate; Pi, inorganic phosphate; LDAO, lauryldimethylamino oxide; PMF, proton motive force.

11.1
Introduction

The F_0F_1 ATP synthase uses a unique rotational mechanism to couple energy from the translocation of protons across the membrane to the synthesis of ATP from ADP and Pi. The movement of the rotor complex is integral to both catalysis and transport, and both mechanisms are designed to drive, or to be driven by, torque applied to the rotor assembly. In addition, each rotor subunit has conformational dynamics that are intimately involved in function. Analyses of mutant enzymes with perturbed function show that the complex is a finely tuned machine and all parts contribute to a system that is optimized for maximal efficiency of proton motive force (PMF)-driven ATP synthesis. Many amino acid substitutions, even in locations away from the catalytic or transport sites, have significant effects on function. Nevertheless, the complex is a robust machine which can tolerate surprisingly invasive manipulations and still retain enough function to support the essential ATP synthesis activity.

The F_0F_1 complex is large, made up of at least eight different protein subunits, most of which have been conserved throughout evolution. The soluble F_1 is easily and reversibly dissociated from membranous F_0 and the division of the two sectors defines the basic functionality. The F_1 sector contains the machinery for synthesis or hydrolysis of ATP. There are more than 3,500 amino acids divided among a minimum of nine polypeptides (for example, $\alpha_3\beta_3\gamma_1\delta_1\epsilon_1$ in *E. coli*) and each subunit contains multiple domains (reviewed in Refs. 1–4). There are six nucleotide binding sites with three interacting in a highly coordinated fashion to carry out the catalytic function. Although the other three sites do not participate directly in the cata-

Handbook of ATPases. Edited by M. Futai, Y. Wada, J. H. Kaplan
Copyright © 2004 WILEY-VCH Verlag GmbH & Co. KGaA, Weinheim
ISBN 3-527-30689-7

lytic mechanism, their occupancy with nucleotides is likely important for the conformational integrity of the complex. The F_O sector contains the transport mechanism and is equally complex, with at least three different subunits and 13 polypeptides ($a_1b_2c_{\sim 10}$ in *E. coli*). While the two sectors are experimentally defined into F_O and F_1, one may also divide the complex into rotor and stator assemblies. For the *E. coli* complex, the rotor consists of single copies of F_1 subunits γ and ε, and an oligomer of approximately ten *c* subunit. In this chapter we use the nomenclature for the *E. coli* complex unless otherwise noted (see Ref. 1).

Until recently, structure–function questions were necessarily addressed indirectly because of the lack of high-resolution structural information. Although becoming a more tractable problem, the structural determination of membrane proteins or large multi-subunit complexes is still tremendously difficult and investigators traditionally use protein chemistry techniques such as chemical cross-linking and accessibility of epitopes to tease out structure–function relationships. For the ATP synthase, such techniques have been used extensively; however, it was also one of the first active transport proteins where high-resolution structural information became available. The X-ray structures of F_1 from bovine heart mitochondria [5–10], rat liver mitochondria [11], and PS3 Bacillus [12] were accomplished at high resolution, and that of *E. coli* at low resolution [13]. An equally remarkable feat was obtaining a low-resolution crystal structure of most of the yeast F_OF_1 complex [14]. In addition, X-ray or NMR structures of some of the *E. coli* subunits have been completed: the F_O *c* subunit [15, 16], a monomeric form of the soluble portion of the F_O *b* subunit [17], a portion of the δ subunit [18], and the ε subunit alone [19] or in complex with a truncated form of γ subunit [20]. Although the entire structure has not been solved from a single crystal form, information from the various structures in combination with that from biochemical approaches has provided a good idea of the subunit–subunit interactions not resolved in the X-ray structures (reviewed in Ref. 21, see also [22, 23]).

A long standing goal in active transport is to achieve a molecular understanding of coupling between the catalytic and transport mechanisms. In other words, how is energy converted from the vectorial translocation of protons across a membrane into the catalysis of ATP synthesis, and in such a manner that has very little slip. Unlike a piston engine, the ATP synthase appears to lose very little energy as estimates of coupling ratios indicate little energy loss to heat [24]. With higher resolution structures and more information on the interactions among subunits, the question of coupling has become better understood. We now understand that we must consider the relationship between the catalytic mechanism and rotation, rotation and the transport mechanism, and the catalytic mechanism and its modulation by transport and other parts of the complex. Each of these relationships is complicated and not independent of the other functions. There are numerous examples of perturbations in one part of the complex that affect functions in another part separated by more than 100 Å. Clearly, proteins do not act as rigid bodies and the dynamics of the subunits influence functionality.

This chapter will discuss the roles of these interactions in the ATP synthase complex, focusing on rotational catalysis.

11.2
Asymmetry of Catalytic Sites

11.2.1
Rotation Through Catalytic Site Conformations

Upon viewing the first high-resolution X-ray structure of the bovine F_1 [5], investigators were immediately struck with the realization that the centrally located γ subunit could act as a shaft that would rotate relative to the $\alpha_3\beta_3$ pseudo-hexamer (Figure 11.1). Subsequently, ATPase-dependent rotation of the γ subunit was demonstrated in several ways [25–27]. The most compelling proof was the direct observation of ATP-dependent rotation of a fluorescently-tagged actin filament covalently attached to the γ subunit while the $\alpha_3\beta_3$ stator complex was fixed to a glass surface [27]. The fluorescent actin filament was easily observed in the fluorescence microscope. Even though the rotation was a relatively rare event, adequate data could be collected to learn several characteristics of the rotational behavior. Later the ε [28, 29] and the c subunit oligomer [30, 31] were shown by similar methods to be a part of the rotor complex. Noji et al. [27] observed that ATP-driven rotation was always in the clock-wise direction (when viewed from the "top" towards the membrane from the cytosolic side) and rotation was stopped upon addition of sodium azide, a specific inhibitor of ATPase activity.

These observations led investigators to ask how the catalytic mechanism drives, or is driven by, rotation of the γ subunit. The binding change mechanism model proposed 20 years earlier by Boyer [32–35] already provided the basic framework. In this model, three sites interact in a cooperative manner, switching among a high-affinity site (tight) where the chemistry of ATP hydrolysis or synthesis is carried out, an intermediate affinity site (loose), and a low-affinity site (open). The tight site holds substrates with extremely high affinity, shielded from solvent in a state allowing reversible hydrolysis–synthesis of ATP. Energy from the translocation of protons down their electrochemical gradient, or from the hydrolysis of ATP, is used to drive changes in affinity of the catalytic sites and the conformational changes are controlled by the cooperative interactions among the nucleotide sites. Occupancy of the open or loose site with substrate triggers a conformational change where (1) the tight site becomes the low-affinity open site and releases products (ATP in synthesis, or ADP and Pi in hydrolysis) and (2) the substrate-bound loose site (ADP and Pi in synthesis, or ATP in hydrolysis) becomes the tight site of chemistry. Boyer also suggested that changes in affinity were driven by the rotation of an asymmetric subunit, possibly γ [34]. Biochemical evidence for such a model has been reviewed by Cross [36] and Boyer [35].

The high-resolution X-ray structure of bovine F_1 [5] and the subsequent proof of rotation [27] provide the long awaited structural basis for the binding change mechanism. A key characteristic is the different conformations of the three β subunits and the catalytic sites within them. Because the conformation of each β subunit must be, at least in part, defined by the specific interface with the asymmetric γ subunit, a catalytic site must proceed sequentially through each of the conformations as the γ subunit rotates (Figure 11.1). In other words, the order of conformations in the catalytic mechanism is implicitly denoted by the relative position of each conformation in the structure and direction of rotation. Using low ATP concentrations, Yasuda et al. [37] observed that the rotor γ subunit dwelled in one of three 120° positions when the enzyme was at rest. We may take the dwell position, which is probably represented structurally by the high-resolution structures of bovine F_1 [5–9], as the ground state of the enzyme, or the low-energy state between catalytic cycles before substrate has bound (this will be further defined below). This is consistent with one ATP hydrolyzed or synthesized per 120° rotation of the γ subunit, or three per one complete revolution. The order of the conformations during hydrolysis is $β_E→β_{TP}→β_{DP}→β_E$, referring to the names of the structurally de-

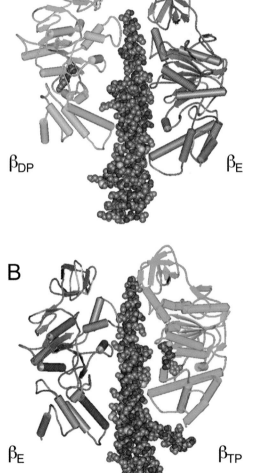

Figure 11.1 Conformational differences between the catalytic β subunits. Views of (A) $β_{DP}$ + $β_E$ and (B) $β_{TP}$ + $β_E$ from the side perpendicular to the long axis of the γ subunit [5]. Bound ADP in $β_{DP}$ or AMPPNP in $β_{TP}$ are in space filling models; α subunits are omitted for clarity.

fined conformations (Figure 11.2; see Ref. 5). Hence, an inescapable conclusion is that the synthesis direction of rotation must be counter-clockwise and the order of the conformations opposite to that of hydrolysis.

The laboratory of Walker and colleagues has made a tremendous effort to obtain high-resolution X-ray structures of the bovine F_1 ATPase with various nucleotides, inhibitors, or transition state analogs in attempts to obtain structural snapshots of catalytic intermediates [5, 8–10]. Under most conditions, two of the sites, termed β_{DP} and β_{TP}, are in "closed" conformations with nucleotides or nucleotide analogs bound, and the third site, β_E, is in an open, solvent-accessible conformation [5, 8, 9]. A different structure is obtained by crystallizing F_1 in the presence of ADP and fluoroaluminate. ADP · AlF$_4^-$ is bound in two closed sites that are very similar in conformation [10]. In addition, the γ subunit is slightly rotated and the third site catalytic site, which is in a "half closed" (β_{HC}) conformation, contains molecules of ADP and sulfate. In the direction of hydrolysis, this conformation is expected to be observed between the $\beta_{DP} \rightarrow \beta_E$ transition just before ADP is released.

The different conformations of the sites and β subunits depend in part upon Mg^{2+} and the presence of the γ subunit. In the absence of Mg^{2+}, nucleotides bind to the catalytic sites with the same affinity [38], and the X-ray structure of the *Bacillus* PS3 $\alpha_3\beta_3$ complex lacking the γ subunit has three-fold symmetry with the three β subunits in the same open conformation [12]. In addition to the influence of the γ subunit on the conformation of the catalytic sites, nucleotide binding in one site triggers activity in the other sites. The site-to-site interactions are transmitted at least in part via interactions with the α subunits. This was demonstrated by several amino acids replacements in the α subunits that block the cooperative interactions [39–43].

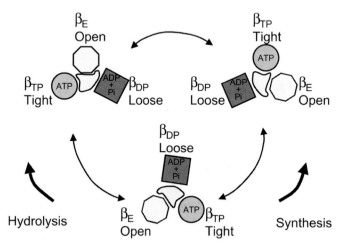

Figure 11.2 Order of the sites in rotational catalysis. β_{TP} is indicated as the "tight" site, β_{DP} as the "loose" site and β_E as the "open" site [63, 103]. In the direction of hydrolysis, the order of the sites is open–tight–loose. In other models (not shown), the assignments are switched; β_{TP} is the "loose" site and β_{DP} is the "tight" site [3, 10]. In the direction of hydrolysis, the order is open–loose–tight.

11.3
Cooperative Interactions and the Promotion of Catalysis

11.3.1
Uni-site versus Multi-site Catalysis

An intrinsic property of the $\alpha_3\beta_3\gamma$ catalytic domain is asymmetry of the three nucleotide binding affinities. Binding of Mg · ATP to the first site is very high affinity ($K_D \approx 10^{-10}$ M for E. coli [44] to 10^{-12} M for bovine mitochondrial enzyme [45]) while binding to the other two sites is much lower affinity. Accurate determinations of the catalytic site affinities were obtained by Weber, Senior and colleagues using an E. coli F₁ complex with a tyrosine to tryptophan substitution in the β subunits, βY331W ([46], see Ref. 47 for a review). The fluorescence of the tryptophan, which is in the binding pocket for the adenine moiety of the nucleotide, is sensitive to occupancy of the site and the fluorescence signal indicates the bound nucleotide is distributed over the three catalytic sites. Titrations with Mg · ATP can be fit to a model with three binding sites with $K_{D1} = 0.0002$ µM (determined using [γ-^{32}P]ATP [44]), $K_{D2} = 0.5$ µM, and $K_{D3} = 25$ µM [46, 48]. The low-affinity site is certainly the β_E conformation, while the assignments of the high (or "tight"; see Refs. 5,35,36) and medium (or "loose") affinity sites are not clear. Assignment of the tight site to the proper β subunit conformer is important because this information forces different possible mechanisms for rotational catalysis (Figure 11.2). Recall that, as the γ subunit rotates, the affinities of the sites will change, depending on the γ subunit position, and the order of the sites is $\beta_E \rightarrow \beta_{TP} \rightarrow \beta_{DP} \rightarrow \beta_E$ during ATP hydrolysis. If β_{DP} is the high-affinity tight site carrying out chemistry, then this site switches from medium to high affinity ($\beta_{TP(loose)}$ to $\beta_{DP(tight)}$). In the other possibility where β_{TP} is the high-affinity site, the site switches from high to medium affinity ($\beta_{TP(tight)}$ to $\beta_{DP(loose)}$). This question will be further addressed below.

At sub-stoichiometric ATP, substrate Mg · ATP binds only to the first high-affinity site on the enzyme, the "uni-site", and is slowly hydrolyzed with a rate constant of ~0.1 s^{-1} {Figure 11.3 (top), [45]}. In this condition, almost all of the rate constants of the elementary reactions steps can be determined because the release of product Pi and ADP are very slow (~1 × 10^{-3} s^{-1}, and ~3 × 10^{-4} s^{-1}, respectively, for E. coli F₁, [44]). Importantly, the rate constants of both hydrolysis (k_2) and synthesis (k_{-2}) may be evaluated. The ratio of the hydrolysis and synthesis reaction rate constants is close to unity, indicating that the step of chemistry occurs with little free energy change. The release of Pi is essentially irreversible because, in the absence of a PMF, the affinity for Pi is extremely low [44, 45]. The affinity of the uni-site for ADP determined by kinetic measurements has a K_D in the range of 10^{-6}–10^{-5} M [44, 49].

Product Pi and ADP release is strongly promoted by addition of Mg · ATP at concentrations high enough to occupy the other two catalytic sites [Figure 11.3 (bottom)]. Through the cooperative "multi-site" mechanism, the turnover of the enzyme is promoted 10^5–10^6-fold and all of the rate constants are increased even though the binding affinities for substrates and products do not change as

Uni-site catalysis

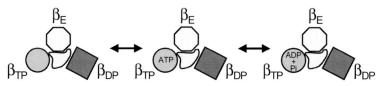

Multi-site catalysis

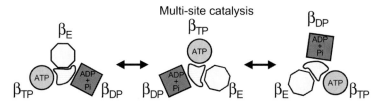

Figure 11.3 Site occupancy in uni-site and multi-site catalysis. Multi-site catalysis requires occupancy of all three catalytic sites and the enzymes are shown just prior to binding substrates in the empty site.

much. During steady-state multi-site ATPase activity, one of the sites has very high affinity for nucleotide and the enzyme still cannot productively bind Pi, during ATP synthesis, in the absence of a PMF. The differences are manifest by the switching of each site for nucleotide affinity and functionality as it cycles through different interactions with the γ subunit. While analysis of the uni-site catalysis behavior has been extremely informative about the chemical mechanism of catalysis and the role of catalytic site residues [43–45, 49–52], uni-site catalysis is not associated with γ subunit rotation [37, 53]. Rotational behavior requires occupancy of the other sites with nucleotide.

11.3.2
Cooperative Promotion of Catalysis

There is a clear correlation between the number of occupied nucleotide sites and rate of ATP hydrolysis. Using strategically placed tryptophans in the nucleotide binding sites that allow real time monitoring of the occupancy of the catalytic [54] or non-catalytic sites [55], Weber et al. found that the K_m for Mg·ATP in steady-state ATP hydrolysis is the same as the K_D for the low-affinity third site, $\sim 3 \times 10^{-5}$ M. The activation of hydrolysis to maximal rates is achieved as the third site is filled with Mg·ATP. Similar results were obtained when Mg·ITP was used as the substrate [56]. The catalytic sites have a much lower affinity for Mg·ITP and the K_m for steady-state ITP hydrolysis is very close to the affinity of the third site for Mg·ITP, which reinforces the model that all three sites have nucleotide bound most of the time and that steady-state activity up to maximal rates is promoted by occupancy of the third site with substrate. In addition, Matsuno-Yagi and Hatefi [57, 58] found that promotion of ATP synthesis by

ADP also requires binding to the low-affinity third site. Importantly, the K_m for ATP for the rotation of the γ subunit is very similar, in the range $2–4 \times 10^{-5}$ M [59], clearly correlating occupancy of the third site, activation of ATPase activity, and γ subunit rotation. This type of mechanism will be referred to as rotational catalysis. Occupancy of the second site may be enough to couple hydrolysis to rotation of the γ subunit [59], but the rate of hydrolysis is very slow at these concentrations of ATP ($\sim 10^{-8}$ M). The occasional rotation events at such low substrate concentrations occur on the rare occasions when two, or more likely three sites, are simultaneously occupied by Mg·ATP [37, 56].

11.3.3
Role of αArg376 in Multi-site Catalysis

The role of ATP binding to the low-affinity site and activation of rotational catalysis is revealed by enzymes with substitutions of catalytic site residue αArg376 (*E. coli* numbering; equivalent of αArg373 in the bovine mitochondrial enzyme). The conserved Arg, which is contributed by the adjacent α subunit, is very close to the γ-phosphate of ATP in the β_{TP} site and slightly closer to the β-phosphate in the β_{DP} site [5]. The Arg guanidinium moiety coordinates the fluoroaluminate in the β_{DP} and β_{TP} sites in the $(ADP \cdot AlF_4^-)_2F_1$ structure of Menz et al. [10]. Consistent with the structures, Nadanaciva et al. [42] used binding of Mg · ADP-fluoroaluminate as an indicator of penta-coordinate transition state formation and found that enzymes with replacements of αArg376 (other than lysine) were unable to form the transition state. From these results, αArg376 appears to participate in stabilizing the penta-coordinate transition state during nucleophilic attack of the γ phosphate in a manner reminiscent of the arginine finger of GTPase-activating proteins, or GAPs. Based on this data, Weber et al. [60] proposed a mechanism where formation of the penta-coordinate transition state is concomitant with rotation of γ subunit. Again like the GTP hydrolysis-driven conformational change of G-proteins, these authors propose that the movement of the phosphate oxygens forces the conformational change via coordination to the active site residues. In this model, because the rotation occurs with the chemistry of hydrolysis and the rotation presumably drives the conformational changes that cause the switching of binding affinities, Pi must be released immediately.

If αArg376 plays a critical role in formation of the transition state, then the enzyme lacking the Arg should not be able to carry out ATP hydrolysis. In contrast to this notion, Le et al. [43] determined that the αR376K or A mutant F_1 carried out uni-site catalysis even though addition of excess ATP was unable to chase ADP and Pi from the site. Analysis of binding energy utilization based on the uni-site rate constants established that the mutant enzymes could utilize substrate-binding energy to coordinate ATP properly for hydrolysis and there was little affect on the step of chemistry where reversible ATP hydrolysis–synthesis occurred. All αArg376 mutant enzymes have <0.1 % of steady-state ATPase activity compared with wild-type, but the three catalytic sites retain the asymmetry of nucleotide binding. The three catalytic sites have different affinities for Mg · ATP (or Mg·AMPPNP), within an order of magnitude of the wild-type dissociation constants [42, 43]. Even though αArg376 is intimately associated with bound nucleotide, it is not essential for the chemistry of hydrolysis. Rather, it is important for the communication among the sites during multi-site catalysis. Mutant enzymes lacking the Arg are unable to promote the release of product ADP and Pi even though the second and third catalytic sites have bound ATP. Based on these results, αArg376 plays a critical role in the catalytic site cooperative mechanism and its interactions with the β and γ-phosphates of bound nucleotide suggest that αArg376 is a sensor of nucleotide and Pi binding. αArg376 may also play the key role in catalysis by assuring that the enzyme releases ADP rather than ATP during hydrolysis, or releases ATP rather than ADP + Pi during synthesis.

Despite the ability of the αArg376 mutant enzymes to carry out uni-site catalysis, Weber et al. [60] have argued that the catalytic mechanism in the uni-site is not the same as in multi-site catalysis. They contend that because the other sites are not occupied by ATP, an essential characteristic of rotational catalysis, uni-site hydrolysis, which is much slower than multi-site catalysis, does not utilize αArg376. These authors suggest that multi-site catalysis is different because αArg376 becomes properly positioned only upon ATP occupancy of the second and third sites, and, in this manner, catalysis is greatly promoted. Conversely, the structure of the $F_1(ADP\text{-}AlF_4^-)_2$ enzyme may not be in the exact conformation of the transition state enzyme. Fluoroaluminate and other transition state analogs are useful because they stabilize the enzyme in a low-energy state that is believed to mimic the transition state; however, a low-energy state is exactly the opposite of the true transition state, which is generally the highest, most unstable energy state in a catalytic pathway. To stabilize such a conformation, it is likely that there will be conformational differences. While αArg376 may contribute to stabilization of a low-energy $(ADP\text{-}AlF_4^-)_2$ structure, which is the same as detected in the nucleotide binding experiments, it may not have the same function in the catalytically active enzyme.

11.4
Identification of a Rotation Step in Steady-state Catalysis

11.4.1
Rate-limiting Transition State Step

In the preceding section, we discussed the formation of the transition state during the chemistry step in the hydrolysis and synthesis of ATP; however, this is not necessarily the transition state of the overall steady-state reaction. The high-energy transition state is, by definition, the rate-limiting step of the reaction. Because the transition state has the largest activation energy it represents the slowest partial reaction. In steady-state rotational catalysis of the F_1 ATPase, the rate-limiting step is difficult to assign because of the involvement of three catalytic sites, which greatly complicates elucidation of the partial reactions. Because the rotation drives the changes in conformation and the affinities of the catalytic sites, the movement of the γ subunit would reasonably be expected to put the enzyme in a high-energy destabilized conformation – a process that requires input of energy. As described in the next section, this expectation is confirmed by the behavior of some rotor–stator interaction mutations.

11.4.2
E. coli γM23K Mutation Perturbs Rotor–Stator Interactions

A mutation which was especially useful in shedding light on the role of rotation in catalysis was replacement of the conserved γ subunit residue Met23 with Lys. In *E. coli*, the mutation caused a temperature-sensitive defect in oxidative phosphorylation-dependent growth [61, 62]. In membrane vesicles, the γM23K substitution caused a temperature-sensitive decrease in efficiency of ATP-dependent H⁺ pumping [62] and NADH-driven ATP synthesis. The lower ATP synthesis was manifested by a seven-fold higher K_m for Pi at 30 °C [63]. In addition, the apparent cooperativity of Pi binding in the ATP synthesis reaction was decreased in the mutant enzyme from a Hill coefficient of 1.6 for the wild-type to 1.1 for the mutant. Consistent with the apparent lower affinity for Pi, the release rate of Pi was increased approximately 50-fold under uni-site conditions [49]. These results indicate that the γM23K enzyme is unable to achieve a conformation that properly coordinates Pi in the catalytic site.

The effects of the γM23K mutation were also notable because it has a uniquely large effect on the temperature dependence of steady-state activity for both ATP hydrolysis and synthesis reactions. From Arrhenius analysis, the mutant enzyme has an exceptionally large increase in the transition state energy of activation, E_A, of 73 compared with 51 kJ mol⁻¹ for the wild-type F₁ [64]. Even larger effects are calculated for F₀F₁ in the membranes (77 compared with 31 kJ mol⁻¹ for the wild type) and ATP synthesis (162 compared with 58 kJ mol⁻¹ for the wild type) [63]. According to transition state theory, this increase indicates that additional bonds must be broken to reach the transition state. The crystal structure [5] revealed that the positively charged Lys replacement of γMet23 could form an extra ionized hydrogen bond with βGlu381 (Figure 11.4) – hence the extra bond detected by the thermodynamic studies. These results strongly suggest that the rate-limiting transition state step involves interactions between the rotor γ subunit and the stator β subunit, and most likely the rotation of the γ subunit.

This hypothesis was supported by the identification of several second-site mutations that suppressed the effects of the original γM23K mutation. All of the second-site suppressor mutations mapped within the regions of the γ subunit that have interactions with the β subunits [62, 65]. One mutation, γR242C, directly interacts with γM23K, while all others are physically separated from γM23K and map near the carboxyl terminus of the γ subunit between residues γGln269 and γVal280 (Figure 11.4). Furthermore, changing βGlu381 to Ala, Asp, or Gln also suppressed the γ subunit mutation [66]. The γ subunit carboxyl terminal and the βGlu381 second-site mutations restored efficient coupling and reduced the activation energy of the γM23K enzyme [64]. These data suggest that the second site suppressors reverse the effects of the primary mutation by decreasing the energy of interaction between rotor and stator. The functional interactions of the γ subunit regions were confirmed by genetic analysis of two primary mutations in the carboxyl terminal region, γQ269E and γT273S [65]. These mutations were suppressed by second site mutations near γMet23 (γLys18–γGln35) and γArg242 (γAla236–γMet246). The compensating effect of second-site suppressors by replacement of residues physically separated suggests that primary mutations do not perturb folding or destabilize the γ subunit coiled-coil region (Figure 11.4, see p. 300).

Other γ subunit mutations had similar effects. Another uncoupling mutation caused by a frameshift mutation near the carboxyl terminus of the γ subunit also affected γ–β interactions [67]. The effects of the frameshift were suppressed by β subunit mutations that mapped near the β subunit amino terminus, replacing βArg52 or βVal77, or in the catalytic site replacing βGly150 [68].

The structural significance of the γM23K substitution is its location proximal to γArg242 (*E. coli* numbering; equivalent of γArg228 in the bovine mitochondrial enzyme) and its direct interactions with βGlu381 of the ^{380}DELSEED^{386} motif of the $β_{DP}$ isoform (Figure 11.4) [5]. The dynamics of the contact of this motif with the γ subunit were demonstrated by several cross-linking studies. Different nucleotides [69, 70], ATP hydrolysis [25, 71], or a PMF in the presence of ADP + Pi [72] caused shifts in the interactions between γCys87 and cysteine residues in place of βAsp380, βGlu381, or the equivalent α subunit residue, αSer411. As expected, the cross-linking between γCys87 and βE381C is specific for a particular conformation of the β subunit [73, 74]. Furthermore, modification of one of the $β^{380}$DELSEED^{386} carboxylic acids with quinacrine mustard inactivated the enzyme [75]. Similar effects were observed for interactions between βE381C and εS108C [76]. In each case, the inter-subunit cross-link prevented activity.

The relative position of the β subunit carboxyl terminal region, which contains the $β^{380}$DELSEED^{386} motif, defines the open versus closed conformations of the catalytic sites [5]. Even though the individual residues of $β^{380}$DELSEED^{386} are not essential for rotation [77], the proper interactions are important for efficient coupling [66]. Molecular dynamics trajectories also show that the $β^{380}$DELSEED^{386} motif plays a key role in driving the conformational changes in the catalytic sites [78]. Uni-site analysis indicates that the added energy of interaction between γM23K and βGlu381 forces the enzyme to utilize excess binding energy to bind substrates, which leaves a deficit for driving catalysis [49]. This defect affects the two energy-requiring steps in ATP synthesis, namely binding of inorganic phosphate and release of ATP, indicating that the γ subunit is critical in transmitting the transport-derived energy to the catalytic sites. The uni-site results are significant because they demonstrate the role of the γ subunit in establishing the behavior of the catalytic sites. Even though γ subunit rotation is not involved in the behavior of uni-site catalysis, the interactions between γ and β subunits contribute to the precise conformation of the catalytic sites.

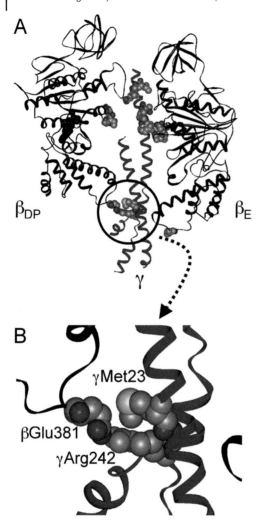

Figure 11.4 Location of γM23K and its suppressor mutations, and proximity of γMet23 to γArg242 and βGlu381 of the β^{380}DELSEED^{386}. The γM23K, γArg242 and βGlu381 are circled in (A) and their direct interactions shown in (B). The γ and β subunit suppressor mutations are in the colored space-filling models (based on Refs. 5, 62, 65, 66, 68). ADP bound in the β$_{DP}$ site is shown in the black space-filling model.

11.4.3
Isokinetic Analysis

Does the effect of the γM23K substitution on the steady-state rate-limiting step alter the structure or the pathway to the transition state? To assess this question, we have developed isokinetic relationships for analysis of the structure of the transition state. We used the methods recommended by Exner [79] (see also Ref. 80 for an enzymological example of the application of the methods utilized). In isokinetic relationships there is a linear free energy relationship between two rate constants (k_1 and k_2) measured at two temperatures ($T_2 > T_1$) which can be given by the following equation [79]:

$$\log k_2 = a + b \log k_1$$

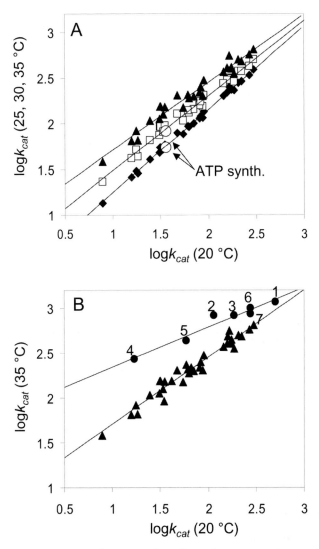

Figure 11.5 Linear free-energy plots of F_1 and F_oF_1 enzymes. (A) Isokinetic correlations of various F_oF_1 and F_1 enzymes carrying different amino acid substitutions or from different sources. See Refs. 44, 63, 64 for lists of the enzymes included in the plot. Arrows point to the data (open circles) for ATP synthesis by wild-type F_oF_1. (B) Enzymes lacking the influence of the ε subunit have a different linear relationship (upper line). Only the 20 versus 35 °C comparison is shown for clarity. The points are (1) wild-type F_oF_1 in 10 mM LDAO; (2) wild-type F_1 without ε subunit; (3) wild-type F_1 with ε subunit in 10 mM LDAO; (4) γM23K F_oF_1 in 10 mM LDAO; (5) γM23K F_1 without ε subunit; (6) γE208K F_oF_1 in 10 mM LDAO; and (7) γY205C F_oF_1. Modified after Ref. 81.

where *a* and *b* are constants. Such a relationship can be used to distinguish transition states of a given reaction pathway. This analysis takes advantage of data from different types of F_1, coupled ATP hydrolysis by F_OF_1 preparations of diverse origins, and a wide range of mutant forms (see Refs. 44, 63, 64, 81). As shown in Figure 11.5, there are excellent linear correlations between the experimental rate constants derived at three different temperatures (25, 30 and 35 °C) compared with the rate constants determined at 20 °C. This is strong evidence that all of these enzymes have the same structure at the transition state. We note that the temperature dependence of *ATP synthesis* also conforms to the isokinetic relationship, indicating that the same transition state structure is used in coupled ATP synthesis as in rotation-coupled ATP hydrolysis. From this analysis, we conclude that cooperative ATP synthesis and hydrolysis follow different directions of the same kinetic pathway [63].

Data for γM23K F_1 and F_OF_1 in ATP hydrolysis and ATP synthesis all conform to the isokinetic relationship, indicating that the mutant enzymes have the same transition state structure. The added rotor–stator interaction and the functional perturbation caused by the amino acid substitution appear to affect the enzyme on the pathway to the transition state, resulting in the increased activation energy, while not altering the structure of the transition state. A perturbation that does affect the structure of the transition state is described next.

11.4.4
Coupling Control and Modulation of Catalysis by the Rotor

Based on the effects of the γM23K mutation and its second-site suppressors, other perturbations in the rotor are also expected to modulate the activation energy to achieve the transition state. As expected, E_A is strongly affected by a number of mutations in the rotor subunits: γ [66, 82], ε (S.-K. Lin and R. K. Nakamoto, unpublished results) and *c* subunits [82, 83], as well as in the stator *a* subunit [83]. Many of these mutations are not directly involved in interactions between γ and α or β subunit.

A different effect on the transition state is obtained with manipulations that alter interactions with ε and *c* subunits. Association of F_1 with the transport F_O sector results in activation of the enzyme [64], while ε subunit association with the $\alpha_3\beta_3\gamma$ catalytic domain inhibits ATP hydrolysis rates [81, 84–90]. The ε subunit also inhibits F_OF_1 ATPase activity [81, 89, 90]. Even though the ε subunit is required for F_1 association with F_O and its effects are complicated [91], manipulations that abrogate the inhibitory property of ε subunit, such as the γY205C mutation [81, 92] or truncation of the ε subunit [90], result in higher ATPase activities. Perturbations in each of these inter-subunit interactions affect the rate-limiting transition state and coupling efficiency, suggesting that the two properties are inter-related [81].

A critical region for the coupling mechanism is the interface of the rotor subunits, γ, ε, and *c* subunits. Although there is a lack of high-resolution structure (see [14, 20]), this region has been partially defined by cross-linking [89, 92–96] and site-directed spin labeling with analysis by electron paramagnetic resonance spectroscopy [97]. Nitroxide spin-labeled γE204C and γY210C show restricted motion upon association with F_O, indicating the role of these residues in interactions with subunit *c*. Even though γE208C is not available to labeling, genetic studies revealed that this residue is also likely to interact with subunit *c*. Several second-site mutations in the hydrophilic loop of subunit *c* suppress the deleterious effects of γE208K [82]. Cysteine replacements of the odd-numbered residues, γK201C, γW203C, γY205C, and γY207C, cause significantly higher F_1 ATPase activities, suggesting that the ε subunit inhibition is blocked, even though ε subunit is still bound [97]. Treatment of the enzyme with the detergent lauryldimethylamino oxide (LDAO) has a similar effect [81, 86, 88]. Some of these mutations also cause reduced coupling efficiency, in particular γY205C [81]. Another uncoupling mutation in this region, *c*Q42E, is suppressed by intergenic second-site mutations of εGlu31 [98]. In addition, a cross-link between γY207C and *c*Q42C causes reduced coupling [89].

The effects of ε subunit are also detected by isokinetic analysis. Significantly, F_1 without the ε subunit, F_OF_1 carrying the γY205C substitution, or F_OF_1 and F_1 enzymes in the presence of high concentrations of LDAO fall on a distinctly different line for each temperature comparison shown in Figure 11.5B (points 1–7; for clarity, only one temperature comparison, 20 versus 35 °C, is shown) [81]. These points conform to a new linear relationship, indicating that the ε subunit affects the rate-limiting transition state structure. The enzymes on the new isokinetic line and lacking the control of the ε subunit also have reduced coupling. For example, the γY205C mutant F_OF_1 has approximately 5-fold lower coupling efficiency compared with wild type [81]. We suggest that the transition state structure is in part defined by the ε subunit and its interactions with the γ and *c* subunits. Achieving the proper transition state structure is a critical feature of efficient coupling.

11.5
Model for Steady-state Rotational Catalysis

Based on the data described above, we propose a model for three-site, steady-state rotational catalysis (Figure 11.6). The major premise of the model is that formation of the rate-limiting transition state for the steady-state reaction occurs during rotation of the γ subunit. Furthermore, progress of the reaction is physically represented by the position of the γ subunit. Results from crystal structures [5, 10] and direct observations of rotational behavior [37, 59] provide clues on the coupling between the γ subunit position and the catalytic reaction. The low-energy ground states occur at the 120° dwell positions, while the high-energy transition state occurs at an intermediate rotation point between the ground states. As described below, the energy-requiring steps of the reaction, Pi binding and ATP release, are likely to be coupled to formation of the transition state. The activation energy is increased in the γM23K mutant, as well as other rotor mutants, because of extra interactions between γ and β subunits or perturbations in the rotor structure, and the enzyme must use more energy to achieve the transition state [1, 63]. Thermodynamic analysis suggests that the transition state of both ATP hydrolysis and synthesis utilize the same structure, implying that the reactions have the same pathway but in reverse of each other.

Another feature of this model is the assignment of the "tight" conformation to β_{TP}. Using free-energy difference simulations for the hydrolysis reaction, ATP + $H_2O \rightarrow$ ADP + Pi, Yang et al. [99] found that the β_{TP} conformation catalyzes hydrolysis with little change in free energy, as expected for the tight site which is defined as the site in which reversible hydrolysis–synthesis occurs. As found in the uni-site conditions, reversible hydrolysis–synthesis occurs in this site while the enzyme is in the ground state (Figure 11.6, state *A*).

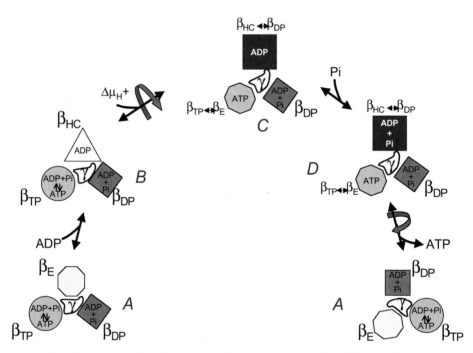

Figure 11.6 Proposed reaction scheme of rotational catalysis. One 120° cycle of rotational catalysis is shown with synthesis from left to right. Reversible hydrolysis–synthesis occurs in β_{TP}, while β_{DP} contains bound ADP and Pi. Thick arrows indicate the partial rotation steps of the γ subunit as described by Yasuda et al. [59]. The two *A* states correspond to ground states after 120° rotation and are represented by the crystal structure [5]. A small rotation occurs with ADP binding to form the β_{HC} conformation (state *B*) [10]. Input of energy from the PMF drives the γ subunit to the 30° position, converting the enzyme into the high-energy transition state (states *C* and *D*). This form of the enzyme productively binds Pi (state *D*), which allows the enzyme to complete the rotation back to the ground state *A*. Structures of states *C* and *D* are not known and are indicated as intermediates between β_{TP} to β_E and β_{HC} to β_{DP}. In hydrolysis, the alternative uncoupled pathway of the γM23K mutant enzyme slips from state *C* back to state *A* (on the right) after Pi is released. In synthesis, the enzyme is likely to slip from state *C* back to state *B*. Modified after Ref. 63.

The binding of ATP and Pi are hypothesized to be controlled by, or to control, γ subunit rotation because these steps have the largest change in free energy. In synthesis, after ATP dissociates from β_E, we propose that a partial rotation of the γ subunit puts this site in an ADP-binding conformation (Figure 11.6, state B). This intermediate conformation may be similar to the ADP-bound β_{HC} conformation found in the $(ADP \cdot AlF_4^-)_2F_1$ structure of Menz et al. [10]. In this structure, the γ subunit was rotated 15° and two $ADP \cdot AlF_4^-$ were bound to the β_{DP} and β_{TP} sites, very similar to the proposed ADP + Pi and ATP/ADP $\cdot$ Pi bound to β_{DP} and β_{TP}, respectively, in state B. Binding of ADP triggers a partial rotation driven by the PMF putting the enzyme in the high-energy transition state C, which creates the Pi binding site. Also during this step, the synthesis reaction occurring in β_{TP} is biased toward ATP so that only ATP is in this site as it converts into β_E. There is no available crystal structure of the transition state and it is represented as an intermediate where β_{TP} is converting into β_E and β_{HC} is converting into β_{DP}. Once Pi has bound (state D), the enzyme relaxes through the remainder of the rotation back to state A and ATP is released from the new β_E. Under the influence of a PMF, Pi binds with positive cooperativity, suggesting that two Pi sites are involved during synthesis [63]. The first site is β_{DP} that has bound phosphate from the previous cycle, and the second site is the transition state intermediate between β_{HC} and β_{DP}.

In hydrolysis, ATP binding to the low-affinity β_E site is required for activation of multi-site catalysis [54, 56] and appears to be associated with a partial 90° γ subunit rotation (Figure 11.6, state A on the right to state D) [59]. With ATP bound to β_E, nucleotide in β_{TP} must be hydrolyzed to ADP + Pi before converting into β_{DP}. The energy from hydrolysis drives this rotation step. As the enzyme relaxes from the transition state, the remaining 30° rotation converts β_{DP} into β_{HC} and Pi is released.

In this model, αArg376 is assigned a specific role in site-to-site cooperativity and monitoring the catalytic sites for bound nucleotide. αArg376 does not participate in the chemical transition state, as demonstrated by the ability of the αR376A and K mutant enzymes to carry out uni-site catalysis [43]. Instead, we propose that αArg376 plays a critical role in creating the high-affinity binding site for ADP and Pi in the transition state between β_{HC} and β_{DP}, and in β_{DP}. In synthesis, ADP and Pi bind to the enzyme as the β_{HC} site converts into β_{DP} and remain bound until the site converts into β_{TP} where synthesis takes place. In hydrolysis, products stay bound to β_{DP} and are not released until the site converts into β_{HC} and β_E. In both cases, the enzyme will have bound, on average, two molecules of ADP during steady-state hydrolysis. This behavior was experimentally confirmed by the observations of Weber et al. [46] using an E. coli tryptophan mutant, βF148W, to monitor occupancy of the catalytic sites with the ability to differentiate bound ATP from ADP. This model is entirely compatible with the role of the low-affinity third catalytic site and takes maximum advantage in utilizing binding energy in the promotion of catalysis.

The model also requires that the uni-site pathway is on the multi-site pathway, which includes a step of reversible ATP hydrolysis–synthesis [45, 100]. The reversible ATP hydrolysis–synthesis step was demonstrated by the incorporation of phosphate oxygens from water as followed by ^{18}O isotopic exchange [101, 102].

The reversible hydrolysis reaction must occur before the rotation step which drives the conformational change that allows release of Pi. Importantly, the level of exchange decreases with increasing concentration of Mg·ATP [101, 102], indicating that all of the rate constants increase in multi-site catalysis compared with uni-site. Even though the extent of exchange decreases to a minimum at maximal activity, the amount of exchange remaining suggests that the reversible hydrolysis–synthesis step, k_2/k_{-2}, is a part of the steady-state rotational catalytic pathway.

The model presented in Figure 11.6 also satisfies the effects of the γM23K mutation. Because of the increased energy barrier to conformational coupling, we propose that an alternative pathway is kinetically available in the mutant enzyme. In ATP hydrolysis, the mutant γ subunit rotates enough to allow release of product Pi but, because of the increased energy of interaction between γ and β subunits, the energy from hydrolysis is not used efficiently to achieve the transition state (Figure 11.6, state C), and so the γ subunit falls back to the starting position. As the rotation is not complete, so is the coupling to transport and heat is produced without translocation of protons. The enzyme can take the pathway above, or the normal coupled pathway if the γ subunit rotation can proceed past the transition state. As indicated by the decreasing coupling efficiency with increasing temperature of the γM23K mutant enzyme [62], the alternate pathway becomes more prevalent with increasing temperature. In ATP synthesis, the γ subunit again is unable to complete the rotation to the point where productive Pi binding can occur. This hypothesis explains the increased K_m for Pi in ATP synthesis [63].

An alternative model has been proposed by Weber et al. [60]. In this case, αArg376 participates in stabilization of the chemical transition state. A step of reversible hydrolysis–synthesis is not used in this model. Rather, the hydrolysis of ATP and the rotation occur concomitantly. The site that carries out chemistry is the same site that releases product because Pi will leave the site as soon as it is formed during the rotation. Because the order of the conformations is $\beta_{TP} \rightarrow \beta_{DP} \rightarrow \beta_E$, the site of chemistry must be β_{DP} and this site will release products before changing to the β_E conformation. In this respect, this model is similar to that presented by Menz et al. [10]. Products cannot be released from β_{TP} because the β_{DP} site would always be open during hydrolysis and this behavior does not satisfy the observation that the three sites are mostly occupied during steady-state activity [54, 56]. This model also does not explicitly provide a pathway for ATP synthesis because there is no discrete step at which Pi can bind. An extra step must be invoked where a PMF-induced conformational change will create the high-affinity Pi binding site. Because of this extra step, the ATP synthesis pathway is not the reverse of the hydrolysis reaction. The important differences between this model and that presented by us [63, 103] and described above are (1) the role of αArg376 and (2) the role of the rotation, whether it drives the step of chemistry as suggested by Weber et al. [60] or couples to affinity changes for the binding of substrates or release of products as delineated in the model in Figure 11.6. Further experimentation is required to elucidate the true mechanism.

11.6
Perspectives

Tremendous advances in our understanding of the molecular mechanism of the ATP synthase have been fueled by a combination of high-resolution structures of the F_1 catalytic domain and extensive structure–function mutagenesis studies. This information extends to understanding of molecular motors and the coupling between the enzymatic mechanism and the generation of torque on a protein rotor. Questions remain pertaining to the assignment of the site that carries out the catalytic reaction and, related to this issue, the role of αArg376 (bovine αArg373) in catalysis and cooperative interactions among the sites. Dissecting the partial reactions of the steady-state rotational catalytic reaction may clarify both issues. In particular, elucidating how the catalytic steps are coordinated with the γ subunit rotation steps is critical. Unfortunately, the direct observation of rotation will not yield such information because no direct observation methods exist to detect the partial reactions of the catalytic reaction. Solution methods are required in this case so that the rate constants of the individual steps may be determined.

Acknowledgments

This work was supported by a US Public Health Services grant from the National Institutes of General Medical Sciences, R01 GM50957.

References

1. R. K. Nakamoto, C. J. Ketchum, P. H. Kuo, Y. B. Peskova, and M. K. Al-Shawi, *Biochim. Biophys. Acta*, **2000**, *1458*, 289–299.
2. D. Stock, C. Gibbons, I. Arechaga, A. G. Leslie, and J. E. Walker, *Curr. Opin. Struct. Biol.*, **2000**, *10*, 672–679.
3. A. E. Senior, S. Nadanaciva, and J. Weber, *Biochim. Biophys. Acta*, **2002**, *1553*, 188–211.
4. M. Yoshida, E. Muneyuki, and T. Hisabori, *Nat. Rev. Mol. Cell. Biol.*, **2001**, *2*, 669–677.
5. J. P. Abrahams, A. G. W. Leslie, R. Lutter, and J. E. Walker, *Nature*, **1994**, *370*, 621–628.
6. J. P. Abrahams, S. K. Buchanan, M. J. Raaij van, I. M. Fearnley, A. G.W. Leslie, and J. E. Walker, *Proc. Natl.*
Acad. Sci. U. S.A., **1996**, *93*, 9420–9424.
7. M. J. Raaij van, J. P. Abrahams, A. G.W. Leslie, and J. E. Walker, *Proc. Natl. Acad. Sci. U. S.A.*, **1996**, *93*, 6913–6917.
8. K. Braig, R. I. Menz, M. G. Montgomery, A. G.W. Leslie, and J. E. Walker, *Structure*, **2000**, *8*, 567–573.
9. C. Gibbons, M. G. Montgomery, A. G.W. Leslie, and J. E. Walker, *Nat. Struct. Biol.*, **2000**, *7*, 1055–1061.
10. R. I. Menz, J. E. Walker, and A. G.W. Leslie, *Cell*, **2001**, *106*, 331–341.
11. M. A. Bianchet, J. Hullihen, P. L. Pedersen, and L. M. Amzel, *Proc. Natl. Acad. Sci. U. S.A.*, **1998**, *95*, 11065–11070.

12. Y. Shirakihara, A. G.W. Leslie, J. P. Abrahams, J. E. Walker, T. Ueda, Y. Sekimoto, M. Kambara, K. Saika, Y. Kagawa, and M. Yoshida, *Structure*, **1997**, *5*, 825–836.

13. A. C. Hausrath, G. Grüber, B. W. Matthews, and R. A. Capaldi, *Proc. Natl. Acad. Sci. U. S.A.*, **1999**, *96*, 13697–13702.

14. D. Stock, A. G.W. Leslie, and J. E. Walker, *Science*, **1999**, *286*, 1700–1705.

15. M. E. Girvin, V. K. Rastogi, F. Abildgaard, J. L. Markley, and R. H. Fillingame, *Biochemistry*, **1998**, *37*, 8817–8824.

16. V. K. Rastogi and M. E. Girvin, *Nature*, **1999**, *402*, 263–268.

17. P. A. Rizzo Del, Y. Bi, S. D. Dunn, and B. H. Shilton, *Biochemistry*, **2002**, *41*, 6875–6884.

18. S. Wilkens, S. D. Dunn, J. Chandler, F. W. Dahlquist, and R. A. Capaldi, *Nat. Struct. Biol.*, **1997**, *4*, 198–201.

19. S. Wilkens, F. W. Dahlquist, L. P. McIntosh, L. W. Donaldson, and R. A. Capaldi, *Nat. Struct. Biol.*, **1995**, *2*, 961–967.

20. A. J.W. Rodgers and M. C.J. Wilce, *Nat. Struct. Biol.*, **2000**, *7*, 1051–1054.

21. S. D. Dunn, D. T. McLachlin, and M. Revington, *Biochim. Biophys. Acta*, **2000**, *1458*, 356–363.

22. J. Weber, S. Wilke-Mounts, and A. E. Senior, *J. Biol. Chem.*, **2003**, *278*, 13409–13416.

23. J. Weber, A. Muharemagic, S. Wilke-Mounts, and A. E. Senior, *J. Biol. Chem.*, **2003**, *278*, 13623–13626.

24. D. G. Nicholls and S. J. Ferguson, *Bioenergetics 2*, Academic Press, London, **1992**.

25. T. M. Duncan, V. V. Bulygin, Y. Zhou, M. L. Hutcheon, and R. L. Cross, *Proc. Natl. Acad. Sci. U. S.A.*, **1995**, *92*, 10964–10968.

26. D. Sabbert, S. Engelbrecht, and W. Junge, *Nature*, **1996**, *381*, 623–625.

27. H. Noji, R. Yasuda, M. Yoshida, and K. Kinosita, *Nature*, **1997**, *386*, 299–302.

28. Y. Kato-Yamada, H. Noji, R. Yasuda, K. Kinosita, and M. Yoshida, *J. Biol. Chem.*, **1998**, *273*, 19375–19377.

29. V. V. Bulygin, T. M. Duncan, and R. L. Cross, *J. Biol. Chem.*, **1998**, *273*, 31765–31769.

30. Y. Sambongi, Y. Iko, M. Tanabe, H. Omote, A. Iwamoto-Kihara, I. Ueda, T. Yanagida, Y. Wada, and M. Futai, *Science*, **1999**, *286*, 1722–1724.

31. S. P. Tsunoda, R. Aggeler, M. Yoshida, and R. A. Capaldi, *Proc. Natl. Acad. Sci. U. S.A.*, **2001**, *98*, 898–902.

32. P. D. Boyer, *FEBS Lett.*, **1975**, *58*, 1–6.

33. P. D. Boyer, *Membrane Bioenergetics*. C. P. Lee, G. Schatz, and L. Ernster, eds., Addison-Wesley, Reading **1979**, pp. 461–479.

34. P. D. Boyer, *FASEB J.*, **1989**, *3*, 2164–2178.

35. P. D. Boyer, *Biochim. Biophys. Acta*, **1993**, *1140*, 215–250.

36. R. L. Cross, *Annu. Rev. Biochem.*, **1981**, *50*, 681–714.

37. R. Yasuda, H. Noji, K. Kinosita, and M. Yoshida, *Cell*, **1998**, *93*, 1117–1124.

38. J. Weber, S. Wilke-Mounts, and A. E. Senior, *J. Biol. Chem.*, **1994**, *269*, 20426–20467.

39. M. B. Maggio, J. Pagan, D. Parsonage, L. Hatch, and A. E. Senior, *J. Biol. Chem.*, **1987**, *262*, 8981–8984.

40. R. Rao, D. S. Perlin, and A. E. Senior, *Arch. Biochem. Biophys.*, **1987**, *255*, 15957–15963.

41. J. Weber, C. Bowman, S. Wilke-Mounts, and A. Senior, *J. Biol. Chem.*, **1995**, *270*, 21045–21049.

42. S. Nadanaciva, J. Weber, S. Wilke-Mounts, and A. E. Senior, *Biochemistry*, **1999**, *38*, 15493–15499.

43. N. P. Le, H. Omote, Y. Wada, M. K. Al-Shawi, R. K. Nakamoto, and M. Futai, *Biochemistry*, **2000**, *39*, 2778–2783.

44. M. K. Al-Shawi and A. E. Senior, *J. Biol. Chem.*, **1988**, *263*, 19640–19648.

45. C. Grubmeyer, R. L. Cross, and H. S. Penefsky, *J. Biol. Chem.*, **1982**, *257*, 12092–12100.

46. J. Weber, C. Bowman, and A. E. Senior, *J. Biol. Chem.*, **1996**, *271*, 18711–18718.

47. J. Weber and A. E. Senior, Catalytic mechanism of F1-ATPase, *Biochim. Biophys. Acta*, **1997**, *1319*, 19–58.

48. S. Löbau, J. Weber, S. Wilke-Mounts, and A. E. Senior, *J. Biol. Chem.*, **1997**, *272*, 3648–3656.

49. M. K. Al-Shawi and R. K. Nakamoto, *Biochemistry,* **1997**, *36,* 12954–12960.

50. M. K. Al-Shawi, D. Parsonage, and A. E. Senior, *J. Biol. Chem.,* **1988**, *263,* 19633–19639.

51. M. K. Al-Shawi, D. Parsonage, and A. E. Senior, *J. Biol. Chem.,* **1989**, *264,* 15376–15383.

52. M. K. Al-Shawi, D. Parsonage, and A. E. Senior, *J. Biol. Chem.,* **1990**, *265,* 4402–4410.

53. J. J. García and R. A. Capaldi, *J. Biol. Chem.,* **1998**, *273,* 15940–15945.

54. J. Weber, S. Wilke-Mounts, R. S.-F. Lee, E. Grell, and A. E. Senior, *J. Biol. Chem.,* **1993**, *268,* 20126–20133.

55. J. Weber, S. Wilke-Mounts, E. Grell, and A. E. Senior, *J. Biol. Chem.,* **1994**, *269,* 11261–11268.

56. J. Weber and A. E. Senior, *J. Biol. Chem.,* **2001**, *276,* 35422–35428.

57. A. Matsuno-Yagi and Y. Hatefi, *J. Biol. Chem.,* **1985**, *260,* 14424–14427.

58. A. Matsuno-Yagi and Y. Hatefi, *J. Biol. Chem.,* **1990**, *265,* 82–88.

59. R. Yasuda, H. Noji, M. Yoshida, K. Kinosita, and H. Itoh, *Nature,* **2001**, *410,* 898–904.

60. J. Weber, S. Nadanaciva, and A. E. Senior, *FEBS Lett.,* **2000**, *483,* 1–5.

61. K. Shin, R. K. Nakamoto, M. Maeda, and M. Futai, *J. Biol. Chem.,* **1992**, *267,* 20835-20839.

62. R. K. Nakamoto, M. Maeda, and M. Futai, *J. Biol. Chem.,* **1993**, *268,* 867–872.

63. M. K. Al-Shawi, C. J. Ketchum, and R. K. Nakamoto, *Biochemistry,* **1997**, *36,* 12961–12969.

64. M. K. Al-Shawi, C. J. Ketchum, and R. K. Nakamoto, *J. Biol. Chem.,* **1997**, *272,* 2300–2306.

65. R. K. Nakamoto, M. K. Al-Shawi, and M. Futai, *J. Biol. Chem.,* **1995**, *270,* 14042–14046.

66. C. J. Ketchum, M. K. Al-Shawi, and R. K. Nakamoto, *Biochem. J.,* **1998**, *330,* 707–712.

67. A. Iwamoto, J. Miki, M. Maeda, and M. Futai, *J. Biol. Chem.,* **1990**, *265,* 5043–5048.

68. C. Jeanteur-DeBeukelaer, H. Omote, A. Iwamoto-Kihara, M. Maeda, and M. Futai, *J. Biol. Chem.,* **1995**, *270,* 22850–22854.

69. R. Aggeler, M. A. Haughton, and R. A. Capaldi, *J. Biol. Chem.,* **1995**, *270,* 9185–9191.

70. M. A. Houghton and R. A. Capaldi, *J. Biol. Chem.,* **1995**, *270,* 20568–20574.

71. Y. Zhou, T. M. Duncan, V. V. Bulygin, M. L. Hutcheon, and R. L. Cross, *Biochim. Biophys. Acta,* **1996**, *1275,* 96–100.

72. Y. Zhou, T. M. Duncan, and R. L. Cross, *Proc. Natl. Acad. Sci. U. S.A.,* **1997**, *94,* 10583–10587.

73. G. Grüber and R. A. Capaldi, *Biochemistry,* **1996**, *35,* 3875–3879.

74. G. Grüber and R. A. Capaldi, *J. Biol. Chem.,* **1996**, *271,* 32623–32628.

75. D. A. Bullough, E. A. Ceccarelli, J. G. Verburg, and W. S. Allison, *J. Biol. Chem.,* **1989**, *264,* 9155–9163.

76. R. Aggeler and R. A. Capaldi, *J. Biol. Chem.,* **1996**, *271,* 13888–13891.

77. K. Y. Hara, H. Noji, D. Bald, R. Yasuda, K. Kinoshita, and M. Yoshida, *J. Biol. Chem.,* **2000**, *275,* 14260–14263.

78. J. Ma, T. C. Flynn, Q. Cui, A. G. W. Leslie, J. E. Walker, and M. Karplus, *Structure,* **2002**, *10,* 921–931.

79. O. Exner, *Prog. Phys. Org. Chem.,* **1973**, *10,* 411–482.

80. O. Smékal, G. A. Reid, and S. K. Chapman, *Biochem. J.,* **1994**, *297,* 647–652.

81. Y. B. Peskova and R. K. Nakamoto, *Biochemistry,* **2000**, *39,* 11830–11836.

82. C. J. Ketchum and R. K. Nakamoto, *J. Biol. Chem.,* **1998**, *273,* 22292–22297.

83. P. H. Kuo and R. K. Nakamoto, *Biochem. J.,* **2000**, *347,* 797–805.

84. J. B. Smith, P. C. Sternweis, and L. A. Heppel, *J. Supramol Struct.,* **1975**, *3,* 248–255.

85. P. C. Sternweis and J. B. Smith, *Biochemistry,* **1980**, *19,* 526–531.

86. H.-R. Lötscher, deJong, C., and Capaldi, R. A. *Biochemistry,* **1984**, *23,* 4140–4143.

87. S. D. Dunn, V. D. Zadorozny, R. G. Tozer, and L. Orr, *Biochemistry,* **1987**, *26,* 4488–4493.

88. S. D. Dunn, R. G. Tozer, and V. D. Zadorozny, *Biochemistry,* **1990**, *29,* 4335–4340.

89. B. Schulenberg, R. Aggeler, J. Murray, and R. A. Capaldi, *J. Biol. Chem.*, **1999**, *274*, 34233–34237.

90. Y. Kato-Yamada, D. Bald, M. Koike, K. Motohashi, T. Hisabori, and M. Yoshida, *J. Biol. Chem.*, **1999**, *274*, 33991–33994.

91. P. C. Sternweis, *J. Biol. Chem.*, **1978**, *253*, 3123–3128.

92. S. D. Watts, C. Tang, and R. A. Capaldi, *J. Biol. Chem.*, **1996**, *271*, 28341–28347.

93. E. N. Skakoon and S. D. Dunn, *Arch. Biochem. Biophys.*, **1993**, *302*, 272–278.

94. Y. Zhang and R. H. Fillingame, *J. Biol. Chem.*, **1995**, *270*, 24609–24614.

95. C. Tang and R. A. Capaldi, *J. Biol. Chem.*, **1996**, *271*, 3018–3024.

96. J. Hermolin, O. Y. Dmitriev, Y. Zhang, and R. H. Fillingame, *J. Biol. Chem.*, **1999**, *274*, 17011–17016.

97. S. H. Andrews, Y. B. Peskova, M. K. Polar, V. B. Herlihy, and R. K. Nakamoto, *Biochemistry*, **2001**, *40*, 10664–10670.

98. Y. Zhang, M. Oldenburg, and R. H. Fillingame, *J. Biol. Chem.*, **1994**, *269*, 10221–10224.

99. W. Yang, Y. Q. Gao, Q. Cui, J. Ma, and M. Karplus, *Proc. Natl. Acad. Sci. U. S.A.*, **2003**, *100*, 874–879.

100. P. Boyer, R. L. Cross, and W. Momsen, *Proc. Natl. Acad. Sci. U. S.A.*, **1973**, *70*, 2837–2839.

101. G. L. Choate, R. L. Hutton, and P. D. Boyer, *J. Biol. Chem.*, **1979**, *254*, 286–290.

102. C. C. O'Neal and P. D. Boyer, *J. Biol. Chem.*, **1984**, *259*, 5761–5767.

103. R. K. Nakamoto, C. J. Ketchum, and M. K. Al-Shawi, *Annu. Rev. Biophys. Biomol. Struct.*, **1999**, *28*, 205–234.

12

ATP Synthase Stalk Subunits b, δ and ε: Structures and Functions in Energy Coupling

Stanley D. Dunn, Daniel J. Cipriano, and *Paul A. Del Rizzo*

12.1
Introduction

The ATP synthase complex is composed of two molecular rotary motors that are connected by two stalks. The central stalk, or rotor, composed of the γ and ε subunits, forms one physical connection between the F_1 and F_o motors and transmits mechanical energy between the two sectors via rotation. The second stalk, composed of two copies of the b subunit, and one copy of the δ subunit, is on the periphery of the ATP synthase complex and acts as a stator, anchoring F_1 to F_o during the rotation of the central stalk. Both stalks are essential to the assembly and the function of ATP synthase.

In this chapter we will focus on the ε subunit of the rotor and the b_2δ stator, concentrating on the bacterial ATP synthase from *Escherichia coli*, ECF_1F_o, but will make comparisons with the homologous ATP synthases of mitochondria (MF_1F_o)

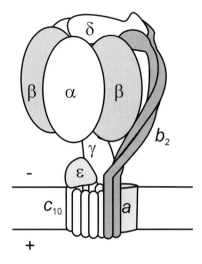

Figure 12.1 Subunit model of the *E. coli* ATP synthase. Model is based on the known high-resolution structures of isolated subunits or complexes, as well as on biochemical evidence.

Handbook of ATPases. Edited by M. Futai, Y. Wada, J. H. Kaplan
Copyright © 2004 WILEY-VCH Verlag GmbH & Co. KGaA, Weinheim
ISBN 3-527-30689-7

and chloroplast (CF_1F_o). The reader may assume that specific comments are directed toward the bacterial enzymes unless otherwise noted. These are the simplest of the ATP synthases and have served as models for understanding the more complex enzymes. A schematic model of the *E. coli* ATP synthase is shown in Figure 12.1. The F_1 sector is composed of five polypeptides in a stoichiometry $\alpha_3\beta_3\gamma\delta\varepsilon$, while the F_o sector is composed of three different polypeptides in a stoichiometry of ab_2c_{10}.

12.2
ε Subunit of ATP Synthase

The ε subunit of the *E. coli* ATP synthase is a 138-residue, 15 kDa, soluble polypeptide. It is a natural inhibitor of F_1-ATPase and is required for assembly of ATP synthase. While the importance of inhibiting the ATPase activity of free F_1 in the cytosol is clear, the existence and significance of the inhibitory function in ATP synthase has been enigmatic [1]. Some evidence in the literature has suggested a role for ε in energy coupling, but until recently this role has been undefined. Recent work from our laboratory implicates ε in ensuring highly efficient coupling by restricting the direction of rotor movement.

Both NMR [2] and X-ray crystallography [3] have been used to solve high-resolution structures of the isolated ε subunit from the *E. coli* ATP synthase. The structures (Figure 12.2) were in good agreement, showing ε as a two-domain protein with an N-terminal β-sandwich domain (residues 1–86, shown in grey) and a C-terminal helix–loop–helix hairpin domain (residues 87–138, shown in red). These will be referred to as the N-domain and C-domain, respectively. The NMR structure showed relatively few long-range NOEs between residues in the N-domain and the C-domain, suggesting that in solution some flexibility exists between the two domains.

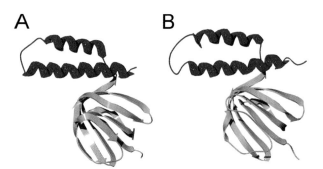

A   B

Figure 12.2 Crystal and NMR structures of the isolated ε subunit from *E. coli*. The N-domain is in gray-blue and the C-domain is in red. (A) Crystal structure (PDB 1AQT) [3]; (B) NMR structure (PDB 1BSN) [2].

12.2.1
Conformations of the ε Subunit in Crystal Structures of ATP Synthase and the γ'ε Subcomplex

While it is known that ε is part of the central stalk, there is currently a discrepancy about the tertiary conformation, particularly of the C-domain, in the ATP synthase complex. Crystal structures of ε complexed with other ATP synthase subunits, and some biochemical evidence, imply that ε can achieve at least two different conformations in ATP synthase. These results suggest that ε is dynamic, undergoing large conformational changes during catalysis that re-orient the two domains as well as possibly unfolding the C-domain.

12.2.1.1   ε in the "Down" Conformation
Two crystal structures from Walker and co-workers have ε in a position and conformation where both domains are located at the base of the rotor. The first study [4] defined a 3.9 Å structure of the mitochondrial $\alpha_3\beta_3\gamma\delta c_{10}$ complex from *Saccharomyces cerevisiae*, while the second study [5] showed the $\alpha_3\beta_3\gamma\delta\epsilon$ complex of the bovine mitochondrial F_1-ATPase at 2.4 Å resolution. Unfortunately, in MF_1F_0 the subunit homologous to bacterial and chloroplast ε subunit is named δ. The mitochondrial homologue of bacterial δ is called OSCP, and the mitochondrial subunit called ε represents an additional subunit with no bacterial homologue.

These crystal structures are compared in Figure 12.3. In both cases mitochondrial δ has a conformation like that of isolated *E. coli* ε, and sits at the base of the rotor with both domains making contact with the polar loops of the *c* oligomer. The N-domain is more centrally located, and the F_1 structure reveals its extensive interaction with γ. In this conformation, commonly referred to as the "down" conformation, no parts of ε come in close proximity to the $\alpha_3\beta_3$ hexamer of F_1. What is not obvious from these two structures is the mechanism by which ε inhibits the ATPase activity of soluble F_1 since it makes no direct contact with the $\alpha_3\beta_3$ hexamer. In the bovine mitochondrial structure, the additional subunit, mitochondrial ε, appears to hold down the C-domain and prevent its interaction with $\alpha_3\beta_3$.

12.2.1.2   ε in the "Up" Conformation
Two crystal structures have shown the *E. coli* ε subunit to be in a different orientation when complexed to either γ or $\alpha_3\beta_3\gamma$. Rodgers and Wilce solved a 2.1 Å resolution structure of a truncated form of the γ subunit (residues 11–258; termed γ') complexed to ε [6]. In this structure (Figure 12.4A), the N-domain of ε is in a slightly different, but basically similar, position as in the yeast and bovine structures, making contacts with γ, and situated adjacent to the expected position of the *c* oligomer. The C-domain, however, is in a drastically different conformation, with the two helices extending up and wrapping around the γ subunit. This has been referred to as the "up" conformation. The location of the very C-terminus of ε is sterically inaccessible when the bovine F_1 crystal structure is overlaid by

A

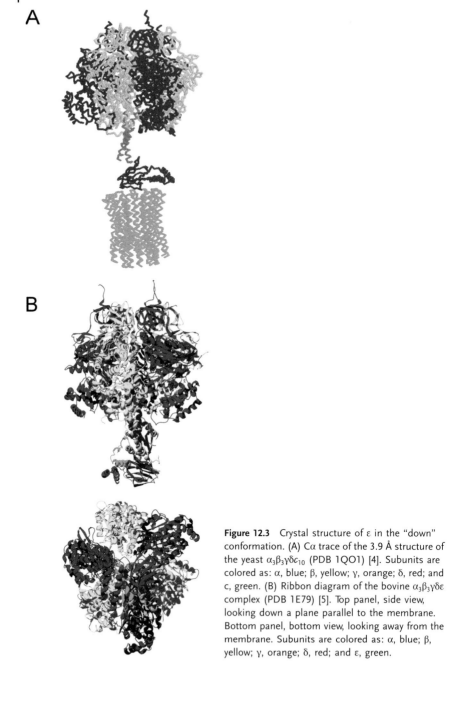

B

Figure 12.3 Crystal structure of ε in the "down" conformation. (A) Cα trace of the 3.9 Å structure of the yeast $\alpha_3\beta_3\gamma\delta c_{10}$ (PDB 1QO1) [4]. Subunits are colored as: α, blue; β, yellow; γ, orange; δ, red; and c, green. (B) Ribbon diagram of the bovine $\alpha_3\beta_3\gamma\delta\varepsilon$ complex (PDB 1E79) [5]. Top panel, side view, looking down a plane parallel to the membrane. Bottom panel, bottom view, looking away from the membrane. Subunits are colored as: α, blue; β, yellow; γ, orange; δ, red; and ε, green.

aligning the appropriate residues of γ. While the relevance of the position of the C-terminus in the absence of $\alpha_3\beta_3$ is questionable, this structure does imply that the C-domain of ε can exist in an alternate conformation. A similar conformation (Figure 12.4b) has been shown in a low-resolution (4.4 Å) structure of the *E. coli* F$_1$-ATPase by Hausrath et al. [7], although the C-terminal helix could not be placed with certainty.

The discrepancy about the C-domain conformation could arise from inherent differences between mitochondrial and bacterial ATP synthase, such as the additional subunit in the central stalk of the mitochondrial enzyme. An alternate explanation is that ε is a dynamic protein, undergoing changes in conformation during normal functioning. A limit of X-ray crystallography is that it only provides a "snapshot" of a protein in one conformation, and if a protein is dynamic we need to rely on biochemical or biophysical studies to characterize its changes. For the bacterial ATP synthase, biochemical evidence implies that ε can achieve multiple conformations.

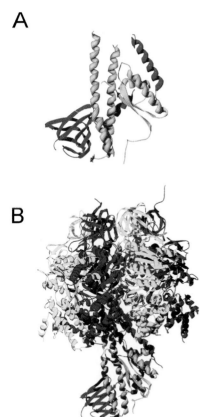

A

B

Figure 12.4 Crystal structure of ε in the "up" conformation. (A) Ribbon diagram of the *E. coli* γ'-ε complex (PDB 1FS0) [6]. (B) ribbon diagram of *E. coli* F$_1$ (PDB 1JNV) [7]. Subunits are colored as: α, blue; β, yellow; γ, orange; and ε, red.

12.2.2

Biochemical Analyses of Conformations and Orientations of ε in ATP Synthase

Before any of the crystal structures described above were determined, biochemical studies had clearly revealed that the major interaction of ε with F_1-ATPase occurs through the γ subunit. In particular, isolated γ and ε form a complex displaying a dissociation constant of 3 nM, just one order of magnitude above the dissociation constant of ε from ECF_1 [8]. In addition, chemical cross-linking studies using standard reagents such as dithiobis(succinimidylpropionate) revealed a prominent γ-ε product [9]. However, these studies also demonstrated formation of a β-ε product, indicating the proximity of the two subunits.

Definitive evidence for the direct interaction of the β and ε subunits was provided through the discovery by Lötscher et al. [10] that ε could be cross-linked to a β subunit in yields approaching 100 %, using a water-soluble carbodiimide. This is particularly significant because carbodiimides are zero-length cross-linking agents, indicating the very close proximity of the cross-linked sites. Furthermore, the reaction showed exquisite specificity, as cross-links between other subunits were formed in trace amounts only. Subsequent analysis of the cross-link showed it to be an ester, rather than the expected amide, and the residues involved were identified through a protein chemical approach as εS108 and βE381 [11]. εS108 is located in the loop (see Figure 12.5A, see p. 318) between the two helices of the C-domain while βE381 lies in the D̲ELSEED (E381 underlined) region of β near the base of the $α_3β_3$ hexamer. This region makes contact with γ and has been suggested to serve as a checkpoint for rotation of that subunit [12]. The cross-linkage of β and ε by carbodiimide was also notable by its nucleotide dependence; the cross-link was favored in the presence of ADP and P_i, but not in the presence of ATP or AMP-PNP, implying that the relationship between β and ε changed depending on nucleotides bounds to the enzyme. Aggeler et al. confirmed the spatial relationship between β by showing that ε mutant S108C can form disulfides to cysteine residues introduced in β at positions 381 and 383 with similar nucleotide dependence [13]. In addition, they found that, in the presence of ATP, the S108C would form a disulfide with a cysteine introduced into the α subunit at position 411 [14], indicating two different orientations of the C-domain with respect to F_1. While it has been expected that catalytic sites are the nucleotide binding sites relevant to these conformational changes, the recent report by Kato-Yamada and Yoshida [15] that isolated ε from the thermophilic bacterium PS3 binds ATP has raised the possibility of a more direct mechanism.

Schulenberg and Capaldi [16] have shown that ε in ATP synthase can be locked into the "down" conformation seen in the isolated subunit. Two sets of ε mutants were created (Figure 12.5B, see p. 318). The first set (εM49C/A126C and εF61C/V130C) was designed to lock the C-domain against the N-domain as in the isolated structure, while the second set of mutants (εA94C/L128C and εA101C/L121C) was designed to lock the C-terminal helices in the hairpin arrangement. In all cases the mutants formed disulfide bonds with 90–100 % efficiency when oxidized with 50 μM Cu^{2+}, suggesting that ε in ATP synthase can achieve a conformation similar to that seen for the isolated subunit. The dynamic nature of ε was also demonstrated by trypsin sensitivity studies. Mendel-Hartvig and Capaldi found that the trypsin sensitivity of ε in the *E. coli* F_1-ATPase [17], and ATP synthase [18], changes depending on whether ATP or ADP was in the medium. Thus, several lines of evidence imply that the conformation of ε changes depending on the nucleotides bound to F_1. In chloroplast F_1F_o similar changes in ε have been observed in response to illumination of thylakoid membranes [19].

Cross-linking studies, including disulfide formation between introduced cysteine residues, have shown that the N-domain of ε also makes contacts with the *c* ring of F_o [20, 21]. In these studies cysteine residues substituted in any of positions 26–33 of ε formed disulfides with cysteines introduced into positions 40, 42, 43, and 44 on the polar loop of subunit *c*. These residues of ε are highlighted blue in Figure 12.5A. This loop corresponds well to the interface between ε and the *c* oligomer in the yeast crystal structure shown in Figure 12.3A.

Tsunoda et al. [22] have trapped ε in ATP synthase in either the up or down conformation by linking the C-domain to either γ or *c* through the formation of disulfides between cysteines introduced into either positions εS118C and γL99C, or else εA117C and *c*Q42C, respectively. The formation of each of these disulfides suggests that ε can exist in either conformation.

In summary, while the N-domain of ε appears to be more or less stationary with respect to γ or *c*, much experimental evidence implies that the C-domain of *E. coli* ε can be in either the "up" or "down" conformations shown in the crystal structures, and that more than one "up" conformation may exist. Biochemical evidence shows that the conformation of ε is dynamic, responding to the nucleotides bound, and the NMR structure supports the potential for flexibility between the two domains. To date, however, no evidence suggests that the "up" conformation exists in the mitochondrial analog, δ, and it is uncertain whether it simply does not occur, or does occur but has yet to be detected.

12.2.3
Role of the ε Subunit in ATP Synthase

Energy transduction from the F_1 to the F_o part of the enzyme is accomplished mechanically through the rotation of the γε rotor. This rotation causes sequential conformational changes in the β subunits of F_1 that are responsible for the synthesis of ATP. Likewise, hydrolysis of ATP is thought to cause conformational changes in the β subunits that drive the rotation of γε. γ forms an asymmetrical spindle that extends up the middle of the $α_3β_3$ hexamer and this spindle acts as a cam, pushing on the C-terminal domains of the β subunits and causing the conformational changes needed for ATP synthesis. In contrast, the function of ε in the rotor is less certain. Certainly it is involved in the binding of F_1 to F_o; several lines of evidence further implicate ε in either regulation of ATP synthase, or in energy coupling.

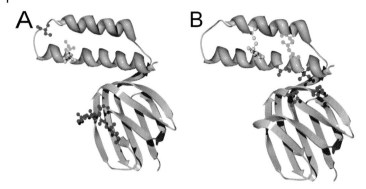

Figure 12.5 Ribbon structures of ε subunit highlighting cross-linkable residues in color as ball and stick. (A) Residues in ε that can be cross-linked to other ATP synthase subunits. Residues ε26–33 (blue) can be cross-linked to the *c* oligomer [20, 21]; residue εS108 (red) can be cross-linked to both β [10, 11, 13] and α [14]; residue εA117 (green) can be cross-linked to cQ42 [22]; and εS118 (yellow) can be cross-linked to γL99 [22]. (B) Cysteine pairs in ε that readily form disulfides. εM49C/A126C, red; εF61C/V130C, blue; εA94C/L128C, green; and εA101C/L121C, yellow [16].

12.2.3.1 Rotation of ε During Hydrolysis

Single molecule fluorescence microscopy has been used to directly visualize the ATP-dependent rotation of fluorescently labeled actin filaments attached to either the γ [23] or ε [24] subunits on F_1. This rotation occurred in a stepwise manner in 120° increments that were later broken down into 30° and 90° sub-steps [25]. This work has been extended to show rotation of the *c* oligomer in membrane bound F_1F_o [26], demonstrating that the rotor of ATP synthase consists of the $γεc_{10}$ complex.

Recently, we obtained the first evidence for the *in vivo* rotation of γε in the F_1F_o complex through a genetic approach whereby a series of proteins were fused to the C-terminus of ε [27]. Rotation of these fusion proteins with the central stalk would sweep the added protein mass through the port formed by central stalk, the b_2 stator, the lower surface of $α_3β_3$ and the polar loops of the *c* subunits, imposing a limit on the size of the fused proteins that might be accommodated before rotation would be sterically blocked. We found that fusion of the 12 kDa cytochrome b_{562} to ε did not disrupt oxidative phosphorylation *in vivo*. Fusing the larger, 20 kDa flavodoxin, or the 28 kDa flavodoxin reductase, prevented oxidative phosphorylation. Analysis of membranes revealed that ATP synthase assembly was normal in all cases, and that ATP-dependant proton pumping was affected in parallel to oxidative phosphorylation. This work supports rotation, particularly of the ε subunit, during oxidative phosphorylation *in vivo*.

12.2.3.2 Role of ε in the Formation of F_1F_o and Inhibition of Soluble F_1-ATPase

During the initial characterization of ε carried out in Leon Heppel's laboratory, Sternweis and Smith discovered that this subunit is required for the binding of *E. coli* F_1 to F_o [28] and is a noncompetitive, dissociable inhibitor of F_1-ATPase [29]. Deletion of the *uncC* gene encoding ε results in an *E. coli* strain with impaired growth and a failure to assemble F_1 onto the membrane [30]. These strains grow more poorly than those with a deletion of the entire ATP synthase, an effect attributed to a high level of ATP hydrolysis by cytoplasmic F_1-ATPase. ε inhibits ATP hydrolysis by F_1-ATPase by reducing the rate of product release from the active site [31]. This is probably accomplished by slowing the rotation-dependent conformational change that reduces the affinity of the active site for ADP and P_i, an interpretation suggesting a role of ε in energy coupling in ATP synthase.

Most evidence suggests that the C-domain is primarily responsible for inhibition. Using deletion analysis Kuki et al. [32] showed that the first 80 residues (i.e., the N-domain) of ε are sufficient for oxidative phosphorylation *in vivo*, but that this domain alone cannot inhibit the ATPase activity of soluble F_1 *in vitro*. Xiong and co-workers [33] have used both deletion analysis and site-directed mutagenesis to confirm and extend this finding. Interestingly, the extent of the C-terminal deletion is related to the severity of the effect on inhibition, as a form of ε truncated at residue 94, well before the 108 position known to interact with β, was still able to inhibit ATP hydrolysis by about 50%. Truncation studies also support a role for the C-domain of chloroplast ε in inhibition of CF_1 [34]. Using a different approach, Mendel-Hartvig and Capaldi [17] demonstrated that proteolytic cleavage in the C-domain of ε in purified ECF_1 results in a concomitant activation of ATP hydrolysis. Using site-directed mutagenesis of TF_1, from the thermophilic bacterium PS3, Hara and co-workers [35] have developed evidence supporting a key role of electrostatic interactions between the acidic residues in the DELSEED sequence of β and basic residues in the C-domain of ε.

12.2.3.3 ε as an Inhibitor in ATP Synthase

Due to the requirement of the ε subunit in the assembly of F_1 onto F_o, it is difficult to determine how ε affects the activity of ATP synthase. In their original characterization, Sternweis and Smith showed that the ATPase activity of ε-replete F_1-ATPase increased when it was bound to F_o, indicating that ε is less inhibitory in the intact ATP synthase than in F_1 [29]. In contrast, several more recent papers have shown that deletion or removal of the inhibitory C-domain of ε, or prevention of this domain from interacting with $\alpha_3\beta_3$, results in a more active ATP synthase complex. Some of these studies are summarized in Table 12.1. In all cases where the normal interactions of the C-domain of ε with $\alpha_3\beta_3$ were disrupted there was an increase in ATP hydrolysis activity. Based on these effects it has thus been suggested that ε acts to regulate the direction of ATP synthase by preventing ATP hydrolysis.

We feel that there are problems with the idea of ε as a regulator in *E. coli*, and will suggest a different primary function for ε in ATP synthase. First, it is not

Table 12.1 Functional consequences of interfering with the interactions of the C-domain of ε with $\alpha_3\beta_3$.

Reference	ε construct	Effect on:		
		ATP hydrolysis	*proton pumping*	*ATP synthesis*
18	Cleavage of C-domain with trypsin	Increase	Not measured	Not measured
80	Delete C-terminal 49 residues	Increase	No change	Not measured
33	Deletion of C-terminal 44 residues	Increase	No change	Not measured
16	Tethered C-domain to N-domain via disulfide bridge	Increase	No change	No change
22	Tethered C-domain to *c* oligomer via disulfide bridge	Increase	Slight decrease	No change
34	Deletion of C-terminal 45 residues	Increase	Slight decrease	No change

clear that a facultative bacterium like *E. coli* would benefit from such regulation. Junge et al. [36] point out that in this type of organism the enzyme should allow the flow of energy either from the proton-motive force to the phosphorylation potential, or vice versa, so that each may be maintained at levels that will support growth. The direction of energy flow would thus be dictated by the physiological conditions of the cell, and inhibition of ATP hydrolysis could be detrimental. Under anaerobic conditions the bacterium uses ATP synthase in the hydrolysis mode to maintain the proton-motive force needed to drive the flagellar motor and various membrane transport processes.

Second, a regulator of a reaction should affect its velocity, but not its efficiency. If ε is regulating the direction of ATP synthase by inhibiting ATP hydrolysis, then the removal of this regulation by any of the methods in Table 12.1 should also lead to an increase in ATP-dependent proton pumping. However, in none of these studies was the rate of proton pumping increased as would be expected if the uncovered ATPase activity was coupled, and more often a small decrease in proton pumping was observed. We conclude that the additional ATPase activity uncovered by these various methods is, in fact, uncoupled from proton pumping. Logically, then, the function of ε that is lost when the C-domain is disrupted or restricted is a coupling function rather than a regulatory function.

12.2.3.4 Bidirectional Ratchet Model for ε Function in ATP Synthase

Tsunoda et al. [22] showed that covalently linking the C-domain of ε to γ, such that it is held in the "up" conformation, strongly reduced the ATPase activity of F_1F_o but had no effect on ATP synthesis. It was suggested that ε, through the interactions of its C-domain with $\alpha_3\beta_3$, acts as a ratchet, allowing rotation only in the direction of ATP synthesis.

More recently, we have proposed that the ratchet is bidirectional [27]. Since ATP synthase can operate in either direction, we hypothesize that ε prevents the rotation of γε in whichever direction is "wrong", depending on the nucleotide, ATP or ADP, that is bound at the active site. Thus ε is postulated to sense nucleotide binding to the catalytic site, to distinguish between ATP and ADP, and to adopt a conformation that prevents rotation in the direction of ATP synthesis if ATP is bound, or in the direction of ATP hydrolysis if ADP is bound. If this restriction of movement is lost, a futile cycle will result, as explained below.

This model of ε function grew from our recent studies of ATP synthases containing ε subunits bearing C-terminal fusions to unrelated proteins [27]. As noted above, strains carrying ATP synthase with C-terminal ε fusions to either flavodoxin or flavodoxin reductase were unable to grow on acetate, indicating loss of oxidative phosphorylation, and membranes prepared from these strains show no ATP-dependent proton pumping. Surprisingly, however, these forms of ATP synthase retained approximately 75 % of the ATPase activity of the wild-type enzyme. Though this activity was uncoupled from proton movements, it remained sensitive to inhibition by dicyclohexylcarbodiimide (DCCD). Work from the Fillingame laboratory [37] has shown that modification of c by DCCD prevents the movement of the c ring past the ab_2. Therefore, the uncoupled ATPase activity must depend on the rotational movement of γε. Since full rotation is sterically blocked, the rotational movements must be less than 360° and reciprocating in nature, as suggested in Figure 12.6.

A model proposing how this might happen is presented in Figure 12.6A. The rotational events linked with ATP hydrolysis in steps 1 and 2 are allowed, because these 120° steps do not move the ε-flavodoxin fusion past the b_2 stator. Step 3, however, is sterically blocked, and we suggest that the rotor can move backwards 120° (step 4) while the site that has just released ADP and phosphate is either empty, or else has already bound ATP. Normally this rotation would be prevented by the C-domain of ε, but it is unable to occupy its normal position because of the bulky protein attached to it. Another 120° rotational step in the forward direction (step 2) can now occur, coupled to the hydrolysis of ATP and the release of ADP and phosphate. Repetition of steps 2 and 4 produces a futile cycle. For each proton that moves through F_o during the forward rotation of the rotor (step 2), another proton will move in the opposite direction while the rotor turns backwards (step 4). It is also possible, if ATP binds before step 4 begins to occur, that a reversal of less than 120° may be sufficient to reach the transition state. The rotor might then return to the position shown in view iii without ever having reached the position shown in view ii. Though the ATPase activity is uncoupled from net proton movement, it still depends on rotational movements of the c oligomer past ab_2 and is therefore inhibited by DCCD. The model presented in Figure 12.6 is based on a hydrolysis scheme where both hydrolysis and product release occur in a single 120° step of the rotor, but it should be noted that the model may be adapted to fit hydrolysis schemes requiring a 240° rotation between binding of ATP and release of ADP and P_i.

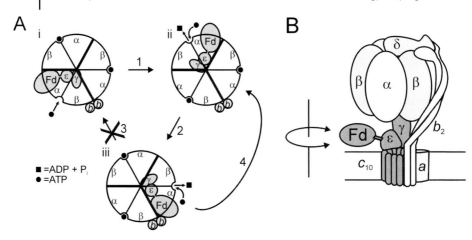

Figure 12.6 Proposed mechanism of uncoupled ATP hydrolysis in ATP synthase containing ε fused to flavodoxin (Fd). (A) Schematic representation modeled after the hydrolytic cycle of F_1 ATPase shown in Figure 12.4b of Leslie and Walker [81]. The view is looking down from the top of F_1 towards F_0. In step 1, ATP binding to an empty site promotes hydrolysis of ATP in an adjacent site and the clockwise rotation of the rotor by 120°. In step 2, the same series of events occur. At this point the ε-flavodoxin fusion comes into close proximity of the b_2 stator and cannot make another clockwise 120° rotation, preventing step 3 (as represented by the large X). Instead, the rotor is proposed to rotate counter-clockwise in the direction of ATP synthesis (step 4), but without ADP and P_i bound, so the step is uncoupled. Uncoupled ATPase activity then consists of repeating cycles of steps 2 and 4. It is proposed that the uncoupled step 4 is normally prevented by the C-domain of ε. (B) Subunit model of ATP synthase illustrating the reciprocating motion of the rotor (shaded dark) that has to occur to unmask the uncoupled hydrolysis.

In our study, we also fused a smaller 4-helix bundle protein, the soluble *E. coli* cytochrome b_{562}, to the C-terminus of ε, and found that ATP synthase bearing this fusion could support growth of cells on acetate, implying that the smaller fused domain was able to pass through the gap between the central and peripheral stalks, allowing full rotation. However, one would expect that the fusion still prevents the ε C-domain from assuming its normal conformation. Analysis of membranes from this strain showed a markedly enhanced ATPase activity, compared to wild type, but a modestly reduced ATP-dependent proton pumping activity. This fusion therefore retained substantial coupled ATPase activity, but also manifested the uncoupled activity noted previously for the larger fusions.

These results are reminiscent of those summarized in Table 12.1 for other mutations or treatments that disrupt the normal properties of the C-domain of ε. ATPase activity was increased in all cases, but, whenever measured, proton pumping was not increased in parallel; usually a modest decline was noted. It seems likely that in all of these cases the disruption of C-domain function generated the uncoupled activity, while having only a small effect on the coupled activity. Until now this had not been obvious, as the uncoupled component always occurred in conjunction with the coupled activity. We recognized the uncoupled ATPase activity initially in our larger fusions because they totally lacked the coupled ATPase activity.

One would expect that cells carrying ATP synthase with impaired C-domain function would give lower growth yields on limiting levels of glucose because of loss of ATP dissipated through the uncoupled activity, and this result is also borne out. The ε-cytochrome b fusions gave yields of 80–85 % of the wild-type growth yield [27], while work in the Vik laboratory [33] showed that truncations of ε at residues 94 and 117 gave 74 % and 81 % of the control, respectively.

The results of Tsunoda et al. [22] imply that ε can prevent rotation in the direction of ATP hydrolysis, while our results [27] imply that ε can prevent rotation in the direction of ATP synthesis. We therefore propose that the ratchet is bidirectional, with ε acting as the switch that determines which direction is allowed. How is the switch set?

The work described above, showing that ε sits in a different position with respect to $\alpha_3\beta_3$ depending on whether ATP or ADP is present, suggests a mechanism. Besides differences in nucleotide-dependent disulfide formation [14], Wilkens and Capaldi [38] have shown, using electron microscopy, that the N-domain of ε is in two different angular positions with respect to $\alpha_3\beta_3$, depending on which of the nucleotides are present. Here the positions are opposite to those suggested by the disulfide cross-linking data, with ε sitting under α when ADP is bound, and under β when ATP is bound, but it is likely that, when the N-domain is under one subunit of $\alpha_3\beta_3$, the C-domain is sitting under an adjacent subunit. The binding of ATP would favor the conformation that allows rotation in the direction of hydrolysis, while the binding of ADP would favor the conformation that allows rotation in the direction of synthesis. Whether these nucleotide-binding events occur at catalytic sites, or on the recently reported site on ε subunit itself [15], is uncertain.

Additional factors that should be considered are the role of proton movements and the proton-motive force. Richter and McCarty [39] showed that the ε subunit of spinach thylakloids is inaccessible to polyclonal antibodies in the dark, but becomes accessible in the light, implying a proton-motive force-dependent conformational change in the subunit. In the *E. coli* system, Mendel-Hartvig and Capaldi [18] showed that DCCD treatment of ATP synthase prevented the nucleotide-dependent conformational change, implying that the change occurs in conjunction with *c* ring rotation. Since the conformation of ε and conformational changes within the subunit are affected by both the nucleotide bound and by *c* ring rotation, we suggest that ε integrates the effects of the phosphorylation potential and the proton-motive force. When ATP is in abundance and proton-motive force is low, ε senses these signals and takes on a conformation that allows ATP hydrolysis. When ATP is less abundant and the proton-motive force is high, ADP is bound and ε undergoes a change in conformation to one that allows the rotor to move in the direction of ATP synthesis, but not ATP hydrolysis. Thus, ε acts as the switch of the ratchet, determining the direction of allowed rotation.

Clearly, the exact mechanism by which ε accomplishes this is not known. One purely physical model is shown in Figure 12.7. In panel A, the enzyme is poised to hydrolyze ATP and act as a proton pump. In this conformation ε allows rotation only in the direction of hydrolysis. The C-terminal helix of the C-domain would have to pivot downward to pass by the individual α and β subunits. This is possible for rotation in the direction of ATP hydrolysis, but rotation in the opposite direction would force the helix into the α/β cleft and jam the motor. Binding of ADP and phosphate to the active site would cause a conformational change in ε, inducing partial rotation of the N-domain, γ, and the *c* ring as shown in Figure 12.7B. This conformation would also be favored by the presence of a high proton-motive force, and it is likely that some protons would move through F_o in conjunction with the conformational change; these protons would not be involved in driving the synthesis of ATP, but rather in poising the enzyme for ATP synthesis. Rotation in the direction of synthesis is now allowed, while rotation in the direction of ATP hydrolysis is blocked. The physical mechanism here is the same as described for panel A. Probably, ionic interactions also play a key role in determining the direction of rotation [35] since they are essential for the inhibition of soluble F_1 by ε. Regardless of the exact mechanism by which ε restricts rotation to one direction, the central features of our model are nucleotide- or proton-motive force-dependent conformational changes in ε and associated partial rotation of the N-domain-γ-*c* ring complex.

The coupling function proposed above provides a good explanation of the easily observed activity of ε, i. e. inhibition of ATP hydrolysis. To maintain coupling, uncoupled activity must be prevented, and certainly ATP hydrolysis by F_1-ATPase is

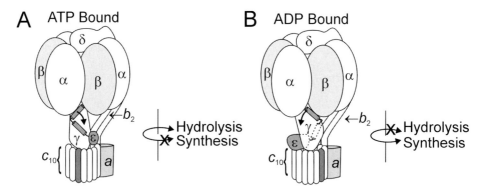

Figure 12.7 Model for the ratchet mechanism of ATP synthase. The orientation of the C-terminal helix is proposed to provide the mechanism by which the ratchet can determine the allowed direction of rotation. One subunit c is shaded in each panel to illustrate the relative angular position of the rotor with respect to the stator. (A) ATP synthase that has bound ATP and is poised to undergo ATP hydrolysis. The α subunit can act as a cam, pushing the C-terminal helix of ε downward, with the connecting loop between the two helices acting as a hinge, to allow rotation in the direction of ATP hydrolysis. Rotation in the opposite direction would force the C-terminal helix further into the cleft between α and β, preventing movement in this direction. (B) ATP synthase that has bound ADP + P_i and is poised to undergo ATP synthesis. Compared to part A, the $γεc_{10}$ complex has rotated 80° upon the binding of ADP + P_i and the C-terminal helix of ε has re-oriented at the same α/β interface. This conformation is similar to that seen in the structure of Rodgers and Wilce [6] and Hausrath et al. [7], and would allow rotation in the direction of ATP synthesis but not in the direction of ATP hydrolysis. Reproduced from [27] with permission.

uncoupled. The model we have proposed has important implications extending beyond just the ε subunit. The most basic of these is that ATP synthesis and ATP hydrolysis occur with the enzyme in different conformations and therefore these two processes are not the exact reverse of each other. Here we have focused on the ε conformation, but other differences will also be expected, such as the way in which the second stalk exerts its force as a stator. One might also expect that the three resting positions of the rotor might be different in ATP synthesis than in ATP hydrolysis, because of the additional proton movements involved in converting the enzyme between the two forms.

12.3
Peripheral Stalk of ATP Synthase

The peripheral stock of ATP synthase functions as a stator, holding the $\alpha_3\beta_3$ catalytic hexamer stationary relative to the $\gamma\varepsilon c_{10}$ rotor during catalysis. Certainly, assembly of the complex depends on the proper interactions among the stalk subunits, as well as between the stalk and $\alpha_3\beta_3$. Although the peripheral stalk is a critical part of the ATP synthase machinery, a high-resolution structure of the intact enzyme including the peripheral stalk complex has not yet been determined, nor has there been a high-resolution structure of the peripheral stalk subassembly in isolation. However, high-resolution structures of some fragments of the stalk have been obtained, and other techniques have provided insight into the structure. Here, we will review the current knowledge of the *E. coli* peripheral stalk, which has been most thoroughly characterized, and will make reference to the relationship between this structure and those of other organisms.

The *E. coli* peripheral stalk is made up of two copies of the *b* subunit of F_o and one copy of the δ subunit of F_1. The *b* subunit has an N-terminal membrane-spanning domain, while the balance of the protein forms a coiled-coil dimer extending into the cytoplasm, up the side of $\alpha_3\beta_3$. The δ subunit has a compact six-helix N-terminal domain which binds near the top of $\alpha_3\beta_3$ and a more extended C-terminal domain that interacts with the C-terminal domain of *b* to form the peripheral connection. A similar architecture is found in the ATP synthases of other bacteria and chloroplasts. In contrast, the peripheral stalks of mammalian mitochondrial ATP synthases are more complex, containing at least four different subunits present in one copy each.

12.3.1
b Subunit Sequences

Examination of *b* sequences reveals an N-terminal membrane spanning segment, while the balance of the sequences are rich in polar and charged residues, and contain many alanines (Figure 12.8). These features are generally maintained in bacterial and chloroplast subunits, and their sequences are easily aligned with few gaps, although *b* is one of the least conserved subunits of ATP synthase. Most eubacteria have single genes for *b* and the resulting subunits form homodimers, but genes for two different *b*-type polypeptides, denoted *b* and *b'*, are encoded by photosynthetic bacteria and by certain other bacterial species, including the mycobacteria. Remarkably, in the mycobacteria, the genes for *b* and δ are fused into a single open-reading frame, separated by a linker of 109 residues, while *b'* is encoded in a different open reading frame. The *b* and *b'* polypeptides from individual organisms

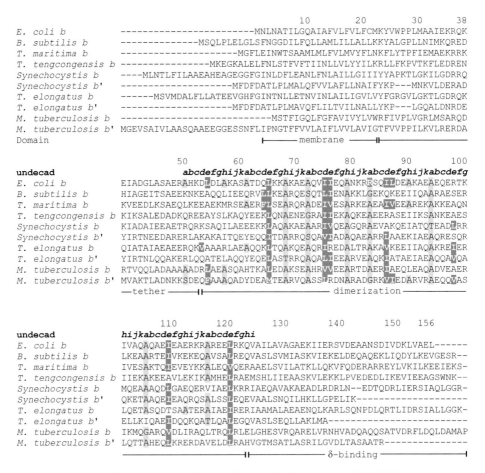

Figure 12.8 Clustal alignment of selected *b*-type subunit sequences. SP/trEMBL entries for the sequences are: *Escherichia coli b*, sp|P00859|ATPF_ECOLI; *Bacillus subtilis b*, sp|P37814|ATPF_BACSU; *Thermotoga maritima b*, tr|Q9X1U9; *Thermoanaerobacter tengcongensis b*,tr|Q8RC19; *Synechocystis 6803 b*, sp|P27181|ATPF_SYNY3; *Synechocystis 6803 b'*, sp|P27183|ATPX_SYNY3; *Thermosynechococcus elongatus b*, tr|BAC07985; *T. elongatus b'*, tr|BACO7984; *Mycobacterium tuberculosis b*, sp|Q10594|ATPD_MYCTU (residues 1–159 of the *b*/δ fusion gene are shown); *M. tuberculosis b'*, sp|Q10596|ATPF_MYCTU. Numbering is by the *E. coli* positions. Small residues (AGST) in *a* and *h* positions are highlighted in yellow; larger hydrophobic residues (VILFMY) in *d* and *e* positions are highlighted in green. Arg-83 of *E. coli b* is highlighted in cyan.

show substantial sequence divergence, and the *b'* subunits are modestly shortened. The expressed polar domains of the *b* and *b'* subunits of the cyanobacterium *Synechocystis* exist individually as monomers, but form a heterodimer when mixed together [40], which is consistent with the expectation that the ATP synthase contains one copy of each. Chloroplast ATP synthases also contain two related *b*-type polypeptides, called subunits I and II [41, 42]. In contrast, mitochondrial *b* subunits (not shown) have two membrane spanning domains near the N-terminus, and the balance of the sequence shows little relationship to bacterial or chloroplast *b*-type subunits, except for sharing their notable polarity. Mitochondrial ATP synthase contains only a single copy of this *b* subunit.

12.3.2
Structure of E. coli b

The two *b* subunits of *E. coli* ATP synthase form a helical, highly elongated 156-residue homodimer. The polar region of the subunit, residues 25–156, has been expressed as a soluble dimer [43] and much studied in that form, though a high-resolution dimeric structure has yet to be obtained. Hydrodynamic studies have revealed a highly extended structure with a frictional ratio of about 1.8. The circular dichroism spectrum is characteristic of an α-helical coiled coil, which would provide the extreme length required to span the distance from the membrane to the top of F_1 [44]. Dimerization of the polar domains of *b* subunits is reversible and of limited strength; the dimer can be converted into monomer by heating to 40°C, as monitored by either analytical ultracentrifugation or circular dichroism [44, 45].

Deletion analysis allowed division of the b_2 structure into four distinct domains that we designate the membrane-spanning domain (residues 1–24), the tether domain (residues 25–52), the dimerization domain (residues 53–122), and the δ-binding domain (residues 123–156) as indicated on the lower line of Figure 12.8 [44]. Each of these domains of *b* will be discussed in turn.

12.3.2.1 Membrane-spanning Domain

The membrane-spanning domain (residues 1–24) crosses the membrane once, with the N-terminus on the periplasmic side. An NMR structure (PDB 1B9U) of the b_{1-34} construct was determined in a membrane mimetic chloroform–methanol–H_2O solvent, showing a hydrophobic α-helix interrupted by a 20° bend through residues 23–26 [46]. Copper-induced disulfide formation studies utilizing *b* subunits containing single cysteines inserted in positions 2–20 of the membrane spanning segment showed that residues on one face of this helix readily form intersubunit cross-links. The current model of this domain is that residues 3–22 are within the membrane and interact through a right-handed crossing at a 23° angle, leaving residues Asn-2 and the bend region, residues 23–26, to flank the membrane on the periplasmic and cytoplasmic sides of the membrane, respectively [46].

12.3.2.2 **Tether Domain**

The tether domain of *b*, residues 25–52, can be deleted from the soluble construct with little effect on dimerization [47]. Although a heptad repeat pattern characteristic of left-handed coiled coils is seen in this region, comparison of CD spectra of various deletion constructs suggests that in solution the region is largely unstructured [47]. The most notably conserved residue in the region, Arg-36, has functional importance in coupling as revealed by mutagenesis studies [48]. The proximity of this residue to the *a* subunit has been demonstrated by cross-linking through an introduced cysteine [49], and it may be that much of the tether domain interacts with the cytoplasmic polar loops of *a*. This region is surprisingly tolerant of residue deletions or insertions; as many as 11 residues can be either added or deleted before *b* will no longer support function or assembly of ATP synthase [50, 51].

12.3.2.3 **Dimerization Domain**

Dimerization of the two *b* subunits is required for interactions with F_1 [52]. The residues essential for the dimerization of the *b* subunit were determined to be between Asp-53 and Lys-122 through a series of truncation mutants that were analyzed by sedimentation equilibrium and sedimentation velocity ultracentrifugation [44]. Circular dichroism spectra showed the b_{53-122} dimer to be highly helical, with a $\theta_{222}/\theta_{208}$ ratio of 0.99, indicative of a coiled-coil structure [47]. Previously, a heptad repeat pattern, characteristic of left-handed coiled coils, had been identified in some parts of the sequence [43, 53]. Experiments designed to confirm such a structure by measuring the tendency to form disulfides between cysteines inserted into positions between 59 and 68 were carried out [54]. In these studies, the *d* positions of the heptad, residues 61 and 68, were expected to be most favorable for disulfide formation, but instead disulfide formation was strongest at positions 59 and 60, leading to the conclusion that this region of *b* forms an atypical coiled coil structure.

More recently, a crystal structure of a monomeric form of b_{62-122} at 1.55 Å resolution revealed an amphipathic helix with a slightly right-handed pattern of hydrophobic residues running along one side [55] as seen in Figure 12.9A and B. Dimerization of two such helices via their hydrophobic surfaces would yield a right-handed, parallel, coiled-coil (Figure 12.9C). The residues making up the hydrophobic surface lie in an 11-residue "undecad" pattern in the sequence, the pattern expected for a right-handed coiled coil (RHCC) [56, 57] with undecad positions designated *a* to *k*. As yet, however, no parallel two-stranded RHCC has been determined at high resolution. Comparison of the *E. coli* sequence with other *b* sequences (Figure 12.8) reveals that the undecad pattern is well conserved. The *a* and *h* positions are usually occupied by small residues, particularly alanine; some of these alanines, such as Ala-79, are among the most conserved residues of *b*. The *a* and *h* positions are suggested to occupy the center of the helix–helix interface (Figure 12.10A). The *d* and *e* positions of the undecad, which are more often occupied by larger hydrophobic residues, would then flank the interface on either side.

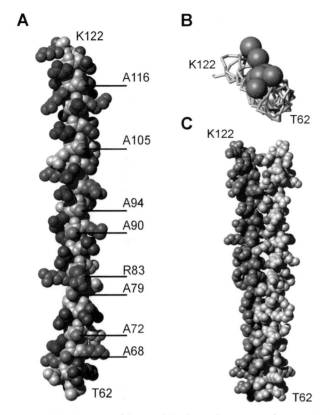

Figure 12.9 Structure of the *E. coli b* subunit dimerization domain. (A) Crystal structure of b_{62-122} (PDB 1L2P). Acidic, basic and hydrophobic side-chains are colored red, blue, and green respectively. Alanine residues thought to be involved in the dimer interface are colored orange. (B) End-on view of the monomer structure, showing the right-handed twist of alanine residues in the proposed interface. (C) Dimer model of b_{62-122}.

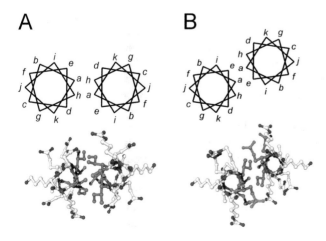

Figure 12.10 Expected residue disposition in the two-stranded RHCC. (A) Pattern in the N-end of the dimerization domain. (B) Pattern in the C-end of the dimerization domain. Upper panels show helical wheel representations; lower panels show a cross-sectional view through one undecad. The *a* and *h* positions are shown in orange while the *d* and *e* positions are in green.

In the b_{62-122} helix, the hydrophobic surface is interrupted by residue Arg-83, which occupies an a position in contrast to the usual alanine. Based on the right-handed model, mutation of this arginine to alanine would be expected to stabilize the dimer by removing the large, charged arginine from the center of the interface. Stability was assessed by sedimentation equilibrium ultracentrifugation and thermal denaturation monitored by circular dichroism [55]. As controls, arginine residues at positions 82 and 98 were also individually changed to alanine. The alanine mutation at position 83 caused a large increase in stability of the dimer, about a 7°C increase in the melting midpoint, whereas the other two mutations destabilized the structure, reducing the melting midpoint. Thus the alanine residues in the a and h positions serve important structural roles.

12.3.2.4 δ-Binding Domain

The C-terminal domain of b (residues 123–156) adopts a more globular structure and is essential for the binding of F_1 through the δ subunit. No well-defined heptad or undecad pattern is apparent in this region, but circular dichroism analysis indicates that it is largely helical [47]. Sedimentation velocity experiments reveal that mutations in this region, including deletion of C-terminal residues [58], introduction of an A128D mutation [59] or even working at low temperature [44] cause substantial decreases in the sedimentation coefficient while not affecting dimerization. We have interpreted these results to mean that the C-terminal region folds back on the protein, possibly through a predicted bend near residue 141, to create a more globular structure, such as a 4-helix bundle. The mutations noted above also cause a loss of ability of the b dimer to bind F_1-ATPase.

The conformation in the region between residues 124 and 132 has also been explored by disulfide formation between introduced cysteines [54]. Disulfide formation was favored at residues 124, 128, and 132. This periodicity of 4 implies that the region is helical, and that the two helices cross in a right-handed manner.

12.3.2.5 Proposed Structural Model of b

As no dimeric high-resolution dimeric structure of b subunit is yet available, we have generated a model (Figure 12.11) of the b_2 dimer based on our monomeric structure and the other experimental data summarized above. In our model, the two helices are arranged in a right-handed coiled-coil, as suggested by the arrangement of hydrophobic residues on the surface of the monomer, and the undecad pattern in the amino acid sequence [55]. To achieve optimal interactions between the monomers, the helices are offset by approximately 3 Å, allowing the β-carbons of the alanines to come in close proximity to the backbone of the adjacent helix (Figure 12.10). The result is a coiled-coil where the two helices are closer together by at least 1 Å, compared with a typical left-handed coiled coil. Ionic or polar interactions on the periphery of the interface, analogous to those between the g and e' positions of a left-handed coiled coil, are possible through interactions between residues d–e', d–i', k–e', or k–i'.

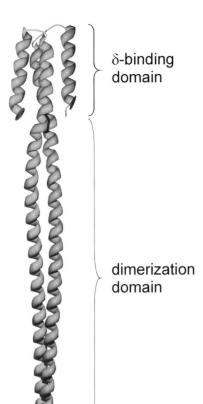

δ-binding
domain

dimerization
domain

Figure 12.11 Current model of residues 53–156 of the b_2 subunit of *E. coli* ATP synthase. Beginning at the dimerization domain, the helices interact by a right-handed coiled-coil that becomes more tightly wound as the C-terminus is approached. We predict that in the absence of δ the polypeptides bend near residue 140, folding back to form a 4-helix bundle.

This right-handed coiled-coil model is further supported by previous cross-linking data, where no disulfides were formed between *d* positions 61 and 68 of the proposed heptad repeat, and mixed disulfides were formed between residues S60C and L65C [54]. The latter sites correspond to positions *c* and *a*, respectively, in the heptad repeat, and positions *e* and *k*, respectively, in the current undecad repeat model. In the heptad repeat, disulfide formation between positions *c* and *a* is unlikely, but in our undecad dimer model with offset helices, the *k* and *e* positions are approximately at the correct distance and orientation to form disulfide bonds.

It is evident from the sequence alignment that the undecad pattern seems to change around residue Ala-90 (*E. coli*), with larger hydrophobic residues such as isoleucine usually occupying the *h* positions, and charged or polar residues in the *d* positions. This suggests a change in the coiled-coil arrangement that is consistent with the helices wrapping around each other with a higher degree of right-handed super-helical twist. This would result in a shift of the helices such that the hydrophobic core of the dimer is formed by the *a* and *e* residues, as the *h* position moves to the periphery of the core (Figure 12.10B).

Based on the sequence of the *b* subunit, a bend is predicted in the α helix around residues 139–141, which would allow the final 15 residues to fold back onto residues 124–138, forming an anti-parallel right-handed four-helix bundle. Such a motif is consistent with the more strongly hydrophobic character of the sequence in this region, the ultracentrifugation studies described above, and the observation that a photoactivatable cross-linking agent incorporated at the C-terminus of *b* can form an internal cross-link [58]. Since many *b'* sequences end near residue 140, the *bb'* heterodimers would form 3-helix bundles at their C-termini.

12.3.3
Structure of δ

E. coli δ is a 177-residue, 19.3 kDa polypeptide, situated at or near the top of the F_1 complex and linking the *b* dimer and F_1. Like *b*, δ has a poorly conserved sequence with few essential residues. In classical work using δ purified from F_1, the circular dichroism spectrum indicated 55–70 % helicity, while size-exclusion chromatography and analytical ultracentrifugation indicated a monomeric, nonspherical protein [60]. δ is currently characterized as a two-domain protein, with residues Ser-1 to Thr-105 forming a compact N-terminal helical bundle, while residues 106 to 177 are more disordered in solution. Both δ in isolation and δ in F_1 are sensitive to proteolytic cleavage in the C-terminal region, with the larger N-terminal fragment remaining bound to F_1 [61]. Similarly, during purification of overexpressed δ, an endogenous protease cleaved the polypeptide after residue 134 [62]. Using the 1–134 fragment, Wilkens and coworkers [63] determined the solution structure of the N-terminal domain to be a compact, 6-helix bundle, with helices 1, 2, 5, and 6 forming an unusual arrangement of intercalating V-shaped pairs. From the hydrodynamic properties of intact δ, and the sensitivity to proteolysis in the C-terminal section, it seems likely that the latter region is largely disordered in solution.

12.3.4
Interaction of *b* and δ

The interaction of isolated *b* and δ has been demonstrated by several techniques [45, 62, 64]. The stoichiometry of the complex was characterized by analytical ultracentrifugation as the expected $b_2δ$, and the shape of the complex is even more extended than either b_2 or δ, implying that the complex is formed in an end-to-end manner [62]. This is consistent with strong evidence that the C-terminal regions of each subunit are involved in formation of $b_2δ$. In particular, C-terminal truncation or proteolysis of either subunit blocks the interaction. Furthermore, a photoactivatable cross-linking agent incorporated at the C-terminus of *b* formed a cross-link with residue Met-158 of δ [58]. Thus, the $b_2δ$ interaction depends on the C-terminal regions that, in isolation, are either loosely folded, for b_2, or apparently unfolded, in the case of δ. It seems very likely that substantial conformational changes accompany the formation of this interaction, but the details are as yet unknown. Interaction between the isolated subunits appears to be weak in solution, with a K_d of 5–10 μM [62], but this determination is limited by the weak dimeriza-tion of the soluble form of b, which is of the same order of magnitude [47]. An en-gineered disulfide between the C-terminus of b and position 158 of δ did not affect ATP hydrolysis or ATP-dependent proton pumping, implying that the b_2–δ interac-tion remains intact during catalytic events [65].

12.3.5
Position of the Peripheral Stalk in ATP Synthase

To function as a crucial part of the stator ($ab_2\alpha_3\beta_3\delta$), the peripheral stalk is respon-sible for holding the $\alpha_3\beta_3$ complex stationary relative to the rotor components ($\gamma\epsilon c_{10}$). To achieve this goal, it must bind the $\alpha_3\beta_3$ hexamer to the a subunit in the membrane. The peripheral stalk was not visible by electron microscopy in early studies, due to its non-globular structure and tendency to be averaged out dur-ing the alignment of single ATP synthase particles that were oriented differently, so it was once expected that b extended from the membrane and interacted with F_1 through the centre of the $\alpha_3\beta_3$ complex. More recently, improved methods of sort-ing images before alignment have allowed visualization of the peripheral structure in bacterial, chloroplast, and mitochondrial enzymes [66–69]. Generally, in these studies, a tapering protrusion is seen rising from the side of F_o, in some cases ex-tending as a narrow filament along the side of $\alpha_3\beta_3$. A more substantial additional density is also apparent alongside the upper outside surface of $\alpha_3\beta_3$ in some views [67].

The b subunit originates at the membrane in close proximity to the other F_o com-ponents. The membrane-spanning domain of b is positioned on the outer surface of the c ring proximal to the a subunit, as determined initially through chemical cross-linking with purified F_o and F_1F_o [70, 71]. Unpublished results (W. Jiang and R. Fillingame) cited by Fillingame et al. [72] further showed that a cysteine in-troduced into position 2 of b will form disulfides with cysteines at positions 227 or 228 of a. These residues lie in the periplasmic polar loop between the fourth and fifth transmembrane helices of that subunit. The same cysteine in b also readily forms disulfides with cysteines introduced into the c subunit at positions 74, 75, or 78 following the second transmembrane helix [37]. Thus, there is good evidence that, on the periplasmic side, the membrane domain of b is located at the edge of the interface between a and the c ring. On the cytoplasmic side of the membrane, interactions between the tether domain of b and polar loops of the a subunit seem likely, although this region has not been studied in detail. Chemical cross-linking using the photoactivatable cross-linking agent benzophenone maleimide, intro-duced into a cysteine at position 36 of b, did link that site to a [49].

At the other end of the peripheral stalk, binding of the δ subunit to F_1 depends on the N-terminal residues of the α subunits, which are located at the top of F_1 structure but are disordered in the F_1 crystal structures lacking OSCP, the mito-chondrial analog of bacterial δ. Proteolytic removal of the first 15 residues of α pre-vents binding of δ [73], and a cysteine inserted into position 2 of α forms disulfides with either δCys-140 or δCys-64, the naturally occurring cysteines in the *E. coli* subunit [74]. K_d of the interaction of δ with δ-depleted F_1 is of the order of 1–2 nM [75]. Recently, Weber and Senior [76] found that a synthetic peptide encompassing residues 1–22 of α bound to δ with a K_d of 130 nM, causing an enhancement in the fluorescence of δTrp-28. Based on the effects of mutations at various sites in δ on binding interactions with F_1 [77] or the α peptide [76], Weber and coworkers postulate that helices 1 and 5 of δ pack with the N-terminal α peptide to form the central part of the interaction.

As noted above, the C-terminal regions of the *b* dimer and of δ interact to form the peripheral connection. Several lines of evidence suggest that this domain complex lies on the upper outside surface of ATP synthase, possibly corresponding to the structure noted in this position in the electron microscopic study by Wilkens and coworkers [67]. Rodgers and Capaldi [78] found that a cysteine introduced at the C-terminus of *b*, position 156, could form a disulfide with Cys-90 of an α subunit, which is located in this region. In the mitochondrial system, Rubinstein and Walker [79] localized the C-terminus of OSCP, the analog of δ, by electron microscopy of ATP synthase using a fused biotinylation tag that was labeled with avidin. These workers found that the C-terminal region may extend down the side of F_1 by almost 10 nm.

The orientation of the *b* dimerization domain along the surface of F_1 has been defined through cross-linking experiments in which photoactivatable cross-linking agents, benzophenone-4-maleimide or *p*-azidophenacyl bromide, were linked to cysteine residues introduced into positions 92 or 109 of *b* [49]. Cross-linked peptides of α or β were then identified by a fragmentation/mass spectrometry approach. The 92 position of *b* was found to cross-link to α within the region between residues 464 and 483, as well as to a nearby region of the β subunit. These peptides are located near the bottom of $\alpha_3\beta_3$ at a noncatalytic interface. Furthermore, the 109 position was found to cross-link farther up the surface of α, probably in the segment between residues 213 and 220. Using these positions to orient the helices of the *b* dimerization domain reveals that residue 53, the beginning of the domain, is located very close to the surface of the membrane, while residue 122, the site of junction with the δ-binding domain, is located about two-thirds of the way up the side of $\alpha_3\beta_3$.

12.3.6
Function of the Peripheral Stalk

The most straightforward functions of the peripheral stalk are to hold F_1 and F_o together and to prevent $\alpha_3\beta_3$ from rotating with $\gamma\varepsilon c_{10}$. These roles can both be considered part of the stator function, and elicit visions of a strong, unyielding structure. However, neither the physical properties of b_2 nor the results of the internal insertion/deletion studies are consistent with this vision. In isolation, the polar domain of b_2 can fold back on itself [58], while the ability of the subunit to tolerate insertions or deletions of up to 11 residues, at least in some regions, suggests that b_2 should be viewed as an elastic rope that can still function when stretched over a substantial range.

Since it seems likely that there are 10 protons pumped across the membrane per 360° rotation (one per *c* subunit), and 3 ATP produced per 360° rotation, the synthesis of one ATP cannot correspond exactly with an integral number of protons pumped. This might pose a problem if the stator was rigid, but not if it was elastic. In the latter case, the proton-driven counter-clockwise rotation of $\gamma\varepsilon$, as viewed from the top, would turn $\alpha_3\beta_3$ in the same direction, with resistance from b_2 as it became more stretched. This would cause the peripheral stalk to wrap around $\alpha_3\beta_3$ in a right-handed manner. For ATP hydrolysis, the wrapping would be in a left-handed manner. In either case, the tension borne elastically by b_2 would serve to absorb and buffer the force exerted by the rotation of $\gamma\varepsilon$ due to individual proton movements, making non-integral H^+/ATP ratios less of a problem. Upon catalysis and product release from the active site, the elastically stretched b_2 would contract, pulling $\alpha_3\beta_3$ back relative to $\gamma\varepsilon$.

In the absence of a high-resolution structure, one can only speculate on the question of how the b_2 dimer can store elastic energy, but our thoughts are guided by the unique right-handed coiled coil structure we propose for the dimerization domain. In our RHCC model, the two helices interact through a relatively flat interaction surface created by the small residues in the *a* and *h* positions. This could allow for the helices to shift relative to each other along the superhelical axis, something that is not possible in a left-handed coiled-coil because of the "knobs into holes" packing. The changes in ionic, or dipole–dipole, interactions relating to such a shift could result in different energy states. Any such shift cannot be more than a few Å, however, as disulfide formation between cysteines introduced into *b* subunits had little or no affect on enzyme function [45].

A second type of conformational change suggested by the RHCC model is an alteration in the type or extent of supercoiling. Twisting motions may be imparted to b_2 by conformational changes in the *a* and *c* subunits associated with proton translocation. Twisting in one direction would tighten the right-handed supercoil, while twisting in the other direction would drive it toward a traditional left-handed structure. Either of these would be energetically unfavorable, resulting in energy storage.

Direct evidence to support any of these possibilities has yet to be obtained, but the hypothesis that the RHCC has functional significance is supported by results recently obtained in our laboratory. In particular, a portion of the dimerization domain of *b* was replaced by the left-handed coiled coil sequence from yeast GCN4. The resultant chimeric *b* supported assembly of ATP synthase, but the strain failed to grow on acetate, indicating a functional impairment in ATP synthesis (S. D. Dunn and Y. Bi, unpublished observations).

12.4
Conclusion

While the rotational behavior of ATP synthase has been well established, a detailed understanding of the mechanism of coupling ATP synthesis and proton translocation at the molecular level is still lacking. It is clear, however, that the transmission of energy between the catalytic sectors depends on the two stalks, the central γε rotor and the peripheral $b_2\delta$ stator. Here we have focussed on the roles of δ, ε and b.

Biochemical evidence shows that the conformation of ε is dynamic and that the subunit has a key function in maintaining a high level of coupling efficiency. Results we have described imply that ε maintains coupling by specifically inhibiting the uncoupled hydrolysis of ATP, and we have proposed a mechanism for this function.

Although the structure of peripheral stalk composed of $b_2\delta$ has yet to be fully described, it appears that this complex contains features that are novel to structural biology. We have proposed a right-handed coiled coil structure for the dimerization domain, based on sequence analysis, a monomeric crystal structure, cross-linking studies, and mutational analyses. The right-handed coiled coil should have properties very different from those of a left-handed coiled coil, and we expect these different properties to be related to the function of second stalk. It seems likely that the conformation of $b_2\delta$, like that of ε, will be dynamic.

References

1. S. D. Dunn, *Nat. Struct. Biol.*, **1995**, 2, 915–918.
2. S. Wilkens, F. W. Dahlquist, L. P. McIntosh, L. W. Donaldson, and R. A. Capaldi, *Nat. Struct. Biol.*, **1995**, 2, 961–967.
3. U. Uhlin, G. B. Cox, and J. M. Guss, *Structure*, **1997**, 5, 1219–1230.
4. D. Stock, A. G. Leslie, and J. E. Walker, *Science*, **1999**, 286, 1700–1705.
5. C. Gibbons, M. G. Montgomery, A. G. Leslie, and J. E. Walker, *Nat. Struct. Biol.*, **2000**, 7, 1055–1061.
6. A. J. Rodgers and M. C. Wilce, *Nat. Struct. Biol.*, **2000**, 7, 1051–1054.
7. A. C. Hausrath, R. A. Capaldi, and B. W. Matthews, *J. Biol. Chem.*, **2001**, 276, 47227–47232.
8. S. D. Dunn, *J. Biol. Chem.*, **1982**, 257, 7354–7359.
9. P. D. Bragg and C. Hou, *Eur. J. Biochem.*, **1980**, 106, 495–503.
10. H. R. Lötscher, *Biochemistry*, **1984**, 23, 4134–4140.
11. H. G. Dallmann, T. G. Flynn, and S. D. Dunn, *J. Biol. Chem.*, **1992**, 267, 18953–18960.
12. J. P. Abrahams, A. G. Leslie, R. Lutter, and J. E. Walker, *Nature*, **1994**, 370, 621–628.
13. R. Aggeler, M. A. Haughton, and R. A. Capaldi, *J. Biol. Chem.*, **1995**, 270, 9185–9191.
14. R. Aggeler and R. A. Capaldi, *J. Biol. Chem.*, **1996**, 271, 13888–13891.
15. Y. Kato-Yamada and M. Yoshida, *J. Biol. Chem.*, **2003**, 278, 36013–36016.
16. B. Schulenberg and R. A. Capaldi, *J. Biol. Chem.*, **1999**, 274, 28351–28355.

17. J. Mendel-Hartvig and R. A. Capaldi, *Biochemistry*, **1991**, *30*, 1278–1284.

18. J. Mendel-Hartvig and R. A. Capaldi, *Biochemistry*, **1991**, *30*, 10987–10991.

19. E. A. Johnson and R. E. McCarty, *Biochemistry*, **2002**, *41*, 2446–2451.

20. Y. Zhang and R. H. Fillingame, *J. Biol. Chem.*, **1995**, *270*, 24609–24614.

21. J. Hermolin, O. Y. Dmitriev, Y. Zhang, and R. H. Fillingame, *J. Biol. Chem.*, **1999**, *274*, 17011–17016.

22. S. P. Tsunoda, A. J. Rodgers, R. Aggeler, M. C. Wilce, M. Yoshida, and R. A. Capaldi, *Proc. Natl. Acad. Sci. U. S.A.*, **2001**, *98*, 6560–6564.

23. H. Noji, R. Yasuda, M. Yoshida, and K. Kinosita, Jr., *Nature*, **1997**, *386*, 299–302.

24. Y. Kato-Yamada, H. Noji, R. Yasuda, K. Kinosita, Jr., and M. Yoshida, *J. Biol. Chem.*, **1998**, *273*, 19375–19377.

25. R. Yasuda, H. Noji, M. Yoshida, K. Kinosita, Jr., and H. Itoh, *Nature*, **2001**, *410*, 898–904.

26. K. Nishio, A. Iwamoto-Kihara, A. Yamamoto, Y. Wada, and M. Futai, *Proc. Natl. Acad. Sci. U. S.A.*, **2002**, *99*, 13448–13452.

27. D. J. Cipriano, Y. Bi, and S. D. Dunn, *J. Biol. Chem.*, **2002**, *277*, 16782–16790.

28. P. C. Sternweis, *J. Biol. Chem.*, **1978**, *253*, 3123–3128.

29. P. C. Sternweis and J. B. Smith, *Biochemistry*, **1980**, *19*, 526–531.

30. D. J. Klionsky, W. S. Brusilow, and R. D. Simoni, *J. Bacteriol.*, **1984**, *160*, 1055–1060.

31. S. D. Dunn, V. D. Zadorozny, R. G. Tozer, and L. E. Orr, *Biochemistry*, **1987**, *26*, 4488–4493.

32. M. Kuki, T. Noumi, M. Maeda, A. Amemura, and M. Futai, *J. Biol. Chem.*, **1988**, *263*, 17437–17442.

33. H. Xiong, D. Zhang, and S. B. Vik, *Biochemistry*, **1998**, *37*, 16423–16429.

34. K. F. Nowak, V. Tabidze, and R. E. McCarty, *Biochemistry*, **2002**, *41*, 15130–15134.

35. K. Y. Hara, Y. Kato-Yamada, Y. Kikuchi, T. Hisabori, and M. Yoshida, *J. Biol. Chem.*, **2001**, *276*, 23969–23973.

36. W. Junge, O. Pänke, D. A. Cherepanov, K. Gumbiowski, M. Muller, and S. Engelbrecht, *FEBS Lett.*, **2001**, *504*, 152–160.

37. P. C. Jones, J. Hermolin, W. Jiang, and R. H. Fillingame, *J. Biol. Chem.*, **2000**, *275*, 31340–31346.

38. S. Wilkens and R. A. Capaldi, *Hoppe-Seyler's Z. Physiol.Chem.*, **1994**, *375*, 43–51.

39. M. L. Richter and R. E. McCarty, *J. Biol. Chem.*, **1987**, *262*, 15037–15040.

40. S. D. Dunn, E. Kellner, and H. Lill, *Biochemistry*, **2001**, *40*, 187–192.

41. U. Pick and E. Racker , *J. Biol. Chem.*, **1979**, *254*, 2793–2799.

42. R. G. Herrmann, J. Steppuhn, G. S. Herrmann, and N. Nelson, *FEBS Lett.*, **1993**, *326*, 192–198.

43. S. D. Dunn, *J. Biol. Chem.*, **1992**, *267*, 7630–7636.

44. M. Revington, D. T. McLachlin, G. S. Shaw, and S. D. Dunn, *J. Biol. Chem.*, **1999**, *274*, 31094–31101.

45. A. J. Rodgers, S. Wilkens, R. Aggeler, M. B. Morris, S. M. Howitt, and R. A. Capaldi, *J. Biol. Chem.*, **1997**, *272*, 31058–31064.

46. O. Dmitriev, P. C. Jones, W. Jiang, and R. H. Fillingame, *J. Biol. Chem.*, **1999**, *274*, 15598–15604.

47. M. Revington, S. D. Dunn, and G. S. Shaw, *Protein Sci.*, **2002**, *11*, 1227–1238.

48. T. L. Caviston, C. J. Ketchum, P. L. Sorgen, R. K. Nakamoto, and B. D. Cain, *FEBS Lett.*, **1998**, *429*, 201–206.

49. D. T. McLachlin, A. M. Coveny, S. M. Clark, S. D. Dunn, and , *J. Biol. Chem.*, **2000**, *275*, 17571–17577.

50. P. L. Sorgen, M. R. Bubb, and B. D. Cain, *J. Biol. Chem.*, **1999**, *274*, 36261–36266.

51. P. L. Sorgen, T. L. Caviston, R. C. Perry, B. D. Cain, *J. Biol. Chem.*, **1998**, *273*, 27873–27878.

52. P. L. Sorgen, M. R. Bubb, K. A. McCormick, A. S. Edison, and B. D. Cain, *Biochemistry*, **1998**, *37*, 923–932.

53. K. A. McCormick, G. Deckers-Hebestreit, K. Altendorf, and B. D. Cain, *J. Biol. Chem.*, **1993**, *268*, 24683–25691.

54. D. T. McLachlin, S. D. Dunn, *J. Biol. Chem.*, **1997**, *272*, 21233–21239.

55. P. A. Rizzo Del, Y. Bi, S. D. Dunn, and B. H. Shilton, *Biochemistry*, **2002**, *41*, 6875–6884.

56. A. Lupas, *Curr. Opin. Struct. Biol.*, **1997**, *7*, 388–393.

57. A. Lupas, *Trends Biochem. Sci.*, **1996**, *21*, 375–382.

58. D. T. McLachlin, J. A. Bestard, and S. D. Dunn, *J. Biol. Chem.*, **1998**, *273*, 15162–15168.

59. S. D. Dunn, Y. Bi, and M. Revington, *Biochim. Biophys. Acta*, **2000**, *1459*, 521–527.

60. P. C. Sternweis and J. B. Smith, *Biochemistry*, **1977**, *16*, 4020–4025.

61. J. Mendel-Hartvig and R. A. Capaldi, *Biochim. Biophys. Acta*, **1991**, *1060*, 115–124.

62. S. D. Dunn and J. Chandler, *J. Biol. Chem.*, **1998**, *273*, 8646–8651.

63. S. Wilkens, S. D. Dunn, J. Chandler, F. W. Dahlquist, and R. A. Capaldi, *Nat. Struct. Biol.*, **1997**, *4*, 198–201.

64. K. Sawada, N. Kuroda, H. Watanabe, C. Moritani-Otsuka, and H. Kanazawa, *J. Biol. Chem.*, **1997**, *272*, 30047–30053.

65. D. T. McLachlin and S. D. Dunn, *Biochemistry*, **2000**, *39*, 3486–3490.

66. S. Karrasch and J. E. Walker, *J. Mol. Biol.*, **1999**, *290*, 379–384.

67. S. Wilkens, J. Zhou, R. Nakayama, S. D. Dunn, and R. A. Capaldi, *J. Mol. Biol.*, **2000**, *295*, 387–391.

68. B. Böttcher and P. Gräber, *Biochim. Biophys. Acta*, **2000**, *1458*, 404–416.

69. B. Böttcher, L. Schwarz, and P. Gräber, *J. Mol. Biol.*, **1998**, *281*, 757–762.

70. J. Hermolin, J. Gallant, and R. H. Fillingame, *J. Biol. Chem.*, **1983**, *258*, 14550–14555.

71. J. P. Aris and R. D. Simoni, *J. Biol. Chem.*, **1983**, *258*, 14599–14609.

72. R. H. Fillingame, W. Jiang, and O. Y. Dmitriev, *J. Exp. Biol.*, **2000**, *203*, 9–17.

73. S. D. Dunn, L. A. Heppel, and C. S. Fullmer, *J. Biol. Chem.*, **1980**, *255*, 6891–6896.

74. I. Ogilvie, R. Aggeler, and R. A. Capaldi, *J. Biol. Chem.*, **1997**, *272*, 16652–16656.

75. J. Weber, S. Wilke-Mounts, and A. E. Senior, *J. Biol. Chem.*, **2002**, *277*, 18390–18396.

76. J. Weber, A. Muharemagic, S. Wilke-Mounts, and A. E. Senior, *J. Biol. Chem.*, **2003**, *278*, 13623–13626.

77. J. Weber, S. Wilke-Mounts, and A. E. Senior, *J. Biol. Chem.*, **2003**, *278*, 13409–13416.

78. A. J. Rodgers, and R. A. Capaldi, *J. Biol. Chem.*, **1998**, *273*, 29406–29410.

79. J. Rubinstein and J. Walker , *J. Mol. Biol.*, **2002**, *321*, 613–619.

80. Y. Kato-Yamada, D. Bald, M. Koike, K. Motohashi, T. Hisabori, and M. Yoshida, *J. Biol. Chem.*, **1999**, *274*, 33991–33994.

81. A. G. Leslie and J. E. Walker, *Philos. Trans. R. Soc. London, B Biol. Sci.*, **2000**, *355*, 465–471.

Part III
V-type ATPases

13

Structure, Mechanism and Regulation of the Yeast and Coated Vesicle V-ATPases

Elim Shao and *Michael Forgac*

Summary

Vacuolar (H^+)-ATPases (or V-ATPases) are ATP-dependent proton pumps that acidify intracellular compartments and, in certain cases, transport protons across the plasma membrane of eukaryotic cells. Acidification of intracellular compartments functions in such processes as receptor-mediated endocytosis, intracellular targeting of lysosomal enzymes, protein processing and degradation, and coupled transport. Plasma membrane V-ATPases are involved in acid–base balance, bone resorption, potassium secretion and tumor metastasis. This chapter will focus on work from our laboratory on the structure, mechanism and regulation of the V-ATPases from clathrin-coated vesicles and yeast. V-ATPases are composed of a peripheral V_1 domain responsible for ATP hydrolysis and an integral V_0 domain responsible for proton transport. V_1 has a molecular mass of 640 kDa and is composed of eight different subunits (subunits A–H) of molecular mass 70–13 kDa, whereas V_0 has a molecular mass of 260 kDa and is composed of five different subunits (subunits a, d, c, c' and c'') of molecular mass 100–17 kDa. Conventional and cysteine-mediated cross-linking as well as electron microscopy have allowed us to define the overall shape of the V-ATPase and the location of subunits within the complex. Site-directed and random mutagenesis together with chemical modification have been used to identify residues involved in nucleotide binding and hydrolysis, proton translocation and the coupling of these two processes. We have also obtained information about the mechanism of intracellular targeting of V-ATPases and regulation of V-ATPase activity by reversible dissociation.

13.1
Function of V-ATPases

V-ATPases are ATP-dependent proton pumps responsible for acidification of intracellular compartments and, in specialized cell types, proton secretion across the plasma membrane (for reviews see Refs. 1–4). V-ATPases are present in such in-

Handbook of ATPases. Edited by M. Futai, Y. Wada, J. H. Kaplan
Copyright © 2004 WILEY-VCH Verlag GmbH & Co. KGaA, Weinheim
ISBN 3-527-30689-7

tracellular compartments as clathrin-coated vesicles, endosomes, lysosomes, Golgi-derived vesicles, secretory vesicles and the central vacuoles of fungi and plants. A major role of V-ATPases within intracellular compartments is in various membrane traffic processes [1]. V-ATPases within early endosomes create the low pH necessary for ligand–receptor dissociation and recycling of receptors to the cell surface and for the formation of endosomal carrier vesicles that move dissociated ligands along the endocytic pathway. Several viruses (such as influenza virus) and toxins (such as diphtheria toxin) gain access to the cytoplasm of target cells via acidic endosomal compartments. Acidification of late endosomes is required for targeting of newly synthesized lysosomal enzymes from the Golgi to lysosomes. In addition to their role in membrane traffic, V-ATPases also provide the driving force for various coupled transport processes. For example, the proton gradient and membrane potential generated by the V-ATPases drives uptake of neurotransmitters into synaptic vesicles. Finally, V-ATPases in lysosomes and the central vacuoles of fungi and plants provide the acidic environment required for macromolecular degradation as well as the driving force for coupled transport. Plasma membrane V-ATPases have been identified in various cell types, including renal intercalated cells, osteoclasts, macrophages, insect goblet cells and tumor cells, where they function in such processes as renal acidification, bone resorption, pH homeostasis, coupled potassium transport and tumor invasion [1, 4].

Table 13.1   Subunit composition of the yeast V-ATPase.

	Subunit	Yeast Gene	Mr (kDa)	Function/location
V_1	A	VMA1	69	Catalytic site, homolog of β of F_1F_0-ATPase
	B	VMA2	58	Non-catalytic site, homolog of α of F_1F_0-ATPase
	C	VMA5	44	Peripheral stalk, released from V_1 complex during dissociation
	D	VMA8	29	Central stalk, homolog of γ of F_1F_0-ATPase
	E	VMA4	26	Peripheral stalk
	F	VMA7	14	Central stalk
	G	VMA10	13	Peripheral stalk
	H	VMA13	54	Peripheral stalk
V_0	a	VPH1	100	Proton translocation, targeting to vacuole, homolog of a of F_1F_0-ATPase
		STV1	100	Proton translocation, targeting to late Golgi, homolog of a of F_1F_0-ATPase
	d	VMA6	40	Cytoplasmic side, non-transmembrane protein
	c	VMA3	17	Proton translocation, DCCD- and concanamycin-binding sites, homolog of c of F_1F_0-ATPase
	c′	VMA11	17	Proton translocation, homolog of c of F_1F_0-ATPase
	c″	VMA16	23	Proton translocation, homolog of c of F_1F_0-ATPase

13.2
Overall Structure of V-ATPases and Relationship to F-ATPases

Our current model of the structure of the V-ATPases is shown in Figure 13.1, and the subunit composition, including the molecular mass of the subunits, the genes encoding these subunits in yeast and information about subunit function, is summarized in Table 13.1. V-ATPases are composed of a peripheral V_1 domain responsible for ATP hydrolysis and an integral V_0 domain responsible for proton translocation [1–4]. V_1 is a 640 kDa complex that contains eight different subunits (A–H) of molecular mass 70–13 kDa that are present in a stoichiometry of $A_3B_3C_1D_1E_1F_1G_2H_{1-2}$ [5, 6]. The nucleotide-binding sites are located on the 70 kDa A subunit and the 60 kDa B subunit, with the catalytic sites located on the three A subunits and so-called "non-catalytic" sites located on the three B subunits [7–9]. V_0 is a 260 kDa complex [10] that contains five different subunits (a, d, c, c' and c'') of molecular mass 100–17 kDa in a stoichiometry $a_1d_1c_{4-5}c'_1c''_1$ [5, 11]. Buried charged residues essential for proton transport are present in both the proteolipid subunits (c, c' and c'') and the 100 kDa a subunit [12, 13].

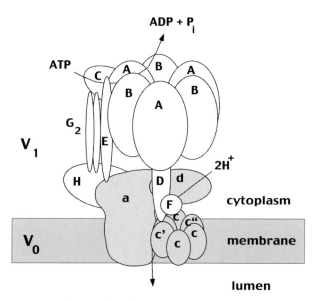

Figure 13.1 Structural model of the V-ATPase complex. The V_1 domain (shown in white) is responsible for ATP hydrolysis whereas the V_0 domain (shaded) is responsible for proton translocation. ATP hydrolysis at the catalytic nucleotide-binding sites (located on the A subunits) is proposed to drive rotation of a central stalk (composed of the D and F subunits) which in turn drives rotation of the ring of proteolipid subunits (c, c', c'') relative to subunit a. Subunit a is held fixed relative to the A_3B_3 head by a peripheral stalk composed of subunits C, E, G, H and the soluble domain of subunit a. Movement of the ring of proteolipid subunits relative to subunit a is postulated to drive unidirectional proton transport across the membrane. Reprinted with permission from Ref. 44, copyright 2002, the American Society for Biochemistry and Molecular Biology.

V-ATPases are structurally similar to the F-ATPases that function in the synthesis of ATP in mitochondria, chloroplasts and bacteria [14–16]. In both cases, the peripheral domain is responsible for nucleotide binding and hydrolysis (or, in the case of the F-ATPases, synthesis), whereas the integral domain is responsible for transport of protons across the membrane. Sequence homology has been identified between the nucleotide-binding subunits of the V and F-ATPases as well as between the proteolipid subunits of these two classes [17–19], suggesting that they have evolved from a common evolutionary ancestor. X-Ray structures have been obtained for F_1 from several sources and they reveal a hexameric arrangement of alternating nucleotide-binding subunits (α and β) surrounding a central cavity occupied by a highly α-helical γ subunit, which extends towards the membrane [20, 21]. More recently, an X-ray structure of F_1 attached to a ring of proteolipid subunits has been obtained from yeast mitochondria that reveals the γ and ε subunits form a central stalk connecting F_1 to a ring of 10 c subunits [22]. F_1 and F_0 have also been shown to be connected by a peripheral stalk containing the δ subunit and the soluble domains of subunit b [23, 24].

The presence of multiple stalks connecting the peripheral and integral domains has been demonstrated for both the V and F-ATPases by electron microscopy [25–27]. The V-ATPase structure, however, appears more complex than the F-ATPase, with many projections emerging from the V_1 domain and a knob present on the lumenal side of the membrane [26]. EM images of theV_0 domain suggest a ring (presumably of proteolipid subunits) adjacent to a membrane embedded mass that probably corresponds to subunit a [28]. This chapter focuses on our work on the structure and function of individual subunits, their arrangement in the V-ATPase complex, the mechanism of ATP-driven proton transport, the mechanism of targeting of V-ATPases to different intracellular compartments and the mechanisms employed in regulating V-ATPase activity *in vivo*.

13.3
Structure and Function of the Nucleotide-binding Subunits

Results from several studies suggest that the catalytic nucleotide-binding sites are located on the 70 kDa A subunits. Modification of the bovine A subunit by sulfhydryl reagents (such as NEM and NBD-Cl) leads to ATP-protectable inhibition of V-ATPase activity [29]. The cysteine residue whose modification is responsible for this inhibition was identified as Cys254 in the bovine A subunit [7]. This cysteine residue is located in a consensus sequence termed the Walker A sequence or P-loop (corresponding to GXGKTV). The X-ray structure of F_1 reveals that this sequence is located at the catalytic nucleotide-binding site on the β subunit, where it surrounds the terminal phosphate groups of ATP [20, 21]. Interestingly, formation of a disulfide bond between Cys254 and Cys532 (located in the C-terminal domain of the A subunit) causes reversible inhibition of V-ATPase activity [30]. It has been proposed that disulfide bond formation between Cys254 and Cys532 leads to inhibition of ATPase activity by preventing this subunit from adopting an open conformation that it must go through during the catalytic cycle [31]. Reversible disulfide bond formation between these cysteine residues in the A subunit has also been proposed to represent an important mechanism of regulating V-ATPase activity *in vivo* [32].

Studies using the photoactivated nucleotide analog 2-azido-ATP also support the idea that the catalytic nucleotide-binding sites are located on the A subunits [8]. This reagent inhibits V-ATPase activity in an ATP-protectable manner, with complete inhibition of ATPase activity occurring after modification of only one of the three A subunit sites per complex. Further support for the location of the catalytic sites on the A subunit comes from mutagenesis studies. Using molecular modeling based on the F-ATPase structure and the homology between the V and F-ATPase nucleotide-binding subunits, A subunit residues were identified that were predicted to play a role in either ATP binding or hydrolysis at the catalytic sites [33]. Mutagenesis of K263 in the Walker A sequence of the A subunit completely inhibited ATP hydrolysis, as did mutagenesis of E286 [34]. K263 is proposed to stabilize the negatively charged phosphate groups of bound ATP whereas E286 is suggested to serve as a proton acceptor from water during ATP hydrolysis. In addition, mutagenesis studies coupled with photochemical modification by 2-azido-ATP have identified four aromatic residues that are thought to form the adenine binding pocket of the catalytic site on the A subunit [33, 35].

More recently, evidence has been obtained for an important role of a novel domain of the V-ATPase A subunit, termed the "non-homologous" region, in coupling of ATP hydrolysis and proton transport and reversible dissociation of the V-ATPase complex [36]. The non-homologous region is a 90 amino acid stretch which is present and conserved in all V-ATPase A subunit sequences but which is absent from the corresponding F-ATPase β subunit [17, 37]. This region is located approximately one-third of the way from the amino terminus of the A subunit and, because the A_3B_3 structure (by analogy with F_1) is predicted to be tightly packed, is likely to be located on the outer surface of the complex. A peripheral location of this domain suggests the possibility that it may contribute to the peripheral stalk connecting the V_1 and V_0 domains [26]. Consistent with this idea, mutations have been identified in this region that lead to significant changes in coupling of proton transport and ATP hydrolysis [36]. Interestingly, one mutation (P217V) dramatically increased the coupling efficiency, suggesting that the wild-type enzyme is not normally optimally coupled. Several mutations were also identified in the non-homologous region that block reversible dissociation of the V-ATPase complex in response to glucose depletion [36], suggesting that this domain may also play a role in regulating V-ATPase activity *in vivo* (see below).

Several lines of evidence suggest that, in addition to subunit A, the 60 kDa B subunit also participates in nucleotide binding. First, subunit B is selectively modified by the photoactivated nucleotide analog BzATP [9]. Second, the B subunit is homologous to the α and β subunits of F_1 and to subunit A of the V-ATPases (approximately 20–25 % amino acid identity), although subunit B lacks the consensus Walker A and B motifs that are often present in nucleotide-binding proteins [18, 38]. Despite the absence of these motifs, the sequences that replace them in the

B subunit are highly conserved among species. Modeling studies of the B subunit based upon sequence homology with the α subunit, the X-ray coordinates of F_1 and energy minimization have been used to predict a structure for this subunit. This model accurately predicts the presence of a number of residues at the nucleotide-binding site on B, based upon nucleotide-protectable chemical modification of mutant proteins [39]. Unique cysteine residues were introduced into a cys-less form of subunit B and their modification by the sulfhydryl reagent biotin maleimide was tested in the intact V-ATPase in the presence and absence of BzATP. All six cysteine residues which reacted with biotin maleimide displayed protection from labeling by BzATP, which is consistent with their presence at the nucleotide-binding site [39]. Although the function of the "non-catalytic" nucleotide-binding sites on the B subunit is not certain, at least one mutation at this site causes a time-dependent change in ATPase activity, suggesting that they may play a role in regulating activity [33]. Other mutations at these sites appear to have less dramatic effects on activity [40]. As with the F-ATPases [20], the nucleotide-binding sites (both catalytic and non-catalytic) appear to be located at the interface of the alternating A and B (or β and α) subunits [33, 40], although, as indicated above, most residues at the catalytic sites are contributed by the A subunits whereas most residues at the non-catalytic sites are contributed by the B subunits.

13.4
Structure and Function of other V_1 Subunits and Arrangement of Subunits in the V-ATPase Complex

Studies from a number of laboratories have begun to shed light on the function of other subunits in the V_1 domain. Mutants have been identified in subunit D (product of the *VMA8* gene) that suggest it plays an important role in coupling of proton transport and ATP hydrolysis [41]. The most dramatic effects on coupling are observed for double mutants in which changes near the amino and carboxyl-terminus appear to act synergistically to cause uncoupling. One possible explanation for these results is that the N and C-terminus of the protein interact. Together with the predicted high α-helical content of subunit D, this has led us to postulate that subunit D acts as the γ subunit homolog in the V-ATPases [41]. Mutations causing uncoupling of the F-ATPase have also been identified in the γ subunit of F_1 [42], which exists as a coiled-coil structure within the central stalk connecting F_1 and F_0 [20, 21].

Support for the hypothesis that subunit D is the γ subunit homolog in the V-ATPases comes from cysteine mutagenesis and covalent cross-linking studies using the bifunctional, photoactivated maleimide reagent maleimido-benzophenone. As indicated above, a molecular model of the A_3B_3 hexamer has been derived using the X-ray coordinates of F_1, the sequence homology between the nucleotide-binding subunits of the V and F-ATPases and energy minimization [35, 39]. This model was used to predict the location of unique cysteine residues introduced into a cys-less form of subunit B by site-directed mutagenesis. These unique cysteine residues were first modified using maleimido-benzophenone followed by photoactivated cross-linking. Cross-linked products were then identified by Western blot analysis using subunit-specific antibodies. Results from these studies indicate that subunits B and D are cross-linked only at sites on the B subunit predicted to be oriented towards the interior of the A_3B_3 hexamer [43]. By contrast, subunits E and G are cross-linked to subunit B at sites predicted to be facing the exterior of the complex [43, 44]. These results suggest that subunit D is located in the central stalk connecting V_1 and V_0 whereas subunits E and G form part of a peripheral connection between these domains.

Conventional cross-linking studies performed on the coated vesicle V-ATPase using a number of bifunctional cross-linking reagents have demonstrated that subunit D makes contact with subunit F whereas subunit E is in contact with a significant number of subunits, including subunits C, G, H and the soluble domain of subunit a [6]. Disruption of V-ATPase assembly in yeast strains lacking certain V_1 subunits gives rise to several subcomplexes, including the heterodimeric species DF and EG [45], providing further support for the structural model shown in Figure 13.1. Dissociation of the bovine coated vesicle V-ATPase also gives rise to a heterodimeric EC subcomplex, as detected by immunoprecipitation [46]. These results suggest that subunits D and F form a central stalk connecting V_1 and V_0 whereas subunits C, E, G, H and the soluble domain of subunit a form the peripheral stalk connecting these domains.

Considerable information about the function of subunits C, G and H has been obtained from studies performed in other laboratories involving mutagenesis and overexpression of these subunits in yeast. Thus, subunit C appears to activate ATP hydrolysis by V_1 whereas subunit H inhibits this activity [47, 48]. Subunit G, which shows some homology to subunit b of the F-ATPases along one helical face [49], has been shown to tolerate short deletions in its sequence without loss of function [50]. This result is similar to those obtained for the F-ATPase b subunit [51], and suggests that subunit G may be the b subunit homolog of the V-ATPases. An X-ray structure of subunit H has recently been obtained [52] and has revealed interesting structural similarities to the importins, proteins involved in nuclear import. Yeast 2-hybrid analysis and co-immunoprecipitation studies have revealed that subunit H makes contact with both subunit A in V_1 and the soluble, cytoplasmic domain of subunit a in V_0 [53], suggesting that it serves an important bridging function. Moreover, subunit H has been shown to serve an important role as the docking site for the NEF protein, which in turn functions in internalization of the HIV receptor CD4 [54].

13.5
Structure and Function of V_0 Subunits

The V_0 domain of the V-ATPases contains five subunits. Three of these correspond to highly hydrophobic proteins termed proteolipids because of their ability to be extracted with organic solvents: subunit c (Vma3p), subunit c' (Vma11p) and subunit c′′ (Vma16p). All three of these proteins are required for proton transport by the V-ATPases [12]. Subunit c and c′ each contain four transmembrane helices with an essential buried glutamate residue present in TM4. Although subunit c′′ was originally predicted to contain five transmembrane helices, recent studies have demonstrated that it instead contains only four, with the critical glutamic acid residue present in TM2 [55]. Interestingly, the region originally predicted to correspond to TM1 has been shown not to be required for function. Topological studies employing expression of epitope-tagged proteins as well as differential modification of unique cysteine residues indicate that the C-terminus of subunit c is present on the lumenal side of the membrane whereas the C-terminus of subunit c′′ is exposed on the cytoplasmic side of the membrane [55, 56]. The buried glutamic acid residues present in the proteolipid subunits are essential for proton transport by both chemical modification by DCCD [31] and site-directed mutagenesis [12]. These residues have therefore been proposed to participate directly in proton translocation. Interestingly, subunit c forms part of the binding site for the specific V-ATPase inhibitor bafilomycin [57].

The three V-ATPase proteolipid subunits are homologous to the F-ATPase subunit c, which is composed of just two transmembrane helices that adopt a helical hairpin structure [58]. The V-ATPase proteolipids appear to have been derived during evolution from a gene duplication and fusion event from an ancestral gene that more closely resembles the F-ATPase c subunit [19]. NMR analysis of the F-ATPase proteolipid suggests that the orientation of the helix containing the essential acidic residue changes upon a change in protonation state [59], leading to the suggestion that this helical swiveling is an important part of the mechanism of proton translocation (see below). The F_0 domain from *E. coli* and yeast mitochondria contain 10 copies of subunit c, which form a ring [22, 60]. Stoichiometry measurements indicate that the V_0 domain contains 5–6 copies of subunits c plus c′ and a single copy of subunit c′′ [5], and epitope-tagging experiments indicate the presence of single copies of both subunits c′ and c′′ [11]. Thus, these results suggest a subunit stoichiometry for the V_0 domain of $c_{4-5} c'_1 c''_1$.

In addition to the proteolipid subunits, the V_0 domain contains two additional subunits. Subunit d (Vma6p) is a 40 kDa hydrophilic polypeptide which appears to lack transmembrane segments [61], but which remains tightly bound to V_0 upon dissociation of V_1 [62]. The fifth V_0 subunit is a 100 kDa integral membrane protein called subunit a. In yeast, subunit a is encoded by two genes (*VPH1* and *STV1*), with Vph1p associated with V-ATPase complexes targeted to the vacuole and Stv1p associated with V-ATPases targeted to the late Golgi [63–65]. In mammals, subunit a is encoded by four genes (a1–a4), which are expressed in a tissue specific manner [66–70]. The a3 isoform appears to be responsible for plasma membrane targeting of the V-ATPase in osteoclasts [68], and a defect in this gene causes the genetic defect autosomal recessive osteopetrosis [69]. The a4 isoform is responsible for targeting of the V-ATPase to the plasma membrane of renal intercalated cells [67], and disruption of this gene causes the disease renal tubule acidosis [70].

The topology of subunit a has been studied by introduction of unique cysteine residues into a cys-less form of Vph1p followed by evaluation of the exposure of these sites in intact vacuolar membranes using differential reactivity towards membrane permeant and impermeant sulfhydryl reagents [71]. These studies indicate that subunit a contains an amino terminal hydrophilic domain of about 50 kDa oriented towards the cytoplasmic side of the membrane and a carboxyl terminal hydrophobic domain containing nine transmembrane segments, with the C-terminus located on the lumenal side of the membrane. Site-directed mutagenesis of Vph1p has identified a buried, charged residue present in TM7 (Arg735) that is absolutely essential for proton translocation [13]. In addition, there are a series of other buried charged residues present in TM7 and TM9, including E789, H743 and R799, whose mutation causes a reduction in proton transport, suggesting that these residues may also play some role in this process [13, 72, 73]. Although no sequence homology exists between the a subunits of V_0 and F_0, we have suggested that the V_0 subunit a plays a similar role in proton transport. For subunit a of F_0, Arg210 located in TM4 is critical for proton transport [74]. It is postulated to directly interact with the buried carboxyl groups of the proteolipid c subunits during proton translocation [75, 76]. The F-ATPase a subunit also contains other buried charged residues that are postulated to form aqueous access channels that allow protons to reach and leave these buried sites [74, 75].

V-ATPase complexes containing different isoforms of subunit a differ in other properties besides intracellular localization [77]. Thus Vph1p-containing complexes show a ca. 10-fold greater assembly of V_1 and V_0 than Stv1p-containing complexes. In addition, Vph1p-containing complexes are more tightly coupled than those containing Stv1p. These differences may reflect the need to maintain a lower lumenal pH in the vacuole (where Vph1p-containing complexes reside) than in the Golgi (where Stv1p-containing complexes are found). Using chimeric constructs, we have shown that whereas intracellular targeting is controlled by the amino terminal domain, the degree of assembly of V_1 and V_0 and the tightness of coupling of proton transport and ATP hydrolysis are controlled by the carboxyl-terminal [65].

13.6
Mechanism of ATP-dependent Proton Transport

Because of their structural similarity, the V and F-ATPases are believed to operate by a similar rotary mechanism [78, 79]. For the F-ATPases, ATP hydrolysis by the β subunits of F_1 causes rotation of a central stalk, composed of the γ and ε subunits, which in turn drives rotation of the ring of proteolipid c subunits relative to subunit a. Rotation of both the γ subunit in F_1 [80–82] and the ring of c subunits in F_1F_0 [83] have been demonstrated. The a subunit is postulated to provide access channels for the protons to reach and leave the buried carboxyl groups on the c subunit ring and to activate release of protons into the egress channel through interaction with a critical arginine residue [74, 75, 16]. Rotation of the c subunit ring relative to subunit a thus causes unidirectional proton transport across the membrane.

For the V-ATPases, we suggest that ATP hydrolysis in V_1 drives rotation of the D and F subunits together with the proteolipid ring relative to subunit a. Rotation of the D and F subunits of the V_1 domain of the V-ATPase from *Thermos thermophilus* was recently demonstrated [84], which is consistent with the placement of these subunits in the central stalk [43]. Interaction of the proteolipid carboxyl groups with Arg735 on subunit a [13] would then lead to release of protons into the egress channel leading to the lumenal side of the membrane. Further rotation of the c subunit ring would bring these deprotonated carboxyl groups into contact with the cytoplasmic aqueous channel. They would need to pick up protons from this channel before re-entering the hydrophobic phase of the bilayer.

13.7
Regulation of V-ATPase Activity *In Vivo*

Control of V-ATPase activity in cells and tissues is critical for the diversity of functions served by these enzymes. Several mechanisms have been proposed to be involved in regulation of V-ATPase activity *in vivo*. As described above, reversible disulfide bond formation between conserved cysteine residues at the catalytic sites on the A subunit causes reversible inhibition of V-ATPase activity [30, 32], and several studies suggest that this represents an important *in vivo* mechanism for controlling activity [32, 85]. Differences in coupling efficiency have been demonstrated for V-ATPases containing different a subunit isoforms [77], and mutations in a number of V-ATPase subunits, including subunit D [41], subunit a [13] and the non-homologous domain of subunit A [36], cause changes in coupling efficiency. Moreover, coupling of proton transport and ATPase activity is readily perturbed [86, 87], suggesting that the enzyme may be poised to change the tightness of coupling *in vivo*.

In addition to the mechanisms described above, reversible dissociation of the V-ATPase into its constituent V_1 and V_0 domains represents an important *in vivo* regulatory mechanism [88, 89]. In yeast, reversible dissociation occurs in response to glucose depletion [88], although many of the signal transduction pathways activated by glucose withdrawal do not appear to be involved in this process [90]. Glucose-dependent dissociation requires a catalytically active enzyme [35, 90] and an intact microtubular network [91]. Interestingly, however, reassembly of the V-ATPase is not dependent upon microtubules. Several mutations have been identified in the non-homologous region which block dissociation without inhibiting activity [36], suggesting that this domain may play a role in controlling dissociation independent of any effects on activity. Similar mutations have been identified in subunit G, where they appear to cause increased stability of the V-ATPase complex [50].

The *in vivo* dissociation behavior of V-ATPase complexes containing different isoforms of the a subunit has proven to be complex [77]. Thus, Vph1p-containing complexes localized to the vacuole undergo dissociation in response to glucose withdrawal whereas Stv1p-containing complexes localized to the Golgi do not. Re-targeting of Stv1p-containing complexes to the vacuole by overexpression of Stv1p results in a re-appearance of the ability to dissociate in response to glucose depletion. Conversely, prevention of Vph1p-containing complexes from reaching the vacuole by disruption of genes involved in intracellular targeting results in V-ATPase complexes that show glucose-dependent dissociation, but less completely than when these complexes have a vacuolar localization. Thus, *in vivo* dissociation appears to be controlled by both the a subunit isoform present in the complex and by the cellular environment in which the complex resides [77]. Recent studies have identified a novel complex (termed RAVE) which includes part of the ubiquitin ligase complex (Skp1p) and which appears to function in both normal and regulated assembly of the V-ATPase [92, 93]. Additional studies will be required to further elucidate the mechanism of regulating *in vivo* dissociation of the V-ATPase complex.

13.8
Conclusions

V-ATPases are multisubunit complexes responsible for ATP-driven proton transport in both intracellular compartments and the plasma membrane. They are composed of a peripheral V_1 domain responsible for ATP hydrolysis and an integral V_0 domain that transports protons. A variety of approaches have begun to elucidate the arrangement and function of subunits in the V-ATPase complex. Regulation of V-ATPase activity *in vivo* is likely to be complex given the great diversity of functions performed by this family of proton pumps.

Acknowledgments

The authors thank Drs Takao Inoue, Tsuyoshi Nishi and Shoko Kawasaki-Nishi for many helpful discussions. This work was supported by NIH Grant GM34478 (to M. F.).

References

1. T. Nishi and M. Forgac, *Nat. Rev. Mol. Cell. Biol.*, **2002**, *3*, 94–103.
2. L. A. Graham, B. Powell, and T. H. Stevens, *J. Exp. Biol.*, **2000**, *203*, 61–70.
3. P. M. Kane and K. J. Parra, *J. Exp. Biol.*, **2000**, *203*, 81–87.
4. H. Wieczorek, D. Brown, S. Grinstein, J. Ehrenfeld, and W. R. Harvey, *Bioessays*, **1999**, *21*, 637–648.
5. H. Arai, G. Terres, S. Pink, and M. Forgac, *J. Biol. Chem.*, **1988**, *263*, 8796–8802.
6. T. Xu, E. Vasilyeva, and M. Forgac, *J. Biol. Chem.*, **1999**, *274*, 28909–28915.
7. Y. Feng and M. Forgac, *J. Biol. Chem.*, **1992**, *267*, 5817–5822.
8. J. Zhang, E. Vasilyeva, Y. Feng, and M. Forgac, *J. Biol. Chem.*, **1995**, *270*, 15494–15500.
9. E. Vasilyeva and M. Forgac, *J. Biol. Chem.*, **1996**, *271*, 12775–12782.
10. J. Zhang, Y. Feng, and M. Forgac, *J. Biol. Chem.*, **1994**, *269*, 23518–23523.
11. B. Powell, L. A. Graham, and T. H. Stevens, *J. Biol. Chem.*, **2000**, *275*, 23654–23660.
12. R. Hirata, L. A. Graham, A. Takatsuki, T. H. Stevens, and Y. Anraku, *J. Biol. Chem.*, **1997**, *272*, 4795–4803.
13. S. Kawasaki-Nishi, T. Nishi, and M. Forgac, *Proc. Natl. Acad. Sci. U. S.A.*, **2001**, *98*, 12397–12402.
14. R. L. Cross, *Biochim. Biophys. Acta*, **2000**, *1458*, 270–275.
15. J. Weber and A. E. Senior, *Biochim. Biophys. Acta*, **2000**, *1458*, 300–309.
16. R. H. Fillingame, W. Jiang, and O. Y. Dmitriev, *J. Exp. Biol.*, **2000**, *203*, 9–17.
17. L. Zimniak, P. Dittrich, J. P. Gogarten, H. Kibak, and L. Taiz, *J. Biol. Chem.*, **1988**, *263*, 9102–9112.
18. B. J. Bowman, R. Allen, M. A. Wechser, and E. J. Bowman, *J. Biol. Chem.*, **1988**, *263*, 14002–14007.
19. M. Mandel, Y. Moriyama, J. D. Hulmes, Y. C. Pan, H. Nelson and N. Nelson, *Proc. Natl. Acad. Sci. U. S.A.*, **1988**, *85*, 5521–5524.
20. J. P. Abrahams, A. G. Leslie, R. Lutter, and J. E. Walker, *Nature*, **1994**, *370*, 621–628.
21. M. A. Bianchet, J. Hullihen, P. L. Pedersen, and L. M. Amzel, *Proc. Natl. Acad. Sci. U. S.A.*, **1998**, *95*, 11065–11070.
22. D. Stock, A. G. Leslie, and J. E. Walker, *Science*, **1999**, *286*, 1700–1705.
23. I. Ogilvie, R. Aggeler, and R. A. Capaldi, *J. Biol. Chem.*, **1997**, *272*, 16652–16656.
24. D. T. McLachlin, J. A. Bestard, and S. D. Dunn, *J. Biol. Chem.*, **1998**, *273*, 15162–15168.
25. E. J. Boekema, T. Ubbink-Kok, J. S. Lolkema, A. Brisson, and W. N. Konings, *Proc. Natl. Acad. Sci. U. S.A.*, **1997**, *94*, 14291–14293.
26. S. Wilkens, E. Vasilyeva, and M. Forgac, *J. Biol. Chem.*, **1999**, *274*, 31804–31810.
27. S. Wilkens and R. A. Capaldi, *Nature*, **1998**, *393*, 29.
28. S. Wilkens and M. Forgac, *J. Biol. Chem.*, **2001**, *276*, 44064–44068.
29. H. Arai, M. Berne, G. Terres, H. Terres, K. Puopolo, and M. Forgac, *Biochemistry*, **1987**, *26*, 6632–6638.
30. Y. Feng and M. Forgac, *J. Biol. Chem.*, **1994**, *269*, 13224–13230.
31. M. Forgac, *J. Biol. Chem.*, **1999**, *274*, 12951–12954.
32. Y. Feng and M. Forgac, *J. Biol. Chem.*, **1992**, *267*, 19769–19772.
33. K. J. MacLeod, E. Vasilyeva, J. D. Baleja, and M. Forgac, *J. Biol. Chem.*, **1998**, *273*, 150–156.
34. Q. Liu, X. H. Leng, P. R. Newman, E. Vasilyeva, P. M. Kane, and M. Forgac, *J. Biol. Chem.*, **1997**, *272*, 11750–11756.
35. K. J. MacLeod, E. Vasilyeva, K. Merdek, P. D. Vogel, and M. Forgac, *J. Biol. Chem.*, **1999**, *274*, 32869–32874.
36. E. Shao, T. Nishi, S. Kawasaki-Nishi, and M. Forgac, *J. Biol. Chem.*, **2003**, *278*, 12985–12991.
37. K. Puopolo, C. Kumamoto, I. Adachi, and M. Forgac, *J. Biol. Chem.*, **1991**, *266*, 24564–24572.

38. K. Puopolo, C. Kumamoto, I. Adachi, R. Magner, and M. Forgac, *J. Biol. Chem.*, **1992**, *267*, 3696–3706.

39. E. Vasilyeva, Q. Liu, K. J. MacLeod, J. D. Baleja, and M. Forgac, *J. Biol. Chem.*, **2000**, *275*, 255–260.

40. Q. Liu, P. M. Kane, P. R. Newman, and M. Forgac, *J. Biol. Chem.*, **1996**, *271*, 2018–2022.

41. T. Xu and M. Forgac, *J. Biol. Chem.*, **2000**, *275*, 22075–22081.

42. K. Shin, R. K. Nakamoto, M. Maeda, and M. Futai, *J. Biol. Chem.*, **1992**, *267*, 20835–20839.

43. Y. Arata, J. D. Baleja, and M. Forgac, *Biochemistry*, **2002**, *41*, 11301–11307.

44. Y. Arata, J. D. Baleja, and M. Forgac, *J. Biol. Chem.*, **2002**, *277*, 3357–3363.

45. J. J. Tomashek, L. A. Graham, M. U. Hutchins, T. H. Stevens, and D. J. Klionsky, *J. Biol. Chem.*, **1997**, *272*, 26787–26793.

46. K. Puopolo, M. Sczekan, R. Magner, and M. Forgac, *J. Biol. Chem.*, **1992**, *267*, 5171–5176.

47. K. J. Parra, K. L. Keenan, and P. M. Kane, *J. Biol. Chem.*, **2000**, *275*, 21761–21767.

48. K. K. Curtis and P. M. Kane, *J. Biol. Chem.*, **2002**, *277*, 2716–2724.

49. I. E. Hunt and B. J. Bowman, *J. Bioenerg. Biomembr.*, **1997**, *29*, 533–540.

50. C. M. Charsky, N. J. Schumann, and P. M. Kane, *J. Biol. Chem.*, **2000**, *275*, 37232–37239.

51. P. L. Sorgen, T. L. Caviston, R. C. Perry, and B. D. Cain, *J. Biol. Chem.*, **1998**, *273*, 27873–27878.

52. M. Sagermann, T. H. Stevens, and B. W. Matthews, *Proc. Natl. Acad. Sci. U. S. A.*, **2001**, *98*, 7134–7139.

53. C. Landolt-Marticorena, K. M. Williams, J. Correa, W. Chen, and M. F. Manolson, *J. Biol. Chem.*, **2000**, *275*, 15449–15457.

54. M. Geyer, H. Yu, R. Mandic, T. Linnemann, Y. H. Zheng, O. T. Fackler, and B. M. Peterlin, *J. Biol. Chem.*, **2002**, *277*, 28521–28529.

55. T. Nishi, S. Kawasaki-Nishi, and M. Forgac, *J. Biol. Chem.*, **2003**, *278*, 5821–5827.

56. T. Nishi, S. Kawasaki-Nishi, and M. Forgac, *J. Biol. Chem.*, **2001**, *276*, 34122–34130.

57. B. J. Bowman and E. J. Bowman, *J. Biol. Chem.*, **2002**, *277*, 3965–3972.

58. M. E. Girvin, V. K. Rastogi, F. Abildgaard, J. L. Markley, and R. H. Fillingame, *Biochemistry*, **1998**, *37*, 8817–8824.

59. V. K. Rastogi and M. E. Girvin, *Nature*, **1999**, *402*, 263–268.

60. W. Jiang, J. Hermolin, and R. H. Fillingame, *Proc. Natl. Acad. Sci. U. S. A.*, **2001**, *98*, 4966–4971.

61. C. Bauerle, M. N. Ho, M. A. Lindorfer, and T. F. Stevens, *J. Biol. Chem.*, **1993**, *268*, 12749–12757.

62. I. Adachi, K. Puopolo, N. Marquez-Sterling, H. Arai, and M. Forgac, *J. Biol. Chem.*, **1990**, *265*, 967–973.

63. M. F. Manolson, D. Proteau, R. A. Preston, A. Stenbit, B. T. Roberts, M. Hoyt, D. Preuss, J. Mulholland, D. Botstein, and E. W. Jones, *J. Biol. Chem.*, **1992**, *267*, 14294–14303.

64. M. F. Manolson, B. Wu, D. Proteau, B. E. Taillon, B. T. Roberts, M. A. Hoyt, E. W. Jones, *J. Biol. Chem.*, **1994**, *269*, 14064–14074.

65. S. Kawasaki-Nishi, K. Bowers, T. Nishi, M. Forgac, and T. H. Stevens, *J. Biol. Chem.*, **2001**, *276*, 47441–47420.

66. T. Nishi and M. Forgac, *J. Biol. Chem.*, **2000**, *275*, 6824–6830.

67. T. Oka, Y. Murata, M. Namba, T. Yoshimizu, T. Toyomura, A. Yamamoto, G. H. Sun-Wada, N. Hamasaki, Y. Wada, and M. Futai, *J. Biol. Chem.*, **2001**, *276*, 40050–40054.

68. T. Toyomura, T. Oka, C. Yamaguchi, Y. Wada, and M. Futai, *J. Biol. Chem.*, **2000**, *275*, 8760–8765.

69. Y. P. Li, W. Chen, Y. Liang, E. Li, and P. Stashenko, *Nat. Gen.*, **1999**, *23*, 447–451.

70. A. N. Smith, J. Skaug, K. A. Choate, A. Nayir, A. Bakkaloglu, S. Ozen, S. A. Hulton, S. A. Sanjad, E. A. Al-Sabban, R. P. Lifton, S. W. Scherer, and F. E. Karet, *Nat. Gen.*, **2000**, *26*, 71–75.

71. X. H. Leng, T. Nishi, and M. Forgac, *J. Biol. Chem.*, **1999**, *274*, 14655–14661.

72. X. H. Leng, M. F. Manolson, Q. Liu, and M. Forgac, *J. Biol. Chem.*, **1996**, *271*, 22487–22493.

73. X. H. Leng, M. Manolson, M. Forgac, *J. Biol. Chem.*, **1998**, *273*, 6717–6723.

74. B. D. Cain, *J. Bioenerg. Biomembr.*, **2000**, *32*, 365–371.

75. S. B. Vik, J. C. Long, T. Wada, and D. Zhang, *Biochim. Biophys. Acta*, **2000**, *1458*, 457–466.

76. W. Jiang and R. H. Fillingame, *Proc. Natl. Acad. Sci.*, **1998**, *95*, 6607–6612.

77. S. Kawasaki-Nishi, T. Nishi, M. Forgac, *J. Biol. Chem.*, **2001**, *276*, 17941–17948.

78. S. B. Vik and B. J. Antonio, *J. Biol. Chem.*, **1994**, *269*, 30364–30369.

79. W. Junge, D. Sabbert, and S. Engelbrecht, *Ber. Bunsenges. Phys. Chem.*, **1996**, *100*, 2014–2019.

80. T. M. Duncan, V. V. Bulygin, Y. Zhou, M. L. Hutcheon, and R. L. Cross, *Proc. Natl. Acad. Sci. U. S.A.*, **1995**, *92*, 10964–10968.

81. D. Sabbert, S. Engelbrecht, and W. Junge, *Nature*, **1996**, *381*, 623–625.

82. H. Noji, R. Yasuda, M. Yoshida, and K. Kazuhiko, *Nature*, **1997**, *386*, 299–302.

83. Y. Sambongi, Y. Iko, M. Tanabe, H. Omote, A. Iwamoto-Kihara, I. Ueda, T. Yanagida, Y. Wada, and M. Futai, *Science*, **1999**, *286*, 1722–1724.

84. H. Imamura, M. Nakano, H. Noji, E. Muneyuki, S. Ohkuma, M. Yoshida, and K. Yokoyama, *Proc. Natl. Acad. Sci. U. S.A.*, **2003**, *100*, 2312–2315.

85. Y. E. Oluwatosin and P. M. Kane, *J. Biol. Chem.*, **1997**, *272*, 28149–28157.

86. H. Arai, S. Pink, and M. Forgac, *Biochemistry*, **1989**, *28*, 3075–3082.

87. I. Adachi, H. Arai, R. Pimental, and M. Forgac, *J. Biol. Chem.*, **1990**, *265*, 960–966.

88. P. M. Kane, *J. Biol. Chem.*, **1995**, *270*, 17025–17032.

89. J. P. Sumner, J. A. Dow, F. G. Early, U. Klein, D. Jager, and H. Wieczorek, *J. Biol. Chem.*, **1995**, *270*, 5649–5653.

90. K. J. Parra and P. M. Kane, *Mol. Cell. Biol.*, **1998**, *18*, 7064–7074.

91. T. Xu and M. Forgac, *J. Biol. Chem.*, **2001**, *276*, 24855–24861.

92. J. H. Seol, A. Shevchenko, A. Shevchenko, and R. J. Deshaies, *Nat. Cell. Biol.*, **2001**, *3*, 384–391.

93. A. M. Smardon, M. Tarsio, and P. M. Kane, *J. Biol. Chem.*, **2002**, *277*, 13831–13839.

14

Role of the V-ATPase in the Cellular Physiology of the Yeast *Saccharomyces cerevisiae*

Laurie A. Graham, Katherine Bowers, Andrew R. Flannery, and *Tom H. Stevens*

14.1
Introduction: the Yeast V-ATPase

The budding yeast *Saccharomyces cerevisiae* serves as an excellent model organism to study the role of ion-pumping ATPases due to the ease of genetic and biochemical manipulation. Specifically, the yeast V-ATPase belongs to a family of V-type ion-pumping ATPases present in all eukaryotic organisms. V-ATPases are large, multi-subunit, membrane associated enzyme complexes that function to acidify cellular compartments. V-ATPases carry out the active transport of protons across the membrane bilayer that is tightly coupled to the hydrolysis of ATP. The disruption of V-ATPase function is lethal in all eukaryotic organisms tested except *Saccharomyces cerevisiae*, making yeast an ideal system to study the structure, function and assembly of this important enzyme complex [1].

When yeast cells are viewed under the light microscope, even at low magnification, the most prominent structure visualized is the vacuole hinting at the importance of this organelle in the cellular physiology of this single celled organism. The yeast vacuole appears as a large multi-lobed organelle in actively growing wild-type cells. It has been observed that the loss of certain proteins, due to the deletion of their genes in the genome, affect either directly or indirectly the appearance or morphology of the yeast vacuole [2]. Characterizing the changes in morphology of the yeast vacuole has proven to be a powerful tool in the identification of proteins involved in similar process or pathways in the cell.

The yeast vacuole is the equivalent of the mammalian lysosome, and contains proteases required for the degradation of proteins sent to the vacuole. The vacuole is the final destination for vesicles derived from several sources including the Golgi complex, the plasma membrane, and even directly from the cytosol. The vacuole is responsible for a variety of physiological processes, including pH regulation, ion regulation, amino acid storage, and metal detoxification [3]. One characteristic of the yeast vacuole is that the lumen or interior of the vacuole is more acidic (pH 6.0) than the surrounding cytosol. Organelle acidification in other eukaryotic systems plays a role in various cellular functions such as receptor-mediated endocyto-

Handbook of ATPases. Edited by M. Futai, Y. Wada, J. H. Kaplan
Copyright © 2004 WILEY-VCH Verlag GmbH & Co. KGaA, Weinheim
ISBN 3-527-30689-7

sis, renal acidification, bone reabsorption, neurotransmitter accumulation, and activation of acid hydrolases. From yeast to humans the acidification of organelles and compartments is due to the action of the ion-pumping V-type ATPase (V-ATPase). The yeast V-ATPase is the best-characterized member of a large family of V-type ATPases present in all eukaryotic cells. The focus of this chapter is the various roles the V-ATPase and organelle acidification plays in the physiology of the yeast cell.

14.2
Role of the V-ATPase in the Acidification of the Yeast Vacuole

Yeast V-ATPase is a membrane-associated enzyme complex that carries out the active transport of protons that is tightly coupled to the hydrolysis of ATP. The yeast V-ATPase is localized primarily to the limiting membrane of the vacuole. A smaller population of V-ATPase complexes can be found localized to the late or last compartment of the Golgi as well as the endosomal network, and these V-ATPases also play a role in acidification [4]. The proton gradient generated by the vacuolar V-ATPase is utilized by other membrane-bound transporters to drive the accumulation of substrates into the lumen of the vacuole. As an example, the V-ATPase helps to maintain low cytosolic calcium levels (100–150 nM) even when the cells are exposed to media containing 10^5–10^6 times the level of calcium in the cytosol [5].

Acidification of the vacuoles in yeast can easily be visualized using the fluorescent dye quinacrine [6]. Incubating live yeast cells for a short time in growth media containing quinacrine allows the dye to diffuse into the cell in its neutral form and accumulate in the protonated form in the lumen of the vacuole. Weak bases such as quinacrine will accumulate in acidic compartments like the vacuole until the pH of the organelle is neutralized. When the cells are viewed using a fluorescent microscope the dye in the vacuole lumen appears as a bright spot. Yeast cells lacking a functional V-ATPase fail to accumulate the dye in the vacuole and no lumenal staining is observed. The intensity of the fluorescence of the quinacrine stain accumulated in the vacuole is directly proportional to the degree of acidification of the vacuole.

Loss of V-ATPase activity not only results in failure to acidify the vacuole but also produces a distinct set of growth phenotypes in yeast. Yeast cells lacking V-ATPase activity are unable to grow on media buffered to a neutral pH (7.5) but can grow on more acidic media buffered to a pH of 5.0. There is not a clear understanding why these cells display this conditional growth phenotype, but it is specific and unique to mutant yeast cells lacking a functional V-ATPase. Additional growth phenotypes, which are shared by other classes of yeast mutants, included the inability to grow in media containing elevated concentrations of calcium (similar to mutants with disrupted calcium ion homeostasis or *cls* mutants) and the inability to grow on non-fermentable carbons sources (similar to mutants lacking mitochondrial function or *pet* mutants).

14.3
Overview of the Function and Structure of the Yeast V-ATPase

V-type ATPases and F-type ATPases belong to a super-family of structurally and functionally related proton translocating multisubunit enzyme complexes. The organization of the yeast V-ATPase subunits is modeled after the well-characterized F-type ATPase. The V-ATPase can be functionally divided into two domains: the catalytic V_1 domain responsible for ATP hydrolysis and the proton-translocating Vo domain (Figure 14.1). Dissociation of the two domains is a means of regulating the function of the V-ATPase complex, suggesting that once separated the domains have limited activity [7].

Two subunits of the V_1 subcomplex possess the sites for ATP binding and ATP hydrolysis, the 69 and 60 kDa proteins, respectively (Table 14.1). These catalytic and nucleotide binding subunits share high homology to the β and α subunits of the F-ATPase (45–47 %). The remaining non-catalytic V_1 subunits form the stalk or the connection between the catalytic subunits and the Vo subcomplex. Unfortunately, the non-catalytic V_1 subunits do not demonstrate sequence homology to any of the well-characterized γ, δ or ε subunits of the F-ATPase stalk.

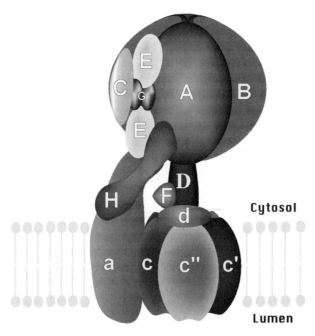

Figure 14.1 Proposed model of the yeast V-ATPase. Thirteen proteins have been identified as subunits of the yeast V-ATPase. Subunits **a**, **c**, **c'**, and **c''** are all predicted to possess transmembrane domains, the remaining subunits are associated peripherally with the membrane. The Golgi/endosome localized V-ATPase is identical to the vacuolar V-ATPase except for the isoforms of subunit **a**, the large 100 kDa protein. Stv1p (subunit **a**) is found to be associated with the Golgi and endosome localized V-ATPase and Vph1p (subunit **a**) is associated with the V-ATPase on the vacuole membrane.

Table 14.1 Subunits of the yeast V-ATPase.

		Universal letter code	Yeast protein	Molecular weight (kDa)	Proposed function
V_0 subunits	a	Vph1p or Stv1p	95/102	H^+ transport, stator
	c	Vma3p	16	H^+ transport, rotor
	c′	Vma11p	17	H^+ transport, rotor
	c″	Vma16p	23	H^+ transport, rotor
	d	Vma6p	36	rotor
V_1 subunits	A	Vma1p	69	ATP catalytic site
	B	Vma2p	60	Nucleotide binding site
	C	Vma5p	42	Stator
	D	Vma8p	32	Rotor
	E	Vma4p	27	Stator
	F	Vma7p	14	Rotor
	G	Vma10p	13	Stator
	H	Vma13p	54	Regulator

The remaining non-catalytic V_1 subunits may participate in forming the stalk or the connection between the catalytic subunits and the Vo subcomplex. In the yeast V-ATPase there are six possible candidates that may serve the function of the well-characterized γ, δ and ε subunits of the F-ATPase stalk. The function of the remaining non-catalytic V_1 subunits is completely unknown except Vma13p, that most likely plays a role in regulation of the complex. Cells lacking Vma13p are able to assemble a twelve subunit complex that is correctly localized but functionally inactive [8] and the Vo subcomplex is stable, able to assemble correctly and is localized to either the vacuole or Golgi membrane [9].

Subunits of the Vo subcomplex play a role in the translocation of protons from the yeast cytosol into the lumen of the vacuole, Golgi or endosomal network. The translocation of protons does not occur through a pore or a channel formed by the Vo subunits but utilizes charged residues buried within the transmembrane domains of several of these subunits [10]. Three Vo subunits share homology to subunit c of the F-ATPase and are proposed to form a hexameric ring, similar to the ring formed by multiple copies of subunit c in the F-ATPase. Each subunit of the ring contains a charged glutamic acid residue that is proposed to accept protons from the cytosol, and then releases the protons into the lumen of the vacuole or Golgi.

Similar to the F-ATPase model, the movement or the rotation of the subcomplexes, V_1 and Vo, relative to each other is believed to link the hydrolysis of ATP coupling to the translocation of protons across the membrane (for review see Ref. [11]). Rotation of the stalk subunits was actually visualized in the F_1 domain when the subcomplex was affixed to a surface via the catalytic subunits [12]. Rotation of the V_1-ATPase was observed using purified V_1 complex from the thermophilic bacterium *Thermus thermophilus*, supporting the rotating mechanism for the V-ATPase complex regardless of the source [13]. These same researchers also observed the rotation of the Vo domain using purified V-ATPase from the same thermophilic bacterium [14]. Recently, the rotation of the yeast V_1 domain was observed when the V_O domain was affixed to a surface and not allowed to rotate [15]. Rotation of the V_1 domain relative to the V_O domain was dependent on ATP and could be inhibited by the presence of concanamycin, a specific inhibitor of the V-type ATPase [15]. The increased number of subunits of the yeast V-ATPase compared with the V-ATPase from a thermophilic bacterium adds an extra level of complexity in understanding the structure and function of the yeast enzyme complex.

14.4
Composition and Assembly of the V₁ Catalytic Domain of the Yeast V-ATPase

The V_1 subcomplex of both the Golgi and vacuolar-localized V-ATPases in yeast is composed of the same eight subunits: Vma1p, Vma2p, Vma4p, Vma5p, Vma7p, Vma8p, Vma10p and Vma13p (Table 14.1). Genes encoding subunits common to both the Golgi and vacuolar yeast V-ATPase are called *VMA* for vacuolar membrane ATPase. Each yeast V-ATPase V_1 complex contains three copies of Vma1p which possess a ATP binding and catalytic site [16]. The sequence of Vma1p is highly conserved across organisms and even between V-ATPases and F-ATPases. Vma1p (subunit A) shares 65–75 % identity with other homologous V-ATPase subunits and 26 % identity with the β subunit of the F-ATPase. Interestingly, in yeast, formation of Vma1p is the result of a unique process called protein splicing [17]. Initially translated as a larger protein of 119 kDa, Vma1p is converted into a 69 kDa V-ATPase subunit by removal of a internal 50 kDa protein sequence (intein) and splicing of the N- and C-terminal halves of the protein together [18].

Three copies of Vma2p (subunit B, Vat2p) are also present in each V-ATPase complex. Vma2p is highly conserved in structure and function within the V-type and F-type ATPase families. Similar to the F-ATPase alpha subunit, Vma2p is proposed to possess a nucleotide binding site that is not the catalytic site [19, 20]. Interestingly, the Vma2p translation product is predicted to possess a 12-amino acid N-terminal leader sequence that is not present in the mature protein [21].

The remaining non-catalytic V_1 subunits possess homologous subunits in V-ATPase complexes in other organisms but none demonstrate homology to any F-ATPase subunits. Vma4p (subunit E) is a 27 kDa subunit of the V_1 complex of the yeast V-ATPase [22]. Vma4p was the first V_1 subunit identified whose stability in the cell is dependent on the presence of another V_1 subunit, Vma10p, suggesting a direct interaction between these two subunits [23]. Cross-linking studies further support an interaction between Vma4p and Vma10p, revealing that both of these proteins interact with the exterior surface of Vma2p [24]. Vma5p (subunit C) is a 42 kDa subunit of the V-ATPase [22].

Vma7p (subunit F) behaves as a typical peripherally associated V_1 subunit since it can be stripped from the membrane by alkaline carbonate treatment and fractionates only with the assembled V-ATPase complex. Cells lacking Vma7p also affect the stability of the Vo complex, unlike the deletion of any other V_1 subunit [25].

In the absence of Vma7p the turnover rate of Vph1p decreases from in excess of 400 min to 60 min (Graham and Stevens, unpublished results; [26]). Vma8p (subunit D) is a 32 kDa V_1 subunit that can be cross-linked to Vma7p, indicating physical association between these two subunits [23, 27]. Vma8p can also be cross-linked to the interior face of Vma2p, suggesting both Vma8p and Vma7p may form part of the rotor connecting the catalytic subunits to the V_O subcomplex [24, 28].

The gene encoding Vma10p is unique for yeast genes because it possesses an intron immediately following the initiating methionine. Vma10p (subunit G) is required to maintain stable levels of Vma4p in the cell. In wild-type cells the half-life of Vma4p is greater than 4 hours, but in cells lacking Vma10p the half-life of Vma4p decreases to about 1 hour, suggesting an association between these two subunits [23]. Stable cross-linked species can be formed between Vma10p and Vma4p, further supporting their close proximity in the V-ATPase complex. The calculated mass of the Vma10p-Vma4p subcomplex identified by native gel electrophoresis is twice that of the Vma10p–Vma4p cross-linked product, suggesting that two copies of each protein (E_2G_2) may be present in each V-ATPase complex [23]. With the exception of Vma10p, Vma4p, Vma1p, and Vma2p, only single copies of all the other V_1 V-ATPase subunits are present in each complex.

Vma13p (subunit H) is thought to play a role in the regulation of the V-ATPase complex. Vma13p is the only subunit not needed for assembly of the complex, but is instead required for activation since in its absence all the other subunits assemble to form a nonfunctional complex residing in the vacuole membrane [8]. V-ATPase and proton-pumping activity can be restored in V-ATPase complexes isolated from cells lacking Vma13p by the addition of recombinantly expressed Vma13p (Flannery and Stevens, unpublished data). There is a physical interaction between Vma13p and the N-terminal cytosolic domain of the Vo subunit Vph1p [29]. Even though it is characterized as a subunit of the V_1 subcomplex, the association of Vma13p with the V-ATPase requires the N-terminal domain of Vph1p, further supporting role for Vma13p in regulation complex activity [29].

The crystal structure for Vma13p has recently been solved [30]. The data reveals that the protein structure consists largely of α helical and contains two domains. The N-terminal domain (amino acids 2–352) consists of 17 consecutive α helices that can be stacked upon one another in a right-handed spiral to form a superhelix. The C-terminal domain (353–478) has a structure that is overall quite similar to the N-terminal domain; however, the α helices are less regularly arranged. A flexible linker connects the domains and creates an interface between the domains. Interestingly, electrostatic analysis suggests that this interface may be a possible site for binding Vma13p to the V-ATPase [30]. There is also a shallow groove in the N-terminal domain, and in the crystal structure the first 10 amino acids are bound in this groove. A structural homology search revealed that the peptide interaction is similar to that seen in the importin family of proteins, and in the importins the binding of the peptide provides a regulatory switch for importin function [30, 31].

When Vma13p was first identified, it was shown that only the C-terminal 299 amino acids (out of 478) are needed for a functional V-ATPase [8]. Since this N-terminal region can be deleted without affecting the function of the V-ATPase, it may play a role in regulating some other functions not related to the complex.

In subsequent studies, Vma13p has been implicated in interactions with other proteins and may have other functions independent of its role in the V-ATPase. Vma13p may play a role in inhibiting the ATPase activity of V_1 subcomplexes that are not associated with the membrane but are found in the cytosol [32]. Yet overexpression of Vma13p is reported to uncouple the yeast V-ATPase, making it functionally inactive [33]. Vma13p has also been implicated in regulating the activity of a Golgi localized ectoapyrases by binding to the protein Ynd1p [34]. In addition to its roles in enzyme regulation, human Vma13p (NBP1) is proposed to play a role in the down regulation of the HIV receptor CD4 by binding to the Nef protein [35, 36].

Analysis of various deletion mutants allowed the identification of specific V_1 subunit complex intermediates in the assembly of the V_1 subcomplex. In most yeast V_1 deletion mutants, the loss of a V_1 subunit prevented the association of the remaining V_1 subunits with the Vo subcomplex, yet they remain present in the cytosol. The largest cytosolic complex observed included Vma1p, Vma2p, Vma4p, Vma7p, Vma8p, and Vma10p [23]. Strong associations where observed in several mutant backgrounds between Vma7p and Vma8p and a second subcomplex between Vma4p and Vma10p. As mentioned previously, association of Vma4p with Vma10p ensures its stability in the cell. Unfortunately, more detailed arrangement of the non-catalytic V_1 subunits relative to Vma1p, Vma2p and the Vo subunits can only be accurately determined from analysis of a crystal structure.

14.5
Composition of the Proton-translocating Domain of the Yeast V-ATPase

The Vo complex is composed of five subunits, Vma3p, Vma11p, Vma16, Vma6p and Vph1p or Stv1p depending on whether the complex is localized to the vacuole or Golgi (Figure 14.2; see following section). Vma3p (subunit c) is a small 16 kDa hydrophobic subunit predicted to possess four transmembrane-spanning domains. Vma11p (subunit c′) is very similar to Vma3p in amino acid sequence, the two proteins share 50 % identical amino acids. Vma11p is also similar in structure to Vma3p; it is predicted to span the membrane four times. Both Vma3p and Vma11p have a highly conserved glutamic acid residue near the center of the fourth transmembrane domain that is essential for proton translocation across the membrane [10]. Protease digestion experiments of epitope-tagged Vma11p and Vma3p support a luminal orientation for both the N-terminus and C-terminus of these proteins (Flannery and Stevens, unpublished results).

Vma16p (subunit c″) is a 23 kDa membrane protein similar to Vma3p (25 % identity) and Vma11p (24 %identity) but predicted to contain five transmembrane domains. Vma16p has two conserved glutamic acid residues located in the third

A

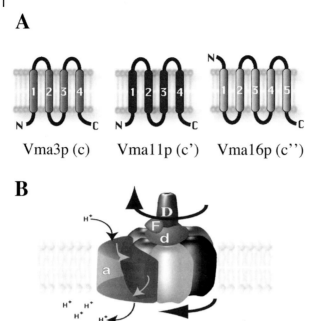

Vma3p (c) Vma11p (c') Vma16p (c'')

B

Figure 14.2   (A) Topology and function of the V_O subunits Vma3p, Vma11p and Vma16p- A. Vma3p (subunit **c**) and Vma11p (subunit **c'**) are predicted to form four membrane-spanning domains with both the C-terminus and N-terminus oriented towards the lumen of the organelle. Vma16p (subunit **c''**) is predicted to have five transmembrane helices, and recent data place the C-terminus in the lumen and the N-terminus in the cytosol. (B) Proton translocation occurs by movement of the rotor subunits (**c, c', c'', d, F, and D**) relative to the membrane domain of the stator subunit **a** (Vph1p or Stv1p). Protonation/deprotonation of glutamic acid residues imbedded in the transmembrane domains of subunits c, c', and c'' allow the transfer of protons from the cytosol to the lumen.

and fifth transmembrane domains that could participate in proton translocation. Mutational analysis of each residue in Vma16p revealed that only the glutamate residue in the third transmembrane domain is required for proton transport activity. Mutation of only one critical glutamic acid residue in either Vma3p, Vma11p, or Vma16p inactivates the V-ATPase by preventing proton translocation, suggesting a high degree of cooperativity between these subunits [10].

Transmembrane modeling programs predict that there are five transmembrane regions formed by Vma16p. An odd number of transmembrane domains suggests that either the N- or C-terminus is located in the lumen, but not both as is the case for Vma3p and Vma11p. We have shown with protease digestion of epitope tags, that the C-terminus lies in the lumen of the vacuole, while the N-terminus is located in the cytosol (Flannery and Stevens, unpublished results). Based on the homology to Vma3p and Vma11p, and predicted packing arrangement, the first helix of Vma16p could be in the center of the proteolipid ring, allowing it to interact with one or more of the stalk subunits.

Our proposed topology for Vma16p is the opposite of that proposed by Nishi et al. [37]. In their model Vma16p forms only four transmembrane domains and has a topology opposite of that of either Vma3p or Vma11p with both the N- and C-terminus facing the cytosol. Our topological model for Vma16p, five transmembrane domains with only the N-terminus facing the cytosol, is further supported by the work of Kim et al. [38] who experimentally determined the topology of several membrane proteins, including Vma16p. Using a reporter protein fused to the C-terminus of Vma16p these researchers determined a cytosolic orientation for the N-terminus and a lumenal orientation for the C-terminus, which is consistent with our model.

Comparison of the amino acid sequences of Vma16p homologous proteins across species reveal a large variability of the length of the first transmembrane domain, questioning its function in the V-ATPase complex. Several deletions have been constructed to determine if the first transmembrane helix of Vma16p is required for V-ATPase function. Deleting amino acids 12–55 of Vma16p disrupted the function of the V-ATPase since the cells carrying this truncated Vma16p exhibited slowed growth on media buffered to neutral pH, reflecting a role for the first TMD in the function of the complex (Graham and Stevens, unpublished results; [39]). However, if only the region of amino acids predicted to be the first transmembrane helix (aa 12–38) were deleted, this form of Vma16p resulted in a completely assembled complex with only slightly reduced ATPase activity, and the cells exhibited no growth defects. A Vma16p/Vma11p chimera, where the first 38 amino acids of Vma16p were placed on the N-terminus of Vma11p, was stable when expressed in yeast cells. Whereas this Vma16p/Vma11p chimera could function in place of Vma11p, it could not replace Vma16p or restore normal growth and V-ATPase activity in *vma16Δ* cells (Flannery and Stevens, unpublished results). Experiments are underway to determine which V-ATPase subunits interact with the Vma16p N-terminus.

Vma6p (subunit d) is the only V_O subunit that is not predicted to span the membrane bilayer, and it behaves as a peripherally associated membrane protein yet it consistently fractionates with the other V_O subunits [9]. In cells unable to assemble a V_O subcomplex, such as *vma3Δ*, Vma6p is no longer associated with the membrane but is found in the cytosol. The apparent reduction of Vma6p levels in *vma3Δ* cells suggested that the protein was destabilized in the absence of a V_O subcomplex [9]. We have yet to determine which specific V_O subunit interacts with Vma6p, mediating its association with the membrane.

Either Vph1p or Stv1p functions as the subunit **a** in the Vo subcomplex depending on where the V-ATPase complex is localized (see following sections). Subunit **a** is a large membrane protein possessing a hydrophilic N-terminal domain and a polytopic transmembrane-spanning C-terminal domain. Analysis of chimeric subunit **a** proteins created by swapping the N-terminal domains of Vph1p with that of Stv1p revealed that the information determining the localization of these proteins is present entirely in the N-terminus (Figure 14.3; [40]). Additional mutational analysis will be aimed toward identifying the specific residues required to localize the V-ATPase to the Golgi and endosomes instead of the vacuole.

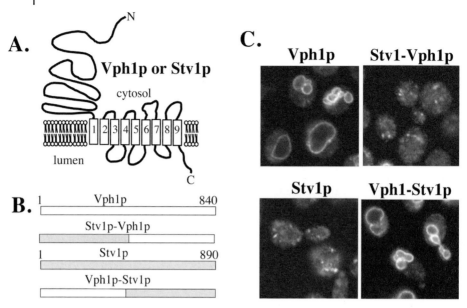

Figure 14.3 The amino-terminal domains of the subunit **a** isoforms contain information for the intracellular targeting of V-ATPase complexes. (A) Model for the topology of Vph1p or Stv1p (subunit **a** isoforms). The N-terminal portion of both Vph1p and Stv1p is hydrophilic and located in the cytosol. The C-terminal portion of each protein is hydrophobic, forming 7–9 transmembrane domains, with the extreme C-terminal tail likely to be located in the lumen. (B) Cartoon of the wild-type Stv1p, Vph1p, and the Stv1p-Vph1p or Vph1p-Stv1p chimeras. (C) *vph1Δstv1Δ* cells were transformed with low copy plasmids expressing HA-tagged versions of Vph1p, Stv1p, or chimeras of the two proteins. Complexes containing Vph1p-HA are localized to the limiting vacuolar membrane by immunofluorescence with an anti-HA monoclonal antibody. Stv1p-HA-continuing complexes are localized to punctate structures distinct from the vacuoles. Other evidence suggests that this is the late Golgi complex [40]. Vph1-Stv1p-HA has the entire N-terminal domain of Vph1p with a HA tag, fused to the C-terminal domain of Stv1p. Stv1-Vph1p-HA has the entire N-terminal domain of Stv1p with a HA tag fused to the C-terminal domain of Vph1p. As shown by immunofluorescence, the N-terminal domain of subunit **a** specifies the subcellular localization of the V-ATPase complexes.

14.6
Assembly of the Vo Domain Occurs in the ER and Requires Dedicated Assembly Factors

Assembly of the Vo subcomplex requires the presence of the subunits Vma3p, Vma6p, Vma11p, Vma16p and both or either Vph1p and Stv1p. The integral membrane Vo subunits are inserted into the ER membrane following their synthesis. Assembly of the Vo subunits forming a subcomplex occurs in the ER and is aided by three ER localized assembly factors. Only a correctly assembled Vo subcomplex is able to exit the ER for the Golgi or vacuolar membrane. The Vo subcomplex is transported to the vacuole or Golgi via the secretory pathway. Where the V_1 subcomplex assembles with the Vo to form an active V-ATPase complex is not known but results suggest it occurs very soon after synthesis and assembly of the Vo subcomplex [41].

A feature very unique to the V-ATPases as compared with the F-ATPase is the presence of three hydrophobic subunits, Vma3p, Vma11p, and Vma16p, that form a hexameric ring containing the charged residues required for proton translocation. In the highly similar F-ATPase, multiple copies (10–14) of a single hydrophobic residue (subunit c), each containing only two transmembrane domains, form a ring. Each V-ATPase subunit c, c', and c'' contains a glutamic acid residue directly involved in proton translocation [10]. Biochemical and genetic analyses confirm that all three proteins, Vma3p, Vma11p, and Vma16p, are required for proton translocation.

Using a biochemical approach we determined the stoichiometry of Vma11p and Vma16p in the yeast V-ATPase complex [42]. Immunoprecipitation of Vo subunits from yeast cells expressing epitope-tagged Vma3p, Vma11p, and Vma16p confirmed that each Vo subcomplex contained all three polypeptides. Further, immunoprecipitation experiments from cells expressing two different epitope-tagged copies of the same subunit (e. g. Vma16p-cmyc and Vma16p-HA) revealed that only one copy of Vma11p and one copy of Vma16p are present in a single Vo subcomplex but multiple copies of Vma3p are present in each V-ATPase complex. The exact number of copies of Vma3p in each yeast V-ATPase complex is yet to be determined, but based on subunit stoichiometry studies the best estimate is four copies of Vma3p [43].

Failure to assemble a Vo subcomplex results in rapid turnover of Vph1p or Stv1p by the components of the ER quality control pathway (ERQC; [44, 45]). In yeast cells lacking an integral membrane Vo subunit and unable to assemble a Vo subcomplex, Vma6p is found not on the membrane but in the cytosol [9]. The association of Vma6p with the membrane is directly dependent on the presence of the Vo subunits. Analysis of the fate of the remaining subunits in the various *vma* mutant cells suggests that failure to assemble a Vo blocks their exit from the ER.

Genetic screens have identified not only the genes encoding subunits of the yeast V-ATPase but also genes encoding non-subunit proteins required for the assembly, stability, and/or targeting of the V-ATPase complex. Three proteins have been identified, Vma12p, Vma21p, and Vma22p, that are required for the formation of a functional V-ATPase in yeast [22]. All of these three proteins have been localized to the membrane of the ER, suggesting they function very early in the biosynthesis and assembly of the V-ATPase. Cells lacking any one of these three non-subunit proteins prevent the assembly of the Vo subunits, and these unassembled Vo subunits fail to exit the ER. Since Vma12p, Vma21p, and Vma22p are required for the proper assembly of the Vo in the ER, these three proteins are referred to as V-ATPase assembly factors. Cells lacking Vma12p, Vma22p, or Vma21p do not display general defects in the processing and targeting of other vacuolar proteins, and thus the function of these assembly factors seems to be dedicated to the assembly of the yeast V-ATPase.

Vma12p is a 25 kDa integral membrane protein predicted to have two transmembrane-spanning domains [46]. The topology of Vma12p was determined by protease digestion studies and is consistent with a model placing both the N- and the C-termini on the cytosolic face of the ER membrane [47]. Cells lacking Vma12p, or any of the other assembly factors, display the same set of phenotypes as cells lacking any one of the Vo subunits. The loss of Vma12p results in the rapid turnover of Vph1p in a process that is independent of vacuolar proteases. Protease digestion experiments of Vph1p ruled out the possibility that the increased degradation of Vph1p in cells lacking Vma12p was due to improper insertion of Vph1p into the ER membrane [47]. The accessibility of Vph1p to proteases in wild-type cells compared with cells lacking Vma12p was identical, suggesting that the topology of Vph1p was not altered when Vma12p is not present.

Vma22p is a 21 kDa hydrophilic protein with no predicted transmembrane-spanning domains. The loss of Vma22p results in an identical phenotype to cells lacking Vma12p. Vma22p was localized to the membranes of the ER except in cells lacking Vma12p, suggesting that Vma12p serves as the anchor for Vma22p to the ER membrane [26]. Vma12p and Vma22p form a core complex that interacts directly but transiently with the V-ATPase subunit a in the ER [45]. The precise role played by the Vma12p-Vma22p complex in V-ATPase assembly is a more difficult issue to address since we can detect associations between Vo subunits even in cells lacking either Vma12p or Vma22p (Graham and Stevens, unpublished results).

The third V-ATPase assembly factor is Vma21p, a small 8.5 kDa hydrophobic protein predicted to possess two transmembrane-spanning domains [48]. Cells lacking Vma21p failed to assemble the Vo subcomplex and displayed phenotypes indistinguishable from cells lacking Vma12p or Vma22p. Fractionation of detergent-solubilized membranes showed that Vma21p was not part of the Vma12p-Vma22p complex, suggesting that Vma21p plays a unique role in V-ATPase assembly in the ER [45].

The carboxy-terminal four residues of Vma21p contain a di-lysine motif (at positions 3 and 4 relative to the C-terminus) that is required to maintain Vma21p in the ER. Mutation of the di-lysine (KK) motif to di-glutamine (QQ) resulted in the localization of Vma21p to the vacuole [48]. Coatomer subunits of COPI vesicles have been shown to bind directly to di-lysine motifs in proteins and that binding is required for retrograde transport, in this case the return of Vma21p to the ER.

An ER retrieval signal in the C-terminal tail of Vma21p suggests the protein may exit the ER with the V-ATPase and be retrieved back from a post-Golgi compartment. Using an *in vitro* ER vesicle budding assay Vma21p was found to be enriched in the transport vesicles formed compared with Vma12p, supporting an additional role for Vma21p in transport of the V-ATPase complex out of the ER (Malkus and Schekman, unpublished results). We are using alanine scanning mutagenesis of the cytosolic tails of Vma21p to identify signals required for packaging and transport out of the ER. So far, substitution of each of the eleven N-terminal amino acids with alanine has had no affect on the assembly, packaging or transport of the V-ATPase, since cells carrying these mutated forms of Vma21p exhibit normal V-ATPase function.

The *de novo* assembly of a functional V-ATPase in yeast not only requires that all the structural subunits are present but also requires the three assembly factors

Vma12p, Vma21p, and Vma22p (Figure 14.4). Vma12p and Vma22p form a stable complex in the ER, and the Vma12p–Vma22p complexes interact directly with the V-ATPase subunit **a**, and play a role in assembly after subunit **a** is synthesized and inserted into the ER membrane. The interaction between Vma12p–Vma22p and subunit **a** occurs transiently and the assembly complex appears to dissociate from subunit **a** as it exits the ER [45].

Vma21p was found not to be part of the Vma12p–Vma22p assembly complex, suggesting it has a separate role in assembly, either in parallel with Vma12p–Vma22p or upstream or even downstream. Since Vma21p is a small hydrophobic membrane protein similar to the Vo subunits Vma3p, Vma11p, and Vma16p it may play a role in the assembly of these proteins. One function of Vma21p may be to ensure the correct stoichiometry of the V-ATPase subunits as they assemble to form a Vo subcomplex, since each V-ATPase contains a single copy of Vma11p and a single copy of Vma16p but multiple copies of Vma3p. The presence of the functional di-lysine ER retrieval motif at the C-terminus of Vma21p strongly suggests that this protein cycles between the ER and Golgi compartments. In addition to its role in assembly, Vma21p may serve to escort the assembled V-ATPase complex out of the ER [49]. According to this model, Vma21p would then separate from the V-ATPase in the cis-Golgi and be retrieved back to the ER for another round of V-ATPase assembly.

Homologues of the V-ATPase structural subunits can easily be identified in a wide range of eukaryotic organisms. The cellular machinery required to assemble this important multisubunit complex would predicted to be conserved across spe-

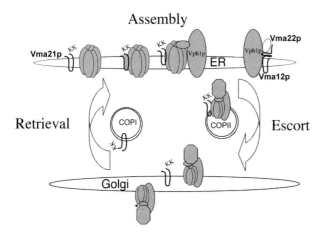

Figure 14.4 Roles of Vma21p in the assembly of the yeast V-ATPase. Assembly: Vma21p aids the assembly of the V-ATPase by association with the hydrophobic Vma3p, Vma11p, Vma16p subunits in the ER. Vma12p/Vma22p forms a core complex that interacts directly with the 100 kDa subunit **a** during assembly in the ER. Escort: unlike Vma12p/Vma22p, Vma21p remains associated with the V-ATPase to escort the complex out of the ER using COPII vesicles. Retrieval-Vma21p separates from the V-ATPase in the cis-Golgi and is retrieved, possibly by COPI vesicles, for another round of assembly and escort.

cies. Currently only two hypothetical proteins appear homologous to yeast Vma12p; one in *Candida albicans* (CaVph2p; 33 % identical) and one from *Schizosaccharomyces pombe* (16 % identity). For Vma22p a homologous protein has also been identified in *C. albicans* (CaVma22; 20 % identical) and a hypothetical human protein (MGC12981) sharing 25 % identity. It may be that the overall structure and function of Vma12p and Vma22p are conserved across species but not the protein sequence, making it more challenging to identify functionally homologous proteins.

An extensive group of small hydrophobic proteins of molecular mass 9.2–9.4 kDa, each predicted to span the membrane twice (similar to Vma21p), have been identified in many higher eukaryotic organisms. Interestingly, unlike Vma21p, these proteins lack a di-lysine ER retrieval motif and instead have been found to purify along with the V-ATPase complex isolated from bovine chromaffin granules (M9.2) and tobacco horn worm [50, 51]. A Vma21p-like protein has been identified in the *Arabidopsis* genome (Atlg05780). Unlike the M9.2 and M9.7 proteins the AtVma21p has a di-lysine motif (-3,-4) suggesting it is also ER localized and may function in the V-ATPase assembly in *Arabidopsis* similar to the yeast Vma21p.

14.7
Regulation of V-ATPase Function in Yeast

One method of regulation of V-ATPase activity is by the binding of the Vma13p subunit to a fully assembled but nonfunctional complex, resulting in an active enzyme complex. Vma13p may work as a switch to activate the V-ATPase complex once it has been targeted to its final location in the cell. A second method of regulation of V-ATPase activity is by the reversible disassembly and reassembly of the V_1 subcomplex from the Vo domain in response to glucose deprivation [7]. The yeast cells sensing starvation conditions prevent the hydrolysis of cytosolic ATP reserves through inactivation the V-ATPase complex by disassembly reassembly of the V-ATPase. This regulation process is highly reversible – following the re-addition of glucose the V_1 subcomplex reassembles on the membrane. Yeast lacking components of several well-characterized glucose-induced response pathways showed no defects in the disassembly/reassembly of the V-ATPase, suggesting the cells utilize a novel glucose sensing mechanism [52].

RAVE, a novel complex of the three proteins Rav1p, Rav2p and Skp1p, plays a role in the reassembly of V_1 complexes released from the membrane following glucose depletion [53]. RAVE binds reversibly to cytosolic V_1 complexes, but not membrane-associated V_1 subcomplexes, and was shown to bind specifically to Vma4p and/or Vma10p. Exactly how the RAVE complex facilitates the reassembly of V_1 subcomplex with Vo is not known. Cells lacking components of the RAVE complex are not affected in their ability to disassemble the V-ATPase complex during glucose starvation [53]. The proteins involved in the glucose sensing and disassembly of V_1 subcomplexes have yet to be identified.

14.8
Targeting and Trafficking of the Yeast V-ATPase (Vph1p vs. Stv1p)

In yeast the 100 kDa subunit (subunit **a**) of the V-ATPase is encoded by two different genes. *VPH1* (vacuolar pH 1) encodes a 95 kDa protein, and *STV1* (similar to
VPH1) encodes a 102 kDa polypeptide [4, 54]. The two isoforms have 54% amino
acid identity and 71% similarity, but are found in different intracellular locations
[4]. Cells lacking both Stv1p and Vph1p fail to grow on medium buffered to neutral
pH, or on medium with 100 mM calcium chloride, and thus show a phenotype typical of cells lacking a functional V-ATPase (see above). However, cells lacking
either Stv1p or Vph1p do not show a Vma$^-$ phenotype, suggesting that the two proteins can functionally compensate for each other [4]. These results led to the idea
that there were two different V-ATPase complexes in the yeast cell, in different locations. One complex contains Vph1p and the other 12 subunits of the complex,
and this is localized predominantly to the limiting membrane of the vacuole
(Table 14.1, Figure 14.3C; [4, 40, 54]). In contrast, the other form of the V-ATPase
contains Stv1p and the other 12 subunits, and is concentrated on membranes of
the late Golgi complex [40].

Stv1p and Vph1p have long N-terminal cytosolic domains of about 50 kDa, and
hydrophobic C-terminal domains containing membrane-spanning segments. However, the exact topology of the 100 kDa subunits remains controversial. Predictions
suggest anywhere from six to nine transmembrane domains, and there is data for
the extreme C-terminus being on the lumenal side of the membrane, or on the cytosolic side (Figure 14.3A; [55, 56]).

The long cytosolic N-terminal domains of Stv1p and Vph1p contain the information for targeting of the V-ATPase complexes to their correct intracellular locations
(Figure 14.3; [40]). Both Vph1p and Stv1p are synthesized on the ER membrane,
and transported through the Golgi. The Vph1p-containing complex travels from
the late Golgi to the vacuole membrane via a prevacuolar compartment, whereas
the Stv1p-containing complex can be transported to the prevacuole but is recycled
back to the late Golgi. Since targeting to the vacuole in yeast is thought to require
no specific signals, it seems likely that Stv1p contains information that allows it to
recycle to the late Golgi and possibly to be retained in that compartment [57, 58].
Stv1p but not Vph1p contains an FXFXD sequence in the amino terminal domain,
a motif required for the recycling of Ste13p (also known as diaminopeptidase A)
from the prevacuole to the late Golgi [59]. However, mutation of this motif in
Stv1p results in no change in its localization, suggesting that this motif either
does not function as a retrieval motif in Stv1p or that Stv1p contains redundant
signals [40].

As well as differences in localization, Stv1p- and Vph1p-containing V-ATPase
complexes have different biochemical properties [40, 60]. Stv1p-containing complexes have a 4–5-fold lower ratio of proton translocation to ATP hydrolysis than
Vph1p-containing complexes, and this is controlled by the C-terminal domains
of Vph1p and Stv1p. V_0 subcomplexes containing Stv1p also show lower levels
of assembly with V_1 subunits than V_0 containing Vph1p, and do not show glu-

cose-dependent dissociation of the V_1 from V_0. Assembly and glucose-dependant dissociation are controlled by the N-terminal, cytosolic regions of the 100 kDa subunits. However, while assembly of the V-ATPase complex appears to depend on the isoform of 100 kDa subunit present (Vph1p or Stv1p), the glucose-dependent dissociation of the complex is dependent on the subcellular location of the V-ATPase complex [60].

The presence of multiple subunit **a** isoforms in the yeast cell may allow tighter control over V-ATPase activity in different subcellular compartments. In mammalian cells, the pH in the lumen of the Golgi is 6.0–6.5, and the pH in the lysosomal lumen is 4.0–5.0 [61–65]. Although the pH of the vacuole is slightly higher than its mammalian counterpart (the lysosome), at pH 5.0–6.0 [21, 66], a similar pH difference is thought to exist. Thus the lower level of assembly and lower coupling of ATP hydrolysis to proton translocation of the Stv1p-containing V-ATPase complex compared with the Vph1p-containing complex may be factors in establishing the lumenal pH difference between the Golgi and the vacuole.

Different 100 kDa V-ATPase isoforms have also been identified in other species. Four different isoforms of the 100 kDa subunit have been identified in mice (a1–a4), and four in *C. elegans* (*vha-5*, *vha-6*, *vha-7*, and *unc-32*), and these isoforms are expressed in different tissues [67–69]. The presence of several 100 kDa subunit isoforms may therefore be a general mechanism for the control of targeting and activity of the V-ATPase.

14.9
Alternative Mechanisms for Intraorganellar Acidification in the Absence of the V-ATPase

Yeast cells lacking a functional V-ATPase are unable to grow at neutral pH, but are able to survive if grown under acidic conditions. One possible explanation is that cells require acidification of one or more intracellular compartments, and in media at neutral pH this acidification depends on the V-ATPase. However, in media at acidic pH, alternative mechanisms of intraorganellar acidification exist. Measurement of vacuolar pH in *vma4Δ* cells shows that cells lacking a functional V-ATPase can acidify their vacuoles to pH 5.9 when grown in medium at pH 5.5 [66]. This is compared with a vacuolar pH of 5.45 in wild-type cells under the same conditions. When grown in medium buffered to pH 7.5, wild-type cells can acidify their vacuoles to pH 5.9, but the vacuoles of *vma4Δ* yeast are now only pH 7.05. Further evidence for acidification of endosomes in the absence of the V-ATPase comes from studies of the sodium/proton exchanger Nhx1p. Nhx1p/Vps44p is located at the prevacuolar endosome compartment (Figure 14.5), and its ion-exchange activity is required for protein transport to the vacuole [70, 71]. The exchanger is thought to move protons out of the endosome in exchange for sodium ions, using the proton gradient across the membrane normally generated by the V-ATPase. Interestingly, *vma2Δ* cells have a distinct and less severe phenotype for vacuolar protein trafficking than cells lacking Nhx1p [70]. For example, cells lacking Nhx1p (unlike *vma2Δ* or *vma3Δ* cells) have a large, aberrant prevacuolar compartment where late Golgi, prevacuolar, and vacuolar proteins accumulate. *nhx1Δvma2Δ* double mutant cells, however, have the same protein transport defects as *nhx1Δ* cells. These results imply that Nhx1p can function in protein trafficking even in the absence of the V-ATPase, and suggest that the endosomes are at least partially acidified in these cells.

The mechanisms of intraorganellar acidification in the absence of the V-ATPase remain controversial. One idea is that protons taken up from acidic medium can reach the intracellular organelles and provide acidification of these compartments [72, 73]. However, following inhibition of endocytosis using an *end4-1* mutant at the non-permissive temperature and in the presence of the V-ATPase inhibitor bafilomycin, cells grown under acidic conditions were still able to acidify their vacuoles [66]. It has also been suggested that mislocalization of the plasma membrane P-type ATPase Pma1p may allow acidification of intracellular compartments in cells lacking the V-ATPase [74]. This seems unlikely, since Pma1p appears to accumulate in the endoplasmic reticulum of *vma8Δ* cells rather than in Golgi, endosomal or vacuolar fractions, and the accumulated Pma1p is inactive [74]. In addition, overexpression of Pma1p in *vma4Δ* cells does not affect either vacuolar acidification at acidic pH or the inability of the cells to grow at neutral pH compared with *vma4Δ* cells [66]. Another explanation of the vacuolar acidification seen in cells lacking the V-ATPase is the presence of high levels of ammonium ions in yeast cul-

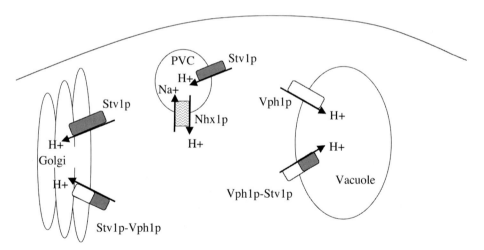

Figure 14.5 Model for the localization and function of the Nhx1p and the V-ATPase complexes. The Stv1p-containing V-ATPase acidifies the Golgi and endosomal compartments, whereas Vph1p-containing V-ATPase acidifies the vacuole. The localization of the Stv1p-Vph1p chimera to the Golgi and the Vph1p-Stv1p chimera to the vacuole suggests that V-ATPase sorting information is present in the N-terminal portion of subunit **a**. Nhx1p localizes to the prevacuolar compartment (PVC) and exchanges protons for Na$^+$ ions. Nhx1p can function in the absence of the V-ATPase, supporting the existence of alternative methods of organelle acidification independent of the V-ATPase.

ture media [66]. Under acidic conditions, ammonium ions can be transported into cells via plasma membrane transporters, causing acidification of the cytosol which in turn promotes vacuolar acidification. The precise mechanism of vacuolar acidification is unknown, but may be due to the entry of ammonium ions, or other weak acids into the vacuole [66].

14.10
Role of the V-ATPase in Protein Trafficking

Experiments in mammalian cells using the V-ATPase inhibitor bafilomycin A suggest that acidification of the endocytic system by the V-ATPase is required for the normal trafficking of proteins [75–78]. In yeast, the loss of a functional V-ATPase causes defects in the trafficking of the soluble vacuolar proteins carboxypeptidase Y (CPY) and proteinase A (PrA), although there is conflicting data regarding the severity of the defects. In our hands, *vma2Δ* and *vma3Δ* strains secrete 20–25 % of newly synthesized CPY, compared with the 4–8 % secreted in wild-type cells [21, 70]. *vma2Δ* cells also secrete 20–25 % of newly synthesized PrA [21]. The CPY or PrA that is not secreted is matured, suggesting only a mild defect in its trafficking. Similar results were found by Umemoto et al. [79]. Other reports suggest much more severe trafficking phenotypes. 53–78 % of CPY was matured after a 60 min chase in *vma2Δ*, *vma1Δ* or *vma3Δ* cells when intact cells were labeled, but only 33–44 % was matured if spheroplasts were labeled [80, 81]. A more recent study in a different strain background showed almost complete lack of CPY processing in a *vma3Δ* strain, but a less severe defect in *vma4Δ* cells [82]. Depletion of Vma3p or Vma4p using the *GAL1* promoter in *vma3Δ* or *vma4Δ* cells showed a kinetic delay in CPY processing, though not an inhibition of maturation. This delay was more pronounced in *vma3Δ* than *vma4Δ* cells [82].

Vacuolar hydrolases like CPY follow a trafficking pathway from the late Golgi via the prevacuolar compartment to the lumen of the vacuole. In contrast, alkaline phosphatase (ALP) follows a distinct route from the late Golgi to the vacuole-limiting membrane that does not pass through the prevacuole. The role of the V-ATPase in ALP trafficking is again controversial. Our results show no effect on ALP trafficking or processing in cells lacking a functional V-ATPase [21, 70]. As for CPY, others have found more severe defects on ALP processing in cells lacking the V-ATPase [80–83]. Unlike CPY processing however, experiments using either the V-ATPase inhibitor bafilomycin A1, or following shutoff of *GAL1*-dependent Vma3p or Vma4p synthesis, suggest that the effects seen on ALP processing may be the secondary physiological result of long-term deacidification [82]. The reasons for the conflicting results obtained by different groups remain unclear, but may well be due either to yeast strain differences, or differences in experimental procedures. Nevertheless, it seems clear that a functional V-ATPase is required for efficient CPY trafficking.

As discussed above, CPY and ALP follow separate routes from the late Golgi to the vacuole. However, proteins can also reach the vacuole following endocytosis from the cell surface (for reviews see Refs. 84,85). Protein internalized from the cell surface meet proteins traveling from the late Golgi along the CPY pathway at or before the prevacuolar compartment. Endocytosis in yeast can be measured using the lipophilic dye FM4-64, which is incorporated in the plasma membrane and internalized through the endocytic system to the vacuole membrane, or by uptake of the sulfonate dye lucifer yellow [86, 87]. Transport of FM4-64 to the vacuole is slowed dramatically in cells lacking a functional V-ATPase (our unpublished results and [79, 88]). We suggest that the V-ATPase is required for an endocytic transport step prior to the prevacuolar compartment, e.g., protein transport from the plasma membrane to the early endosome, or from early endosome to prevacuolar compartment. This would explain why in our experiments more of an effect is seen on endocytic protein trafficking than on CPY or ALP transport. In addition, this hypothesis agrees with studies in mammalian cells that suggest the V-ATPase is required for early to late endosome transport, or for recycling from the early endosome to the plasma membrane [75–77].

Currently, in yeast it is unclear how the V-ATPase (or indeed a proton gradient within intracellular compartments) may be required for efficient protein trafficking. In mammalian cells it has been shown that neutralization of the early endosome lumenal pH with bafilomycin inhibits the binding of a subset of cytosolic coat proteins (α, β, β', ε, and ζ COP) to the outer membrane [75, 89]. A transmembrane pH sensor protein has been proposed that would monitor the pH within the early endosome lumen and relay this information to the cytosolic side of the membrane, allowing COP proteins to bind [75, 90]. The binding of the COP proteins to the membrane is required for protein transport from early to late endosomes, though their precise role in this process remains unclear. To date, there is no evidence that the yeast homologues of the COP proteins (Ret1p/Sec33p, Sec26p, Sec27p, Sec28p, and Ret3p) play a role in endosomal protein trafficking.

In mammalian cells, many proteases are transported as inactive zymogens that require activation by proteolytic processing in the lysosome. The acidic pH of the lysosome is required for the activation of these enzymes. However, in yeast cells lacking an active V-ATPase, zymogen activation was normal [21]. The precursor forms of ALP and CPY seen in some studies of V-ATPase deficient yeast (see above) were not present in the vacuole, but held in other intracellular compartments [83]. The CPY and ALP found in the vacuole were fully mature, suggesting that even in cells lacking a functional V-ATPase the vacuolar proteases were able to function.

14.11
Conclusion

The yeast V-ATPase complex is an ion-pumping V-type ATPase functioning to acidify cellular compartments by driving the translocation of protons from the cytosol into the lumen of various organelles. Organelle acidification is essential for life in all eukaryotic organisms tested, with the exception of the budding yeast *S. cerevi-*

siae, making yeast an ideal model system to study this important complex. The V-ATPase complex in yeast is present on the membranes of several organelles and its specific localization is determined by varying the isoforms of the **a** subunit in the complex. Yeast cells lacking a functional V-ATPase are able to grow only under acidic conditions and also display defects in the trafficking of other proteins within the cell, supporting a global role for V-ATPase dependent organelle acidification.

Extensive biochemical and genetic analysis has identified a large collection of proteins required to form a functional V-ATPase in yeast. With the exception of the catalytic and nucleotide-binding subunits (Vma1p and Vma2p), the function of most peripherally associated V_1 subunits is unknown. By analogy with the structurally and functionally related F-ATPases, some subunits are likely to play a structural role, linking the catalytic domain to the membrane V_O domain. The remaining V_1 subunits may serve to transfer the energy released from ATP hydrolysis to drive the translocation of protons across the membrane bilayer. While structural studies of the V-ATPase are in their infancy, the Vma13p structure is beginning to shed some light on how this protein functions in the complex. Until details of the structure of the V-ATPase complex are available the actual arrangement of the individual subunits within the V_1 subcomplex will remain unclear.

The stoichiometry and topology of the V_O subunits has become an active area of investigation. The conclusions from our results support a model of the V-ATPase complex containing a single copy each of Vma11p and Vma16p, but multiple copies of Vma3p per complex. One role of the assembly factor proteins may be to regulate the stoichiometry and arrangement of the V_O subunits as the Vo subcomplex assembles in the ER. Vma3p, Vma11p and Vma16 each contain an acidic residue within a transmembrane domain that is intimately involved in the translocation of protons across the membrane, and these three subunits are proposed to assemble to form a proton-translocating "proteolipid" ring. We have found that the topology of Vma16p is the same as Vma3p and Vma11p, with the C-terminus of Vma16p facing the lumen and the extra transmembrane domain of Vma16p spanning the membrane. Details of the precise arrangement and function of these hydrophobic proteins within the V_O subcomplex must await future investigations.

References

1. T. Nishi and M. Forgac, *Nat. Rev. Mol. Cell. Biol.*, **2002**, *3*, 94–103.
2. C. K. Raymond, I. Howald-Stevenson, C. A. Vater, and T. H. Stevens, *Mol. Biol. Cell*, **1992**, *3*, 1389–1402.
3. T. H. Stevens and M. Forgac, *Annu. Rev. Cell. Dev. Biol.*, **1997**, *13*, 779–808.
4. M. F. Manolson, B. Wu, D. Proteau, B. E. Taillon, B. T. Roberts, M. A. Hoyt, and E. W. Jones, *J. Biol. Chem.*, **1994**, *269*, 14064–14074.
5. D. Halachmi and Y. Eilam, *FEBS Lett.*, **1993**, *316*, 73–78.
6. E. Conibear and T. H. Stevens, *Methods Enzymol.*, **2002**, *351*, 408–432.
7. P. M. Kane, *J. Biol. Chem.*, **1995**, *270*, 17025–17032.
8. M. N. Ho, R. Hirata, N. Umemoto, Y. Ohya, A. Takatsuki, T. H. Stevens, and Y. Anraku, *J. Biol. Chem.*, **1993**, *268*, 18286–18292.
9. C. Bauerle, M. N. Ho, M. A. Lindorfer, and T. H. Stevens, *J. Biol. Chem.*, **1993**, *268*, 12749–12757.
10. R. Hirata, L. A. Graham, A. Takatsuki, T. H. Stevens, and Y. Anraku, *J. Biol. Chem.*, **1997**, *272*, 4795–4803.
11. R. A. Capaldi and R. Aggeler, *Trends Biochem. Sci.*, **2002**, *27*, 154–160.
12. H. Noji, R. Yasuda, M. Yoshida, and K. Kinosita, Jr., *Nature*, **1997**, *386*, 299–302.
13. H. Imamura, M. Nakano, H. Noji, E. Muneyuki, S. Ohkuma, M. Yoshida, and K. Yokoyama, *Proc. Natl. Acad. Sci. U. S. A.*, **2003**, *100*, 2312–2315.
14. K. Yokoyama, M. Nakano, H. Imamura, M. Yoshida, and M. Tamakoshi, *J. Biol. Chem.*, **2003**, *278*, 24255–24258.
15. T. Hirata, A. Iwamoto-Kihara, G. H. Sun-Wada, T. Okajima, Y. Wada, and M. Futai, *J. Biol. Chem.*, **2003**, *278*, 23714–23719.
16. Y. Anraku, N. Umemoto, R. Hirata, and Y. Wada, *J. Bioenerg. Biomembr.*, **1989**, *21*, 589–603.
17. P. M. Kane, C. T. Yamashiro, D. F. Wolczyk, N. Neff, M. Goebl, and T. H. Stevens, *Science*, **1990**, *250*, 651–657.
18. R. Hirata, Y. Ohsumk, A. Nakano, H. Kawasaki, K. Suzuki, and Y. Anraku, *J. Biol. Chem.*, **1990**, *265*, 6726–6733.
19. J. Zhang, E. Vasilyeva, Y. Feng, and M. Forgac, 5, *J. Biol. Chem.*, **1995**, *270*, 15494–15500.
20. E. Vasilyeva, Q. Liu, K. J. MacLeod, J. D. Baleja, and M. Forgac, *J. Biol. Chem.*, **2000**, *275*, 255–260.
21. C. T. Yamashiro, P. M. Kane, D. F. Wolczyk, R. A. Preston, and T. H. Stevens, *Mol. Cell. Biol.*, **1990**, *10*, 3737–3749.
22. M. N. Ho, K. J. Hill, M. A. Lindorfer, and T. H. Stevens, *J. Biol. Chem.*, **1993**, *268*, 221–227.
23. J. J. Tomashek, L. A. Graham, M. U. Hutchins, T. H. Stevens, and D. J. Klionsky, *J. Biol. Chem.*, **1997**, *272*, 26787–26793.
24. Y. Arata, J. D. Baleja, and M. Forgac, *Biochemistry*, **2002**, *41*, 11301–11307.
25. L. A. Graham, K. H. Hill, and T. H. Stevens, *J. Biol. Chem.*, **1994**, *269*, 25974–25977.
26. K. J. Hill and T. H. Stevens, *J. Biol. Chem.*, **1995**, *270*, 22329–22336.
27. L. A. Graham, K. J. Hill, and T. H. Stevens, *J. Biol. Chem.*, **1995**, *270*, 15037–15044.
28. Y. Arata, J. D. Baleja, and M. Forgac, *J. Biol. Chem.*, **2002**, *277*, 3357–3363.
29. C. Landolt-Marticorena, K. M. Williams, J. Correa, W. Chen, and M. F. Manolson, *J. Biol. Chem.*, **2000**, *275*, 15449–15457.
30. M. Sagermann, T. H. Stevens, and B. W. Matthews, *Proc. Natl. Acad. Sci. U. S. A.*, **2001**, *98*, 7134–7139.
31. B. Kobe, *Nat. Struct. Biol.*, **1999**, *6*, 388–397.
32. K. J. Parra, K. L. Keenan, and P. M. Kane, *J. Biol. Chem.*, **2000**, *275*, 21761–21767.
33. K. Keenan Curtis and P. M. Kane, *J. Biol. Chem.*, **2002**, *277*, 2716–2724.
34. X. Zhong, R. Malhotra, and G. Guidotti, *J. Biol. Chem.*, **2000**, *275*, 35592–35599.

35. X. Lu, H. Yu, S. H. Liu, F. M. Brodsky, and B. M. Peterlin, *Immunity,* **1998**, *8,* 647–656.

36. R. Mandic, O. T. Fackler, M. Geyer, T. Linnemann, Y. H. Zheng, and B. M. Peterlin, *Mol. Biol. Cell.,* **2000**, *12,* 463–473.

37. T. Nishi, S. Kawasaki-Nishi, and M. Forgac, *J. Biol. Chem.,* **2003**, *278,* 5821–5827.

38. H. Kim, K. Melen, and G. Heijne Von, *J. Biol. Chem.,* **2003**, *278,* 10208–10213.

39. L. C. Gibson, G. Cadwallader, and M. E. Finbow, *Biochem. J.,* **2002**, *366,* 911–919.

40. S. Kawasaki-Nishi, K. Bowers, T. Nishi, M. Forgac, and T. H. Stevens, *J. Biol. Chem.,* **2001**, *276,* 47411–47420.

41. P. M. Kane, M. Tarsio, and J. Liu, *J. Biol. Chem.,* **1999**, *274,* 17275–17283.

42. B. Powell, L. A. Graham, and T. H. Stevens, *J. Biol. Chem.,* **2000**, *275,* 23654–23660.

43. H. Arai, G. Terres, S. Pink, and M. Forgac, *J. Biol. Chem.,* **1988**, *263,* 8796–8802.

44. K. Hill and A. A. Cooper, *Embo J.,* **2000**, *19,* 550–561.

45. L. A. Graham, K. J. Hill, and T. H. Stevens, *J. Cell. Biol.,* **1998**, *142,* 39–49.

46. R. Hirata, N. Umemoto, M. N. Ho, Y. Ohya, T. H. Stevens, and Y. Anraku, *J. Biol. Chem.,* **1993**, *268,* 961–967.

47. D. D. Jackson and T. H. Stevens, *J. Biol. Chem.,* **1997**, *272,* 25928–25934.

48. K. J. Hill and T. H. Stevens, *Mol. Biol. Cell,* **1994**, *5,* 1039–1050.

49. J. M. Herrmann, P. Malkus, and R. Schekman, *Trends Cell Biol.,* **1999**, *9,* 5–7.

50. J. Ludwig, S. Kerscher, U. Brandt, K. Pfeiffer, F. Getlawi, D. K. Apps, and H. Schägger, *J. Biol. Chem.,* **1998**, *273,* 10939–10947.

51. H. Merzendorfer, M. Huss, R. Schmid, W. R. Harvey, and H. Wieczorek, *J. Biol. Chem.,* **1999**, *274,* 17372–17378.

52. K. J. Parra and P. M. Kane, *Mol. Cell Biol.,* **1998**, *18,* 7064–7074.

53. A. M. Smardon, M. Tarsio, and P. M. Kane, *J. Biol. Chem.,* **2002**, *277,* 13831–13839.

54. M. F. Manolson, D. Proteau, R. A. Preston, A. Stenbit, B. T. Roberts, M. A.

Hoyt, D. Preuss, J. Mulholland, D. Botstein, and E. W. Jones, *J. Biol. Chem.,* **1992**, *267,* 14294–14303.

55. X. H. Leng, T. Nishi, and M. Forgac, *J. Biol. Chem.,* **1999**, *274,* 14655–14661.

56. J. L. Urbanowski and R. C. Piper, *J. Biol. Chem.,* **1999**, *274,* 38061–38070.

57. C. J. Roberts, S. F. Nothwehr, and T. H. Stevens, *J. Cell. Biol.,* **1992**, *119,* 69–83.

58. C. A. Wilcox, K. Redding, R. Wright, and R. S. Fuller, *Mol. Biol. Cell.,* **1992**, *3,* 1353–1371.

59. S. F. Nothwehr, C. J. Roberts, and T. H. Stevens, *J. Cell. Biol.,* **1993**, *121,* 1197–1209.

60. S. Kawasaki-Nishi, T. Nishi, and M. Forgac, *J. Biol. Chem.,* **2001**, *276,* 17941–17948.

61. S. Ohkuma and B. Poole, *Proc. Natl. Acad. Sci. U. S.A.,* **1978**, *75,* 3327–3331.

62. M. M. Wu, M. Grabe, S. Adams, R. Y. Tsien, H. P. Moore, and T. E. Machen, *J. Biol. Chem.,* **2001**, *276,* 33027–33035.

63. O. Seksek, J. Biwersi, and A. S. Verkman, *J. Biol. Chem.,* **1995**, *270,* 4967–4970.

64. J. H. Kim, C. A. Lingwood, D. B. Williams, W. Furuya, M. F. Manolson, and S. Grinstein, *J. Cell. Biol.,* **1996**, *134,* 1387–1399.

65. I. Mellman, R. Fuchs, and A. Helenius, *Annu. Rev. Biochem.,* **1986**, *55,* 663–700.

66. P. J. Plant, M. F. Manolson, S. Grinstein, and N. Demaurex, *J. Biol. Chem.,* **1999**, *274,* 37270–37279.

67. T. Nishi and M. Forgac, *J. Biol. Chem.,* **2000**, *275,* 6824–6830.

68. T. Oka, Y. Murata, M. Namba, T. Yoshimizu, T. Toyomura, A. Yamamoto, G. H. Sun-Wada, N. Hamasaki, Y. Wada, and M. Futai, *J. Biol. Chem.,* **2001**, *276,* 40050–40054.

69. T. Oka, T. Toyomura, K. Honjo, Y. Wada, and M. Futai, *J. Biol. Chem.,* **2001**, *276,* 33079–33085.

70. K. Bowers, B. P. Levi, F. I. Patel, and T. H. Stevens, *Mol. Biol. Cell.,* **2000**, *11,* 4277–4294.

71. R. Nass and R. Rao, *J. Biol. Chem.,* **1998**, *273,* 21054–21060.

72. A. L. Munn and H. Riezman, *J. Cell. Biol.,* **1994**, *127,* 373–386.

73. H. Nelson and N. Nelson, *Proc. Natl. Acad. Sci. U. S.A.*, **1990**, *87*, 3503–3507.

74. N. Perzov, H. Nelson, and N. Nelson, *J. Biol. Chem.*, **2000**, *275*, 40088–40095.

75. F. Aniento, F. Gu, R. G. Parton, and J. Gruenberg, *J. Cell. Biol.*, **1996**, *133*, 29–41.

76. M. J. Clague, S. Urbe, F. Aniento, and J. Gruenberg, *J. Biol. Chem.*, **1994**, *269*, 21–24.

77. L. S. Johnson, K. W. Dunn, B. Pytowski, and T. E. McGraw, *Mol. Biol. Cell.*, **1993**, *4*, 1251–1266.

78. A. W. Weert van, K. W. Dunn, H. J. Gueze, F. R. Maxfield, and W. Stoorvogel, *J. Cell. Biol.*, **1995**, *130*, 821–834.

79. N. Umemoto, T. Yoshihisa, R. Hirata, and Y. Anraku, *J. Biol. Chem.*, **1990**, *265*, 18447–18453.

80. D. J. Klionsky, H. Nelson, and N. Nelson, *J. Biol. Chem.*, **1992**, *267*, 3416–3422.

81. D. J. Klionsky, H. Nelson, N. Nelson, and D. S. Yaver, *J. Exp. Biol.*, **1992**, *172*, 83–92.

82. K. A. Morano and D. J. Klionsky, *J. Cell. Sci.*, **1994**, *107*, 2813–2824.

83. D. S. Yaver, H. Nelson, N. Nelson, and D. J. Klionsky, *J. Biol. Chem.*, **1993**, *268*, 10564–10572.

84. N. J. Bryant and T. H. Stevens, *Microbiol. Mol. Biol. Rev.*, **1998**, *62*, 230–247.

85. E. Conibear and T. H. Stevens, *Biochim. Biophys. Acta*, **1998**, *1404*, 211–230.

86. H. Riezman, *Cell*, **1985**, *40*, 1001–1009.

87. T. A. Vida and S. D. Emr, *J. Cell. Biol.*, **1995**, *128*, 779–792.

88. N. Perzov, V. Padler-Karavani, H. Nelson, and N. Nelson, *J. Exp. Biol.*, **2002**, *205*, 1209–1219.

89. F. Gu, F. Aniento, R. G. Parton, and J. Gruenberg, *J. Cell. Biol.*, **1997**, *139*, 1183–1195.

90. F. Gu and J. Gruenberg, *FEBS Lett.*, **1999**, *452*, 61–66.

15

Vacuolar-Type Proton ATPases:
Subunit Isoforms and Tissue-Specific Functions

Ge-Hong Sun-Wada, Yoh Wada, and *Masamitsu Futai*

Vacuolar-type H^+-ATPases (V-ATPase), a family of multi-subunit ATP-dependent proton pumps, are one of the ubiquitous eukaryotic enzymes. They are present in endomembrane organelles such as vacuoles, lysosomes, endosomes, the Golgi apparatus, chromaffin granules and coated vesicles [1–4], and acidify the luminal pH of these intracellular compartments. They also pump protons across the plasma membranes of specialized cells, including osteoclasts and epithelial cells in the kidneys and male genital tracts [4–6]. Therefore, V-ATPases are required for diverse cellular processes, including receptor-mediated endocytosis, protein processing and degradation, targeting of lysosomal enzymes, renal acidification, bone resorption, neurotransmitter accumulation, and activation of acid hydrolases. One of the most important questions is how ubiquitous V-ATPases can function in a wide variety of physiological processes. Recent studies indicated that the diverse functions of V-ATPases are possibly due to the utilization of a specific subunit isoform(s), the basic functional structure being maintained. In this chapter, we will focus on recently obtained knowledge on the subunit isoforms, including their tissue-specific functions in mammals, and inherited diseases caused by mutations in specific isoforms.

15.1
Similarities Between V- and F-ATPase

V-ATPases from fungi, plants and animals are structurally similar [7, 8], consisting of two major functional sectors known as V_1 and V_o (Figure 15.1, see p. 382). The V_1 sector consists of at least eight different subunits (*A–H*). This sector contains catalytic sites formed from the *A* and *B* subunits and is responsible for ATP hydrolysis. The V_o sector contains up to 5 subunits (*a, c, c', c''* and *d*) and is mainly responsible for proton translocation across the membrane in which it is anchored [9].

V-ATPases are structurally and evolutionarily related to F-ATPase (ATP synthase), which is responsible for ATP synthesis in mitochondria, chloroplast and bacteria [10]. F-ATPase can synthesize ATP when coupled with an electrochemical proton

Handbook of ATPases. Edited by M. Futai, Y. Wada, J. H. Kaplan
Copyright © 2004 WILEY-VCH Verlag GmbH & Co. KGaA, Weinheim
ISBN 3-527-30689-7

gradient, or hydrolyze ATP forming the gradient, whereas it is considered that V-ATPases can only perform the latter function. It is of interest to determine whether V-ATPases can synthesize ATP from ADP and phosphate coupled with an electrochemical proton gradient. This possibility was examined using engineered yeast vacuolar membranes, in which both yeast V-ATPase and a plant proton pumping pyrophosphatase are expressed [11]. Although the activity is low when compared with that of F-ATPase, the ATPase synthesis was observed coupled with the electrochemical proton gradient generated by pyrophosphatase, indicating that V-ATPases are also reversible enzymes [11].

The catalytic mechanism of F-ATPase has been studied extensively and a binding-change mechanism has been proposed by Boyer [12]. Consistent with this mechanism, the γ subunit occupying the central space of the F-ATPase $\alpha_3\beta_3$ hexamer rotates, interacting alternately with the three β subunits [13, 14]. We [15], and Junge and coworkers [16] have shown continuous rotation of a complex of the γ subunit and c ring of a purified F-ATPase, observing rotation of an actin filament connected to the c subunit, when the $\alpha_3\beta_3$ hexamer was immobilized on a glass surface. Similarly, rotation of the $\alpha_3\beta_3$ hexamer was observed when the c ring was immobilized [17]. Finally, the rotation of the a subunit or $\alpha_3\beta_3$ hexamer relative to the c ring has been shown in membranes [18]. These results clearly indicate that F-ATPase is a motor enzyme of which the rotor and stator are interchangeable.

Although V-ATPases are significantly different from F-ATPase in structure, kinetics and physiological roles, it has been speculated that V-ATPases have a rotary mechanism [1, 3]. We have examined this possibility, introducing a histidine-tag and a biotin-tag to the yeast c and G subunits, respectively, to immobilize a V-ATPase on a glass surface [19]. Upon the addition of ATP, we observed continuous counter-clockwise rotation of an actin filament connected to the G subunit of the V-ATPase. This rotation was inhibited by concanamycin, a specific V-ATPase inhibitor, but not by azide, an inhibitor of F-ATPase. Since bafilomycin A, a similar antibiotic to concanamycin, has been shown to bind to V_o [20], possibly its subunit a [21], the rotation was obviously blocked by the tightly bound antibiotic. These results suggest that V-ATPases and F-ATPases carry out similar rotational catalysis. However, the different structures of the membrane sectors (ab_2c_{10-14} and $ac_4c'c''d$ for *E. coli* Fo and yeast V_o, respectively) and stators of the two enzymes (Figure 15.1) suggest that their individual rotation mechanisms may be different.

So far, the assembly, structure and enzymatic characteristics of V-ATPases have been studied using the yeast enzyme; however, the physiological functions of the enzyme in higher eukaryotes, especially in mammals, are not as simple as those in uni-cellular organisms. In addition, various diseases associated with V-ATPases imply complicated mechanisms are required for functional expression of these ubiquitous but specific enzymes.

15.2
V-ATPase is Essential for Mammalian Development

We first addressed the obvious but pertinent question of whether this ubiquitous proton pump is required during early mammalian embryogenesis. In yeast, deletion mutants of the *VMA* genes encoding the V-ATPase subunits can not grow at neutral pH but can at acidic pH [22]. The viability of yeast cells at acidic pH is probably due to their capability to acidify vacuoles by uptake of the acidic medium through the endocytic pathway. Consistently, double mutants of *end* (defective in endocytosis) and *vma* exhibit a lethal phenotype even at acidic pH [23]. In mammals, the acidic compartments exist at very early developmental stages.

We have examined the distribution of acidic compartments in mouse pre-implantation embryos using a fluorescent dye, acridine orange, that accumulates in acidic compartments [24]. At the one cell-stage, the acidic compartments are distributed as a diffuse granular pattern throughout the cytoplasm. This staining pattern is maintained until the 8-cell stage. After compaction, much of the cortical staining is aggregated into coarse clump. In blastocysts, we found that the staining in the inner cell mass appeared to be diffusely distributed throughout the cells, whereas it was primarily localized around the nucleus in trophoectoderm cells. The orange staining pattern was not observed when embryos were exposed to bafilomycin. These results indicate that the intracellular acidic compartments in pre-implantation embryos are generated by V-ATPase.

How far can a mammalian embryo develop without the function of V-ATPase? The *c* subunit proteolipid is encoded by a single gene (*Atp6Voc* or PL16) in the whole mouse genome, although several pseudogenes have been found [25]. Therefore, deletion of the *c* subunit gene could disrupt the function of the enzyme. Knockout of the proteolipid *c* subunit in mouse caused an embryonic lethal phenotype [26]. However, PL16$^{-/-}$ embryos could develop up to the blastocyst stage, the latest stage of preimplantation development, and could maintain acidic compartments, possibly utilizing maternal mRNA for the V-ATPase proteolipid (Figure 15.2a–d). Histochemical analysis revealed that a null mutant of the *c* subunit is defective in postimplantation development [24]. On embryonic day 5.5, we found abnormal implantation sites in which the embryonic structures were not detectable. The ratio of abnormal implantation sites of PL16$^{+/-}$ × PL16$^{+/-}$ crosses was close to a Mendelian ratio expected for the appearance of abnormal embryos. Thus, the high frequency of abortive postimplantation development (in PL16$^{+/-}$ × PL16$^{+/-}$ crosses) suggests that the PL16$^{-/-}$ mutants attached to the basement membrane of the uterus but failed to differentiate into egg cylinders [24].

Acidification of organelles was impaired in cells of outgrowths of blastocysts after culture *in vitro* for 3 days (Figure 15.2). Then we examined endocytosis by means of uptake of extracellular fluorescein isothiocyanate-labeled dextran (FITC-dextran). The fluorescent FITC-dextran was accumulated within cells as a punctate lysosome-like pattern when cells of outgrowths of wild-type blastocysts were incubated with the probe. However, the fluorescence intensity in the null mutant cells was significantly lower than that in the wild-type ones: the fluorescence

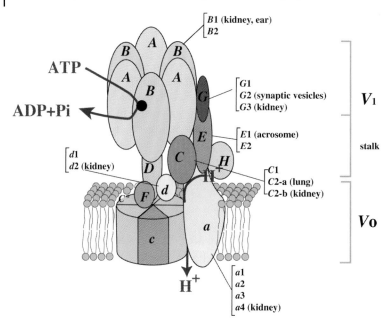

Figure 15.1 Subunit organization of V-ATPases. Structure of V-ATPases is schematically shown together with catalysis and proton transport. The membrane intrinsic (V_o), and peripheral (V_1) sectors, catalytic hexamer, stalk regions, proton pathway, and subunit isoforms are indicated. Isoforms with tissues or organelles in parentheses are expressed specifically, and the others are ubiquitous.

often remained at the cell surface, and the lysosomal staining pattern was not observed, indicating that endocytosis is defective in the mutant cells. In addition, the morphology of the Golgi apparatus in null mutant cells was disrupted, especially in the *trans* region. These results demonstrate that the impaired luminal acidification causes defects in the intracellular vesicle trafficking required for protein processing/degradation and the internalization of growth factors essential for early embryonic development [24].

Schoonderwoert and Martens reported the generation of a null embryonic stem cell of chromaffin granular V-ATPase-associated protein Ac45, which is a subunit only found in the mammalian enzymes. However, they could not obtain living mice [27]. The embryonic stem cells containing the null mutation appeared to affect the normal development of blastocysts.

In conclusion, the acidic compartments generated by V-ATPases are essential for mammalian development immediately after implantation.

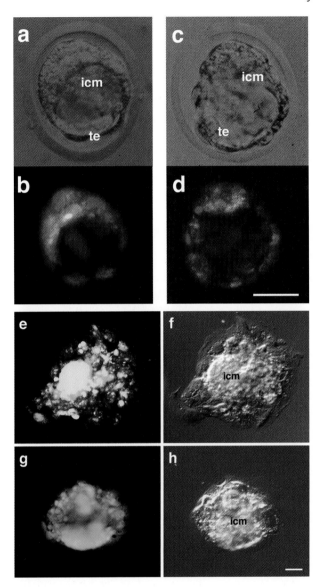

Figure 15.2 Acidic compartments in blastocysts and the outgrowth of a *c* subunit null mutant. Top panel: The morphology of wild-type blastocysts (a, b) was found to be similar to that of null embryos (c, d) on Nomarski microscopy (a, c) and acridine orange staining (b, d). Inner cell mass (icm) and trophectoderm (te) are indicated. Bar, 50 μm. Bottom panel: Acidification of wild-type and *c* subunit null blastocysts cultured *in vitro*. Blastocysts cultured for three days were stained with acridine orange (e and f, wild-type; g and h, null mutant). Respective Nomarski images are shown in f and h. No acidic compartments (red fluorescence) were observed in the outgrowths from null blastocysts (g, h) (24). Bar, 10 μm.

15.3
Subunit Isoform Involved in Tissue-specific Functions of V-ATPase

As indicated above, V-ATPases are localized in the plasma membranes of specialized cells, including osteoclasts and renal epithelial cells, in addition to the intracellular organellar membranes. What is the mechanism involved in targeting the enzyme to various membranes? We were interested in membrane-embedded V_o subunits, because their unique isoforms may play significant roles in the localization of V-ATPases to specific membranes. The *S. cerevisiae* genome encodes two different *a* subunits (Vph1p and Stv1p) for V-ATPases that differ in proton pump energy-coupling efficiency, subcellular localization, and subunit association regulation under cellular metabolic conditions [28–30]. The Vph1p-containing V-ATPase was localized in the vacuolar membrane and required for acidification of the central vacuoles, whereas the Stv1p-containing enzyme was needed for acidification of the lumens of the Golgi apparatus and prevacuolar compartments [28–30]. The *C. elegans* genome encodes four *a* subunits that differ in cell type localization and expressions during developmental stages [31]. In mouse, four *a* subunit isoforms have been identified, *a1*, *a2* and *a3* being expressed ubiquitously [32–35]. As discussed below, in addition to the *a* isoforms, isoforms of the V_1 subunit may also be involved in tissue-specific functions of the enzymes.

15.3.1
V-ATPase and Osteoclasts

Bone homeostasis is dependent on the two opposing processes of bone formation and resorption in vertebrates and is regulated throughout adult life [36]. Defective bone resorption or osteopetrosis [37, 38] results in the accumulation of mineralized bone and cartilage due to a lack of bone remodeling activity. This activity is normally provided by osteoclasts [39, 40], and fully differentiated multinucleated cells formed through the fusion of myeloid cells of the monocyte-macrophage lineage [41].

The bone resorption process requires the secretion of protons by osteoclasts. This allows the dissolution of bone minerals and maintenance of the acidic environment required for proteolytic enzymes to degrade the bone matrix [42]. The proton pumps highly concentrated at the osteoclast raffled border (apical surface) membrane are V-ATPases [43].

We have found that the *a3* isoform is specifically localized in the plasma membrane and its vicinity of osteoclasts derived from bone marrow cells [32]. Consistent with this localization, disruption of the *Atp6i* gene for mouse *a3* causes severe osteopetrosis. Osteoclasts isolated from such animals failed to generate resorption pits on bone chips [44]. A spontaneous mouse osteosclerosis (*oc*) mutation [45] has been mapped to the same gene [46]. Mutations in the *a3* gene account for ~50% of human kindreds with recessive osteopetrosis, nearly all of them being frameshift or termination mutations [47, 48]. These results indicate that the *a3* isoform is an essential subunit of the osteoclast plasma membrane V-ATPase.

However, the *a*3 isoform is expressed in all tissues so far examined, and it is not restricted to osteoclasts [32, 33]. We found that the V-ATPases with the *a*3 isoform were localized to late endosomes and lysosomes in NIH3T3 and other cell lines [49]. What is the regulatory mechanism that targets V-ATPase with the *a*3 to the osteoclast plasma membrane? Does the osteoclast progenitor have a V-ATPase with the *a*3 isoform in the plasma membrane? Osteoclasts are known to differentiate from progenitor cells. During the differentiation, activation by RANKL (receptor activator of nuclear factor κB ligand) through interaction with osteoblasts is essential [50]. The established murine macrophage cell line RAW 264.7 can form multi-nuclear cells when cultured with sRANKL (the extra-cellular domain of RANKL) [51]. The differentiated cells express osteoclast markers such as tartrate-resistant acid phosphatase, calcitonin receptor, *c-src* and cathepsin K. Thus, this cell line can be used as a good model for examining the localization of V-ATPases with the *a*3 isoform during osteoclast differentiation.

Upon stimulation with sRANKL, the *a*3 isoform together with a lysosome marker, lamp2 (lysosomal associated membrane protein 2), were localized in the dot-like organelles associated with the filamentous structures of microtubules extending to the cell surface, and resided on the plasma membranes of mature multi-nuclear osteoclast-like cells (Figure 15.3) [49]. Immunoelectron microscopy confirmed that the *a*3 signal was highly concentrated in the cell periphery and plasma membrane facing the bone matrix. These results suggest that lysosomal V-ATPases are targeted to the ruffled border membrane together with other lysosomal markers, including lamp2, upon osteoclast differentiation. This is consistent with the fact that various lysosomal enzymes, including tartrate-resistant acid phosphatase and cathepsin K, are found in the resorption lacunas between bone and the ruffled membranes of osteoclasts [42].

The molecular machinery underlying the lysosome fusion with plasma membranes in osteoclasts remains unclear. Several genetic diseases, which are characterized by varying degrees of hypopigmentation, prolonged bleeding and immunological deficiency, are the results of impaired lysosome exocytosis in cell types including cytotoxic T cells and melanocytes [52, 53]. It is thus of interest to examine the bone metabolism in patients and mutant mice.

The transport of vesicles containing V-ATPases with the *a*3 isoform to plasma membranes in osteoclasts depends on microtubules. *c-src*, which plays a pivotal role in osteoclast differentiation, is colocalized with intracellular microtubules through direct interaction with tubulin [54]. A V-ATPase is also colocalized with microtubules in osteoclasts [32, 54], suggesting that organelles containing molecules involved in bone resorption are transported to the bone-apposed cell surface through microtubules. Consistently, the depolymerization of microtubules disturbed the *a*3 localization with the filamentous structures. Holiday and colleagues have shown that the *B* subunit of V-ATPase contains a binding site for filamentous actin [55]. It is possible that the luminal acidic organelles (vesicles) delivered to the vicinity of the plasma membrane through microtubules may switch tracks, and be concentrated to form an active resorption domain through interaction with microfilaments.

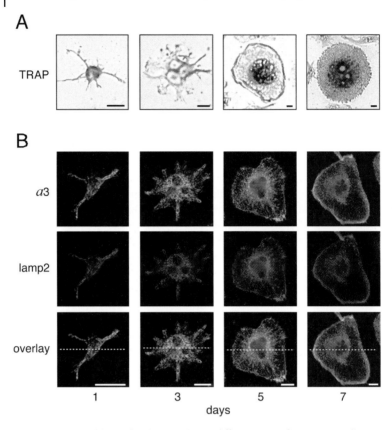

Figure 15.3 *a*3 and lamp2 localization during differentiation of RAW 264.7 cells into osteoclast-like cells. (A) RAW 264.7 cells were cultured on a plastic surface for seven days in medium containing sRANKL and M-CSF. Osteoclast-like cells were identified as multi-nuclear cells exhibiting positive staining for tartrate-resistant acid phosphatase (TRAP). (B) Localization of *a*3 and lamp2. Cells were fixed, and then stained with antibodies against *a*3 and lamp2 (49).

An alternative spliced transcript of the *a*3 gene encodes the lymphocyte protein TIRC7, an amino-terminally truncated version beginning from amino acid 217 of *a*3. Anti-TIRC7 antibodies prevent T cell proliferation triggered by an alloantigen or mitogen. TIRC7 is considered to be a lymphocyte costimulatory molecule that binds to a surface receptor, because its soluble fragment (region without the membrane-span domains) can produce T cell unresponsiveness that can be rescued by interleukin-2 [56].

15.3.2
Kidney-specific Isoforms and Distal Renal Tubular Acidosis

The human body fueled by an average western diet daily generates 1–1.5 mmol of mineral acid per kg of body weight, and this acid load must be excreted by the kidneys [57]. The acid is secreted by the type A intercalated cells of the collecting segment and collecting duct via a V-ATPase in the apical membrane, cytoplasmic carbonic anhydrase CAII, and the AE1 Cl^-/HCO_3^- exchanger of the basolateral membrane (Figure 15.4). Therefore, it has been estimated that the kidneys have the largest amount and activity of plasma membrane V-ATPase of any mammalian tissues. Failure of the kidneys to generate sufficiently acidic urine is diagnosed as distal renal tubular acidosis (dRTA), the symptoms of dRTA including inappropriately alkaline urine, low serum potassium ions, and elevated urinary calcium. If untreated, the disease may result in osteomalacia, rickets and renal stone formation. It has been shown that autosomal-dominant and recessive forms of dRTA are caused by mutations in the AE1 Cl^-/HCO_3^- exchanger of the basolateral membrane and at least two subunits of the apical membrane V-ATPase, V_1 subunit $B1$ and V_o subunit $a4$ [58, 59].

Karet et al. have found that fifteen mutations were distributed throughout the 513 amino acid $B1$ sequence in 19 of 62 evaluated kindreds, which included 7 missense mutations in addition to premature terminations, frameshifts and splice-site mutations [58]. Bilateral sensorineural deafness was found in 87 % of the evaluated families. As expected, immunocytochemical analysis revealed that the $B1$ subunit is expressed in the mouse cochlea and endolymphatic sac. The V-ATPase with the $B1$ isoform is thought to have the same function in both intercalated cells and cochlea. The mouse $B1$ gene encodes a protein that is 93 % identical to human $B1$ gene. Unlike humans with $B1$ mutations, $B1$ subunit knockout mice were born in a Mendelian ratio, and grew at a normal rate. The mice did exhibit failure of acidification of the urine in response to oral acid loading [60].

In addition to $B1$, mutations in a kidney-specific $a4$ subunit isoform also cause dRTA [59]. The mouse $a4$ isoform has also been cloned and shows 85 % identity to the human ortholog [34, 35]. Unlike human $a4$ mainly localized in type A (or type α) cells, mouse $a4$ is localized in the basolateral membranes of type B (or type β) intercalated cells, as well as in the apical membranes of type A cells.

In addition to $B1$ and $a4$, subunits C, d and G also have isoforms, C2-b, d2 and G3, respectively, which are expressed specifically in kidneys [61, 62]. We have shown that these kidney-specific isoforms predominantly form the intercalated cell proton pump localized in the apical or basolateral membranes of intercalated cells [62a]. These isoforms are candidate dRTA disease genes, although the mutations have not been identified yet.

The $B1$ and $a4$ isoforms are also expressed in the apical membranes of occasional cells of the epydidimis and *vas deferens*. V-ATPases with these subunit isoforms are required for luminal acidification of the spermatic duct, which is essential for sperm maturation [35, 63]. Other kidney-specific isoforms may be involved in the pump function in the *vas deferens*, because these organs have the same developmental origin.

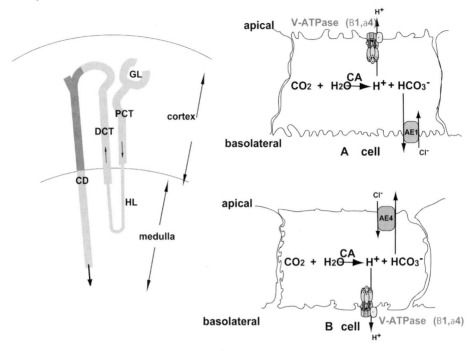

Figure 15.4 Schematic representation of the two major subtypes of intercalated cells in the cortical collecting duct of the nephron. Both cell types contained cytosolic carbonic anhydrase II (CA), which is responsible for the intracellular production of protons and bicarbonate. Type A cells, proton-secreting cells, have an apical V-ATPase with $a4$ and $B1$ isoforms, and a basolaterally located chloride–bicarbonate exchanger (AE1). Type B cells V-ATPase with $a4$ and $B1$ is of the opposite polarity. The apically located anion exchanger is AE4 (for details see Ref. 34). The cortical collecting duct in the nephron is also shown schematically. PCT, proximal convoluted tubules; DCT, distal convoluted tubules; GL, glomeruli, CD, collecting duct; HL, Henle loop.

15.3.3
Brain-specific Isoform *G2*

In neurons, V-ATPases are present in the membranes of synaptic vesicles, besides the housekeeping organelles [64–66]. The pH gradient generated by these enzymes, pH 5.2 to 5.5 inside, is utilized by specific vesicular transporters to accumulate neurotransmitters [67, 68]. Therefore, V-ATPase activity is essential not only for intracellular membrane traffic, processing and degradation of proteins, and receptor-mediated endocytosis in the neuronal cell bodies, but also for the storage and release of the neurotransmitters in nerve endings. One important question is how the V-ATPases localized in synaptic vesicles reach the axon terminals. There is evidence suggesting that synaptic vesicles are assembled at the axon terminals, and that their proteins are carried to endosomes and the plasmalemma of nerve terminals in precursor membranes (for a review, see Ref. 69). It is not known whether fully functional V-ATPases are transported to the axonal terminals. In this regard, Morel et al. observed that the B and c subunits exhibited different transport rates in *Torpedo* axons [70]. Their results may suggest that the V_1 and V_o domains are transported separately in axons. It is also of interest to know whether neurons or synaptic vesicles have a unique V-ATPase for their targeting and function.

The bovine G subunit was shown to have two isoforms [71], although the expression pattern and localization of each isoform were not shown. We have identified *Apt6V1g2* encoding the G2 isoform in mouse [72]. Therefore, together with the kidney specific isoform G3 as described above, there are three G isoforms so far found in mice. The three G isoforms consist of 118 amino acid residues, and show high homology (over 50% identity). The amino-terminal regions of the three isoforms are highly conserved (95% identity), whereas their carboxyl-terminals are variable (40% identity).

G1 was distributed ubiquitously in the tissues examined, whereas G2 was specifically distributed in central nervous system neurons. G1 was expressed at an early embryonic stage, whereas G2 transcription was significantly induced at 10.5 dpc (embryonic day 10.5, i.e., 2 days before axon outgrowth). Both G1 and G2 were strongly expressed in cortical and hippocampal neurons, cerebellar granule cells and Purkinje cells. Immunohistochemistry with isoform-specific antibodies revealed that G2 was localized not only in the cell bodies and dendrites, but also in axons where G2 is colocalized with synaptophysin, a marker of synaptic vesicles (Figure 15.5, see p. 390). In addition, electron microscopy and subcellular fractionation indicated that G2 was localized in synaptic vesicles where G1 was not detectable. These results suggest that the G2 isoform may be involved in not only the acidification of synaptic vesicles, but also in the assembly of a functional proton pump for the neuron-specific organelles.

15.3.4
E1 Isoform and Acidification of Acrosome

The acrosome is an acidic secretory organelle containing hydrolytic enzymes that are involved in sperm passage across the *zona pellucida*. Thus, the assembly of a proton pump is an essential step for the biogenesis of this unique organelle. Both the human and mouse E subunits contain a testis-specific isoform, E1, whereas E2 is expressed ubiquitously [73, 74]. The E1 transcript appears about three weeks after birth, corresponding to the start of meiosis, and is expressed specifically in round spermatids in seminiferous tubules. Immunohistochemical analysis with isoform-specific antibodies revealed that the V-ATPase with the E1 and a2 isoforms is located specifically in developing acrosomes of spermatids and acrosomes in mature sperm (Figure 15.6). In contrast, the E2 isoform was expressed in all tissues examined, and present in the perinuclear compartments of spermatocytes.

The E1 isoform exhibits 70% identity with E2, and both of them functionally complemented a null mutation of the yeast counterpart *vma4*. The chimeric enzymes showed slightly lower K_m^{ATP} than yeast V-ATPase. Consistent with the tem-

synaptophysin   **G2**   **merged**

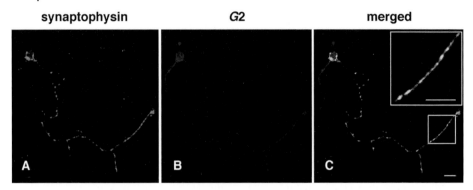

Figure 15.5 Co-localization of G2 with synaptophysin. Hippocampal neurons cultured for 14 days were stained with anti-synaptophysin (A) and anti-G2 IgG (B). A merged image is also shown (C), and the boxed region is magnified [72]. Scale bar, 15 μm.

perature-sensitive growth of *Δvma4* expressing the *E*1 isoform, vacuolar membrane vesicles exhibited temperature-sensitive coupling between ATP hydrolysis and proton transport. This temperature-sensitivity may be consistent with the lower optimal temperature (~33°C) for the culture of spermatid cells [75, 76]. These results suggest that the *E*1 isoform is essential for the energy coupling involved in the acidification of acrosomes. It is possible that mutations in the *E* subunit may cause male sterility.

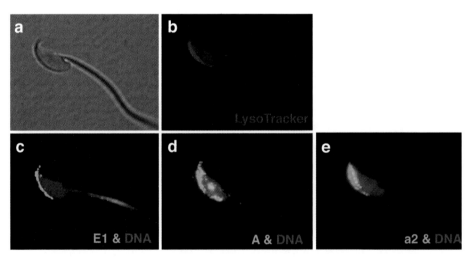

Figure 15.6 Localization of the *E*1 isoform in sperm. A phase-contrast image (a) is shown together with a living epididymal sperm stained with LysoTracker Red (b). Sperm were fixed and labeled with anti-*E*1 (c), anti-A (d), and anti-*a*2 (e) antibodies. Nuclei were visualized using the fluorescent dye DAPI [73].

15.3.5
C2-a Subunit Isoform and Acidification of Lamellar Bodies in Type II Alveolar Cells

Pulmonary alveoli are the terminal sac-like extensions of the distal respiratory tree that are specialized for gas exchange. The alveolar surface is covered by a phospholipid film (pulmonary surfactant), which is synthesized in type II epithelial cells and stored in specialized secretory granules, lamellar bodies [77]. Lamellar bodies range in size from 1.0 to 2.0 μm, and contain lysosomal markers, including acid phosphatase, cathepsins, and CD63. In addition, lamellar bodies have an acidic interior (pH ≈ 5.5), which is maintained by V-ATPases [78]. The luminal acidic pH may be required for the packaging of surfactant phospholipids, processing of surfactant proteins, and surfactant protein-dependent lipid aggregation [79].

We have found that C2 contained two splicing variants (C2-a and C2-b) exhibiting differential expressions in kidneys and lungs [62a]. C2-a was specifically expressed in type II alveolar epithelial cells (type II cells) and localized to the lamellar bodies (Figure 15.7), whereas C2-b was found predominantly in the plasma membranes of renal intercalated cells. C2-a contains 46 additional amino acid residues (^{276}P to ^{321}E) compared with C2-b. A survey of genomic databases revealed that the 138 bp insertion in C2-a was encoded exactly by the exon 12 among the total 15

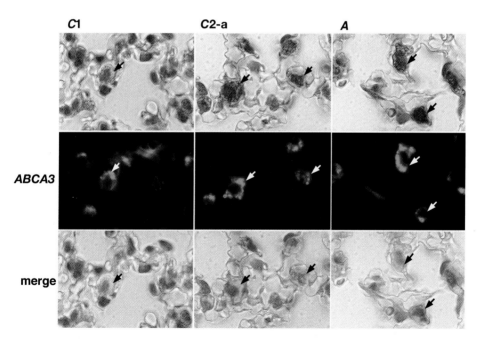

Figure 15.7 Localization of C2 in lamellar bodies in type II alveolar cells. Rabbit polyclonal antibodies against the C1, C2 and A subunits were used to label paraffin sections of lung. Each section was double-labeled with a mouse monoclonal antibody against ABCA3, a lamellar body membrane specific transporter (*ABCA3*). Signals of C2-a and A were detected at high levels in AE2 cells (arrows). C1 antibodies labeled all the cell types in lung.

exons of *C2* isoforms, and that *C2*-b lacked this exon. This insertion was not found in the *C1* isoform, or in the *C* subunits of other organisms, including yeast [80], plant [81] and nematode [82]. The 46 amino acid sequence contained potential sites for phosphorylation by protein kinase C and casein kinase, suggesting that phosphorylation of this region may be required for lung-specific functions.

15.4
Future Perspectives

In high eukaryotic organisms, in addition to the acidification of organellar lumen, V-ATPases are targeted to specific membranes of highly differentiated cells and required for various tissue-specific functions. Various lines of evidence have revealed that the diverse functions of V-ATPase in unique acidic compartments are established and regulated, at least in part, through the utilization of distinct subunit isoforms. It is possible that tissue-specific isoforms confer a targeting signal that differs from those of ubiquitous forms, leading to the V-ATPase localization in the cell; alternatively, specific isoforms may interact with different proteins, leading to differences in pump assembly and targeting. Therefore, characterization of the unique enzymes with specific isoforms will probably provide evidence of how these specialized pumps are assembled and targeted to the surface membranes, giving a molecular explanation for the diseases that are caused by disruption of these subunits. Studies in this direction will facilitate the development of specific drugs for diseases associated with V-ATPase or acidic compartments. In conclusion, detailed elucidation of the mechanism involved in determining the localization and activity of the enzymes at specific destinations will be an important future issue.

References

1. M. Futai, T. Oka, G.-H. Sun-Wada, Y. Moriyama, H. Kanazawa, and Y. Wada, *J. Exp. Biol.*, **2000**, *203*, 107–116.
2. N. Nelson and W. R. Harvey, *Physiol. Rev.*, **1999**, *79*, 361–385.
3. T. Nishi and M. Forgac, *Nat. Rev. Mol. Cell. Biol.*, **2002**, *3*, 94–103.
4. G.-H. Sun-Wada, Y. Wada, and M. Futai, *J. Bioenerg. Biomembr.*, **2003**, *35*, 347–358.
5. D. Brown and S. Breton, *J. Exp. Biol.*, **2000**, *203*, 137–145.
6. W. R. Harvey and H. Wieczorek, *J. Exp. Biol.*, **1997**, *200*, 203–216.
7. T. H. Stevens and M. Forgac, *Annu. Rev. Cell. Dev. Biol.*, **1997**, *13*, 779–808.
8. H. Sze, K. Schumacher, M. L. Muller, S. Padmanaban, and L. Taiz, *Trends Plant Sci.*, **2002**, *7*, 157–161.
9. Y. Anraku, *Handbook of Biological Physics*. N. W. Koning, H. R. Kaback, and J. S. Lolkema, eds., **1996**, Vol. 2. pp. 93–109.
10. M. Futai, T. Noumi, and M. Maeda, *Annu. Rev. Biochem.*, **1989**, *58*, 111.
11. T. Hirata, N. Nakamura, Y. Wada, and M. Futai, *J. Biol. Chem.*, **2000**, *275*, 386–389.

12. P. D. Boyer, *Biochim. Biophys. Acta*, **1993**, *1140*, 215–250.

13. H. Noji, R. Yasuda, M. C. Yoshida, and K. J. Kinosita, *Nature*, **1997**, *386*, 299–302.

14. H. Omote, N. Sambonmatsu, K. Saito, Y. Sambongi, A. Iwamoto-Kihara, T. Yanagida, Y. Wada, and M. Futai, *Proc. Natl. Acad. Sci. U. S. A.*, **1999**, *96*, 7780–7784.

15. Y. Sambongi, Y. Iko, M. Tanabe, H. Omote, A. Iwamoto-Kihara, I. Ueda, T. Yanagida, Y. Wada, and M. Futai, *Science*, **1999**, *286*, 1722–1724.

16. O. Pänke, K. Gumbiowski, W. Junge, and S. Engelbrecht, *FEBS Lett.*, **2000**, *472*, 34–38.

17. M. Tanabe, K. Nishio, Y. Iko, Y. Sambongi, A. Iwamoto-Kihara, Y. Wada, and M. Futai, *J. Biol. Chem.*, **2001**, *276*, 15269–15274.

18. K. Nishio, A. Iwamoto-Kihara, A. Yamamoto, Y. Wada, and M. Futai, *Proc. Natl. Acad. Sci. U. S. A.*, **2002**, *99*, 13448–13452.

19. T. Hirata, A. Iwamoto-Kihara, G. H. Sun-Wada, T. Okajima, Y. Wada, and M. Futai, *J. Biol. Chem.*, **2003**, *278*, 23714–23719.

20. H. Hanada, Y. Moriyama, M. Maeda, and M. Futai, *Biochem. Biophys. Res. Commun.*, **1990**, *170*, 873–878.

21. J. Zhang, Y. Feng, and M. Forgac, *J. Biol. Chem.*, **1994**, *269*, 23518–23523.

22. Y. Anraku, N. Umemoto, R. Hirata, and Y. Wada, *J. Bioenerg. Biomembr.*, **1989**, *21*, 589–603.

23. A. L. Munn and H. Riezman, *J. Cell. Biol.*, **1994**, *127*, 373–386.

24. G.-H. Sun-Wada, Y. Murata, A. Yamamoto, H. Kanazawa, Y. Wada, and M. Futai, *Dev. Biol.*, **2000**, *228*, 315–325.

25. K. Hayami, T. Noumi, H. Inoue, G. Sun-Wada, T. Yoshimizu, and H. Kanazawa, *Gene*, **2001**, *273*, 199–206.

26. H. Inoue, T. Noumi, M. Nagata, H. Murakami, and H. Kanazawa, *Biochem. Biophys. Acta*, **1999**, *1413*, 130–138.

27. V. T. Schoonderwoert and G. J. Martens, *Mol. Membr. Biol.*, **2002**, *19*, 67–71.

28. M. F. Manolson, D. Proteau, and E. W. Jones, *J. Exp. Biol.*, **1992**, *172*, 105–112.

29. M. F. Manolson, B. Wu, D. Proteau, B. E. Taillon, B. T. Roberts, M. A. Hoyt, and E. W. Jones, *J. Biol. Chem.*, **1994**, *269*, 14064–14074.

30. S. Kawasaki-Nishi, T. Nishi, and M. Forgac, *J. Biol. Chem.*, **2001**, *276*, 17941–17948.

31. T. Oka, T. Toyomura, K. Honjo, Y. Wada, and M. Futai, *J. Biol. Chem.*, **2001**, *276*, 33079–33085.

32. T. Toyomura, T. Oka, C. Yamaguchi, Y. Wada, and M. Futai, *J. Biol. Chem.*, **2000**, *275*, 8760–8765.

33. T. Nishi and M. Forgac, *J. Biol. Chem.*, **2000**, *275*, 6824–6830.

34. T. Oka, Y. Murata, M. Namba, T. Yoshimizu, T. Toyomura, A. Yamamoto, G.-H. Sun-Wada, N. Hamasaki, Y. Wada, and M. Futai, *J. Biol. Chem.*, **2001**, *276*, 40050–40054.

35. A. N. Smith, K. E. Finberg, C. A. Wagner, R. P. Lifton, M. A. Devonald, Y. Su, and F. E. Karet, *J. Biol. Chem.*, **2001**, *276*, 42382–42388.

36. G. Karsenty, *Gene Dev.*, **1999**, *13*, 3037–3051.

37. R. Felix, W. Hofstetter, and M. G. Cecchini, *Eur. J. Endocrinol.*, **1996**, *134*, 143–156.

38. F. Lazner, M. Gowen, D. Pavasovic, and I. Kola, *Hum. Mol. Genet.*, **1999**, *8*, 1839–1849.

39. G. D. Roodman, *Exp. Hematol*, **1999**, *27*, 1229–1241.

40. H. K. Vaananen, H. Zhao, M. Mulari, and J. M. Halleen, *J. Cell. Sci.*, **2000**, *113*, 377–381.

41. T. Suda, N. Takahashi, N. Udagawa, E. Jimi, M. T. Gillespie, and T. J. Martin, *Endocrine Rev.*, **1999**, *20*, 345–357.

42. R. Baron, L. Neff, D. Louvard, and P. J. Courtoy, *J. Cell. Biol.*, **1985**, *101*, 2210–2222.

43. H. C. Blair, S. M. Teitelbaum, R. Ghiselli, and L. S. Gluck, *Science*, **1989**, *245*, 855–857.

44. Y. Li, W. Chen, Y. Liang, E. Li, and P. Stashenko, *Nat. Gen.*, **1999**, *23*, 447–451.

45. H. Nakamura, Y. Moriyama, M. Futai, and H. Ozawa, *Arch. Histol. Cytol.*, **1994**, *57*, 535–539.

46. J.-C. Scimeca, A. Franchi, C. Trojani, H. Parrinello, J. Grosgeorge, C. Robert, O. Jaillon, C. Poirier, P. Gaudray, and G. F. Carle, *Bone*, **2000**, 26.

47. C. Sobacchi, A. Frattini, P. Orchard, O. Porras, and I. Tezcan, *Hum. Mol. Genet.*, **2001**, *10*, 1767–1773.

48. U. Kornak, et al., *Hum. Mol. Gen.*, **2000**, *9*, 2059–2063.

49. T. Toyomura, Y. Murata, A. Yamamoto, T. Oka, G.-H. Sun-Wada, Y. Wada, and M. Futai, *J. Biol. Chem.*, **2003**, *278*, 22023–22030.

50. H. Takayanagi, S. Kim, and T. Taniguchi, *Arthritis Res.*, **2002**, *4*, 227–232.

51. L. Huang, J. Xu, D. J. Wood, and M. H. Zheng, *Am. J. Pathol.*, **2000**, *156*, 761–767.

52. N. W. Andrews, *Trends Cell Biol.*, **2000**, *10*, 316–321.

53. E. J. Blott and G. Griffiths, *Nat. Rev. Mol. Cell. Biol.*, **2002**, *3*, 122–131.

54. Y. Abu-amer, F. P. Ross, P. Schlesinger, M. M. Tondravi, and S. L. Teitelbaum, *J. Cell. Biol.*, **1997**, *137*, 247–258.

55. B. S. Lee, S. L. Gluck, and L. S. Holiday, *J. Biol. Chem.*, **1999**, *274*, 29164–29171.

56. T. Heinemann, G.-C. Bulwin, J. Randall, B. Schneiders, and K. Sandhoff, et al., *Genomics*, **1999**, *57*, 398–406.

57. M. D. Penney and D. A. Oleesky, *Ann. Clin. Biochem.*, **1999**, *36*, 408–422.

58. F. E. Karet, K. E. Finberg, R. D. Nelson, A. Nayir, H. Mocan, S. A. Sanjad, J. Rodriguez-Soriano, F. Santos, C. W. Cremers, A. Di Pietro, B. I. Hoffbrand, J. Winiarski, A. Bakkaloglu, S. Ozen, R. Dusunsel, P. Goodyer, S. A. Hulton, D. K. Wu, A. B. Skvorak, C. C. Morton, M. J. Cunningham, V. Jha, and R. P. Lifton, *Nat. Gen.*, **1999**, *21*, 84–90.

59. A. N. Smith, J. Skaug, K. A. Choate, A. Nayir, A. Bakkaloglu, S. Ozen, S. A. Hulton, S. A. Sanjad, E. A. Al-Sabban, R. P. Lifton, S. W. Scherer, and F. E. Karet, *Nat. Gen.*, **2000**, *26*, 71–75.

60. K. E. Finberg, T. Wang, C. A. Wagner, J. P. Geibel, H. Dou, and R. P. Lifton, *J. Am. Soc. Nephrol.*, **2001**, *12*, 3–4.

61. G.-H. Sun-Wada, T. Yoshimizu, Y. Imai-Senga, Y. Wada, and M. Futai, *Gene*, **2003**, *302*, 147–153.

62. A. N. Smith, K. J. Borthwick, and F. E. Karet, *Gene*, **2002**, *297*, 169–177.

62a. G.-H. Sun-Wada, Y. Murata, A. Yamamoto, Y. Wada, and M. Futai, *J. Biol. Chem.*, **2003**, *278*, 44843–44851.

63. S. Breton, P. J. S. Smith, B. Lui, and D. Brown, *Nat. Med.*, **1996**, *2*, 470–472.

64. Y. Moriyama, M. Maeda, and M. Futai, *J. Exp. Biol.*, **1992**, *172*, 171–178.

65. Y. Moriyama, M. Maeda, and M. Futai, *J. Biochem. (Tokyo)*, **1990**, *108*, 689–693.

66. Y. Moriyama and M. Futai, *Biochem. Biophys. Res. Commun.*, **1990**, *173*, 443–448.

67. D. M. Michaelson and I. Angel, *Life Sci.*, **1980**, *27*, 39–44.

68. H. H. Fuldner and H. Stadler, *Eur. J. Biochem.*, **1982**, *121*, 519–524.

69. M. J. Hannah, A. A. Schmidt, and W. B. Huttner, *Annu. Rev. Cell. Dev. Biol.*, **1999**, *15*, 733–798.

70. N. Morel, V. Gérard, and G. Shiff, *J. Neurochem.*, **1998**, *71*, 1702–1708.

71. B. P. Crider, P. Andersen, A. E. White, Z. Zhou, X. Li, J. P. Mattsson, L. Lundberg, D. J. Keeling, X. S. Xie, D. K. Stone, and S. B. Peng, *J. Biol. Chem.*, **1997**, *272*, 10721–10728.

72. Y. Murata, G.-H. Sun-Wada, T. Yoshimizu, A. Yamamoto, Y. Wada, and M. Futai, *J. Biol. Chem.*, **2002**, *277*, 36296–36303.

73. G.-H. Sun-Wada, Y. Imai-Senga, A. Yamamoto, Y. Murata, T. Hirata, Y. Wada, and M. Futai, *J. Biol. Chem.*, **2002**, *277*, 18098–18105.

74. Y. Imai-Senga, G.-H. Sun-Wada, Y. Wada, and M. Futai, *Gene*, **2002**, *289*, 7–12.

75. M. Glassner, J. Jones, I. Kligman, M. J. Woolkalis, G. L. Gerton, and G. S. Kopf, *Dev. Biol.*, **1991**, *146*, 438–450.

76. L. J. Romrell, A. R. Bellve, and D. W. Fawcett, *Dev. Biol.*, **1976**, *49*, 119–131.

77. A. Chander and A. B. Fisher, *Am. J. Physiol.*, **1990**, *258*, 241–253.

78. A. Chander, R. G. Johnson, J. Reicherter, and A. B. Fischer, *J. Biol. Chem.*, **1986**, *261*, 6126–6131.

79. S. J. Wadsworth and A. Chander, *J. Membr. Biol.*, **2000**, *174*, 41–51.

80. M. N. Ho, K. J. Hill, M. A. Lindorfer, and T. H. Stevens, *J. Biol. Chem.*, **1993**, *268*, 221–227.

81. K. Schumacher, D. Vafeados, M. McCarthy, H. Sze, T. Wilkins, and J. Chory, *Gen. Dev.*, **1999**, *13*, 3259–3270.

82. T. Oka, R. Yamamoto, and M. Futai, *J. Biol. Chem.*, **1998**, *273*, 22570–22576.

Part VI

Cell Biology and Pathophysiology of ATPases and their Compartments

Handbook of ATPases. Edited by M. Futai, Y. Wada, J. H. Kaplan
Copyright © 2004 WILEY-VCH Verlag GmbH & Co. KGaA, Weinheim
ISBN 3-527-30689-7

16
Physiological Role of Na,K-ATPase Isoforms

Jerry B Lingrel, Jonathan Neumann, Iva Dostanic, and *Amy E. Moseley*

16.1
Introduction

Na,K-ATPase is an integral membrane protein which transports Na^+ out of cells and K^+ in moving each ion against its concentration gradient. The energy for this translocation is supplied by the hydrolysis of ATP with three Na^+ ions being transported out of the cell for every two K^+ ions transported in. The electrical gradient produced maintains the resting potential of cells and is essential for the excitable activity of muscle and nerve tissue. The Na^+ gradient is responsible for the uptake of many sugars, amino acids and other nutrients as well as some vitamins and ions transported largely through a Na^+ coupled process. While the enzymatic activity of the Na,K-ATPase is to transport Na^+ and K^+ across the cell membrane, the physiological processes which the resulting gradients drive is varied and complex. This raises the question of whether the Na,K-ATPase plays a passive role in these processes or whether the enzyme is integrally involved in specific physiological processes. As isoforms of both the α and β subunits of the Na,K-ATPase exist, it is possible that these isoforms contribute a unique property to the enzyme related to a specific physiological role. A particular isoform could possess different cation, ATP or ligand affinities, reside in a particular plasma membrane location, be coupled with other transporters or be regulated differentially by modifications such as phosphorylation. The observation that the α and β isoforms of the Na,K-ATPase are expressed differently during development and exhibit unique tissue distributions is compatible with differing functions.

This chapter will concentrate on the α and β isoforms, including their existence, distribution, activity and possible physiological roles. The γ protein which associates with the Na,K-ATPase in certain tissues and alters Na,K-ATPase activity will not be covered but has been reviewed recently [1, 2].

Handbook of ATPases. Edited by M. Futai, Y. Wada, J. H. Kaplan
Copyright © 2004 WILEY-VCH Verlag GmbH & Co. KGaA, Weinheim
ISBN 3-527-30689-7

16.2
α Isoforms

In 1957 Skou discovered the Na,K-ATPase and in the ensuing years much of the work concentrated on understanding the properties and mechanism of action of this enzyme. In 1979 Sweadner [3] observed two α subunit bands on SDS gels from brain. One of these bands corresponded to the α subunit found in kidney while the other, termed α(+), exhibited an apparent higher molecular weight. The two proteins differed in their cardiac glycoside sensitivity [3], their N-terminal amino acid sequence [4], and Na^+ affinity [5]. The nature of the α (+) subunit and the discovery and basis of four α isoforms came from cDNA cloning. In particular, Shull et al. [6] screened a brain cDNA library and found three unique clones corresponding to different α isoforms of the Na,K-ATPase. Similar findings were reported by other laboratories [7–9]. One of these cDNA clones corresponded to the renal or α1 isoform that had been cloned by Shull et al. [10] earlier. Of the two other cDNA clones one corresponded to the α(+) isoform and was termed α2. The third cDNA clone represented a new isoform and was named α3. Subsequently a fourth α isoform, α4, was discovered, again by cDNA cloning [11].

The α1 isoform appears to be ubiquitously expressed while the α2 isoform is limited to fewer tissues, including brain, skeletal and smooth muscle, heart, adipocytes, osteoblasts, retina, chororid plexus, lung, and optic nerve [12–22]. This isoform is also found at low levels in other tissues such as erythrocytes [23]; however, the functional significance is unknown. The α3 isoform is limited largely to neural tissue [12, 16–22] in most animals but is present in the heart of humans [24]. The α4 isoform appears limited to spermatogonia and sperm [11, 25, 26].

16.3
β Isoforms

Three β isoforms occur in mammalian cells. These are the ubiquitous isoform β1 [27–29], the β2 isoform, which is expressed mostly in neural tissue, and the β3 isoform, which is more widely expressed. The β2 isoform was originally identified as an adhesion protein that facilitates interaction between neurons and astrocytes in culture and was initially called AMOG (adhesion molecule on glia) [30]. This protein was shown to associate with the Na,K-ATPase [31] and it became the second identified β isoform.

Expression of the β2 isoform in L cells, and using these cells as substrates for neurite outgrowth of cerebellar and hippocampal neurons, demonstrated that this isoform promoted an increase in neurite length [32]. Expression of the β1 isoform did not promote neurite extension. These studies demonstrate that the β1 and β2 isoforms play different roles, at least with respect to signaling pathways. The amino acid sequences of the β1 and β2 isoforms are similar with 40% identity [31, 33].

The β3 isoform was first identified in *Xenopus* [34] and then cloned from human and rat [35]. The human β3 isoform shows 38% and 48% amino acid identity with the human β1 and β2 isoforms [35]. The β3 isoform gene is expressed in several tissues, including testes, brain, kidney, lung, heart, stomach, small intestine, colon, skeletal muscle, spleen, and liver in rat [35], and immunological studies revealed that the β3 isoform protein is expressed in lung, testes, liver, skeletal muscle, kidney, heart, and brain of rat [36].

16.4
Physiological Role of Na,K-ATPase Isoforms

16.4.1
Duplication and Divergence of α and β Isoform Genes

During evolution the α and β isoforms of the Na,K-ATPase, each of which is coded for by a single gene, most likely arose by gene duplication. The α isoform genes of both human and mouse are shown in Figure 16.1. In humans the α1 isoform gene is located on chromosome 1p, α2 on 1q, α3 on chromosome 19 and α4 on chromosome 1q. In mice the α1 isoform gene is located on chromosome 3, α2 on chromosome 1, α3 on chromosome 7 and α4 on chromosome 1. The α2 and 4 genes are found directly next to one another on chromosome 1 and this arrangement has been conserved in mammals. Each α isoform gene consists of 23 exons and the

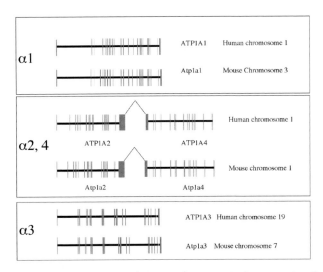

Figure 16.1 Exons 3–22 in human and mouse α isoforms are virtually identical in size. Scale for the above genes is arbitrary, with variation in absolute gene size ranging from 23–35 kB. The last exon, number 23, varies in length within the 3′ untranslated region but all encode the final 9 amino acids. The first two exons vary in size but maintain a consistent pattern with a large intron (4–10 kbp) in between. Maps were drawn from data available from Ensembl web site [98].

intron exon arrangement is highly conserved between mouse and human. The β1, β2 and β3 genes are located on chromosomes1, 17 and 3, respectively, for humans, and on 1, 11 and 7 for mouse. The 7-exon arrangement of the β genes is also highly conserved.

Conservation of these genes must have conferred an advantage for survival. One possibility is that as an early α1 subunit gene duplicated the resulting genes were conserved to provide sufficient Na,K-ATPase activity for transport. That is, if a particular tissue needed more Na,K-ATPase this could be accomplished by expressing two genes rather than a single gene. This does not appear likely as one of the tissues which requires the greatest amount of Na,K-ATPase, the kidney, expresses only one α isoform gene, namely the α1 isoform gene.

Another possibility is that the duplicated genes came under different regulation, either tissue specific expression or developmental regulation, and some tissues utilized a single gene while others used a combination of α isoform genes to provide the correct amount of Na,K-ATPase activity. Multiple α and β genes could make it easier to regulate the levels of Na,K-ATPase by hormones, cations, etc. In this case different α or β isoform genes could respond to different effectors or a combination of effectors rather than the complex regulation of Na,K-ATPase built into a single α or β subunit gene. Thus regulatory elements could vary among the isoform genes to provide precise expression in various cells and during development. This hypothesis in its simplest form would assume that the kinetic properties of the Na,K-ATPase remain unchanged and that the complex regulation of the isoforms provides the correct amount of Na,K-ATPase in that particular tissue or at a specific development stage. It is almost certain that organisms have taken advantage of the multiple α and β isoform genes to regulate the amount of Na,K-ATPase, as the duplicated genes show differences in developmental and tissue specific regulation as well as the response to hormones and other affectors.

It is possible that during evolution amino acid substitutions occurred that altered the enzymatic properties of the Na,K-ATPase and these differences have been conserved as they offer some advantage to a particular cell or can respond to a particular need. Such properties could include cation affinity, ATP affinity, ligand affinity, cellular location, modulation by covalent modification, etc. Most likely, differences in gene regulation and altered enzymatic activities evolved together to provide both fine regulation of the amount of Na,K-ATPase as well as the enzymatic properties of Na,K-ATPase produced.

16.4.2
Enzymatic Properties of α and β Isoform Genes

Several studies have shown that there are differences in the Na^+, K^+, ATP, and cardiac glycoside affinity between the different isoforms. Expression of the α1, α2, and α3 isoforms of the rat Na,K-ATPase in HeLa cells has shown that the Na^+ affinity of enzyme containing the α1 and α2 isoforms is approximately equal and higher than the enzyme carrying the α3 isoform, while the K^+ affinity of the α1 isoform is higher than the α2 and α3 Na,K-ATPase, which are approximately equal [37, 38].

Similar results have been obtained by Blanco and Mercer [39] who have expressed the rat α isoforms in insect cells. Here various combinations of α and β isoforms were introduced and the kinetic properties of each of these Na,K-ATPases determined. In general the Na^+ affinity of the α2 Na,K-ATPase is higher than α1 which is higher than the α3 isoform. The K^+ affinity of the α1 isoform Na,K-ATPase is higher than that containing α2 and α3 isoforms.

Horisberger and Kharoubi-Hess [40] have expressed the rat α isoforms in *Xenopus* oocytes and have found generally similar but somewhat different results than observed in HeLa and insect cells. In this case, the Na^+ affinity of the α2 isoform is higher than that of the α1 and α3 isoforms, while the K^+ affinity of α3 isoform is higher than that of the α2 and α3 isoforms, which are approximately equal. These studies reveal that the α2 isoform is more sensitive to voltage-dependent inhibition by Na^+. It is postulated that the activity of the α2 isoform would increase approximately three-fold between the resting potential and depolarized potential and therefore may play a unique role in heart, skeletal muscle, neurons, and glial cells as they become depolarized following depletion of their ionic stores.

Various combinations of the human α and β isoforms have been expressed in *Xenopus* oocytes and the α2 β2 enzyme exhibits a lower apparent K^+ and Na^+ affinity compared with the other combinations of isoforms [41]. Similar findings were observed with the human isoforms expressed in yeast [42]. This expression system also demonstrated that the human α1, α2 and α3 isoforms confer approximately equal affinities for ouabain. Although the β subunit lacks catalytic activity itself, this subunit clearly plays a role in determining Na^+ and K^+ affinities [39, 43, 44], along with its role of delivering the α subunit to the plasma membrane [45, 46].

Because the systems and assay conditions vary among the studies the differences observed may not always reflect intrinsic properties of the isoforms. In addition, there are differences in cation affinities with the same isoform measured in different cells [47]. Never the less there seems to be differences in cation affinities among the Na,K-ATPases.

Of particular interest is the difference in turnover observed among the α isoforms [38]. The maximum turnover rate of α2β1 Na,K-ATPase is approximately one-half that observed for α1β1 Na,K-ATPase. The significance of this difference must be viewed in relation to the amounts of each enzyme in a particular cell.

Additional studies are required to determine if these differences play a physiological role. One could argue, however, that under conditions where a higher Na^+ affinity of Na,K-ATPase is required, a particular combination of α and β isoforms would be expressed. Unfortunately, the actual kinetic parameters for a particular Na,K-ATPase isoform combination in a specific cell, under the conditions occurring in that cell, are unknown. Never the less, it seems reasonable that the differences in the various isoform specific Na,K-ATPases will be found to play a physiological role in particular tissues.

Little work has yet been carried out on the α4 isoform, although indirect studies using ouabain competition show that the Na^+ and K^+ affinities are in line with the other α isoforms [25]. Therefore, this isoform is likely to exhibit similar characteristics in Na^+ or K^+ affinity compared with that of the other rat α isoforms.

16.4.3
Differential Expression of α and β Isoform Genes

The α and β isoforms are regulated differentially by a variety of affectors, including hormones. Thyroid hormone regulates both the expression of the α and β isoform genes and posttranscriptional regulation occurs as well [48–51]. Alteration in both mRNA and protein levels in various rat tissues was observed following induction of a hypothyroid state. The α2 isoform mRNA and protein decreased significantly in most muscle tissues, but the non-equivalent decrease in mRNA and protein suggest both transcriptional and translational mechanisms play a role. This is also true for the β isoform where the β1 isoform protein decreases in the hypothyroid state while mRNA levels remain unchanged.

Similarly aldosterone, which acts both at the transcriptional level and protein level, exerts its affect in an isoform-specific manner. Of particular interest is the aldosterone-mediated increase in the activity of the rat Na,K-ATPase α1 isoform, which does not occur with the α2 Na,K-ATPase [52]. In other studies, aldosterone has been shown to up-regulate the expression of the α3 isoform gene in dendate gyrus cells of the hippocampus, while not increasing expression in the α1 or α2 isoforms. Possibly, aldosterone will have its affect largely on specific neuronal populations while the α1 isoform is regulated by aldosterone only in epithelial cells [53].

Hypokalemia also reduces the levels of α2β2 Na,K-ATPase selectively [54]. This suggests that the α2β2 Na,K-ATPase plays an important role in conserving K^+ in muscle following prolonged contractions.

For the α1 isoform, dopamine-induced phosphorylation of the rat α1 isoform initiates endocytosis in kidney epithelial cells [55], thus causing a decrease in total plasma membrane Na,K-ATPase. In contrast, dopamine induces exocytosis of the Na,K-ATPase in lung epithelial cells [56]. Enzyme containing both α1 and α2 isoforms is involved in insulin-stimulated translocation of the Na,K-ATPase [57]. Whether tissues with more than one isoform translocate their isoform specific Na,K-ATPases differentially is unknown.

Protein kinase regulation of Na,K-ATPase is varied [58–63] and appears to differ from tissue to tissue [39]. Different protein kinases also act differently on the Na,K-ATPase. Most studies to date have concentrated on the α1 isoform Na,K-ATPase and studies are only beginning on the other isoforms.

These and other studies point to the differential regulation of the α and β isoforms of the Na,K-ATPase but the significance of these findings is not entirely clear. The regulation observed may simply represent the manner of producing the right amount of Na,K-ATPase in a specific tissue without regard to intrinsic differences in enzymatic activity of the enzyme. Alternatively, hormones and other affectors may be directly regulating isoforms with a specific enzymatic property to cope with particular transport requirements at a specific time or in a specific tissue.

16.5
Non-transport Function of the Na,K-ATPase

While transport of Na$^+$ and K$^+$ is considered the major function of the Na,K-ATPase, other roles unrelated to transport have been suggested. Exposure of cells to low concentrations of ouabain, which do not alter intracellular ion concentrations, activates signaling pathways resulting in increased cell growth [64–68]. Low concentrations of ouabain activate two Ras-dependent signaling cascades [69–72], the first includes Ras-induced generation of reactive oxygen species from mitochondria leading to activation of NF-κB. The second is a ouabain induced Ras-dependent pathway that activates the Raf/MEK/MAPK cascade.

It is unknown whether ouabain induces these signaling pathways through a specific α isoform but as the effect occurs in kidney and HeLa cells, which only express the α1 isoform, signaling must act through this isoform [68, 69]. Ouabain-induced signaling and cell growth are also observed in cardiac myocytes and vascular smooth muscle where α1 and α2 isoforms are expressed. Thus, it is possible that both α isoforms of the Na,K-ATPase can induce ouabain mediated signaling, but this is yet to be established. The effect of ouabain on signaling pathways is of pharmacological importance as the digitalis class of drugs might play a role in cardiac hypertrophy using these unique signaling functions of the Na,K-ATPase.

16.6
In Vivo Studies of Differential Isoform Function

The differential regulation of isoforms in a tissue specific and developmental regulated manner as well as changes which occur in response to hormones, ionic changes, etc. are suggestive of isoforms playing differential roles. Such studies can be interpreted as (1) a way of regulating the total amount of Na,K-ATPase required at a particular time, (2) delivering a Na,K-ATPase enzyme having a qualitative difference in enzymatic properties, (3) an enzyme capable of responsiveness to changes in modifications such as phosphorylation, or (4) changes in cellular localization.

16.6.1
Gene Knock-out of the β2 Isoform

One approach for determining the unique function of a particular isoform is to develop animals lacking the gene coding for it. This has been accomplished with the β2 isoform gene of the Na,K-ATPase [73, 74]. In this case animals lacking the β2 isoform lack motor coordination early in life and develop subsequent tremor and paralysis of the extremities during the later neonatal period, with the animals dying around 17 to 18 days following birth. Several morphological abnormalities are observed in the central nervous system of these animals, including enlarged ventricles and swollen astrocytic end feet in the brain stem, thalamus, and spinal cord. In addition, there is photoreceptor cell death in the retina in the second week after birth. Such studies suggest a unique function for the β2 isoform, but another explanation may be the lack of sufficient Na,K-ATPase to the particular cell type or tissue.

16.6.2
Gene Replacement of the β2 Isoform with β1

Another approach for answering the question of isoform function is to replace one isoform with another. This can be accomplished by using gene replacement and this approach has been used to determine whether the function of the β2 isoform of the Na,K-ATPase can be replaced by the β1 isoform [75]. These studies utilize so-called knock-in technology where the coding region, including both introns and exons, of the β2 isoform gene is substituted with the β1 isoform cDNA.

The β2 isoform replacement targeting vector used two pieces of DNA for recombination, one homozygous piece of DNA downstream of the coding region on the 3' end and the other a region on the 5' end of the gene. This latter fragment produced a targeted gene with a small amount of β2 being retained in the β1 isoform replacement. This produced a chimeric protein with 18 amino acids of the β2 isoform protein at the amino terminus of the β1 isoform. This fusion protein, which is under regulation of the mouse β2 isoform gene promoter, rescues animals lacking the β2 isoform gene. These animals do not exhibit the juvenile lethality observed previously. The overall morphological structure of the brain was normal; however, progressive degeneration of photo-receptor cells occurred. Thus, while degeneration of photoreceptor cells was reduced in the β2 deficient animals, substitution by β1 did not completely restore this defect. These data indicate that the β1 isoform can substitute reasonably well for the β2 isoform, although not completely. In these studies only about 15 % of the normal expression of the β2 isoform gene occurred, raising the possibility that if levels of β1 expression comparable to that of the β2 isoform normally produced had been attained the β2 deficient phenotype could have been completely corrected by the introduction of the β1 isoform. It would have been assumed that because the sequences surrounding the β1 coding sequences are that of the β2 isoform, the β1 isoform would have had similar expression to that of the β2 isoform. It is possible that regulatory regions occur in the introns of the β2 genes which are lacking in the replacement gene. The small amount of β2 isoform conserved in the chimeric protein may have affected the outcome of the studies, although this seems unlikely.

16.6.3
Overexpression of the α2 Isoform

A recent study used overexpression of the α2 isoform in lung to study its role in lung liquid clearance [76]. Detailed immunohistochemistry revealed the α2 isoform is abundantly expressed in Type I cells in mouse lung, whereas it was previously thought the α1 isoform provided most of the Na,K-ATPase activity. This is significant because type I cells are thought to provide most of the fluid movement in the lung. Using an adenoviral transfection system, the overproduction of the α2 isoform in lung correlated with increased fluid clearance. This approach has revealed a novel role for the Na,K-ATPase α2 isoform involved in fluid clearance, with obvious implications in treatment of lung disease.

16.6.4
Gene Knock-out of the α1 and α2 Isoform Genes

Gene knockout studies have also been utilized to address the functional differences in the α isoform genes. James et al. [77] have developed mice lacking either one or both copies of the gene coding for the α1 and α2 isoforms of the Na,K-ATPase. Animals lacking both copies of the α1 isoform are not born and die in early embryonic development. Previous studies have shown that using ouabain to inhibit Na,K-ATPase activity will block blastocyst formation which occurs at embryonic day 3.5 [78]. While the concentrations of ouabain used would inhibit the α1 insensitive isoform, which is the main isoform expressed in mouse at this stage [79], it is clear that the activity of the Na,K-ATPase α1 isoform is essential for driving water movement during the formation of blastocoel cavity [80, 81]. Mice lacking both copies of the α1 isoform cannot be found as early as embryonic day 8.5 and die sometime before then (Moseley and Lingrel, unpublished).

Mice lacking both copies of the α2 isoform, unlike the α1 isoform-deficient mice, develop normally through all stages of embryonic growth and are born. However, these mice die immediately after birth even though they lack any apparent gross morphological abnormalities. Further examination of these mice revealed the cause of death is primarily due to asphyxia [82]. At birth, the α2 isoform is expressed in diaphragm and brain. Interestingly, in wild-type mice high levels of expression of the α2 isoform in neurons were observed throughout the brain at this time in development and lower levels in astrocytes. This finding of α2 isoform expression in neurons has uncovered a previously unrecognized role of α2 isoform activity directly in neurons. The absence of α2 isoform expression in the brain altered neural activity, particularly in the respiratory center (Pre-Bötzinger Complex), since these animals do not breathe. Both depressed amplitude and altered periodicity of synchronized neural firing occurred in the respiratory center and this could contribute the asphyxia in mice that lack the α2 isoform. Since the α2 isoform was expressed in neurons and astrocytes throughout the brain, and not limited to Pre-Bötzinger Complex, in wild-type mice, it is possible that the neuronal deficiency observed extends into most regions of the brain. It is also unknown whether the lack of synchronized firing of the neurons is due to a direct defect of neurons or astrocytes or both. Interestingly, no compensatory changes in expression occurred with the absence of the α2 isoform, but clearly the α1 and/or α3 isoforms remaining in the cell cannot support the Na,K-ATPase activity required. While such studies can be interpreted as a unique function of the α2 isoform in brain, it is uncertain whether this has to do with a particular property of the α2 isoform or relates to the total amount of Na,K-ATPase activity.

Studies of diaphragm function were also performed in mice lacking the α2 isoform. While maximal force contraction is normal [83] the rate of relaxation is altered in mice homozygous for the α2 isoform (Moseley, Paul, Lingrel, Heiny, unpublished). Although the mechanism for altered relaxation is not clear, relaxation in muscle occurs as a result of removal of cytosolic calcium via several transporters; the sarcoplasmic reticulum Ca-ATPase, which transports calcium into the sarcoplasmic reticulum, as well as by several plasma membrane transporters which transport Ca^{2+} out of the cell, the plasma membrane Ca-ATPase and the Na/Ca exchanger. Several studies suggest a functional coupling of the Na,K-ATPase α2 isoform with the Na/Ca exchanger. Future studies are required to determine the mechanism by which the absence of the α2 isoform alters calcium transport in the diaphragm muscle.

Animals lacking only one copy of the α1 isoform or α2 isoform gene of the Na,K-ATPase are viable, appear normal, and produce normal numbers of offspring. However, a phenotype is observed in both heart and skeletal muscle with respect to contractility [77, 83]. Animals lacking one copy of the α2 isoform gene express approximately one-half of the α2 isoform protein compared with wild-type animals. The hearts of these animals exhibit an increase in the force of contraction as assayed using a working heart preparation. This is expected based on the known action of cardiac glycosides, i. e. when the Na,K-ATPase is inhibited by these compounds intracellular Na^+ increases resulting in an increase in intracellular Ca^{2+} via the Na/Ca exchanger. The increase in Ca^{2+} increases the force of contraction. An increase in transient Ca^{2+} levels in cardiac myocytes isolated from these animals was indeed observed compared with myocytes from wild-type animals.

Hearts from animals lacking one copy of the α1 isoform gene have approximately one-half of the corresponding protein but show an opposite phenotype, namely a reduction in the force of cardiac contraction. Interestingly, when ouabain is administered to these animals at concentrations which only inhibit the α2 isoform, the contractibility is increased, demonstrating that the α2 isoform plays a key role in regulating cardiac contractility, while the α1 isoform must have some other function, possibly a general transport role. The possibility that the hypocontractility observed in the hearts of animals lacking one copy of the α1 isoform gene is the result of Ca^{2+} overload is not responsible for the decreased force of contraction, as lower Ca^{2+} levels do not correct the reduced contractile force (James, Cougnon, Moseley and Lingrel, unpublished data).

A possible explanation for the differential function of the α1 and α2 isoforms may relate to different cellular localizations of the two isoforms. Juhaszova and Blaustein [84] have shown that the α2 isoform localizes to microdomains within the plasma membrane in close proximity to the sacrcoplasmic reticulum, while the α1 isoform is evenly distributed within the plasma membrane (illustrated in Figure 16.2). The Na/Ca exchanger localizes to similar domains as the α2 isoform [85]. It is possible that the close proximity of the α2 isoform and the Na/Ca exchanger, both of which are near the underlying sarcoplasmic reticulum, regulates Ca^{2+} concentrations related to muscle contraction while the α1 isoform maintains ionic concentrations in the cell as a whole.

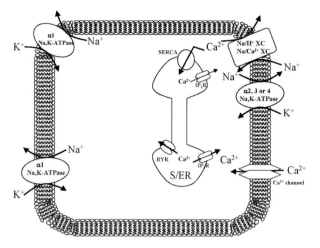

Figure 16.2 Model of α isoform function. Na,K-ATPase isoforms may provide unique functional roles in cells based on their differential localization in the plasma membrane and colocalization with other transporters such as Ca^{2+} handling transporters.

Additional studies have been carried out using cultured astrocytes from animals lacking the α2 isoform gene [86] and have demonstrated that, while the lack of both copies of the α2 isoform has only a small affect on total cytoplasmic Na^+ concentrations, a significant elevation in both resting cytoplasmic Ca^{2+} concentration as well as elevated stores of calcium occurs. This again supports the model that the Na,K-ATPase α2 isoform plays a role in regulating calcium levels.

A difference in α1 and α2 isoform function in heart has also been presented as related to the differential location of these two isoforms [87]. Inhibition of the α1 isoform, which occurs in endocardial endothelial cells, may contribute to negative inotrophy, while inhibiting the myocyte α2 isoform causes a positive inotrophy.

A differential response of skeletal muscle contractility also exists in mice heterozygous for the α1 or α2 isoforms. Extensor digitalis longus muscle from animals lacking one copy of either the α1 or α2 isoform genes show a similar pattern of contraction [83]. The force of contraction is lower in animals lacking one copy of the α1 isoform gene but a stronger force is observed in muscle lacking one copy of the α2 isoform gene. Thus, the differential role of the α1 and α2 isoforms in muscle may be a general phenomenon.

The α4 isoform of the Na,K-ATPase is found in spermatogonia and sperm [25, 26] and is required for sperm motility [88, 89]. When sperm are treated with ouabain, at concentrations that inhibit the α4 Na,K-ATPase but not the α1 Na,K-ATPase, motility is inhibited. In contrast, the administration of nigericin, a H/K ionophore, and monesin, a H/Na ionophore, reinitiates motility in ouabain-inhibited sperm. These studies are compatible with the ouabain inhibiting proton accumulation. Possibly, hydrogen ions accumulating from mitochondria inhibit sperm motility if they are not removed. Sperm motility is quite sensitive to pH and without removal of protons motility would be inhibited. It is possible that the α4 isoform functions in concert with the Na/H exchanger to remove accumulated protons.

16.7
Model for Differential Function of the α1 Isoform and the α2, α3 and α4 Isoforms

Studies of the α2 and α4 isoforms, and possibly the α3 isoform, are compatible with the hypothesis that the Na,K-ATPase isoforms play specialized roles in cells. The α2 isoform may be functionally coupled to the Na/Ca exchanger, and thus play a critical role in regulating Ca^{2+} levels in small intracellular compartments (Figure 16.2). There is evidence that the α4 Na,K-ATPase is coupled to the Na/H exchanger, at least in sperm [88, 89]. Thus the α2, α4 and possibly the α3 isoforms of the Na,K-ATPase may be functionally coupled to exchangers, revealing an important functional aspect of each of these isoforms. Further work will be required along these lines to determine whether both co-localization and functional interaction exists.

16.8
Na,K-ATPase Isoforms and Disease

16.8.1
Essential Hypertension and the α1 Isoform

The α1 isoform is the most abundant isoform in the kidney, providing the majority of Na,K-ATPase activity. The Dahl salt-sensitive and salt-insensitive rat is an animal model for essential hypertension, displaying abnormal Na^+ handling in the kidney [90]. These rats contain an α1 isoform with a gln-276-leu mutation resulting in an enzyme with reduced K^+ transport capability [91, 92]. Re-expression of the wild-type α1 isoform, introduced into the Dahl salt-sensitive rat using a transgenic approach, produced rats with less salt-sensitive hypertension and reduced hypertensive renal disease compared with non-transgenics [93]. In the same study, genetic linkage analysis of the Dahl salt-sensitive rats showed that the α1 isoform gene co-segregated with salt-sensitive hypertension. In fact, a recent report identified that the Na,K-ATPase α1 isoform and the Na,K-2Cl co-transporter interactively increase the Dahl salt-sensitive hypertension [94]. Interestingly, a similar mechanism was identified in human essential hypertension in that same study.

16.8.2
Migraine Headache and the α2 Isoform

Na,K-ATPase is abundantly expressed in the brain, where three α isoforms are expressed: the α1, α2 and α3 isoforms. The α2 isoform is expressed mainly in astrocytes, whereas the α3 isoform is almost exclusively expressed in neurons in adults

[20, 95, 96]. A recent report describes two different human mutations in the α2 iso-form gene in which either mutation independently segregates with familial hemi-plegic migraine-2 [97]. These individuals are heterozygous for the mutation, carry-ing one normal copy and one mutated copy of the α2 isoform gene. Although the protein is translated and inserted into the plasma membrane, it is not functionally active. Interestingly, although the phenotype of humans carrying the mutation manifest with neurological problems, there are no overt defects observed in cardiac or skeletal muscle performance, as these tissues also abundantly express the α2 iso-form. The mechanisms by which the absence of one copy of the α2 isoform con-tributes to the pathophysiology is unclear. However, the identification of altered Na,K-ATPase activity in migraine disease opens new avenues for exploring the etiology and ultimate treatment of this disease.

Acknowledgments

This work was supported by NIH Grants RO1 HL 28573 and R01 HL 66062.

References

1. A. G. Thieren, H. X. Pu, S. J. Karlish, and R. Blostein, *J. Bioenerg. Biomembr.*, **2001**, *33*, 407–414.

2. G. Crambert and K. Geering, *Sci STKE* **2003**, REI, 1–9.

3. K. J. Sweadner, *J. Biol. Chem.*, **1979**, *254*, 6060–6067.

4. J. Lytton, *Biochem. Biophys. Res. Com-mun.*, **1985**, *132*, 764–769.

5. J. Lytton, *J. Biol. Chem.*, **1985**, *260*, 10075–10080.

6. G. E. Shull, J. Greeb, and J. B. Lingrel, *Biochemistry*, **1986**, *25*, 8125–8132.

7. V. L.M. Herrera, J. R. Emanuel, N. Ruiz-Opazo, R. Levenson, and B. Nadal-Ginard, *J. Cell. Biol.*, **1987**, *105*, 1855–1865.

8. Y. Hara, O. Urayama, K. Kawakami, H. Nojima, H. Nagumune, T. Kojima, T. Ohta, K. Nagano, and M. Nakao, *J. Biochem.*, **1987**, *102*, 43–58.

9. J. W. Schneider, R. W. Mercer, M. Caplan, J. R. Emanuel, K. J. Sweadner, E. J. Benz, Jr., and R. Levenson, *Proc. Natl. Acad. Sci. U. S.A.*, **1985**, *82*, 6357–6361.

10. G. E. Shull, A. Schwartz, and J. B. Lingrel, *Nature*, **1985**, *316*, 691–695.

11. O. L. Shamraj and J. B. Lingrel, *Proc. Natl. Acad. Sci. U. S.A.*, **1994**, *91*, 12952–12956.

12. J. W. Schneider, R. W. Mercer, M. Gil-more-Herbert, M. F. Utset, C. Lai, A. Greene, and E. Benz, Jr., *Proc. Natl. Acad. Sci. U. S.A.*, **1988**, *85*, 284–288.

13. J. R. Emanuel, S. Garetz, L. Stone, and R. Levenson, *Proc. Natl. Acad. Sci. U. S.A.*, **1987**, *84*, 9030–9034.

14. J. Lytton, J. C. Lin, and G. Guidotti, *J. Biol. Chem.*, **1985**, *260*, 1177–1184.

15. M. J. Francis, R. L. Lees, E. Trujillo, P. Martin-Vasallo, J. N. Heersche, and A. Mobasheri, *Int. J. Biochem. Cell. Biol.*, **2002**, *34*, 459–476.

16. P. S. Biser, K. A. Thayne, J. Q. Kong, W. W. Fleming, and D. A. Taylor, *Brain Res. Dev. Brain Res.*, **2000**, *123*, 165–172.

17. A. Hibiba, G. Blanco, and R. W. Mercer, *Brain Res.*, **2000**, *875*, 1–13.

18. R. K. Wetzel, E. Arystarkhova, and K. J. Sweadner, *J. Neurosci.*, **1999**, *19*, 9878–9889.

19. B. V. Zlokovic, J. B. Mackic, L. Wang, J. G. McComb, and A. McDonough, *J. Biol. Chem.*, **1993**, *268*, 8019–8025.

20. K. M. McGrail, J. M. Phillips, and K. J. Sweadner, *J. Neurosci.*, **1991**, *11*, 381–391.

21. K. M. McGrail and K. J. Sweadner, *Eur. J. Neurosci. Assoc.*, **1990**, *2*, 170–176.

22. J. Orlowski and J. B. Lingrel, *J. Biol. Chem.*, **1988**, *263*, 10436–10442.

23. M. K. Stengelin and J. F. Hoffman, *Proc. Natl. Acad. Sci. U. S.A.*, **1997**, *94*, 5943–5948.

24. O. I. Shamraj, D. Melvin, and J. B. Lingrel, *Biochem. Biophys. Res. Commun.*, **1991**, *179*, 1434–1440.

25. A. L. Woo, P. F. James, and J. B. Lingrel, *J. Membr. Biol.*, **1999**, *169*, 39–44.

26. G. Blanco, G. Sanchez, R. J. Melton, W. G. Tourtellotte, and R. W. Mercer, *J. Histochem. Cytochem.*, **2000**, *48*, 1023–1032.

27. S. Noguchi, M. Noda, H. Takahashi, K. Kawakami, T. Ohta, K. Nagano, T. Hirose, S. Inayama, M. Kawamura, and S. Numa, *FEBS Lett.*, **1986**, *196*, 315–320.

28. K. Kawakami, H. Nojima, T. Ohta, and K. Nagano, *Nucl. Acids Res.*, **1986**, *14*, 2833–2844.

29. G. E. Shull, L. K. Lane, and J. B. Lingrel, *Nature*, **1986**, *321*, 429–431.

30. H. Antonicek, E. Persohn, and M. Schachner, *J. Cell. Biol.*, **1987**, *104*, 1587–1595.

31. S. Gloor, H. Antonicek, K. J. Sweadner, S. Pagliusi, R. Frank, M. Moos, and M. Schachner, *J. Cell. Biol.*, **1990**, *110*, 165–174.

32. G. Muller-Husmann, S. Gloor, and M. Schachner, *J. Biol. Chem.*, **1993**, *268*, 26260–26267.

33. P. Martin-Vasallo, W. Dackowski, J. R. Emanuel, and R. Levenson, *J. Biol. Chem.*, **1989**, *264*, 4613–4618.

34. P. J. Good, K. Richter, and I. B. Dawid, *Proc. Natl. Acad. Sci. U. S.A.*, **1990**, *87*, 9088–9092.

35. N. Malik, V. A. Canfield, M. C. Beckers, P. Gros, and R. Levenson, *J. Biol. Chem.*, **1996**, *271*, 22754–22758.

36. E. Arystarkhova and K. J. Sweadner, *J. Biol. Chem.*, **1997**, *272*, 22405–22408.

37. E. A. Jewell and J. B. Lingrel, *J. Biol. Chem.*, **1991**, *266*, 16925–16930.

38. L. Segall, S. E. Daly, and R. Blostein, *J. Biol. Chem.*, **2001**, *276*, 31535–31541.

39. G. Blanco and R. W. Mercer, *Am. J. Physiol.*, **1998**, *275*, 633–650.

40. J. D. Horisberger and S. Kharoubi-Hess, *J. Physiol.*, **2002**, *539*, 669–680.

41. G. Crambert, U. Hasler, A. T. Beggah, C. L. Yu, N. N. Modyanov, J. D. Horisberger, L. Lelievre, and K. Geering, *J. Biol. Chem.*, **2000**, *275*, 1976–1986.

42. J. Muller-Ehmsen, P. Juvvadi, C. B. Thompson, L. Tumyan, M. Croyle, J. B. Lingrel, R. H. Schwinger, A. A. McDonough, and R. A. Farley, *Am. J. Physiol. Cell. Physiol.*, **2001**, *281*, 1355–1364.

43. S. Lutsenko and J. H. Kaplan, *Biochemistry*, **1993**, *32*, 6737–6743.

44. K. A. Eakle, M. A. Kabalin, S. G. Wang, and R. A. Farley, *J. Biol. Chem.*, **1994**, *269*, 6550–6557.

45. A. A. McDonough, K. Geering, and R. A. Farley, *FASEB J.*, **1990**, *4*, 1598–1605.

46. K. Geering, *FEBS Lett.*, **1991**, *285*, 189–193.

47. A. G. Therien, N. B. Nestor, W. J. Ball, and R. Blostein, *J. Biol. Chem.*, **1996**, *271*, 7104–7112.

48. J. Orlowski and J. B. Lingrel, *J. Biol. Chem.*, **1990**, *265*, 3462–3470.

49. K. J. Sweadner, K. M. McGrail, and B. A. Khaw, *J. Biol. Chem.*, **1992**, *267*, 769–773.

50. B. Horowitz, C. B. Hensley, M. Quintero, K. K. Azuma, D. Putnam, and A. A. McDonough, *J. Biol. Chem.*, **1990**, *265*, 14308–14314.

51. C. B. Book, X. Sun, and Y. C. Ng, *Biochim. Biophys. Acta*, **1997**, *1358*, 172–180.

52. R. Pfeiffer, J. Beron, and F. Verrey, *J. Physiol.*, **1999**, *516*, 647–655.

53. N. Farman, J. P. Bonvalet, and J. Seckl, *Am. J. Physiol. Cell. Physiol.*, **1994**, *266*, C423-C428.

54. C. B. Thompson and A. A. McDonough, *J. Biol. Chem.*, **1996**, *271*, 32653–32658.

55. A. V. Chibalin, G. Ogimoto, C. H. Pedemonte, T. A. Pressley, A. I. Katz, E. Feraille, P. O. Berggren, and A. M. Bertorello, *J. Biol. Chem.*, **1999**, *274*, 1920–1927.

56. K. M. Ridge, L. Dada, E. Lecuona, A. M. Bertorello, A. I. Katz, D. Mochly-Rosen, and J. I. Sznajder, *Mol. Biol. Cell.*, **2002**, *13*, 1381–1389.

57. L. Al-Khalili, M. Yu, and A. V. Chibalin, *FEBS Lett.*, **2003**, *536*, 198–202.

58. P. L. Jorgensen, *Cell. Mol. Biol.*, **2001**, *47*, 231–238.

59. G. Blanco, G. Sanchez, and R. W. Mercer, *Arch. Biochem. Biophys.*, **1998**, *359*, 139–150.

60. M. S. Feschenko, E. Stevenson, and K. J. Sweadner, *J. Biol. Chem.*, **2000**, *275*, 34693–34700.

61. E. Feraille, P. Beguin, M. L. Carranza, S. Gonin, M. Rousselot, P. Y. Martin, H. Favre, and K. Geering, *Mol. Biol. Cell.*, **2000**, *11*, 39–50.

62. K. A. Buhagiar, P. S. Hansen, N. L. Bewick, and H. H. Rasmussen, *Am. J. Physiol. Cell. Physiol.*, **2001**, *281*, 1059–1063.

63. J. M. Capasso, C. J. Rivard, and T. Berl, *Am. J. Physiol. Renal Physiol.*, **2001**, *280*, F768-F776.

64. G. Scheiner-Bobis and W. Schoner, *Nat. Med.*, **2001**, *7*, 1288–1289.

65. Z. Xie and A. Askari, *Eur. J. Biochem.*, **2002**, *269*, 2434–2439.

66. P. Kometiani, J. Li, L. Gnudi, B. B. Kahn, A. Askari, and X. Xie, *J. Biol. Chem.*, **1998**, *273*, 15249–15256.

67. A. Xie, P. Kometiani, J. Lui, J. Li, J. I. Shapiro, and A. Askari, *J. Biol. Chem.*, **1999**, *274*, 19323–19328.

68. M. Haas, A. Askari, and Z. Xie, *J. Biol. Chem.*, **2000**, *275*, 27832–27837.

69. O. Aizman, P. Uhlen, M. Lal, H. Brismar, and A. Aperia, *Proc. Natl. Acad. Sci. U.S.A.*, **2001**, *98*, 13420–13424.

70. J. Tian, X. Gong, and Z. Xie, *Am. J. Physiol. Heart Circ. Physiol.*, **2001**, *281*, H1899–H1907.

71. A. Aydemir-Koksoy, J. Abramowitz, and J. C. Allen, *J. Biol. Chem.*, **2001**, *276*, 46605–46611.

72. M. Haas, H. Wang, J. Tian, and Z. Xie, *J. Biol. Chem.*, **2002**, *277*, 18694–18702.

73. J. P. Magyar, U. Bartsch, Z. Q. Wang, N. Howells, A. Aguzzi, E. F. Wagner, and M. Schachner, *J. Cell. Biol.*, **1994**, *127*, 835–845.

74. M. Molthagen, M. Schachner, and U. Bartsch, *J. Neurocytol.*, **1996**, *25*, 243–255.

75. P. Weber, U. Bartsch, M. Schachner, and D. Montag, *J. Neurosci.*, **1998**, *18*, 9192–9203.

76. K. M. Ridge, W. G. Olivera, F. Saldias, Z. Azzam, S. Horowitz, D. H. Rutschman, V. Dumasius, P. Factor, and J. I. Sznajder, *Circ. Res.*, **2003**, *92*, 453–460.

77. P. F. James, I. L. Grupp, G. Grupp, A. L. Woo, G. R. Askew, M. L. Croyle, R. A. Walsh, and J. B. Lingrel, *Mol. Cell.*, **1999**, *3*, 555–563.

78. L. M. Wiley, *Dev. Biol.*, **1984**, *105*, 330–342.

79. D. J. MacPhee, D. H. Jones, K. J. Barr, D. H. Betts, A. J. Watson, and G. M. Kidder, *Dev. Biol.*, **2000**, *222*, 486–498.

80. D. H. Betts, L. C. Barcroft, and A. J. Watson, *Dev. Biol.*, **1998**, *197*, 77–92.

81. L. J. Winkle van and A. L. Campione, *Dev. Biol.*, **1991**, *146*, 158–166.

82. A. E. Moseley, S. P. Lieske, R. K. Wetzel, P. F. James, S. He, D. A. Shelly, R. J. Paul, G. P. Boivin, D. P. Witte, J. M. Ramirez, K. J. Sweadner, and J. B. Lingrel, *J. Biol. Chem.*, **2003**, *278*, 5317–5324.

83. S. He, D. A. Shelly, A. E. Moseley, P. F. James, J. H. James, R. J. Paul, and J. B. Lingrel, *Am. J. Physiol. Regul. Integr. Comp. Physiol.*, **2001**, *281*, 917–925.

84. M. Juhaszova and M. P. Blaustein, *Proc. Natl. Acad. Sci. U.S.A.*, **1997**, *94*, 1800–1805.

85. M. Juhaszova, H. Shimizu, M. L. Borin, R. K. Yip, E. M. Santiago, G. E. Lindenmayer, and M. P. Blaustein, *Ann. New York Acad. Sci.*, **1996**, *779*, 318–335.

86. V. A. Golovina, H. Song, P. F. James, J. B. Lingrel, and M. P. Blaustein, *Am. J. Physiol. Cell. Physiol.*, **2002**, *284*, 475–486.

87. P. Fransen, J. Hendrickx, D. L. Brutsaert, and S. U. Sys, *Cardiovasc. Res.*, **2001**, *52*, 487–499.

88. A. L. Woo, P. F. James, and J. B. Lingrel, *J. Biol. Chem.*, **2000**, *275*, 20693–20699.

89. A. L. Woo, P. F. James, and J. B. Lingrel, *Mol. Reprod. Dev.*, **2002**, *62*, 348–356.

90. L. K. Dahl, M. Heine, and L. Tassinari, *Nature*, **1962**, *194*, 480–482.

91. V. L. Herrera and N. Ruiz-Opazo, *Science*, **1990**, *249*, 1023–1026.

92. N. Ruiz-Opazo, F. Barany, K. Hirayama, and V. L. Herrera, *Hypertension*, **1994**, *24*, 260–270.

93. V. L. Herrera, H. X. Xie, L. V. Lopez, N. J. Schork, and N. Ruiz-Opazo, *J. Clin. Invest.*, **1998**, *102*, 1102–1111.

94. N. Glorioso, F. Filigheddu, C. Troffa, A. Soro, P. P. Parpaglia, A. Tsikoudakis, R. H. Myers, V. L. Herrera, and N. Ruiz-Opazo, *Hypertension*, **2001**, *38*, 204–209.

95. L. Peng, P. Martin-Vasallo, and K. J. Sweadner, *J. Neurosci.*, **1997**, *17*, 3488–3502.

96. A. G. Watts, G. Sanchez-Watts, J. R. Emanuel, and R. Levenson, *Proc. Natl. Acad. Sci. U. S. A.*, **1991**, *88*, 7425–7429.

97. M. Fusco De, R. Marconi, L. Silvestri, L. Atorino, L. Rampoldi, L. Morgante, A. Ballabio, P. Aridon, and G. Casari, *Nat. Genet.*, **2003**, *33*, 192–196.

98. M. Clamp, D. Andrews, D. Barker, P. Bevan, G. Cameron, Y. Chen, L. Clark, T. Cox, J. Cuff, V. Curwen, T. Down, R. Durbin, E. Eyras, J. Gilbert, M. Hammond, T. Hubbard, A. Kasprzyk, D. Keefe, H. Lehvaslaiho, V. Iyer, C. Melsopp, E. Mongin, R. Pettett, S. Potter, A. Rust, E. Schmidt, S. Searle, G. Slater, J. Smith, W. Spooner, A. Stabenau, J. Stalker, E. Stupka, A. Ureta-Vidal, I. Vastrik, and E. Birney, Ensembl 2002: accommodating comparative genomics, *Nucl. Acids Res.*, **2003**, *31*, 38–42.

17

Renal V-ATPase: Physiology and Pathophysiology

Dennis Brown and *Vladimir Marshansky*

17.1
Introduction

Vacuolar H^+-ATPase (V-ATPase) is a large, multi-subunit enzyme that plays a dual role in the kidney. Conversely, it is involved in the acidification of several intra-cellular vesicular compartments such as endosomes, parts of the Golgi, and lysosomes, similar to its well-established functions in most cell types [1–3]. In addition, renal V-ATPase is also critical for systemic acid–base homeostasis via its effects on bicarbonate reabsorption and net acid secretion in the kidney [4]. The discovery that this enzyme plays a key role in the acidification of organelles in mammalian cells, as well as lower organisms such as yeast, resulted in the terminology "vacuolar ATPase". However, many studies over the past several years have clearly shown that the V-ATPase can be inserted into the plasma membrane of many cell types in the kidney and other tissues, where it is involved in different physiological processes, including bicarbonate reabsorption, sperm maturation and storage, bone reabsorption and hearing [3, 5, 6]. This chapter will concentrate on the distribution and role of the V-ATPase as it relates to renal function, but other tissues will be introduced for comparative purposes when appropriate. Finally, the role of intracellular V-ATPase and vesicle acidification on renal function, notably on proximal tubule reabsorptive processes, has begun to be elucidated at the molecular level, and these findings will be addressed in some detail.

17.2
Distribution of V-ATPase in the Kidney

V-ATPase is ubiquitously expressed in all cells, where it plays a role in the acidification of intracellular organelles. In the kidney, several cell types contain abundant V-ATPase both on intracellular vesicles and at the cell surface. This high level of expression can, in many cases, be correlated with established physiological func-

Handbook of ATPases. Edited by M. Futai, Y. Wada, J. H. Kaplan
Copyright © 2004 WILEY-VCH Verlag GmbH & Co. KGaA, Weinheim
ISBN 3-527-30689-7

tions of specific cell types and tubule segments in acid–base transport processes [4, 7]. Furthermore, regulation of acid–base transport often correlates with changes in the level of plasma membrane V-ATPase expression. The physiological regulation of V-ATPase trafficking to and from the plasma membrane as a mechanism of regulating proton and bicarbonate transport across renal epithelia will be discussed more extensively below. More recently, the identification of some V-ATPase subunit mutations associated with several disease conditions, including renal tubular acidosis and neurisensorial deafness, have led to a deeper appreciation of the physiological role of the V-ATPase in different cell types [8–10].

17.2.1
Proximal Tubule

The proximal tubule is responsible for reabsorbing most of the water, ions, and proteins that are filtered and that subsequently appear in the tubule lumen. These active reabsorptive processes are dependent upon the polarized insertion of many transport proteins, channels, enzymes and receptors into the apical and/or basolateral plasma membrane domains of proximal tubule epithelial cells [11]. The initial concentration of bicarbonate at the apical pole of the initial portion (S1 segment) of the proximal tubule is around 25 mM. At the end of the proximal tubule, the bicarbonate concentration has fallen to about 5 mM. Thus, around 80 % of the filtered load of bicarbonate is reabsorbed by the proximal tubule [4]. This massive reabsorptive capacity (about 3500 meq per day of bicarbonate) is achieved by the integrated action of several transport proteins, notably sodium/hydrogen exchangers (apical NHE3), apical V-ATPase, a basolateral sodium-bicarbonate co-transporter (NBC1) and two carbonic anhydrase isoforms, CAII (cytosolic) and CAIV (with its catalytic site in the tubule lumen). All of these proteins have been localized by immunocytochemistry, and their physiological significance has been shown by functional studies [4], production of knock-out mice, or naturally-occurring gene mutations in man and in animal models.

17.2.2
Role of Proton Secretion in Bicarbonate Reabsorption

The general mechanism by which apical proton secretion drives bicarbonate reabsorption in the proximal tubule is illustrated in Figure 17.1. Protons are extruded through the apical membrane by the Na/H exchanger NHE3 [12] and by the apical V-ATPase [13]. The relative contribution of these two proteins varies among species, with the V-ATPase being more important in carnivores than in herbivores [14]. The secreted protons combine with luminal HCO_3^- under the influence of apical CAIV [15] to produce CO_2 and H_2O. The carbon dioxide diffuses into the proximal tubule cytoplasm, where it recombines with H_2O (catalyzed by cytosolic CAII) to produce protons and HCO_3^-. The protons are recycled into the lumen by the apical NHE3 and V-ATPase, while the bicarbonate is reclaimed into the blood via a basolateral Na/HCO_3 co-transporter [16, 17]. Thus, apical secretion of one proton results in

Proximal tubule bicarbonate reabsorption

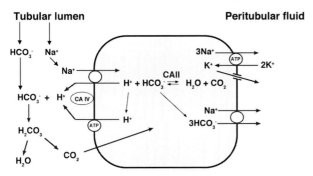

Figure 17.1 Mechanism of bicarbonate reabsorption in the proximal tubule. An apical Na/H exchanger (NHE3) and an apical V-ATPase secrete H^+ into the tubule lumen. These protons combine with filtered bicarbonate to produce CO_2 and H_2O, a reaction that is catalyzed by luminal carbonic anhydrase IV. The CO_2 diffuses into the cytoplasm and combines with water to produce HCO_3^- and H^+. This reaction is catalyzed by a cytosolic isoform of carbonic anhydrase, CAII. This newly-produced HCO_3^- passes across the basolateral plasma membrane into the interstitial space, and ultimately into the blood, through an electrogenic Na/HCO_3 cotransporter. The net result of apical proton secretion in the proximal tubule is, therefore, the reclamation of filtered bicarbonate. (Fig. provided by Dr. S. Breton)

the reclamation of one bicarbonate. No net acid secretion occurs during bicarbonate reabsorption in the proximal tubule.

17.2.3
Role of the V-ATPase in Vesicle Recycling in the Proximal Tubule

As mentioned above, vectorial transport processes in the proximal tubule depend critically on the regulated, polarized insertion of a multitude of membrane proteins into apical and basolateral plasma membranes of the tubular epithelial cells. Many of these proteins are continually recycled between intracellular vesicles and the cell surface [11, 18], a process that requires passage through one or more V-ATPase-containing, acidic compartments within the cell. The high rate of endocytosis in this tubule segment can be readily visualized by introducing fluid phase markers such as fluorescent FITC-dextran or horseradish peroxidase into the tubule lumen [19]. This endocytotic process, as well as the subsequent recycling of proteins such as the scavenger receptor megalin back to the apical plasma membrane, can be disrupted by inhibiting endosomal acidification [20, 21]. Thus, pharmacological interventions or pathophysiological events that lead to defective endosomal acidification invariably result in defective proximal tubule function. This phenomenon, which results in a generalized Fanconi syndrome (i. e., a diminished proximal tubule reabsorptive capacity), is discussed in much greater depth below. Some potential molecular mechanisms that relate defective vesicle acidification to proximal tubule pathophysiologies are beginning to emerge and are also addressed below.

17.2.4
Localization of V-ATPase in the Proximal Tubule

Functional studies and immunocytochemical data have shown that a bafilomycin- and NEM-sensitive V-ATPase is present in the apical membrane (brush border) and in intracellular organelles of proximal tubules [4, 22, 23]. These data are consistent with the physiological role of the V-ATPase in proximal tubule bicarbonate reabsorption, and with the involvement of the V-ATPase in endocytosis and vesicle trafficking in these cells. When immunostaining is performed using antibodies against a C-terminal peptide from the 31 kDa E-subunit of the V-ATPase (based on the bovine sequence), intense staining of the base of the microvilli is seen in the proximal tubule (Figure 17.2A), although brush border staining is also detectable in some tubules. Intercalated cells in adjacent collecting ducts are brightly stained with this antibody. The proximal tubule expresses the B_2 ("brain") isoform

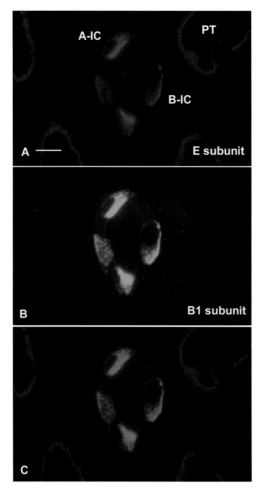

Figure 17.2 The B_1 56 kDa V-ATPase subunit is present in intercalated cells but not in proximal tubules. The 31 kDa E-subunit (A, red – chicken polyclonal antibody) has a sub-brush border apical localization in proximal tubules (PT), an apical localization in A-IC, and a basolateral distribution in B-IC. In contrast, the 56 kDa B_1 subunit (B, green – rabbit polyclonal antibody) can be found only in A-IC and B-IC in this image of the kidney cortex. It is absent from proximal tubules. (C) Merged image in which the yellow intercalated cell staining indicates colocalization of the E and B_1 subunits in these cells. Bar = 10 µm.

of the 56 kDa B-subunit, but the B_1 ("kidney") isoform, which is amplified in intercalated cells, is undetectable in proximal tubules (Figure 17.2B) [24, 25]. High-resolution immunogold staining revealed that the V-ATPase is concentrated at the base of the microvilli in proximal tubules, but is not concentrated in clathrin-coated pits that are present between the microvilli.

Studies on isolated, perfused proximal tubules have shown that exocytosis of vesicles containing the V-ATPase is stimulated by the basolateral application of CO_2 at a partial pressure of 25 mmHg [26]. Fusion was inhibited by prior treatment of the tubules with colchicine, a microtubule depolymerizing agent that inhibits vesicle recycling and exocytotic events. Increased proton secretion by the proximal tubule is an appropriate homeostatic physiological response to partially compensate for systemic acidosis (increase in pCO_2). Increased proton secretion due to the exocytotic insertion of more V-ATPase pumps into the plasma membrane has been reported in other tissues and cell types, including proton secreting cells in the toad and turtle urinary bladders, the epididymis, and in renal intercalated cells (see below) [5].

17.2.5
Distal Tubule and Loop of Henle

Plasma membrane and intracellular V-ATPase has been localized in various segments of the loop of Henle, as well as in the distal convoluted tubule. The initial segment of the descending thin limb of Henle, which follows the S3 segment of the proximal tubule, has both apical and basolateral V-ATPase immunoreactivity [22]. Recent functional data have shown that the bafilomycin-sensitive V-ATPase is involved in intracellular pH regulation in thin limbs of Henle [27], but whether it also plays a role in aspects of systemic pH regulation is unknown. In contrast, the thick ascending limb of Henle (TAL) is responsible for reabsorbing most of the HCO_3^- that escapes the proximal tubule [28]. As for the proximal tubule, the apical transporters involved in this process are NHE3 and the V-ATPase. By immunocytochemistry, a substantial amount of V-ATPase is located in cells of the TAL, where it is present in many subapical vesicles as well as on the apical plasma membrane [22]. The major B-subunit isoform present in the TAL is the B_2-subunit, although recent data from our laboratory have shown co-expression of the B_1 isoform in some regions of the TAL, notably the outer medullary and cortical TAL (Paunescu et al., see Ref. 196). The presence of functional V-ATPase in TAL cells has also been demonstrated, and bafilomycin inhibits pHi recovery from an acid load by 36 % in isolated TAL suspensions [29]. While the presence of the V-ATPase both on vesicles and at the cell surface indicates the existence of a vesicular shuttling mechanism similar to that existing in some other cells, this has not been formally demonstrated in this tubule segment.

The distal convoluted tubule (DCT) shows a consistently high level of apical membrane expression of the V-ATPase [22]. The B_1 56 kDa subunit isoform is expressed in this segment. Very little vesicular staining is detectable in the DCT, in marked contrast to the TAL, where vesicular staining is abundant. This may imply that the apical V-ATPase in the DCT is recycling less than in some other segments.

The role of the V-ATPase in this segment is unclear. DCT are mainly involved in sodium reabsorption and potassium secretion, whereas bicarbonate reabsorption is negligible under baseline conditions. A significant increase in distal tubule bicarbonate reabsorption has been reported in acidotic animals, implying that some regulation can take place in this segment [30]. Bicarbonate reabsorption in acidotic animals was not, however, prevented by the V-ATPase inhibitor bafilomycin, but it was sensitive to the carbonic anhydrase inhibitor acetozolamide [31], indicating that the V-ATPase does not play a critical role in this process. Finally, *in vitro* perfusion data on acid–base transport in this segment are sometimes complicated by the inclusion in the assays of late portions of distal tubules and even connecting segments, which contain intercalated cells. Relatively few data have been obtained on early distal tubules, which are devoid of intercalated cells and in which the role of apical V–ATPase in DCT cells alone could be examined.

17.2.6
Connecting Segment and Collecting Duct

The DCT is linked to the collecting duct system by the connecting segment – a tubule segment of variable length, depending on the species examined, and of complex cellular composition. Both connecting segments and collecting ducts contain a population of specialized proton and HCO_3^- transporting cells known as intercalated cells [5, 32, 33]. These cells contain abundant vesicular and plasma-membrane-associated V-ATPase and high levels of cytosolic CAII [5]. In connecting segments, many of the adjacent non-intercalated cells express apical V-ATPase at an

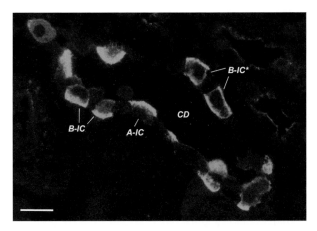

Figure 17.3 Immunocytochemical staining for V-ATPase in a rat cortical collecting duct, showing different manifestations of the intercalated cell (IC) phenotype. A-IC have an exclusive apical staining. This reflects V-ATPase in the apical plasma membrane as well as sub apical cytoplasmic staining that consists of vesicular staining and free, cytosolic V-ATPase subunits. B-IC (operationally identified as being AE1 negative – not shown here) are more variable in their appearance. Some B-IC have a predominant basolateral V-ATPase staining pattern (B-IC) whereas others have a clear bipolar V-ATPase distribution (B-IC*).

intensity similar to that seen in the DCT. In the collecting duct, most principal cells (non-intercalated cells) express very little if any apical V-ATPase [22].

The role of the intercalated cells in acid–base transport is relatively well-established and understood. The cortical collecting duct can achieve either net acid or net base secretion under different physiological conditions [34, 35]. These opposing functions are carried out by distinct intercalated cell phenotypes, known as A-cells and B-cells, respectively. A-cells can express high levels of V-ATPase on their apical plasma membrane, as well as in subapical vesicles, whereas B-cells often have V-ATPase located on their basolateral plasma membrane [22, 36]. However, B-cells can also show an apical, diffuse cytoplasmic, or even a bipolar membrane distribution of the V-ATPase [36, 37]. These different intercalated cell phenotypes are shown in Figure 17.3. The intracellular vesicles that are involved in transporting the V-ATPase to and from the cell surface are highly-characteristic structures, sometimes referred to as tubulovesicles, that have an electron-dense coat formed by the V_1 sector of the V-ATPase [5], which is described in more detail below. The basolateral plasma membrane of all A-IC contains the Cl^-/HCO_3^- exchanger AE1, whereas AE1 is not detectable in B-cells in any membrane domain [38, 39]. Recently, pendrin has been identified as the most likely apical HCO_3^- extrusion pathway in B-cells [40–44]. One group reported apical expression of another anion exchanger, AE4, in rabbit B-IC [45] but it was subsequently localized in the basolateral membranes of rat and mouse A-IC, and in both apical and basolateral membrane of rabbit B-IC by a second group [46]. Therefore, the role of this protein in transepithelial Cl^-/HCO_3^- movement remains unclear.

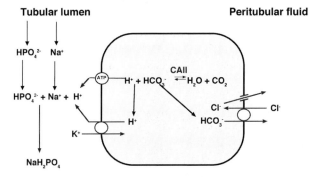

Acid secretion by A-intercalated cell

Figure 17.4 Proton secretion by type A-intercalated cells in the collecting duct. Protons are produced in the cytosol by the catalytic activity of CAII, and are secreted apically through the V-ATPase. An apical H^+,K^+-ATPase may also be involved in apical proton secretion under some conditions. Bicarbonate exits the cell basolaterally via the AE1 Cl^-/HCO_3^- exchanger. Cl^- recycles from the cytosol to the extracellular fluid through a basolateral chloride conductance pathway. In the lumen, protons combine with buffers such as HPO_4^{2-} to form $H_2PO_4^-$, which is excreted as "titratable acid". Protons also combine with luminal NH_3 to form NH_4^+, which also acts as a luminal buffer and is excreted (not shown here). In type B-intercalated cells, basolateral V-ATPase moves protons into the blood, and HCO_3^- exits apically through a different anion exchanger, probably pendrin (see text for details). (Fig. provided by Dr. S. Breton)

A diagram of the mechanism by which acid–base transport is achieved in A-IC is presented in Figure 17.4. In A-IC, cytosolic CAII catalyzes the hydration of CO_2 to produce H^+ and HCO_3^-. Protons are secreted apically by the V-ATPase, and also by the action of an apical H,K-ATPase. HCO_3^- exits the cell through basolateral AE1 and Cl^- that enters the cell recycles back to the interstitium via a chloride conductance pathway. The net effect of proton secretion by A-IC is to replenish HCO_3^- levels in the blood. In the lumen, the H^+ combines with luminal buffers such as HPO_4^{2-} to form $H_2PO_4^-$, which is secreted as titratable acid, and with NH_3, to form NH_4^+, which is excreted into the urine (not shown in Figure 17.4).

17.3
Modulation of Intercalated Cell Phenotypes

17.3.1
Regulation of V-ATPase Polarity in Intercalated Cells by Acid–Base Balance

Proton secretion by intercalated cells in the kidney is regulated mainly by modulating the amount of V-ATPase at the cell surface. In these cells, specialized intracellular acidic vesicles have been identified that shuttle the V-ATPase to and from the plasma membrane in response to physiological acid–base conditions [5, 32, 33]. Systemic acidosis induced in rats and rabbits increases the number of cortical intercalated cells with apical membrane V-ATPase expression and decreases the number of cells with basolateral V-ATPase immunocytochemical staining [37, 47]. Conversely, alkalosis increases the number of cells with basolateral V-ATPase expression and decreases apical V-ATPase staining. These cellular changes are predicted to increase net acid secretion (to compensate for acidosis) and increase net bicarbonate secretion (to compensate for metabolic alkalosis). Much work has been performed to uncover the mechanisms by which the relative number of apparent "A-IC" and "B-IC" changes during this adaptive response, which can occur in just a few hours after the acid–base perturbation is induced. These issues will be discussed briefly in the sections below.

17.3.2
Modulation of the Intercalated Cell Phenotype

Schwartz and Al-Awqati were the first to propose that A- and B-cells were functionally distinct manifestations of a single type of intercalated cell [48]. They hypothesized that, in acidosis, cells have an apical V-ATPase and a basolateral anion exchanger (AE1), which can repolarize, presumably by transcytosis to produce B-cells that have the opposite polarity of these key membrane proteins, i.e., a basolateral V-ATPase and apical AE1. This attractive hypothesis is supported by several studies from the Al-Awqati laboratory showing that the novel matrix protein, hensin, reverses the functional phenotype of intercalated cells in culture [49]. Bicarbonate-secreting intercalated cells could be converted into proton-secreting cells by growing them on a matrix containing hensin [49]. In addition, anti-hensin antibodies applied to the basolateral bathing medium of collecting ducts incubated *in vitro* inhibited the induction of acid secretion and bicarbonate reabsorption that normally occurs following incubation of the tubules in acidic medium [50]. Hensin is proposed to induce terminal differentiation in intercalated cells [51], reflected by the A-cell phenotype, but this phenotypic change cannot occur when hensin function is blocked by antibodies. This change in functional activity of intercalated cells involves the concerted action of microtubules and microfilaments, and also requires *de novo* protein synthesis [50, 52]. According to the model described above, A-IC and B-IC should interconvert repeatedly *in vivo* in response to systemic acid–base variations, implying that the "terminal differentiation" that is induced by hensin must be reversible as A-IC reconvert back into B-IC during alkalosis. The mechanism by which this matrix protein-induced differentiation process is reversed remains undetermined.

17.3.3
Membrane Protein Polarity in Intercalated Cells

As detailed above, cortical collecting ducts contain intercalated cells with predominantly apical (A-IC) or basolateral (B-IC), V-ATPase expression. However, intermediate phenotypes are always found, ranging from bipolar V-ATPase expression to diffuse cytoplasmic expression with little or no apparent membrane staining [36, 37]. This supports the idea that all intercalated cells are variants of the same cell type, and that the precise cellular location of the V-ATPase is determined by prevailing physiological acid–base status. In addition, apical and basolateral anion exchangers can be regulated and/or internalized under appropriate acid–base conditions [53].

While A-IC and B-IC are functional mirror-images, the change in phenotype from A-IC to B-IC clearly involves more than a simple transcytosis-mediated reversal in the polarity of membrane transporters. This is supported by the recent data discussed above, showing that protein synthesis is required to elicit these functional changes in isolated perfused collecting tubules [50]. Among the observations that need to be taken into account when mechanisms underlying this transition are being considered are the following.

(A) Only A-IC are found the medulla. This could indicate that the A-cell phenotype is determined by the prevailing acidic interstitial environment [54] in the medulla. It would be an important observation if medullary A-IC could be shown to convert into the B-cell phenotype under some physiological or experimental conditions. This is likely to be a complex issue because, even under apparently identical acid–base and perfusion conditions in the cortex, A-IC and B-IC with opposite membrane polarities of the V-ATPase can co-exist in close proximity in the same tubule segment at any one point in time [22].

(B) The basolateral anion exchanger in A-IC is AE1 [38, 39, 55, 56], whereas a functionally important apical anion exchanger in B-IC has been identified as pendrin [40, 41]. Apical pendrin is relocated to the cytosol of B-intercalated cells in acid-loaded mouse kidney [42], which is consistent with previous reports of acid-induced apical anion exchanger internalization in rabbit collecting ducts [53].

(C) A- and B-intercalated cells have been distinguished by their differential expression of other proteins, including NHERF [57], in addition to the anion exchangers described above.

17.4
Structural Organization of the V-ATPase in Proton Secreting Cells

Intercalated cells of the kidney, as well as several other proton-secreting cells in other mammalian and non-mammalian tissues, express high levels of the V-ATPase on their plasma membranes as well as on intracellular vesicles. Because of its unusually high density in these membranes, the V-ATPase is readily detectable by various electron microscopic procedures.

17.4.1
Conventional and Immunogold Thin-section Electron Microscopy

The V-ATPase can be detected as an electron-dense array of 10×20 nm stud-like projections on the plasma membrane and on cytoplasmic vesicles of various proton-secreting cells, including kidney intercalated cells (Figure 17.5A), mitochondria-rich cells in the toad and turtle urinary bladders, and ion transporting epithelia in insects [5, 58]. This morphology is similar to that of the F_oF_1 ATP synthase [59]. The V-ATPase studs are located on the apical membrane of A-IC and on the basolateral plasma membrane of B-IC. They were shown to relocate from cytoplasmic vesicles to the apical membrane of A-IC during acute metabolic and respiratory acidosis, providing early evidence that the V-ATPase was actively shuttled between a vesicular compartment and the cell surface in response to physiological conditions [32, 60, 61]. Similar data had been reported previously using the turtle bladder as a model proton-translocating epithelium [62, 63]. Definitive evidence that the studs were formed from the V_1 sector of the V-ATPase came from immunogold labeling studies with specific antibodies against V_1 subunits [22, 64].

17.4.2
Rapid-freeze, Deep-etch Imaging of the V-ATPase

Toad urinary bladder was used as a model, highly-accessible tissue to examine the ultrastructure of the V-ATPase. Fragments of apical plasma membrane from the specialized proton-secreting cells in this epithelium were examined by rapid-freeze, deep-etch microscopy [64]. Clusters of V-ATPase molecules were detectable as arrays of hexagonally-packed projections on the cytoplasmic side of the membranes (Figure 17.5B). Similar projections were also found on the membranes of small vesicles, presumably those that are involved in shuttling large numbers of V-ATPase molecules to and from the cell surface to modulate proton secretion (Figure 17.5C). Isolated, purified V-ATPase from bovine kidney medulla yielded similar images of packed molecules, which represent the large V_1 sector of the enzyme (Figure 17.5D). Proton pumps on contractile vacuoles of unicellular organisms were subsequently shown to have a similar morphological appearance [65].

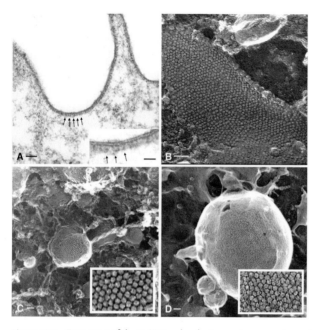

Figure 17.5 Detection of the V-ATPase by electron microscopy in proton-secreting epithelial cells. (A) Apical plasma membrane of an A-type intercalated cell from rat kidney. The cytoplasmic side of the membrane is decorated by many electron dense projections (arrows) that are about 10 nm wide and 20 nm long. These represent the V_1 sector of the V-ATPase, illustrating the high density of membrane-associated proton pumps in these specialized cell types. Bar = 50 nm (inset, bar = 20 nm). (B) Underside of the apical plasma membrane isolated from a toad bladder mitochondria-rich, proton-secreting cell. The membrane was prepared by the rapid-freeze, deep-etch procedure, which reveals structures present on the cytoplasmic face of the membrane. The large V_1 sectors of the V-ATPase form an array of tightly-packed projections in this membrane domain, again illustrating the high level of expression of the V-ATPase in these membranes. (Bar = 50 nm.) (C) Cytoplasmic surface of an intracellular vesicle that is also decorated with V-ATPase projections, forming a "coat" on the vesicle. These specialized vesicles are involved in the exo- and endocytosis of proton pumps as they recycle between the cytoplasm and the cell surface in response to physiological stimuli (Bar = 50 nm). (D) Rapid-freeze, deep-etch image of a phospholipid vesicle that was reconstituted with affinity-purified bovine medullary V-ATPase molecules. Purified protein assumes the same hexagonal packing and appearance as the projections shown in panels B and C, confirming their identity as the V_1 sector of the enzyme. (Bar = 50 nm). The insets in panels C and D show the hexagonally-packed V-ATPase native protein (C) and purified V-ATPase (D) at higher magnification – each V_1 sector is about 10 nm in diameter in both cases.

17.5
Acidification of Intravesicular Compartments by the V-ATPase

Vacuolar H^+-ATPase was originally isolated from intracellular organelles, and has a key function in many facets of organelle function. Both immunocytochemical and functional studies have demonstrated the presence of proton pumps on endosomal membranes and parts of the Golgi/TGN [1, 13, 66–69] as well as on lysosomal membranes [70]. These mammalian intracellular organelles have an acidic lumen generated by a V-ATPase in conjunction with a parallel chloride conductance [1, 2]. In kidney proximal tubules, individual intra-vesicular compartments also showed considerable pH heterogeneity [71], suggesting that endocytotic trafficking is associated with a differential acidification of these organelles: early, late, common, recycling endosomes and finally lysosomes. While it has been well-documented that perturbation of the intraluminal pH of intracellular compartments adversely affects several aspects of protein and vesicle trafficking, the molecular mechanisms underlying these effects remain largely unknown.

17.5.1
Regulation of V-ATPase Activity and Vesicle Acidification by Protein Interactions

The major endosomal acidification machinery consists of: (i) V-ATPase; (ii) chloride channels and (iii) NHE3. Acidification of proximal tubule intravesicular compartments including endosomes is driven by the V-ATPase [7, 20, 72–76]. Increasing evidence over the last decade has shown that the V-ATPase has the capacity to bind to numerous regulatory proteins. It, therefore, appears that the cascade of events that ultimately modulate the V-ATPase (assembly, targeting and activity) and subsequent intravesicular acidification might be controlled at different levels via protein–protein interactions. Some early reports indicated direct interaction between the V-ATPase and low-molecular-weight inhibitory [77] or stimulatory [78] cytosolic proteins. No further identification of these proteins has emerged since the initial reports, however. Recent studies have demonstrated a direct interaction of V-ATPase with SNARE proteins [79–81] and with the actin cytoskeleton [82, 83]. In the male reproductive tract, apical V-ATPase mediated acidification is inhibited by exposing tissue to cholera toxin, which cleaves the v-SNARE cellubrevin and thereby prevents the delivery of V-ATPase to the apical membrane [84]. Both the B_1 and B_2 56 kDa subunits of the V-ATPase contain N-terminal actin-binding domains [83], and the C subunit of the V_1 sector has also been shown to bind actin [85]. In addition, the B_1 subunit has a C-terminal DTAL motif that is involved in indirect binding of the V-ATPase to actin via the PDZ protein NHERF [57]. The respective roles of direct and indirect actin binding to the V-ATPase remain to be determined.

An alternative mechanism for modulation of V-ATPase function is by reversible assembly-disassembly of the V_o and V_1 sectors of the enzyme. This was shown to occur in the larval insect midgut during molting [86, 87]. In yeast, glucose deprivation induces disassembly of the V_1 domain from the plasma membrane V_o domain

[88, 89] and the role of novel RAVE (<u>R</u>egulator of the <u>A</u>TPase of <u>V</u>acuolar and <u>En</u>-dosomal membranes) proteins in glucose-dependent and reversible assembly–dis-assembly of V_1 and V_o domains has been recently uncovered [90, 91]. The role of assembly–disassembly of V_1 and V_o domains to regulate lysosomal acidification has been also demonstrated during maturation of dendritic cells [92].

A potential coupling of V-ATPase activity to glycolysis is suggested by the direct binding of the glycolytic enzyme, aldolase, to subunit E [93]. This direct association suggests the possibility that aldolase may be involved in regulating the activity, dis-tribution, or assembly of the V-ATPase. In addition, evidence for a direct interaction of the C-terminal domain of the V_o subunit a4 with phosphofructokinase I, another enzyme of the glycolytic pathway [94], further indicates a potential regulatory me-chanism(s) between V-ATPase function and energy supply.

17.5.2
Regulation of Acidification by Chloride Channels

It is generally accepted that the electrogenic V-ATPase translocates protons from the cytoplasm to the endosomal lumen and rapidly generates a positive electrical potential ($\Delta\Psi$) within endosomal vesicles, which reduces the net pH gradient (ΔpH) formation. This problem is overcome by the presence of Cl^- channels to-gether with the V-ATPase in endosomal membranes [20, 67, 72, 73, 95, 96]. The nature of the endosomal chloride conductance is a matter of some debate, since channels with different characteristics have been reported in endosomal mem-branes, including a PKA-sensitive Cl^- conductance pathway [97]. However, the CLC5 chloride channel has also been located in endocytotic vesicles of S1, S2 and S3 proximal tubule segments, and its apparent colocalization with the V-ATPase in situ has been reported in rat [98, 99] and mouse [100] kidneys. It is also found in apical vesicles in proton-secreting cells in the epididymis, although in this tissue many intracellular vesicles contain either the V-ATPase or ClC5, but not both [101]. Recently, the role of CLC4 in endosomal acidification in proximal tubules has been also demonstrated [102]. Regardless of the nature of the specific protein, a single Cl^- channel can conduct 10^6–10^8 ions s^{-1}, while purified V-ATPase has a turnover of no more than 100 ions s^{-1} per molecule. Hence, the membrane poten-tial generated by 1000 or more V-ATPases per vesicle can be collapsed by a single Cl^- channel that is open less than 10 % of the time. Thus, chloride channels could be important players in differential acidification of distinct intra-vesicular compart-ments.

Even though the precise role and identity of the chloride channels that are in-volved in endosomal acidification is still subject to considerable discussion [21, 103], there is no doubt that a chloride conductance pathway is important in the re-ceptor-mediated endocytosis pathway and, therefore, for protein reabsorption by kidney proximal tubules [20, 73, 96, 104]. V-ATPase driven endosomal acidification is greatly diminished in the absence of chloride when the proton-motive force (pmf) is predominantly present in the form of a membrane potential ($\Delta\Psi$). Addi-tion of chloride significantly accelerates endosomal acidification by converting $\Delta\Psi$

into a ΔpH [72] (Figure 17.6). Similarly, the acidification capacity of cortical kidney endosomes isolated from CLC5 knock-out mice was somewhat diminished in comparison to control mice [105]. As mentioned above, the presence of a different Cl⁻ channel, ClC4, in proximal tubule endosomes could explain the lack of a severe acidification defect in ClC5 knock-out mice [102]. Direct measurement of chloride accumulation during endosomal acidification has been recently reported *in vivo* using J774 and CHO cell lines [106]. Endosomal [Cl⁻] increased from 17 to 53 mM in J774 cells and from 28 to 73 mM in CHO cells, as the endosomal pH decreased from 6.95 to 5.30 (J774) and 6.92 to 5.60 (CHO). These experiments provide the first direct assay for endosomal [Cl⁻] and indicate that endosomal acidification is accompanied by significant Cl⁻ entry.

$$pmf = \frac{\Delta\mu H}{F} = \Delta\Psi + \frac{RT}{F}\Delta pH$$

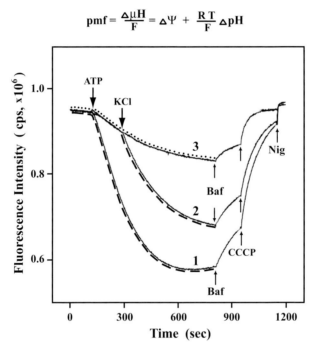

Figure 17.6 Chemiosmotic principle of intra-endosomal acidification and effect of chloride on V-ATPase-driven acidification. Acidification of endosomes purified from kidney proximal tubule was followed using the acridine orange *in vitro* acidification assay. Acidification of endosomes in the presence of 100 mM KCl (Curve 1) or 100 mM K-gluconate (Curve 3) is shown. The effect of 50 mM of KCl added to the medium containing 50 mM K-gluconate during acidification is indicated by the arrow labelled 'KCl' (Curve 2). Dotted line indicates acidification in absence while dashed lines indicate acidification in presence of chloride ions. The effects of 10^{-7}M bafilomycin (Baf), 10^{-5}M carbonylcyanide-m-chlorophenylhydrazone (CCCP) and 2×10^{-5} M nigericin (Nig) are also indicated by the arrows. Adapted from [72] with permission.

17.6
Physiological Role of Acidification in Proximal Tubule Function

In this section, the cell biology of vesicle acidification processes will be related to post-filtration protein handling by proximal tubules. In particular, the role of V-ATPase driven acidification of organelles in health and in proximal tubulopathies will be addressed, with emphasis on normal and abnormal vesicular trafficking processes. Human diseases related to protein and vesicle trafficking are prevalent [107, 108], and the kidney is one of many organs that is affected by this category of disease. Kidney disorders related to defective vesicular trafficking in proximal tubule epithelial cells include: (i) some forms of Fanconi syndrome (such as Dent's disease, cystinosis and cadmium intoxication); (ii) diabetic nephropathy and (iii) autosomal dominant polycystic kidney disease. After a general introduction to some aspects of acidification and the regulation of vesicle trafficking, the role of receptor-mediated uptake of protein in the proximal tubule will be discussed. Finally, acquired and inherited disease states related to the acidification process, and which perturb proximal tubule function, will be presented.

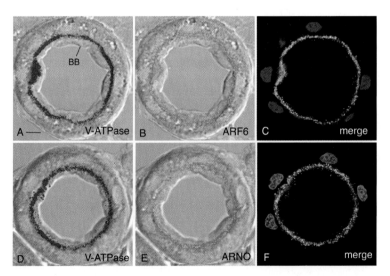

Figure 17.7 Localization of the V-ATPase and two associated "regulatory" proteins at the apical pole of proximal tubule cells. Panels A, B, D, E show double immunofluorescence staining for V-ATPase (red), the small GTPase Arf6 (panel B – green) or the GTP/GDP exchange factor ARNO (panel E – green) superimposed over DIC images of the corresponding proximal tubule. All of the proteins are colocalized at the base of the brush border (BB), the region of extensive endocytotic activity in these tubules. Panels C and F show merged red/green fluorescence images without DIC. Some vesicles contain only the V-ATPase (red) or only Arf6/ARNO (green), whereas other are colabeled (yellow). Nuclei are counterstained in red. (Bar = 5 μm.)

17.6.1
Acidification is Required for the Normal Function of Many Intracellular Vesicular Compartments

Many effects of V-ATPase inhibition or neutralization of luminal pH on intracellular transport processes and protein trafficking have been described. Among the effects that are best understood are the requirement of an acidic pH for the dissociation of many receptor–ligand complexes in endosomes, and the acidic pH requirement of lysosomal hydrolases [1, 2, 109–112]. However, even though functional effects of collapsing pH gradients have been repeatedly found for many vesicular trafficking events, the mechanism(s) underlying these observations remain poorly understood. For example, late endosome to lysosome transport and endosome carrier vesicle formation, as well as a block of cell surface to trans-Golgi (TGN) recycling in some cell types is inhibited by bafilomycin [113–115]. Recycling of specific membrane proteins, including aquaporin 2 [116] and GLUT4 [117], is also blocked by bafilomycin at the level of the TGN. The effect of bafilomycin on AQP2 trafficking bears a striking similarity to the effect of low temperature (20°C) [116], perhaps implying a mechanistic similarity.

In the proximal tubule of the kidney, endocytosis of proteins such as albumin and immunoglobulin light chains is inhibited by bafilomycin treatment [20, 21, 118]. Inhibition of vesicle acidification by other factors, including the heavy metal cadmium and NH_4Cl, also inhibits endocytosis by proximal tubule cells both *in vivo* and *in vitro* [119–122].

17.6.2
Role of Acidification in Vesicle "Coat" Protein Recruitment

Coatomer (COP) coated transport vesicles are involved in many of the acidification-dependent steps described above [123–125]. Increasing evidence is now linking the cytosol-to-membrane recruitment of functionally important vesicle coat proteins to intravesicular acidification of organelles. It has been clearly shown that the association of COP proteins, in particular β-COP and small GTPases of the Arf (ADP-ribosylation factor) family, with some vesicles depends on the establishment of an acidic luminal pH [73, 126–128]. A current hypothesis is that collapsing the normal intraluminal pH in membrane-bounded organelles such as endosomes and the TGN inhibits vesicle trafficking by preventing the recruitment of key coat proteins that are required for vesicle formation and fission.

Several proteins that are important for regulation of the endocytotic and recycling pathways are located at the apical pole of renal proximal tubules (Figure 17.7, see p. 427). These include the scavenger receptor proteins megalin and cubilin, the V-ATPase, the small GTPases Arf1, Arf6, rab5 and rab 11, and an Arf6 GEF (GDP/GTP exchange factor) known as ARNO (ADP-ribosylation factor nucleotide binding site opener) [73]. Arf6 and ARNO, but not Arf1, are recruited from the cytosol to proximal tubule endosomal membranes in response to acidification of the endosomal lumen (Figure 17.8) [73]. In contrast, it was also reported that Arf1 but not Arf6 was recruited to endosomes in a pH-dependent manner in a cell culture model [127]. The extent and specificity of pH-dependent vesicle coat recruitment may vary under different cellular conditions. However, both Arf1 and Arf6 can be recruited to the same endosomal membranes in a GTP-dependent manner [73]. This distinction between pH-dependent/GTP-independent recruitment, compared with GTP-dependent/pH-independent recruitment may allow functionally distinct populations of vesicles to recruit different coat components at different stages of their intracellular trafficking process. For example, clathrin-coated vesicles budding from the apical membrane of proximal tubule epithelial cells (as well as other cell types, including collecting duct principal cells and hepatocytes) do not seem to be acidic compartments [22, 129–131]. Therefore, the earliest stage of clathrin-mediated endocytosis may involve the GTP-dependent recruitment of Arf1 to the membrane. Subsequently, Arf6 may be recruited to early endosomes in a pH-dependent step, since these vesicles contain a functional V-ATPase that generates an acidic luminal pH.

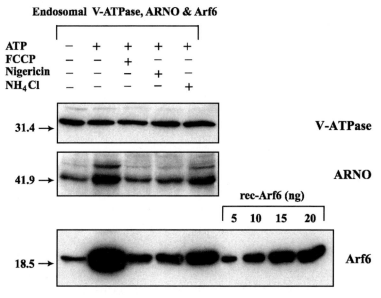

Figure 17.8 Acidification-dependent recruitment of ARNO and Arf6 to proximal tubule endosomes. Purified endosomes were incubated with proximal tubule cytosol in the presence of ATP (to stimulate intravesicular acidification via the V-ATPase), or in the presence of various uncouplers of acidification. The endosomes were then pelleted, washed and subjected to Western blotting with anti-V-ATPase, ARNO or Arf6 antibodies. While V-ATPase protein in the endosomal preparation remained constant (top panel), ARNO (middle panel, lane 2) and Arf6 (lower panel, lane 2) were recruited to the endosomal membranes in an acidification dependent manner. All of the uncouplers reduced the level of ARNO and Arf6 recruitment to membranes, with FCCP being the most effective in this experiment and NH_4Cl the least effective. The amount of Arf6 recruited to endosomes was quantified by comparing with a standard curve using known amounts of recombinant Arf6 (lower panel, 4 lanes on the right). Adapted from [73] with permission.

17.6.3
V-ATPase a2-Isoform as a Putative pH-sensing Protein

The sensor that detects luminal pH in endosomes is probably a transmembrane protein that detects a drop in endosomal pH and in some way transmits this information to its cytosolic domain. [21, 73, 126]. A pH-induced change in the conformation of the cytosolic domain of the putative pH-sensing protein (PSP) would be the signal for the recruitment of key components of the endosomal coat, including Arf proteins and regulators of Arf function such as ARNO. Recently, the transmembrane "a" subunit from the V_0 sector of the V-ATPase has emerged as a potential candidate PSP protein. The a2-isoform is enriched in proximal tubule endosomes, and its cytosolic N-terminal domain binds directly to ARNO both *in vitro* and *in vivo* [132, 133]. Furthermore, immunoprecipitation and pull-down assays have shown that the V-ATPase, ARNO and Arf6 form a multiprotein complex. Thus, while regulation of the a2-isoform/ARNO/Arf6 interaction by pH remains to be demonstrated, it is now clear that the a2-isoform of the V-ATPase has binding characteristics that are predicted for the PSP protein in the endosomal membrane [132, 133]. Clearly, transmembrane signaling is a common phenomenon that regulates various cellular processes, most notably receptor–ligand induced cascades. A pH sensor would, therefore, add to the growing list of signals that can be transmitted from one side of a biological membrane to the other. In the present case, the V-ATPase itself might be a multifunctional protein that is responsible for both the generation and detection of intravesicular pH in the endocytotic and biosynthetic pathways.

17.6.4
Endocytosis and Protein Reabsorption by the Proximal Tubule

Proximal tubules are specialized to reabsorb ions, glucose, amino acids and low-molecular weight proteins from the glomerular ultrafiltrate. This is achieved by the concerted interaction of many ion channels, enzymes, pumps and receptors that are present on the apical (and on the basolateral) plasma membranes. The reabsorption and degradation of filtered proteins occurs through an extensive apical endocytotic apparatus, which is also involved in the internalization and subsequent recycling of membrane proteins [11, 20, 21]. Two multiligand-binding receptors, megalin and cubilin, are involved in protein reabsorption via receptor-mediated endocytosis through clathrin-coated pits [134, 135]. Megalin is a 600 kDa transmembrane glycoprotein of the LDL-receptor family [136–138] while cubilin is a 460 kDa glycoprotein with no transmembrane domain, but which may be anchored to the membrane via palmitoylation of its cysteine moiety [139, 140]. In the proximal tubule, both megalin and cubilin are localized on brush-border microvilli (BBM), in coated pits, apical endosomes and dense apical tubules [135]. A role for both cubilin [141] and megalin [142] receptors in albumin reabsorption has been reported. Many other ligands also interact with megalin and cubilin, including (i) low-molecular weight plasma proteins, (ii) peptide hormones such as insulin, (iii) vitamin-binding proteins such as retinol-binding protein, (iv) apolipoproteins, (v) enzymes, and (vi) polybasic drugs [135, 143, 144]. Endocytosis of proteins by these receptors involves passage through acidic intracellular compartments. As discussed below, normal protein reabsorption is severely impaired (leading to Fanconi syndrome) when acidification of these pathways is disrupted.

17.6.5
Normal Albumin Handling by the Proximal Tubule

Until recently, it was believed that 25 mg or less of albumin are excreted daily in man. Most of the filtered albumin was assumed to be internalized by endocytosis in the proximal tubule, and delivered to lysosomes for complete degradation [145, 146]. However, only intact albumin is detected in the usual radioimmunoassay for urine albumin, whereas partially degraded albumin fragments escape detection [147]. In rats, up to 90 % of the albumin found in their urine is in the form of fragments of various sizes. Normal excretion of albumin by humans is also significantly (50 times) higher than previously thought [148–150], and most of this is also fragmented. Both intact albumin and a 55 kDa albumin fragment were detected in purified early endosomes from proximal tubules [20, 151] and the appearance of the 55 kDa albumin fragment was attributed to specific cleavage by endosomal proteases. The lumen of these endosomes has an acidic pH generated by the V-ATPase and contains cathepsin D [21, 72]. One group has also postulated a mechanism that involves generation of protein fragments in lysosomes, followed by regurgitation from lysosomes into the tubular lumen [150, 152]. A "retrieval pathway" through which intact filtered albumin is returned to the blood stream by transcytosis *in vivo* has also been identified [153, 154]. Thus, albumin handling by the proximal tubule is complex, and can involve internalization followed by degradation (including partial degradation into fragments) in endosomes and lysosomes, as well as transcytosis of intact albumin from the tubule lumen to the blood-facing side of the epithelium.

17.7
Relationship of Defective Acidification to Proximal Tubule Dysfunction

Intracellular vesicle trafficking involves the sequential passage of transported molecules through a series of acidified compartments. It is to be expected, therefore, that perturbation of the acidification process will lead to tubule dysfunction and potentially to pathophysiological states. The following sections will focus on the importance of protein trafficking in some types of proximal tubule dysfunction. Available data implicate a role of defective acidification in some cases, while this has not been examined in other disease conditions.

17.7.1
Inherited Fanconi Syndrome – Dent's Disease

This disease is characterized by a Fanconi-like syndrome and nephrolithiasis. It results from a loss of function mutation in the ClC5 chloride channel (see above) [155–158]. Variable pathophysiological conditions have been duplicated in different strains of ClC5 knock-out mice [104, 105, 159, 160]. Defective endocytosis of filtered proteins and defective recycling of apical membrane proteins is the most likely cause of the proximal disease seen in man and in the mouse models. Absence of the ClC5 chloride channel limits the extent of endosomal acidification, and, as discussed earlier, cortical endosomes ClC5 knock-out mice have a reduced capacity to acidify in response to ATP addition by comparison with endosomes from normal mice [105]. While ClC5 knock-out does have some of the predicted effects on renal tubule function, further work will be required to understand more fully the precise mechanisms underlying these functional defects in view of the potential expression of other chloride channels in acidic intracellular compartments.

17.7.2
Perturbation of the Megalin/Cubilin Pathway

Intra-endosomal acidification-dependent recycling of megalin and cubilin plays a role in the pathogenesis of several diseases. A tubular reabsorption deficiency has been demonstrated in animal models, including megalin knock-out mice [161], and cubilin-deficient dogs [135, 141, 162, 163]. Megalin constitutes a low affinity but high capacity receptor pathway for scavenging proteins from the glomerular filtrate [144], but megalin knock-out mice do not have albuminuria, suggesting an alternative mechanism for albumin reabsorption. Indeed, cubilin is also important for albumin reabsorption both in animal models and in humans [141, 162, 164–167]. Significant albuminuria associated with decreased proximal tubule albumin endocytosis was observed in cubilin-deficient dogs [162, 164, 165]. Patients with Imerslund–Grasbecks disease (caused by mutations in the human cubilin gene) [166]) also develop various degrees of albuminuria [167].

Megalin is present in normal human urine, probably due to shedding from proximal tubule apical membranes into the kidney tubular lumen. However, megalin is virtually undetectable in the urine of patients with Dent's disease. These data could be explained by: (i) diminished expression of megalin in Dent's disease patients and/or by (ii) inhibition of its recycling due to diminished intra-endosomal acidification [168]. Finally, a recent report demonstrates reduced expression of the endocytic receptor proteins megalin and cubilin in a ClC5 knock-out mouse model, which could partly explain the endocytotic defect observed in humans [169].

17.7.3
Acquired Fanconi Syndrome – Cadmium Nephrotoxicity

Both in man and in experimental animals, chronic exposure to the heavy metal cadmium results in proximal tubule dysfunction with an associated Fanconi syndrome [122, 170–176]. Cadmium affects various cellular components, including microtubules [177, 178] and the Na^+,K^+-ATPase [176]. However, it also directly inhibits V-ATPase activity, leading to defective endosomal acidification [122]. Interestingly, cadmium intoxication also inhibits V-ATPase function and luminal acidification in the male reproductive tract [179–181]. Thus, by inhibiting V-ATPase activity, cadmium may adversely affect proximal tubule function in a manner similar to that observed in Dent's disease. In support of this, cadmium intoxicated rats show impaired recycling of apical membrane proteins [177] and defective endocytosis of filtered proteins [122]. Because cadmium affects other cellular components, including microtubules, the precise contribution of its effect on V-ATPase activity to proximal tubule dysfunction is difficult to quantify *in vivo*.

17.7.4
Endocytotic Defect in Polycystic Kidney Disease

Autosomal-dominant polycystic kidney disease (ADPKD) is the prevalent hereditary renal tubulo-interstitial disease, which affects 1 in 800 individuals and is caused mainly by mutations in the PKD1 [182] and PKD2 genes [183]. The disease is characterized by the formation of multiple cysts followed by significant enlargement of both kidneys. Proteinuria and albuminuria have been implicated in severe progression of the disease leading to chronic renal failure. Recently, a significant impairment of proximal tubule endocytotic function was demonstrated in the rat *in vivo* [184, 185]. These data indicate that proteinuria and albuminuria in an ADPKD rat model are due to a loss of endocytic machinery proteins (including CLC5 and megalin) in the epithelial cells lining proximal tubular cysts. Up to now, the potential role of a loss of endosomal acidification in the defective endocytosis seen in this disease has not been investigated.

17.7.5
Endosomal and Lysosomal Acidification in Diabetic Nephropathy

Diabetes is one of the leading causes of a slow deterioration of the kidneys (nephropathy), leading to end-stage renal disease (ESRD) and finally to kidney failure. Approximately 35 % of patients in the United States who develop chronic renal failure have diabetes. Persistent albuminuria (>300 mg d^{-1} or 200 µg min^{-1}) is the hallmark of diabetic nephropathy and is used for clinical diagnosis in both Type I and Type II diabetes [186]. Until recently it was believed that only a very small amount of albumin is normally filtered by glomeruli. The development of albuminuria in diabetes was, therefore, predominantly attributed to changes in the structure and function of the glomerular filtration barrier [186]. In contrast, the role of proximal tubules in the pathogenesis of diabetic nephropathy was underestimated and is largely unknown.

Changes in the normal post-filtration handling of albumin by proximal tubules may lead to: (i) an increase of intact albumin excretion, if the "fragmentation–degradation" pathway is inhibited; or (ii) changes in net albumin excretion (in the form of intact or albumin-derived fragments) if the "retrieval-transcytosis" pathway is disrupted. In the early stages of the disease in a rat model (4–16 weeks) of diabetes, changes in the "fragmentation–degradation" pathway for albumin were observed [150, 187] that were attributed to lysosomal changes. In the later stages (24 weeks) of progression, changes in net albumin excretion reflected changes in the "retrieval–transcytosis" pathway. A change in glomerular permeability was not responsible for the observed albuminuria [187]. Similar changes in albumin excretion have also been observed in human Type I diabetic patients [148]. The role of lysosomes and endosomes in the "fragmentation–degradation" of filtered low-molecular weight proteins including albumin and the role of protein fragments in the development of diabetic nephropathy is currently under investigation in our laboratory. Furthermore the potential role of V-ATPase-dependent endosomal and lysosomal acidification in the pathogenesis of diabetic nephropathy now need to be addressed.

17.7.6
Cystinosis is a Lysosomal-dependent Inherited Fanconi Syndrome

Cystinosis is inherited autosomal recessive disorder that causes infantile and juvenile renal Fanconi syndrome. The most severe form is infantile cystinosis which manifests by proximal renal tubulopathy (accompanied by aminoaciduria, glycosuria, phosphaturia, renal tubular acidosis), leading to ESRD at an early age. Cystinosis is a lysosomal storage disease characterized by the accumulation of cysteine in proximal tubule lysosomes [188, 189]. The CTNS gene which is mutated in cystinosis encodes the cystinosin protein. The protein consists of seven transmembrane domains and is localized in the lysosomal membrane. Cystinosin is a cystine/H^+-symporter, which cotransports cysteine and protons out of lysosomes. Thus, its function is directly related to the acidification of the lysosomal lumen by the V-ATPase [188–193]. Novel cell biological functions for lysosomes as well as their associated proteins, including V-ATPases, are now emerging. For example, "secretory lysosomes" may be involved in the trafficking and delivery of V-ATPase to the plasma membrane during osteoclast maturation [194]. Similarly, it is possible that albumin and other protein fragments may be delivered to the proximal tubule lumen either directly by fusion of secretory lysosomes with the plasma membrane, or indirectly after transfer from lysosomes to another vesicular compartment.

17.8
Role of the Intercalated Cell V-ATPase in Distal Tubule Acidosis

As discussed above, acid–base disturbances result in phenotypic alterations in A- and B-intercalated cells as part of a physiological homeostatic response. Loss or perturbation of intercalated cell V-ATPase activity results in distal tubular acidosis, which is characterized by an inability to acidify the urine below pH 5.5 in the context of acidemia (inappropriately acidic blood pH). In the past few years, mutations in some V-ATPase subunits have been associated with distal renal tubular acidosis (RTA) by genetic analysis of individuals and families.

17.8.1
Mutations in the B_1 (56 kDa) V-ATPase Subunit Cause Distal RTA and Sensorineural Deafness

Families were identified in which sensorineural hearing loss cosegregated with distal RTA [8]. The data showed that all affected individuals had mutations in the gene *ATP6V1B* which encodes the B_1 subunit of the V-ATPase. These mutations were either inherited as a recessive trait, or were classified as spontaneous when neither parent had an affected allele. The mutation causes distal RTA by preventing the function of intercalated cells, and causing a loss of the ability to secrete titratable acid in the collecting duct system. The B_1 subunit had previously been localized to distinct cell types in the inner ear, where it was proposed to be involved in maintaining the characteristic high K^+ composition of the endolymph [195]. Again, the B_1 mutation causes this function to be lost, resulting in deafness. At present, no information on the fertility status of these affected individuals is available because they were all diagnosed and examined at a young age. The relatively restricted pathophysiology caused by B_1 subunit mutations reflects its relatively narrow tissue distribution (see above). In contrast, the B_2 isoform is virtually ubiquitous, and defects in this subunit would be much more severe, if not embryologically lethal.

A model B_1 knock-out mouse has been developed recently, and surprisingly, this mouse does not develop distal RTA when given ad libitum access to normal food and water. However, recent data from our laboratory have shown that the B_2 subunit is expressed in rodent intercalated cells in addition to the B_1 subunit, and that it can colocalize with other V-ATPase subunits on the plasma membrane under some experimental conditions [196]. Expression of the B_2 subunit in B_1 knock-out mice will be examined in future studies. Nevertheless, the fact that human subjects do develop distal RTA in the absence of a functional B_1 subunit suggests that compensatory expression and membrane insertion of the B_2 subunit does not occur in man.

17.8.2
Mutations in the a4 (110 kDa) V-ATPase Subunit Cause Distal RTA

A subsequent study on individuals and families with distal RTA but normal hearing resulted in the identification of mutations in the a4 subunit gene (*ATP6V0A-4*) as the causative factor [9]. This transmembrane subunit is part of the V_o sector of the V-ATPase, and has been localized in intercalated cells as well as in proton-secreting clear cells in the epididymis. In intercalated cells, it colocalized at the apical pole of A-cells with the B_1 subunit [197]. However, it is also present in the apical domain of proximal tubular epithelial cells, yet the patients with mutations in a4 subunit are indistinguishable from patients with the B_1 subunit mutations. This again could be due to compensatory expression of other "a" subunit isoforms in the proximal tubule.

17.9
Summary

By modulating transmembrane proton transport, the V-ATPase plays a key role in many aspects of kidney function both at the cellular level and with respect to whole body acid–base homeostasis. Acidification of organelles in the endocytotic, biosynthetic, recycling, transcytotic and degradative pathways within renal cells is essential for their normal function. Pathophysiological changes in kidney function occur when this process is perturbed. These include low-molecular weight proteinuria and other reabsorptive defects in the proximal tubule (Fanconi syndrome). Furthermore, mutations in "kidney-specific" V-ATPase subunits lead to systemic acidification and distal tubule acidosis, although it is known that some of these subunits are also expressed in a limited number of other organs, including the male reproductive tract and the inner ear. The V-ATPase is a multisubunit protein complex, and the functions of many of its components are not well understood. An increasing number of accessory and interacting proteins are being identified, including cytoskeletal proteins, SNARE proteins, small GTPases and glycolytic enzymes. These may play a key role in the regulation of proton-pumping function, as well as in aspects of intracellular trafficking of the V-ATPase. Its potential role as a transmembrane pH sensor has also been proposed. Thus, the V-ATPase seems to be a multifunctional protein that is involved in the regulation of vesicular trafficking, in energizing membranes for the transport of other ionic species and molecules [6], and in regulating the pH of various extracellular fluids, including plasma, endolymph, seminal fluid and perhaps others. By comparing and contrasting its role in the kidney that of other organ systems and cell types, a more complete understanding of the physiology and pathophysiology of the V-ATPase will undoubtedly emerge in the years ahead.

Acknowledgments

Work from our laboratory described in this chapter received continuous support from the National Institutes of Heath, grants DK38452 and DK42956. We thank many colleagues for their contributions to these studies, particularly Seth Alper, Sylvie Breton, Steve Gluck, John Hartwig and Ivan Sabolic.

References

1. I. Mellman, R. Fuchs, and A. Helenius, *Annu. Rev. Biochem.*, **1986**, *55*, 663–700.
2. I. Mellman, *J. Exp. Biol*, **1992**, *172*, 39–45.
3. T. Nishi and M. Forgac, *Nat. Rev. Mol. Cell. Biol.*, **2002**, *3*, 94–103.
4. L. L. Hamm and R. J. Alpern, *The Kidney: Physiology and Pathophysiology, Vol. I, 3rd Edition.* G. Giebisch and D. W. Seldin, ed., Lippincott Williams & Wilkins, Philadelphia **2000**, pp. 1935–1980.
5. D. Brown and S. Breton, *J. Exp. Biol.*, **1996**, *199*, 2345–2358.
6. H. Wieczorek, D. Brown, S. Grinstein, J. Ehrenfeld, and W. R. Harvey, *Bioessays*, **1999**, *21*, 637–648.
7. D. Brown and S. Breton, *The Kidney: Physiology and Pathophysiology, Vol. 1.* G. Giebisch and D. W. Seldin, ed., Lippincott, Philadelphia **2000**, pp. 171–192.
8. F. E. Karet, K. E. Finberg, R. D. Nelson, A. Nayir, H. Mocan, S. A. Sanjad, J. Rodriguez-Soriano, F. Santos, C. W. Cremers, A. Di Pietro, B. I. Hoffbrand, J. Winiarski, A. Bakkaloglu, S. Ozen, R. Dusunsel, P. Goodyer, S. A. Hulton, D. K. Wu, A. B. Skvorak, C. C. Morton, M. J. Cunningham, V. Jha, and R. P. Lifton, *Nat. Gen.*, **1999**, *21*, 84–90.
9. A. N. Smith, J. Skaug, K. A. Choate, A. Nayir, A. Bakkaloglu, S. Ozen, S. A. Hulton, S. A. Sanjad, E. A. Al-Sabban, R. P. Lifton, S. W. Scherer, and F. E. Karet, *Nat. Gen.*, **2000**, *26*, 71–75.
10. K. J. Borthwick and F. E. Karet, *Curr. Opin. Nephrol. Hypertens.*, **2002**, *11*, 563–568.
11. D. Brown and J. L. Stow, *Phys. Rev.*, **1996**, *76*, 245–297.
12. D. Biemesderfer, J. Pizzonia, A. Abu-Alfa, M. Exner, R. Reilly, P. Igarashi, and P. S. Aronson, *Am. J. Physiol.*, **1993**, *265*, 736–742.
13. I. Sabolic, W. Haase, and G. Burkhardt, *Am. J. Physiol.*, **1985**, *248*, 835–844.
14. M. Duplain, J. Noel, A. Fleser, V. Marshansky, A. Gougoux, and P. Vinay, *Am. J. Physiol.*, **1995**, *269*, 104–112.
15. D. Brown, X. L. Zhu, and W. S. Sly, *Proc. Natl. Acad. Sci. U. S.A.*, **1990**, *87*, 7457–7461.
16. M. F. Romero and W. F. Boron, *Annu. Rev. Physiol.*, **1999**, *61*, 699–723.
17. M. Soleimani, *J. Nephrol.*, **2002**, *15 (Suppl 5)*, 32–40.
18. D. Brown and S. Breton, *Kidney Int.*, **2000**, *57*, 816–824.
19. W. I. Lencer, P. Weyer, A. S. Verkman, D. A. Ausiello, and D. Brown, *Am. J. Physiol.*, **1990**, *258*, 309–317.
20. V. Marshansky, S. Bourgoin, I. Londono, M. Bendayan, B. Maranda, and P. Vinay, *Electrophoresis*, **1997**, *18*, 2661–2676.
21. V. Marshansky, D. A. Ausiello, and D. Brown, *Curr. Opin. Nephrol. Hypertens.*, **2002**, *11*, 527–537.
22. D. Brown, S. Hirsch, and S. Gluck, *J. Clin. Invest.*, **1988**, *82*, 2114–2126.
23. F. Turrini, I. Sabolic, Z. Zimolo, B. Moewes, and G. Burckhardt, *J. Membr. Biol.*, **1989**, *107*, 1–12.
24. R. D. Nelson, X. L. Guo, K. Masood, D. Brown, M. Kalkbrenner, and S. Gluck, *Proc. Natl. Acad. Sci. U. S.A.*, **1992**, *89*, 3541–3545.
25. K. Puopolo, C. Kumamoto, I. Adachi, R. Magner, and M. Forgac, *J. Biol. Chem.*, **1992**, *267*, 3696–3706.
26. G. J. Schwartz and Q. Al-Awqati, *Annu. Rev. Physiol.*, **1986**, *48*, 153–161.
27. T. L. Pannabecker, O. H. Brokl, Y. K. Kim, D. E. Abbott, and W. H. Dantzler, *Pflugers Arch.*, **2002**, *443*, 446–457.
28. G. Capasso, R. Unwin, S. Agulian, and G. Giebisch, *J. Clin. Invest.*, **1991**, *88*, 430–437.
29. M. Froissart, P. Borensztein, P. Houillier, F. Leviel, J. Poggioli, E. Marty, M. Bichara, and M. Paillard, *Am. J. Physiol.*, **1992**, *262*, 963–970.
30. D. Z. Levine, M. Iacovitti, S. Buckman, D. Vandorpe, V. Harrison, D. M. Boisvert, and S. P. Nadler, *Am. J. Physiol.*, **1994**, *267*, 737–747.

31. D. H. Vandorpe and D. Z. Levine, *Clin. Invest. Med.*, **1989**, *12*, 224–229.

32. K. M. Madsen, J. W. Verlander, J. Kim, and C. C. Tisher, *Kidney Int Suppl*, **1991**, *33*, 57–63.

33. Q. Al-Awqati, *Am. J. Physiol.*, **1996**, *270*, 1571–1580.

34. T. D. McKinney and M. B. Burg, *J. Clin. Invest.*, **1977**, *60*, 766–768.

35. W. E. Lombard, J. P. Kokko, and H. R. Jacobson, *Am. J. Physiol.*, **1983**, *244*, 289–296.

36. D. Brown, S. Hirsch, and S. Bluck, *Nature*, **1988**, *331*, 622–624.

37. B. Bastani, H. Purcell, P. Hemken, D. Trigg, and S. Gluck, *J. Clin. Invest.*, **1991**, *88*, 126–136.

38. S. L. Alper, J. Natale, S. Gluck, H. F. Lodish, and D. Brown, *Proc. Natl. Acad. Sci. U. S.A.*, **1989**, *86*, 5429–5433.

39. V. L. Schuster, G. Fejes-Toth, and A. Naray-Fejes-Toth, and S. Gluck, *Am. J. Physiol.*, **1991**, *260*, 506–517.

40. Y. H. Kim, T. H. Kwon, S. Frische, J. Kim, C. C. Tisher, K. M. Madsen, and S. Nielsen, *Am. J. Physiol. Renal Physiol.*, **2002**, *283*, 744–754.

41. I. E. Royaux, S. M. Wall, L. P. Karniski, L. A. Everett, K. Suzuki, M. A. Knepper, and E. D. Green, *Proc. Natl. Acad. Sci. U. S.A.*, **2001**, *98*, 4221–4226.

42. C. A. Wagner, K. E. Finberg, P. A. Stehberger, R. P. Lifton, G. H. Giebisch, P. S. Aronson, and J. P. Geibel, *Kidney Int.*, **2002**, *62*, 2109–2117.

43. S. M. Wall, K. A. Hassell, I. E. Royaux, E. D. Green, J. Y. Chang, G. L. Shipley, and J. W. Verlander, *Am. J. Physiol. Renal Physiol.*, **2003**, *284*, 229–241.

44. S. Petrovic, Z. Wang, L. Ma, and M. Soleimani, *Am. J. Physiol. Renal Physiol.*, **2003**, *284*, 103–112.

45. H. Tsuganezawa, K. Kobayashi, M. Iyori, T. Araki, A. Koizumi, S. Watanabe, A. Kaneko, T. Fukao, T. Monkawa, T. Yoshida, D. K. Kim, Y. Kanai, H. Endou, M. Hayashi, and T. Saruta, *J. Biol. Chem.*, **2001**, *276*, 8180–8189.

46. S. B. Ko, X. Luo, H. Hager, A. Rojek, J. Y. Choi, C. Licht, M. Suzuki, S. Muallem, S. Nielsen, and K. Ishibashi, *Am. J. Physiol. Cell. Physiol.*, **2002**, *283*, 1206–1218.

47. I. Sabolic, D. Brown, S. L. Gluck, and S. L. Alper, *Kidney Int.*, **1997**, *51*, 125–137.

48. G. J. Schwartz, J. Barasch, and Q. Al-Awqati, *Nature*, **1985**, *318*, 368–371.

49. J. Takito, C. Hikita, and Q. Al-Awqati, *J. Clin. Invest.*, **1996**, *98*, 2324–2331.

50. G. J. Schwartz, S. Tsuruoka, S. Vijayakumar, S. Petrovic, A. Mian, and Q. Al-Awqati, *J. Clin. Invest.*, **2002**, *109*, 89–99.

51. Q. Al-Awqati, *Annu. Rev. Physiol.*, **2003**, *65*, 567–583.

52. S. Tsuruoka and G. J. Schwartz, *Am. J. Physiol.*, **1998**, *275*, 982–990.

53. L. M. Satlin and G. J. Schwartz, *J. Cell. Biol.*, **1989**, *109*, 1279–1288.

54. G. Kuramochi, U. Kersting, W. H. Dantzler, and S. Silbernagl, *Pflugers Arch.*, **1996**, *432*, 1062–1068.

55. D. Drenckhahn, K. Schluter, D. P. Allen, and V. Bennett, *Science*, **1985**, *230*, 1287–1289.

56. G. Fejes-Toth, W. R. Chen, E. Rusvai, T. Moser, and A. Naray-Fejes-Toth, *J. Biol. Chem.*, **1994**, *269*, 26717–26721.

57. S. Breton, T. Wiederhold, V. Marshansky, N. N. Nsumu, V. Ramesh, and D. Brown, *J. Biol. Chem.*, **2000**, *275*, 18219–18224.

58. W. R. Harvey, M. Cioffi, J. A. Dow, and M. G. Wolfersberger, *J. Exp. Biol.*, **1983**, *106*, 91–117.

59. T. L. Chan, J. W. Greenawalt, and P. L. Pedersen, *J. Cell. Biol.*, **1970**, *45*, 291–305.

60. K. M. Madsen and C. C. Tisher, *Lab. Invest.*, **1984**, *51*, 268–276.

61. K. M. Madsen and C. C. Tisher, *Am. J. Physiol.*, **1983**, *245*, 670–679.

62. D. L. Stetson and P. R. Steinmetz, *Am. J. Physiol.*, **1985**, *249*, 553–565.

63. D. L. Stetson and P. R. Steinmetz, *Pflugers Arch.*, **1986**, *407 (Suppl)*, 80–84.

64. D. Brown, S. Gluck, and J. Hartwig, *J. Cell. Biol.*, **1987**, *105*, 1637–1648.

65. J. Heuser, Q. Zhu, and M. Clarke, *J. Cell. Biol.*, **1993**, *121*, 1311–1327.

66. J. Glickman, K. Croen, S. Kelly, and Q. Al-Awqati, *J. Cell. Biol.*, **1983**, *97*, 1303–1308.

67. S. A. Hilden, C. A. Johns, and N. E. Madias, *Am. J. Physiol.*, **1988**, *255*, F885–F897.

68. T. E. Machen, M. J. Leigh, C. Taylor, T. Kimura, S. Asano, and H. P. Moore, *Am. J. Physiol. Cell. Physiol.*, **2003**, *285*, c205–c214.

69. Y. Moriyama and N. Nelson, *J. Biol. Chem.*, **1989**, *264*, 18445–18450.

70. Y. Moriyama and N. Nelson, *Biochim. Biophys. Acta*, **1989**, *980*, 241–247.

71. L. B. Shi, K. Fushimi, H. R. Bae, and A. S. Verkman, *Biophys. J.*, **1991**, *59*, 1208–1217.

72. V. Marshansky and P. Vinay, *Biochim. Biophys. Acta*, **1996**, *1284*, 171–180.

73. B. Maranda, D. Brown, S. Bourgoin, J. E. Casanova, P. Vinay, D. A. Ausiello, and V. Marshansky, *J. Biol. Chem.*, **2001**, *276*, 18540–18550.

74. I. Sabolic and G. Burckhardt, *Am. J. Physiol.*, **1986**, *250*, 817–826.

75. I. Sabolic and G. Burckhardt, *Biochim. Biophys. Acta*, **1988**, *937*, 398–410.

76. I. Sabolic and G. Burckhardt, *Methods Enzymol.*, **1990**, *191*, 505–520.

77. K. Zhang, Z. Q. Wang, and S. Gluck, *J. Biol. Chem.*, **1992**, *267*, 14539–14542.

78. X. S. Xie, B. P. Crider, and D. K. Stone, *J. Biol. Chem.*, **1993**, *268*, 25063–25067.

79. G. Li, E. A. Alexander, and J. H. Schwartz, *J. Biol. Chem.*, **2003**, *278*, 1991–19797.

80. A. Banerjee, G. Li, E. A. Alexander, and J. H. Schwartz, *Am. J. Physiol. Cell. Physiol.*, **2001**, *280*, 775–781.

81. A. Banerjee, T. Shih, E. A. Alexander, and J. H. Schwartz, *J. Biol. Chem.*, **1999**, *274*, 26518–26522.

82. B. S. Lee, S. L. Gluck, and L. S. Holliday, *J. Biol. Chem.*, **1999**, *274*, 29164–29171.

83. L. S. Holliday, M. Lu, B. S. Lee, R. D. Nelson, S. Solivan, L. Zhang, and S. L. Gluck, *J. Biol. Chem.*, **2000**, *275*, 32331–32337.

84. S. Breton, N. N. Nsumu, T. Galli, I. Sabolic, P. J. Smith, and D. Brown, *Am. J. Physiol. Renal Physiol.*, **2000**, *278*, 717–725.

85. O. Vitavska, H. Wieczorek, and H. Merzendorfer, *J. Biol. Chem.*, **2003**, *278*, 18499–18505.

86. J. P. Sumner, J. A. Dow, F. G. Earley, U. Klein, D. Jager, and H. Wieczorek, *J. Biol. Chem.*, **1995**, *270*, 5649–5653.

87. W. Zeiske, H. Meyer, and H. Wieczorek, *J. Exp. Biol.*, **2002**, *205*, 463–474.

88. K. J. Parra and P. M. Kane, *Mol. Cell. Biol.*, **1998**, *18*, 7064–7074.

89. P. M. Kane, *FEBS Lett.*, **2000**, *469*, 137–141.

90. J. H. Seol, A. Shevchenko, and R. J. Deshaies, *Nat. Cell. Biol.*, **2001**, *3*, 384–391.

91. A. M. Smardon, M. Tarsio, and P. M. Kane, *J. Biol. Chem.*, **2002**, *277*, 13831–13839.

92. E. S. Trombetta, M. Ebersold, W. Garrett, M. Pypaert, and I. Mellman, *Science*, **2003**, *299*, 1400–1403.

93. M. Lu, L. S. Holliday, L. Zhang, and W. A. Dunn, Jr. and S. L. Gluck, *J. Biol. Chem.*, **2001**, *276*, 30407–30413.

94. Y. Su, A. Zhou, R. S. Al-Lamki, and F. E. Karet, *J. Biol. Chem.*, **2003**, *278*, 20013–20018.

95. S. Uchida, *Am. J. Physiol. Renal Physiol.*, **2000**, *279*, 802–808.

96. T. J. Jentsch, V. Stein, F. Weinreich, and A. A. Zdebik, *Physiol. Rev.*, **2002**, *82*, 503–568.

97. H. R. Bae and A. S. Verkman, *Nature*, **1990**, *348*, 637–639.

98. W. Gunther, A. Luchow, F. Cluzeaud, A. Vandewalle, and T. J. Jentsch, *Proc. Natl. Acad. Sci. U.S.A.*, **1998**, *95*, 8075–8080.

99. V. A. Luyckx, F. O. Goda, D. B. Mount, T. Nishio, A. Hall, S. C. Hebert, T. G. Hammond, and A. S. Yu, *Am. J. Physiol.*, **1998**, *275*, 761–769.

100. H. Sakamoto, Y. Sado, I. Naito, T. H. Kwon, S. Inoue, K. Endo, M. Kawasaki, S. Uchida, S. Nielsen, S. Sasaki, and F. Marumo, *Am. J. Physiol.*, **1999**, *277*, 957–965.

101. C. Isnard-Bagnis, N. Silva Da, V. Beaulieu, A. S. Yu, D. Brown, and S. Breton, *Am. J. Physiol. Cell. Physiol.*, **2003**, *284*, 220–232.

102. R. Mohammad-Panah, R. Harrison, S. Dhani, C. Ackerley, L. J. Huan, Y. Wang, and C. E. Bear, *J. Biol. Chem.*, **2003**, *278*, 29267–29277.

103. N. K. Wills and P. Fong, *News Physiol. Sci.*, **2001**, *16*, 161–166.

104. N. Piwon, W. Gunther, M. Schwake, M. R. Bosl, and T. J. Jentsch, *Nature*, **2000**, *408*, 369–373.

105. W. Gunther, N. Piwon, and T. J. Jentsch, *Pflugers Arch.*, **2003**, *445*, 456–462.
106. N. D. Sonawane, J. R. Thiagarajah, and A. S. Verkman, *J. Biol. Chem.*, **2002**, *277*, 5506–5513.
107. M. Aridor and L. A. Hannan, *Traffic*, **2000**, *1*, 836–851.
108. M. Aridor and L. A. Hannan, *Traffic*, **2002**, *3*, 781–790.
109. P. P. Breitfeld, C. F. Simmons, Jr., G. J. Strous , H. J. Geuze, and A. L. Schwartz, *Int. Rev. Cytol.*, **1985**, *97*, 47–95.
110. C. G. Davis, J. L. Goldstein, T. C. Sudhof, R. G. Anderson, D. W. Russell, and M. S. Brown, *Nature*, **1987**, *326*, 760–765.
111. D. J. Yamashiro and F. R. Maxfield, *J. Cell. Biochem.*, **1984**, *26*, 231–246.
112. J. E. Zijderhand-Bleekemolen, A. L. Schwartz, J. W. Slot, G. J. Strous, and H. J. Geuze, *J. Cell. Biol.*, **1987**, *104*, 1647–1654.
113. M. J. Clague, S. Urbe, F. Aniento, and J. Gruenberg, *J. Biol. Chem.*, **1994**, *269*, 21–24.
114. B. Reaves and G. Banting, *J. Cell. Biol.*, **1992**, *116*, 85–94.
115. A. W. Weert van, K. W. Dunn, H. J. Gueze, F. R. Maxfield, and W. Stoorvogel, *J. Cell. Biol.*, **1995**, *130*, 821–834.
116. C. E. Gustafson, T. Katsura, M. McKee, R. Bouley, J. E. Casanova, and D. Brown, *Am. J. Physiol. (Renal Physiol.)*, **1999**, *278*, 317–326.
117. B. Thorens and J. Roth, *J. Cell. Sci.*, **1996**, *109*, 1311–1323.
118. V. Batuman and S. Guan, *Am. J. Physiol.*, **1997**, *272*, 521–530.
119. J. S. Choi, K. R. Kim, D. W. Ahn, and Y. S. Park, *Toxicol. Appl. Pharmacol.*, **1999**, *161*, 146–152.
120. M. Gekle, S. Mildenberger, R. Freudinger, and S. Silbernagl, *Am. J. Physiol.*, **1995**, *268*, 899–906.
121. M. Gekle, S. Mildenberger, R. Freudinger, and S. Silbernagl, *J. Am. Soc. Nephrol.*, **1998**, *9*, 960–968.
122. C. M. Herak-Kramberger, D. Brown, and I. Sabolic, *Kidney Int.*, **1998**, *53*, 1713–1726.
123. L. Orci, D. J. Palmer, M. Ravazzola, A. Perrelet, M. Amherdt, and J. E. Rothman, *Nature*, **1993**, *362*, 648–652.
124. J. Ostermann, L. Orci, K. Tani, M. Amherdt, M. Ravazzola, Z. Elazar, and J. E. Rothman, *Cell*, **1993**, *75*, 1015–1025.
125. M. Bremser, W. Nickel, M. Schweikert, M. Ravazzola, M. Amherdt, C. A. Hughes, T. H. Sollner, J. E. Rothman, and F. T. Wieland, *Cell*, **1999**, *96*, 495–506.
126. F. Aniento, F. Gu, R. G. Parton, and J. Gruenberg, *J. Cell. Biol.*, **1996**, *133*, 29–41.
127. F. Gu and J. Gruenberg, *J. Biol. Chem.*, **2000**, *275*, 8154–8160.
128. S. Zeuzem, P. Feick, P. Zimmermann, W. Haase, R. A. Kahn, and I. Schulz, *Proc. Natl. Acad. Sci. U. S.A.*, **1992**, *89*, 6619–6623.
129. W. I. Lencer, A. S. Verkman, M. A. Arnaout, D. A. Ausiello, and D. Brown, *J. Cell. Biol.*, **1990**, *111*, 379–389.
130. I. Sabolic, F. Wuarin, L. B. Shi, A. S. Verkman, D. A. Ausiello, S. Gluck, and D. Brown, *J. Cell. Biol.*, **1992**, *119*, 111–122.
131. R. Fuchs, A. Ellinger, M. Pavelka, I. Mellman, and H. Klapper, *Proc. Natl. Acad. Sci. U. S.A.*, **1994**, *91*, 4811–4815.
132. A. Hurtado-Lorenzo, J. El-Annan, S. Bechoua, M. Futai, S. Bourgoin, J. Casanova, D. Brown, D. A. Ausiello, and V. Marshansky, *J. Am. Soc. Nephrol.*, **2003**, *14*, 317a Abstract.
133. A. Hurtado-Lorenzo, J. El-Annan, S. Bechoua, M. Futai, S. Bourgoin, J. Casanova, D. Brown, D. A. Ausiello, and V. Marshansky, *Mol. Biol. Cell*, **2003**, *14*, 237a–238a Abstract.
134. E. I. Christensen, H. Birn, P. Verroust, and S. K. Moestrup, *Int. Rev. Cytol.*, **1998**, *180*, 237–284.
135. E. I. Christensen and H. Birn, *Am. J. Physiol. Renal Physiol.*, **2001**, *280*, 562–573.
136. D. Kerjaschki and M. G. Farquhar, *Proc. Natl. Acad. Sci. U. S.A.*, **1982**, *79*, 5557–5581.
137. R. Raychowdhury, J. L. Niles, R. T. McCluskey, and J. A. Smith, *Science*, **1989**, *244*, 1163–1165.
138. A. Saito, S. Pietromonaco, A. K. Loo, and M. G. Farquhar, *Proc. Natl. Acad. Sci. U. S.A.*, **1994**, *91*, 9725–9729.

139. S. K. Moestrup, R. Kozyraki, M. Kristiansen, J. H. Kaysen, H. H. Rasmussen, D. Brault, F. Pontillon, F. O. Goda, E. I. Christensen, T. G. Hammond, and P. J. Verroust, *J. Biol. Chem.*, **1998**, *273*, 5235–5242.

140. D. Sahali, N. Mulliez, F. Chatelet, R. Dupuis, P. Ronco, and P. Verroust, *J. Exp. Med.*, **1988**, *167*, 213–218.

141. H. Birn, J. C. Fyfe, C. Jacobsen, F. Mounier, P. J. Verroust, H. Orskov, T. E. Willnow, S. K. Moestrup, and E. I. Christensen, *J. Clin. Invest.*, **2000**, *105*, 1353–1361.

142. S. Cui, P. J. Verroust, S. K. Moestrup, and E. I. Christensen, *Am. J. Physiol.*, **1996**, *271*, 900–907.

143. M. Marino, D. Andrews, D. Brown, and R. T. McCluskey, *J. Am. Soc. Nephrol.*, **2001**, *12*, 637–648.

144. E. I. Christensen and H. Birn, *Nat. Rev. Mol. Cell Biol.*, **2002**, *3*, 258–268.

145. B. E. Sumpio and T. Maack, *Am. J. Physiol.*, **1982**, *243*, 379–392.

146. T. Maack, C. Park, and M. Camargo, *The Kidney: Physiol.ogy and Pathophysiology, Vol. V.3, Second Edition* Ed. D. W. Seldin and G. Giebisch, eds., Raven Press, New York **1992**, pp. 3005–3038.

147. M. E. Rosenberg and T. H. Hostetter, *The Kidney: Physiology and Pathophysiology, Vol. v.3, Second Edition* Ed. D. W. Seldin and G. Giebisch, ed., Raven Press, New York **1992**, pp. 3039–3062.

148. T. M. Osicka, C. A. Houlihan, J. G. Chan, G. Jerums, and W. D. Comper, *Diabetes*, **2000**, *49*, 1579–1584.

149. K. A. Greive, N. D. Balazs, and W. D. Comper, *Clin. Chem.*, **2001**, *47*, 1717–1719.

150. L. M. Russo, G. L. Bakris, and W. D. Comper, *Am. J. Kidney Dis.*, **2002**, *39*, 899–919.

151. V. Marshansky, A. Fleser, J. Noel, S. Bourgoin, and P. Vinay, *J. Membr. Biol.*, **1996**, *153*, 59–73.

152. M. J. Burne, T. M. Osicka, and W. D. Comper, *Kidney Int.*, **1999**, *55*, 261–270.

153. G. A. Eppel, T. M. Osicka, L. M. Pratt, P. Jablonski, B. O. Howden, E. F. Glasgow, and W. D. Comper, *Kidney Int.*, **1999**, *55*, 1861–1870.

154. G. A. Eppel, T. M. Osicka, L. M. Pratt, P. Jablonski, B. Howden, E. F. Glasgow, and W. D. Comper, *Ren Fail*, **2001**, *23*, 347–363.

155. S. E. Fisher, G. C. Black, S. E. Lloyd, E. Hatchwell, O. Wrong, R. V. Thakker, and I. W. Craig, *Hum. Mol. Gen.*, **1994**, *3*, 2053–2059.

156. S. E. Fisher, I. Bakel van, S. E. Lloyd, S. H. Pearce, R. V. Thakker, and I. W. Craig, *Genomics*, **1995**, *29*, 598–606.

157. K. Steinmeyer, B. Schwappach, M. Bens, A. Vandewalle, and T. J. Jentsch, *J. Biol. Chem.*, **1995**, *270*, 31172–31177.

158. S. E. Lloyd, S. H. Pearce, S. E. Fisher, K. Steinmeyer, B. Schwappach, S. J. Scheinman, B. Harding, A. Bolino, M. Devoto, P. Goodyer, S. P. Rigden, O. Wrong, T. J. Jentsch, I. W. Craig, and R. V. Thakker, *Nature*, **1996**, *379*, 445–449.

159. S. S. Wang, O. Devuyst, P. J. Courtoy, X. T. Wang, H. Wang, Y. Wang, R. V. Thakker, S. Guggino, and W. B. Guggino, *Hum. Mol. Gen.*, **2000**, *9*, 2937–2945.

160. I. V. Silva, V. Cebotaru, H. Wang, X. T. Wang, S. S. Wang, G. Guo, O. Devuyst, R. V. Thakker, W. B. Guggino, and S. E. Guggino, *J. Bone Miner. Res.*, **2003**, *18*, 615–623.

161. J. R. Leheste, B. Rolinski, H. Vorum, J. Hilpert, A. Nykjaer, C. Jacobsen, P. Aucouturier, J. O. Moskaug, A. Otto, E. I. Christensen, and T. E. Willnow, *Am. J. Pathol.*, **1999**, *155*, 1361–1370.

162. J. C. Fyfe, K. S. Ramanujam, K. Ramaswamy, D. F. Patterson, and B. Seetharam, *J. Biol. Chem.*, **1991**, *266*, 4489–4494.

163. X. Y. Zhai, R. Nielsen, H. Birn, K. Drumm, S. Mildenberger, R. Freudinger, S. K. Moestrup, P. J. Verroust, E. I. Christensen, and M. Gekle, *Kidney Int.*, **2000**, *58*, 1523–1533.

164. J. C. Fyfe, U. Giger, C. A. Hall, P. F. Jezyk, S. A. Klumpp, J. S. Levine, and D. F. Patterson, *Pediatr. Res.*, **1991**, *29*, 24–31.

165. D. Xu, R. Kozyraki, T. C. Newman, and J. C. Fyfe, *Blood*, **1999**, *94*, 3604–3606.

166. M. Aminoff, J. E. Carter, R. B. Chadwick, C. Johnson, R. Grasbeck, M. A. Abdelaal, H. Broch, L. B. Jenner, P. J.

Verroust, S. K. Moestrup, A. Chapelle de la, and R. Krahe, *Nat. Gen.*, **1999**, *21*, 309–313.

167. H. Broch, O. Imerslund, E. Monn, T. Hovig, and M. Seip, *Acta Paediatr. Scand.*, **1984**, *73*, 248–253.

168. A. G. Norden, M. Lapsley, T. Igarashi, C. L. Kelleher, P. J. Lee, T. Matsuyama, S. J. Scheinman, H. Shiraga, D. P. Sundin, R. V. Thakker, R. J. Unwin, P. Verroust, and S. K. Moestrup, *J. Am. Soc. Nephrol.*, **2002**, *13*, 125–133.

169. E. I. Christensen, O. Devuyst, G. Dom, R. Nielsen, P. Smissen Van der, P. Verroust, M. Leruth, W. B. Guggino, and P. J. Courtoy, *Proc. Natl. Acad. Sci. U. S. A.*, **2003**, *100*, 8472–8477.

170. R. G. Adams, J. F. Harrison, and P. Scott, *Q. J. Med.*, **1969**, *38*, 425–443.

171. H. Gonick, S. Indraprasit, H. Neustein, and V. Rosen, *Curr. Probl. Clin. Biochem.*, **1975**, *4*, 111–118.

172. C. M. Herak-Kramberger, B. Spindler, J. Biber, H. Murer, and I. Sabolic, *Pflugers Arch.*, **1996**, *432*, 336–344.

173. K. R. Kim, H. Y. Lee, C. K. Kim, and Y. S. Park, *Toxicol. Appl. Pharmacol.*, **1990**, *106*, 102–111.

174. H. Y. Lee, K. R. Kim, and Y. S. Park, *Pharmacol. Toxicol.*, **1991**, *69*, 390–395.

175. K. Nogawa, A. Ishizaki, M. Fukushima, I. Shibata, and N. Hagino, *Environ. Res.*, **1975**, *10*, 280–307.

176. F. Thevenod and J. M. Friedmann, *FASEB J.*, **1999**, *13*, 1751–1761.

177. I. Sabolic, M. Ljubojevic, C. M. Herak-Kramberger, and D. Brown, *Am. J. Physiol. Renal Physiol.*, **2002**, *283*, 1389–1402.

178. I. Sabolic, C. M. Herak-Kramberger, and D. Brown, *Toxicology*, **2001**, *165*, 205–216.

179. C. R. Caflisch and T. D. DuBose, Jr., *J. Toxicol. Environ. Health*, **1991**, *32*, 49–57.

180. C. R. Caflisch and J. Toxicol, *Environ. Health*, **1994**, *42*, 323–330.

181. C. M. Herak-Kramberger, I. Sabolic, M. Blanusa, P. J. Smith, D. Brown, and S. Breton, *Biol. Reprod.*, **2000**, *63*, 599–606.

182. T. I. P. K. D Consortium, *Cell*, **1995**, *81*, 289–298.

183. T. Mochizuki, G. Wu, T. Hayashi, S. L. Xenophontos, B. Veldhuisen, J. J. Saris, D. M. Reynolds, Y. Cai, P. A. Gabow, A. Pierides, W. J. Kimberling, M. H. Breuning, C. C. Deltas, D. J. Peters, and S. Somlo, *Science*, **1996**, *272*, 1339–1342.

184. N. Obermuller, B. Kranzlin, W. F. Blum, N. Gretz, and R. Witzgall, *Am. J. Physiol. Renal Physiol.*, **2001**, *280*, 244–253.

185. R. Witzgall, B. Kranzlin, N. Gretz, and N. Obermuller, *Kidney Int. Suppl.*, **2002**, *61 (Suppl)*, 132–137.

186. H.-H. Parving, R. Osterby, and E. Ritz, *The Kidney, Vol. V.II (Sixth Edition Ed)*. B. M. Brenner, ed., W. B. Saunders Company, Philadelphia **2000**, pp. 1731–1773.

187. L. M. Russo, T. M. Osicka, G. C. Brammar, R. Candido, G. Jerums, and W. D. Comper, *Am. J. Nephrol.*, **2003**, *23*, 61–70.

188. V. Kalatzis and C. Antignac, *Pediatr. Nephrol.*, **2003**, *18*, 207–215.

189. E. L. Eskelinen, Y. Tanaka, and P. Saftig, *Trends Cell Biol.*, **2003**, *13*, 137–145.

190. A. J. Jonas, M. L. Smith, W. S. Allison, P. K. Laikind, A. A. Greene, and J. A. Schneider, *J. Biol. Chem.*, **1983**, *258*, 11727–11730.

191. A. J. Jonas, *Biochem. J.*, **1986**, *236*, 671–677.

192. V. Kalatzis, S. Cherqui, C. Antignac, and B. Gasnier, *EMBO J.*, **2001**, *20*, 5940–5949.

193. M. L. Smith, A. A. Greene, R. Potashnik, S. A. Mendoza, and J. A. Schneider, *J. Biol. Chem.*, **1987**, *262*, 1244–1253.

194. T. Toyomura, Y. Murata, A. Yamamoto, T. Oka, G. H. Sun-Wada, Y. Wada, and M. Futai, *J. Biol. Chem.*, **2003**, *278*, 22023–22030.

195. K. M. Stankovic, D. Brown, S. L. Alper, and J. C. Adams, *Hear Res.*, **1997**, *114*, 21–34.

196. T. G. Paunescu, N. Da Silva, V. Marshansky, M. McKee, S. Breton, and D. Brown, *Am J. Physiol.*, in press, epub: 10.1152/ajpcell. 00464. 2003 (March 10, **2004**).

197. A. N. Smith, K. E. Finberg, C. A. Wagner, R. P. Lifton, M. A. Devonald, Y. Su, and F. E. Karet, *J. Biol. Chem.*, **2001**, *276*, 42382–42388.

18

Lytic Function of Vacuole, Molecular Dissection of Autophagy in Yeast

Yoshinori Ohsumi

18.1
Introduction

Proteolysis has become one of the most interesting fields of biology. Every cellular event is maintained by a balance of synthesis and breakdown of the related proteins. It has become a general concept that proteins are not degraded spontaneously, but rather are degraded by active processes. Each protein has its own wide-ranging lifetime, but we do not yet know the determinants of the lifetime of each protein and its modulation by physiological conditions. There are two major pathways of intracellular protein degradation. First, the ubiquitin-proteasome system in the cytoplasm is involved in degradation of short-lived, damaged or misfolded proteins [1, 2]. Proteins targeted for degradation are first tagged with a small protein of 76 amino acid, ubiquitin, and then digested by a large proteinase complex, the proteasome. Both the sophisticated ubiquitin ligation system and the cleavage by proteasome ensure the specificity of targets to be degraded depend upon ATP hydrolysis.

The second major pathway performs the degradation of long-lived proteins within a lytic compartment, the lysosome/vacuole. Almost half a century has passed since the lytic organelle, the lysosome, was reported by Christian de Duve using cell fractionation procedures [3]. Since then, many electron microscopic studies have revealed autophagy in various cells from different organs and in cultured cells. However, the molecular basis of the autophagy has remained unclear because of the dynamic and complicated nature of mammalian lysosomes. Moreover, a quantitative monitoring system for autophagy and a specific marker protein in mammalian cells were lacking. Recently, genes specifically involved in autophagy were discovered using yeast, *Saccharomyces cerevisiae*.

Handbook of ATPases. Edited by M. Futai, Y. Wada, J. H. Kaplan
Copyright © 2004 WILEY-VCH Verlag GmbH & Co. KGaA, Weinheim
ISBN 3-527-30689-7

Table 18.1 Yeast vacuolar enzymes.

Protein	Function
Pep4	Proteinase A (PrA/yscA/saccharopepsin)
Prb1	Proteinase B (yscB/PrB/cerevisin)
Prc1	Carboxypeptidase Y (CPY/yscY)
Ape3	Aminopeptidase Y (yscIII, APY)
Lap4	Aminopeptidase I (yscI, API)
Dap2	Dipeptidyl aminopeptidase B (DPAP B)
Pho8	Alkaline phosphatase (ALP)
Ams1	α-Mannosidase
Sga1	Glucoamylase
Ath1	Acid trehalase
?	RNase

18.2
Vacuolar Protein Degradation in Yeast

The yeast vacuole had been postulated as a lytic organelle, since it is acidic inside and contains various hydrolytic enzymes, such as mammalian lysosomes [4]. Intensive genetic studies on yeast proteinases have been undertaken in the laboratories of Wolf and of Jones [5, 6]. The enzymology of vacuolar enzymes and their biosynthetic pathways have also been studied quite extensively [7, 8]. Table 18.1 lists the major yeast vacuolar enzymes. Interestingly, major proteinases in yeast vacuole are serine proteases instead of cysteine proteases in mammalian lysosome and plant vacuole. Previous work indicated that bulk protein turnover is induced upon nitrogen starvation and is dependent on vacuolar enzyme activities [9, 10], suggesting that the vacuole is actually a lytic compartment. However, little was known about the mechanism of transport of substrates to the vacuole. What kinds of substrates are degraded in the vacuole? What is the mechanism of transporting those substrates into the vacuole? How is the process regulated?

18.3
Intracellular Bulk Degradation – Autophagy

In various organisms several delivery routes to the lytic compartment are proposed. The degradation of intracellular components in lysosomes is generally called autophagy, in contrast to heterophagy of extracellular materials [11]. Macroautophagy, the major autophagic pathway, begins with an isolation membrane enclosing a portion of the cytoplasm to form an autophagosome, a double membrane structure containing the engulfed cytoplasm [12] (Figure 18.1). The outer membrane of an autophagosome subsequently fuses with the lysosome, and, thereby, the inner membrane and cytoplasmic contents are digested by the lysosomal hydrolases for reuse. Several other processes deliver cytoplasmic components or organelles

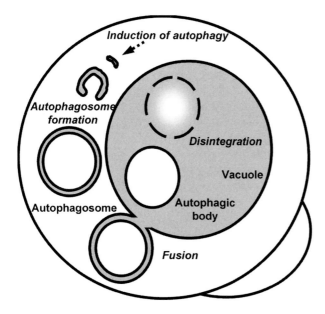

Figure 18.1 Scheme of autophagy in yeast.

to the lysosome. Microautophagy is a process in which the lysosomal or vacuolar membrane engulfs a portion of the cytoplasm by invagination. Chaperone-mediated-autophagy is the direct transport of proteins across the lysosomal membrane, with the help of the chaperones Hsc73 and Lgp96 [13, 14]. The physiological relevance of these systems is still not so clear. This chapter focuses on macroautophagy, which will be referred to simply as autophagy. Autophagy is characterized as the nonselective and bulk degradation of cytoplasmic components, including cytosolic proteins, large complexes such as ribosomes, and organelles. More than 90 % of cellular proteins are long-lived proteins; thus the turnover of long-lived proteins via autophagy is important in understanding cell growth physiology.

18.4
Discovery of Autophagy in Yeast

We started work on the lytic function of vacuole of yeast in 1988. At that time we had realized that mutants defective in various vacuolar functions, including H^+-ATPases and Ca^{2+} homeostasis, showed sporulation-negative phenotype. This cell differentiation is triggered by nitrogen starvation and, therefore, must strictly depend upon the degradation of preexisting proteins and nucleic acids. Thus, we predicted that nitrogen starvation evokes the bulk protein degradation in the vacuole. We then simply observed the morphological change of vacuole during starvation. When vacuolar proteinase-deficient mutants grown in a rich medium were transferred to nitrogen-depleted medium or sporulation medium, spherical structures appeared in the vacuole after 30–40 min, and accumulated and almost filled the vacuole for up to 10 h [15]. These structures were assumed to be an intermediate structure of delivery to the vacuole. Electron microscopy proved that these, named autophagic bodies, were mostly single membrane-bound structures containing a portion of the cytoplasm [15]. Subsequently autophagosomes, double membrane structures, were found in the cytoplasm of the starved cells. Autophagosomes in yeast are about 300–900 nm in diameter and contain cytosolic enzymes, ribosomes and, occasionally, other cellular structures, including mitochondria and rough endoplasmic reticulum (rER) [15]. Cytosolic enzymes such as PGK, ADH, Glu6PDH are equally distributed in the cytoplasm, autophagosomes, and autophagic bodies. Biochemical and immuno-electron microscopic analyses indicated that autophagosomes enclose cytoplasmic components non-selectively. Freeze–fracture electron microscopy clearly demonstrated fusion of the outer membrane of autophagosomes with the vacuolar membrane [16, 17].

Exactly the same membrane phenomena were induced under not only nitrogen- but also carbon-, sulfate-, phosphate-, and auxotrophic single amino acid-starvation. However, other stress conditions such as heat or osmotic shock did not induce

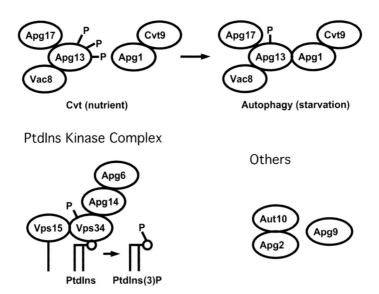

Figure 18.2 Functions of Apg proteins. Various types of starvation trigger signal transduction to induce autophagy. A small membrane sac, termed the isolation membrane, appears in the cytoplasm and elongates to enwrap a portion of the cytoplasm. Once a double membrane structure, the autophagosome (AP), is formed it immediately fuses with the vacuole. Upon fusion of the outer membrane of the autophagosome with the vacuole (V), the inner membrane vesicle of the autophagosome is released into the vacuolar lumen, giving rise to the autophagic body (AB). In wild-type cells, the autophagic body is rapidly disintegrated and its contents digested for reuse.

autophagy at all. This strongly indicated that, in yeast cells, autophagy is primarily a nutrient starvation response and under adverse conditions functions to transport a portion of cytoplasm to the lytic compartment via autophagosomes. The membrane dynamics of yeast autophagy are essentially the same as macroautophagy in mammals, though the yeast vacuole is much larger than a mammalian lysosome. The autophagy in yeast is shown schematically in Figure 18.1.

18.5
Monitoring of Autophagy

Autophagy has been studied mainly by electron microscopy, which requires much skill and time, and sometimes causes ambiguity in distinguishing autophagosome from other membrane structures. Another way to follow bulk degradation is the release of amino acids from pre-labeled proteins. Generally it is more difficult to quantify the decrease than an increase. We developed a method to monitor autophagic activity using molecular genetic trick. Alkaline phosphatase (Pho8) is a vacuolar enzyme whose active site is in vacuolar lumen, and is synthesized as a proenzyme and activated by the proteinases in the vacuole. When the cell expressing the cytosolic form, Pho8Δ60, was transferred to starvation medium, the proform was transferred to the vacuole via autophagy and became an active enzyme [18]. This ALP assay system provided a reliable estimation of autophagy and has been used for various analyses.

18.6
Genetic Approaches to Yeast Autophagy

The progress of autophagy in yeast can be monitored in real time as the accumulation of autophagic bodies under a light microscope. Using this advantage, genetic approaches were taken to screen for autophagy-defective mutants. Cells were screened for a defect in the vacuolar accumulation of autophagic bodies, and only one mutant, *apg1*, was obtained [19]. This mutant failed to induce protein degradation in response to starvation, and could not maintain viability upon prolonged nitrogen starvation. The latter phenotype, assumed to be due to a defect in autophagy, was applied to screen for more mutants. A total of fifteen *apg* mutants were thus obtained. Another approach, taken by Thumm and colleagues, involved immunoscreening for cells that retained a cytosolic enzyme, fatty acid synthase, after starvation [20]. By this method, originally 6 *aut* mutants were obtained. Later, two hybrid screens using Apg proteins as bait identified two more *APG* genes [21, 22]. Klionsky's group, isolating mutants defective in maturation of the vacuolar enzyme aminopeptidase I (API), identified the Cvt (Cytoplasm to vacuole targeting) pathway [23, 23a]. The *CVT* genes heavily overlapped with the autophagy-defective *APG* and *AUT* genes [23, 24]. The Cvt and autophagy pathways use quite similar machinery for membrane dynamics [25], although the two pathways are quite different. The autophagy pathway is degradative and induced by starvation whereas the Cvt pathway is biosynthetic and active under growing conditions.

18.7
Characterization of Autophagy Defective Mutants

All *apg* and *aut* mutants display similar growth phenotypes. They grow normally despite failing to induce bulk protein degradation under various nutrient-starvation conditions. Homozygous diploids with any *apg* or *aut* mutation fail to sporulate [19], suggesting that bulk protein degradation via autophagy is requisite for this intracellular remodeling process triggered by nitrogen starvation. Another characteristic feature of autophagy-defective mutants is loss of viability during nitrogen star-

Table 18.2 Apg, Aut, Cvt and Gsa proteins and of their mammalian orthologues.

Apg	Aut	Cvt	Gsa	Mammals	Function etc.
1	3	10	10	ULK1	Protein kinase, interacts with Apg13, Apg17 and Cvt19, PAS localization
2	8		11	Apg2	PAS localization
3	1		20	Apg3	Apg8-conjugating enzyme, E2
4	2			Autophagin1-4	C-terminal processing and deconjugating enzyme of Apg8, cysteine protease
5				Apg5	Target of Apg12, required for elongation of isolation membranes
6				Beclin-1	Subunit of PtdIns 3-kinase complex, Vps30p
7		2	7	Apg7	Apg12- and Apg7-activating enzyme, E1
8	7	5		LC3, GATE16 GABARAP	Ubl, conjugates to PE, localizes on PAS and autophagosome
9	9	7	14	Apg9?	Membrane protein
10				Apg10	Apg12-conjugating enzyme, E2
12				Apg12	Ubl, conjugate to Apg5
13				?	Subunit of the Apg1 protein complex, phosphorylated under growing
14		12		?	Subunit pf PtdIns 3-kinase complex (autophagy-specific)
16	11			Apg16L	Bind to the Apg12-apg5 conjugate
17				?	Interacts with Apg1, not required for the Cvt pathway
	10	18	12	?	Function in an early step of macroautophagy and pexophagy
	4			?	Required for degradation of autophagic bodies
	5			?	Required for degradation of autophagic bodies, putative lipase

vation. These mutants start to die after two days of starvation and almost all cells lose viability within one week [19]. The *pep4* mutant, which has pleiotropic defects in vacuolar hydrolases, shows a similar loss of viability phenotype as *apg*. All the isolated *apg* mutants behave like *apg* null mutants. They do not display any significant differences from wild-type cells in their response to various types of stress such as heat, osmotic and salt stress. Furthermore, vacuolar functions, secretion, and endocytosis in these mutants are almost completely normal. As a consequence of the mutant isolation strategies, mutants with abnormal vacuole morphology, partially autophagy-defective mutants, and mutants of genes overlapping with certain essential functions were eliminated. Molecular biological studies by us, Thumm and Klionsky's group revealed that a total of sixteen genes are necessary specifically for autophagy in yeast (Table 18.2). Typically autophagy-specific genes seem to be nearly saturated, although it is becoming clear that more genes are required for normal levels of autophagy.

18.8
Proteins Involved in Autophagy

Cloning and identification of the autophagy genes revealed that almost all were previously uncharacterized genes. The sole exception was *APG6*, which turned out to be allelic to *VPS30*, a gene required for vacuolar protein sorting [26]. The genes related to autophagy are shown in Table 18.1. In yeast, autophagy is almost completely shut off under nutrient-rich, growth promoting conditions. Autophagy is strictly induced upon nutrient starvation, but every *APG* gene is expressed under growing conditions. This constitutive expression of the *APG* genes may be in anticipation of emergencies or may be partly due to the Cvt pathway being active in growing cells. Transcription of several *APG* genes, such as *APG8* and *APG14*, is up-regulated by starvation and is under negative regulation of TOR kinase [27–29]. Molecular biological and biochemical analyses of the *APG* genes and their products have revealed genetic and physical interactions among Apg proteins. We now know that the Apg proteins have four functional subgroups: the Apg12 conjugation system, the Apg8 lipidation system, the Apg1 protein kinase complex, and a PI 3-kinase complex.

18.8.1
Apg12 Conjugation System

Analysis of the Apg proteins revealed the remarkable discoveries of two ubiquitin-like conjugation systems [30] (Figure 18.3). Actually, more than half of the *APG* genes are involved in these novel conjugation systems. First, Apg12, a small hydrophilic protein of 186 amino acids with no apparent homology to ubiquitin, forms a covalently-linked conjugate with Apg5 [31]. This conjugate formation is essential for autophagy and, like ubiquitination, is mediated by consecutive reactions. The C-terminal Gly residue of Apg12 is activated by an activating enzyme, Apg7 (E1),

Apg12 conjugation system

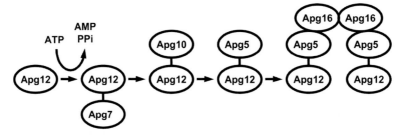

Apg8 lipidation system

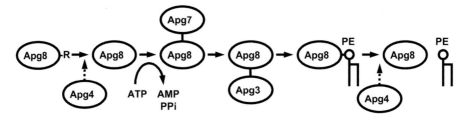

Figure 18.3 Apg 12 conjugation and Apg8 lipidation systems.

and then attached via a thioester linkage to a conjugating enzyme, Apg10 (E2) [32, 33]. Finally Apg12 forms a conjugate with Apg5 via an isopeptide bond between the C-terminal Gly of Apg12 and Lys149 of Apg5. Apg12 and Apg5 form a conjugate immediately after synthesis; free forms of Apg12 and Apg5 are hardly detectable. Apg5 seems to be the only target of the Apg12 modification. So far, no protease activity to de-conjugate Apg12-Apg5 has been found, suggesting that the conjugation reaction is irreversible. The conjugate further forms a complex with Apg16. Apg16 was originally isolated in a two-hybrid screen using Apg12 as bait, although Apg16 binds Apg5 directly, but not Apg12 [21]. Apg16 has a coiled-coil region in its C-terminal half and forms an oligomer through this region. Four copies of the Apg12-Apg5-Apg16 trimeric complex probably form a supercomplex that is essential for autophagy [34].

18.8.2
Apg8 Lipidation System

The second ubiquitin-like protein essential for autophagy is Aut7/Apg8, an 117 amino acid protein. Apg8 was the first marker protein for membrane dynamics during autophagy because it resides on the membrane of the isolated membrane, the autophagosome, and the autophagic body [27]. Cell fractionation studies indicated that about half of the Apg8 is peripherally bound to membrane whereas the other half behaves like an integral membrane protein [35] (Figure 18.3).

Epitope tagging proved that nascent Apg8 is processed at its C-terminus [35]. Apg4, a novel cysteine protease, removes a terminal Arg, thus exposing a Gly at the C-terminus of Apg8. The processed form of Apg8 is in turn activated by Apg7 and then transferred to a conjugating enzyme, Apg3 [36]. Apg7 is a unique enzyme that activates two different ubiquitin-like proteins, Apg12 and Apg8, and assigns them to proper E2 enzymes, Apg10 and Apg3, respectively. Apg3 is somewhat homologous to Apg10, but has no overall significant homology to E2 enzymes of the ubiquitin system. An obvious question was what is the target of Apg8? In conventional SDS-PAGE, Apg8 gave a single band at the expected molecular size. Mass spectrometry on the tightly membrane bound form of Apg8 revealed that Apg8 forms a conjugate not with a protein but with phosphatidylethanolamine (PE), an abundant membrane phospholipid [36]. This lipidated Apg8 is necessary for the membrane dynamics of autophagy. Furthermore, for normal autophagosome formation, Apg8-PE needs to be deconjugated, by the deconjugating enzyme Apg4, suggesting that Apg8 is cycled between a lipidated and a nonlipidated form [36]. The two Apg8-PE and Apg12-Apg5 conjugation reactions are somehow related, as the level of Apg8-PE is significantly reduced in cells lacking the Apg12-Apg5 conjugate.

18.8.3
Apg1 Protein Kinase Complex

The third protein complex required for autophagy is the Apg1 protein kinase complex. Apg1 is a serine/threonine protein kinase [37, 38]. The protein kinase domain of Apg1 resides in the N-terminal region, but C-terminal extension is also necessary for the function. The kinase activity can be detected *in vitro* using artificial substrate. Apg1 kinase activity increases during the induction of autophagy and kinase-negative *apg1* mutations block the induction of autophagy, suggesting that Apg1 kinase activity is essential for the regulation of autophagosome formation [22, 37]. Apg1 is physically associated with Apg13, Apg17 and Cvt9.

Apg13 is highly phosphorylated under nutrient-rich conditions. Upon starvation or addition of the TOR inhibitor rapamycin, Apg13 is dephosphorylated by an as-yet-unknown phosphatase [22]. Conversely, upon addition of nutrients to starved cells, Apg13 is rapidly hyperphosphorylated. Overproduction of Apg1 partially suppresses the autophagy-defect of an *apg13* mutation [39]. Consistent with this genetic interaction, Apg13, via a central region, physically associates with Apg1 [22]. Under starvation conditions, Apg13 is tightly associated with Apg1, while under nutrient-rich conditions the affinity becomes lower [22]. In addition, in an *apg13Δ* mutant, the kinase activity of Apg1 is low. These results suggest that Apg13 is a positive regulator for the Apg1 protein kinase. The above observations also suggest that Apg13 and, ultimately, Apg1 kinase activity are controlled by the TOR signaling pathway in response to nutrient conditions.

Transport of API to the vacuole is completely blocked when the *apg13* null mutant is grown in a nutrient-rich medium, but the block can be partially overcome by incubation in starvation conditions [40]. Furthermore, in an *apg13* mutant in which

Apg13 lacks most of the Apg1-binding region, transport of API is normal but autophagy is completely absent [22]. Thus, Apg13 may regulate both autophagy and the Cvt pathway through the Apg1 protein kinase.

18.8.4
PI 3-kinase Complex

The fourth complex is an autophagy-specific PI 3-kinase complex. Cloning and characterization of *APG6* revealed that it is identical to *VPS30*. Vps30/Apg6 has at least two functions in vacuolar protein sorting and autophagy. Apg14 is a possible coiled-coil protein, and associated with Vps30. Overexpression of Apg14 partially suppresses the autophagic defect of a mutant expressing a truncated form of Vps30, but does not suppress a complete deletion allele of *VPS30*. This suggests that Apg14 binds Vps30 to exert its autophagic function. In contrast to a *vps30* mutant, an *apg14* mutant has no effect on vacuolar protein sorting.

Recently, it was found that Vps30 forms two distinct protein complexes [41]. One consists of Vps30, Apg14, Vps34, Vps15, and the other of Vps30, Vps38, Vps34, Vps15. Vps34 is the sole phosphatidyl inositol 3-kinase in yeast and Vps15 is a regulatory protein kinase of Vps34. The former PI 3-kinase complex is responsible for the autophagy, whereas the latter mediates vacuolar protein sorting. Vps30 is a possible coiled-coil protein and is peripherally membrane associated. Lack of Vps34 or Vps15 results in solubilization of Vps30. Apg14 is a specific factor in the autophagy-specific PI 3-kinase complex and, therefore, it may play an important role in determining the specificity of the PI 3-kinase complex [41].

18.9
Function of the Apg Proteins

All *apg* mutants fail to accumulate autophagosomes during starvation, indicating that all the *APG* genes products function at or before the step of autophagosome formation. So far, studies on the Apg proteins have reached the conclusion that all the Apg proteins function at the autophagosome formation step. There are many fundamental problems to be solved. What is the origin of the autophagosomal membranes? How does the autophagosome assemble to form a spherical structure? What is the machinery mediating closure of the autophagosome and fusion of the autophagosomal outer membrane with the vacuolar membrane.

The membrane dynamics of autophagy are entirely different from classical vesicular membrane trafficking. For a long time, the origin of autophagosomal membrane was thought to be the endoplasmic reticulum (ER). The autophagosomal membranes of yeast seem thinner than any other cellular membranes, and the outer and inner membranes stick together with almost no lumenal space [15, 16]. In freeze–fracture images, the inner membrane completely lacks intramembrane particles, and the outer membrane contains sparse but significant particles [17]. This indicates that both membranes are somehow differentiated and specialized, presumably for delivery of a portion of the cytoplasm to the lytic compartment. We never obtained evidence that membrane vesicles are involved in the elongation of the isolation membrane which leads to autophagosome formation. We proposed that autophagosome formation is not simply the enwrapping of cytoplasm by preexisting large membrane structures such as the ER, but rather assembly of a new membrane from its constituents.

As mentioned above, all Apg proteins function at or just before the autophagosome formation step. Recently, we have shown that many Apg proteins are localized in a small structure, named the preautophagosomal structure (PAS) next to the vacuole [42]. PAS, detected by GFP-Aut7, colocalizes with (Apg12-)Apg5, the Apg1 kinase complex, Apg2, and presumably Apg14. Recent studies demonstrated that 13 Apg proteins localized in the PAS. Thus, PAS seems to be an organizing center of the autophagosome. The lipidation of Apg8 is a prerequisite for the recruitment of Apg8 to the PAS. Furthermore, in *apg14* or *apg6* mutants, Apg8 and Apg5 do not form a dot structure in the cytoplasm, suggesting that the autophagy specific PI 3-kinase complex plays an important role in the organization of PAS [42]. In contrast, defects in the Apg1 kinase complex cause little effect on PAS structure.

18.10
Disintegration of Autophagic Bodies in the Vacuole

As discussed above, in wild-type cells autophagic bodies are disintegrated so quickly. Recently we visualized this process by selective cargo, API-GFP, showing that it takes less than 1 min. The screening strategy for *aut* and *cvt* mutants included mutants that cannot digest autophagic bodies and Cvt bodies. Cvt17/Aut5 shows complete and Aut4 shows partial defect in this process [42a, 42b]. We also tried to isolate mutants which accumulate autophagic bodies in spite of active proteinases. Nakamura isolated 4 *abd* mutants which are defective in autophagic bodies' degradation. Among them three turned out to be *vma* mutants [43]. Furthermore all *vma* mutants showed *abd* phenotype, indicating that an acidic pH of the vacuole is important for disintegration of autophagic bodies. The mechanism of effective and selective degradation of inner membrane structure is unknown. Aut5 has a putative lipase domain, but its activity has not been confirmed. It is likely that a certain activation by vacuolar proteinases A and B is required. In mammals V-type ATPase inhibitor is shown to block a fusion of autophagosome with lysosomes, resulting in the accumulation of autophagosomes. Currently, we do not know the significance of this difference.

18.11
Physiological Roles of Autophagy

Autophagy is induced not only by nitrogen starvation, but also by starvation for other nutrients, which means that autophagy is a general physiological response to nutrient limitation. In yeast cells growing in nutrient-rich conditions, the autophagic level is negligible. Conditions that promote cell growth, such as high cAMP levels or mutations that activate protein kinase A (PKA), block the induction of autophagy [29], indicating that cell growth and the induction of autophagy are oppositely regulated. Furthermore, when rapamycin is added to cells growing in a nutrient-rich medium, the cells behave as if in starvation medium and autophagy is induced [29]. Thus, TOR kinase negatively regulates autophagy as a master regulator. At present, the upstream regulators and downstream effectors of TOR controlling autophagy are not well understood, but may involve an interaction between the TOR pathway and PKA.

So far it has been shown that autophagy is a non-selective degradation process. However, one cannot exclude the possibility that certain molecules are selectively sequestered by autophagosomes. In fact, API and α-mannosidase are selectively enclosed in autophagosomes under starvation conditions though as part of a biosynthetic process. Whether autophagy targets certain molecules for selective degradation is an open question, but it is possible that autophagy functions as a system to eliminate obstacle proteins from cytoplasm.

Several mutants accumulate autophagosomes in the cytoplasm under starvation conditions. These mutants display aberrant vacuolar morphology, suggesting that the processes of autophagosome-vacuole fusion and vacuolar biogenesis share common components, such as SNARE molecules.

In yeast, *S. cerevisiae*, autophagy is not essential for growth. However, autophagy-defective mutants cannot form spores upon nitrogen starvation, indicating that this cell differentiation process requires bulk protein turnover. Another common feature of autophagy-defective mutants is a loss of viability upon prolonged starvation. Mutants start to die after two days of starvation and most lose viability within one week. It is still unclear why autophagy is essential for the maintenance of cell viability. A simple explanation is that the minimal nutrient supply provided by autophagy is essential for synthesis of proteins to survive. Another possibility is that a reduction in cellular activity, such as a down-regulation of ribosomal activity and cell growth, is a prerequisite for survival during adverse conditions. The most frequent stress in nature, especially for microorganisms, is likely to be nutrient limitation. Thus, there must be strong selective pressure to acquire systems, such as the autophagic system, that allow survival under severe starvation conditions. The ubiquitin-proteasome system alone would not be sufficient to satisfy this selective pressure as it does not mediate bulk protein degradation. Autophagy is a bulk degradation process during which significant amounts of cytoplasm and large structures are degraded en masse, and is therefore energetically less costly than the ubiquitin-proteasome system.

18.12
Autophagy in Higher Eukaryotes

The identification of autophagy genes in yeast has facilitated the study of autophagy in higher eukaryotic cells. Many Apg orthologues were found in many eukaryotes. The two ubiquitin-like conjugation systems are highly conserved from yeast to mammals and plants [31, 44–46, 49]. Mouse Apg12-Apg5 is necessary for autophagosome formation, in particular for elongation of the isolation membrane [47]. LC3, an Aut7/Apg8 orthologue, is a good marker for monitoring autophagic membrane dynamics in mammals [48]. It is an interesting question as to why higher eukaryotes have so many Apg8 orthologues. The finding that the basic mechanism of autophagy is well conserved from yeast to human suggests that autophagy is a fundamental cellular activity in eukaryotic organisms. However, the physiological role of autophagy must be much broader in higher eukaryotes, where autophagy is involved in development, differentiation, cellular homeostasis and even cell death. In higher eukaryotes, autophagy may be induced not only upon starvation but also in response to various physiological demands. The molecular dissection of autophagy in higher eukaryotes is currently an exciting area of biology.

References

1. M. Hochstrasser, Ubiquitin-dependent protein degradation, *Annu. Rev. Gen.*, **1996**, *30*, 405–439.
2. A. Hershko and A. Ciechanover, The ubiquitin system, *Annu. Rev. Biochem.*, **1998**, *67*, 425–479.
3. C. Duve de, *Subcellular Particles*, T. Hayashi, ed., Ronald, New York **1959**, pp. 128–159.
4. D. J. Klionsky, P. K. Herman, and S. D. Emr, The fungal vacuole: composition, function, and biogenesis, *Microbiol. Rev.*, **1990**, *54*, 266–292.
5. T. Achstetter and D. H. Wolf, Proteinases, proteolysis and biological control in the yeast Saccharomyces cerevisiae, *Yeast*, **1985**, *1*, 139–157.
6. E. W. Jones, G. C. Webb, and M. A. Hiller, Biogenesis and function of the yeast vacuole. In: *The Molecular and Cellular Biology of the Yeast Saccharomuces: Cell Cycle and Cell Biology*, J. Pringle, J. Broach, and E. Jones, eds., Cold Spring Harbor Laboratory Press, New York, **1997**, pp. 363–470.
7. C. K. Raymond, C. J. Roberts, K. E. Moore, I. Howald, and T. H. Stevens, Biogenesis of the vacuole in Saccharomyces cerevisiae, *Int. Rev. Cytol.*, **1992**, *139*, 59–120.
8. J. H. Stack, B. Horazdovsky, and S. D. Emr, Receptor-mediated protein sorting to the vacuole in yeast: roles for a protein kinase, a lipid kinase and GTP-binding proteins, *Annu. Rev. Cell. Dev. Biol.*, **1995**, *11*, 1–33.
9. G. S. Zubenko and E. W. Jones, Protein degradation, meiosis and sporulation in proteinase-deficient mutants of Saccharomyces cerevisiae, *Genetics*, **1981**, *97*, 45–64.
10. R. Egner, M. Thumm, M. Straub, A. Simeon, H. J. Schuller, and D. H. Wolf, Tracing intracellular proteolytic pathways. Proteolysis of fatty acid synthase and other cytoplasmic proteins in the yeast Saccharomyces cerevisiae, *J. Biol. Chem.*, **1993**, *268*, 27269–27276.
11. G. E. Mortimore and A. R. Poso, Intracellular protein catabolism and its

control during nutrient deprivation and supply, *Annu. Rev. Nutr.*, **1987**, *7*, 539–564.

12. P. O. Seglen and P. Bohley, Autophagy and other vacuolar protein degradation mechanisms, *Experientia*, **1992**, *48*, 158–172.

13. S. R. Terlecky, H. L. Chiang, T. S. Olson, and J. F. Dice, Protein and peptide binding and stimulation of in vitro lysosomal proteolysis by the 73-kDa heat shock cognate protein, *J. Biol. Chem.*, **1992**, *267*, 9202–9209.

14. A. M. Cuervo and J. F. Dice, A receptor for the selective uptake and degradation of proteins by lysosomes, *Science*, **1996**, *273*, 501–503.

15. K. Takeshige, M. Baba, S. Tsuboi, T. Noda, and Y. Ohsumi, Autophagy in yeast demonstrated with proteinase-deficient mutants and conditions for its induction, *J. Cell. Biol.*, **1992**, *119*, 301–311.

16. M. Baba, K. Takeshige, N. Baba, and Y. Ohsumi, Ultrastructural analysis of the autophagic process in yeast: detection of autophagosomes and their characterization, *J. Cell. Biol.*, **1994**, *124*, 903–913.

17. M. Baba, M. Osumi, and Y. Ohsumi, Analysis of the membrane structures involved in autophagy in yeast by freeze-replica method, *Cell. Struct. Funct.*, **1995**, *20*, 465–471.

18. T. Noda, A. Matsuura, Y. Wada, and Y. Ohsumi, Novel system for monitoring autophagy in the yeast Saccharomyces cervisiae, *Biochem. Biophys. Res. Commun.*, **1995**, *210*, 126–132.

19. M. Tsukada and Y. Ohsumi, Isolation and characterization of autophagy-defective mutants of Saccharomyces cerevisiae, *FEBS Lett.*, **1993**, *333*, 169–174.

20. M. Thumm, R. Egner, B. Koch, M. Schlumpberger, M. Straub, M. Veenhuis, and D. H. Wolf, Isolation of autophagocytosis mutants of Saccharomyces cerevisiae, *FEBS Lett.*, **1994**, *349*, 275–280.

21. N. Mizushima, T. Noda, and Y. Ohsumi, Apg16p is required for the function of the Apg12p-Apg5p conjugate in the yeast autophagy pathway, *EMBO J.*, **1999**, *18*, 3888–3896.

22. Y. Kamada, T. Funakoshi, T. Shintani, K. Nagano, M. Ohsumi, and Y. Ohsumi, Tor-mediated induction of autophagy via an Apg1 protein kinase complex, *J. Cell. Biol.*, **2000**, *150*, 1507–1513.

23. T. M. Harding, A. Hefner-Gravink, M. Thumm, and D. J. Klionsky, Genetic and phenotypic overlap between autophagy and the cytoplasm to vacuole protein targeting pathway, *J. Biol. Chem.*, **1996**, *271*, 17621-17624.

23a. T. M. Harding, K. A. Morano, S. V. Scott, and D. J. Klionsky, 1995. Isolation and characterization of yeast mutants in the cytoplasm to vacuole protein targeting pathway. *J. Cell. Biol.*, **131**, 591–602.

24. S. V. Scott, A. Hefner-Gravink, K. A. Morano, T. Noda, Y. Ohsumi, and D. J. Klionsky, Cytoplasm-to-vacuole targeting and autophagy employ the same machinery to deliver proteins to the yeast vacuole, *Proc. Natl. Acad. Sci. U. S.A.*, **1996**, *93*, 12304–12308.

25. S. V. Scott, M. Baba, Y. Ohsumi, and D. J. Klionsky, Aminopeptidase I is targeted to the vacuole by a nonclassical vesicular mechanism, *J. Cell. Biol.*, **1997**, *138*, 37–44.

26. S. Kametaka, T. Okano, M. Ohsumi, and Y. Ohsumi, Apg14p and Apg6/Vps30p form a protein complex essential for autophagy in the yeast, Saccharomyces cerevisiae, *J. Biol. Chem.*, **1998**, *273*, 22284–22291.

27. T. Kirisako, M. Baba, N. Ishihara, K. Miyazawa, M. Ohsumi, T. Yoshimori, T. Noda, and Y. Ohsumi, Formation process of autophagosome is traced with Apg8/Aut7p in yeast, *J. Cell. Biol.*, **1999**, *147*, 435–446.

28. T. F. Chan, P. G. Bertram, W. Ai, and X. F. Zheng, Regulation of APG14 expression by the GATA-type transcription factor Gln3p, *J. Biol. Chem.*, **2001**, *276*, 6463–6467.

29. T. Noda and Y. Ohsumi, Tor, a phosphatidylinositol kinase homologue, controls autophagy in yeast, *J. Biol. Chem.*, **1998**, *273*, 3963–3966.

30. Y. Ohsumi, Molecular dissection of autophagy: two ubiquitin-like systems,

Nat. Rev. Mol. Cell. Biol., **2001**, *2*, 211–216.

31. N. Mizushima, T. Noda, T. Yoshimori, Y. Tanaka, T. Ishii, M. D. George, D. J. Klionsky, M. Ohsumi, and Y. Ohsumi, A protein conjugation system essential for autophagy, *Nature*, **1998**, *395*, 395–398.

32. T. Shintani, N. Mizushima, Y. Ogawa, A. Matsuura, T. Noda, and Y. Ohsumi, Apg10p, a novel protein-conjugating enzyme essential for autophagy in yeast, *EMBO J.*, **1999**, *18*, 5234–5241.

33. I. Tanida, N. Mizushima, M. Kiyooka, M. Ohsumi, T. Ueno, Y. Ohsumi, and E. Kominami, Apg7p/Cvt2p: A novel protein-activating enzyme essential for autophagy, *Mol. Biol. Cell.*, **1999**, *10*, 1367–1379.

34. A. Kuma, N. Mizushima, N. Ishihara, and Y. Ohsumi, Formation of the approximately 350-kDa Apg12-Apg5 · Apg16 multimeric complex, mediated by Apg16 oligomerization, is essential for autophagy in yeast, *J. Biol. Chem.*, **2002**, *277*, 18619–18625.

35. T. Kirisako, Y. Ichimura, H. Okada, Y. Kabeya, N. Mizushima, T. Yoshimori, M. Ohsumi, T. Takao, T. Noda, and Y. Ohsumi, The reversible modification regulates the membrane-binding state of Apg8/Aut7 essential for autophagy and the cytoplasm to vacuole targeting pathway, *J. Cell. Biol.*, **2000**, *151*, 263–276.

36. Y. Ichimura, T. Kirisako, T. Takao, Y. Satomi, Y. Shimonishi, N. Ishihara, N. Mizushima, I. Tanida, E. Kominami, M. Ohsumi, T. Noda, and Y. Ohsumi, A ubiquitin-like system mediates protein lipidation, *Nature*, **2000**, *408*, 488–492.

37. A. Matsuura, M. Tsukada, Y. Wada, and Y. Ohsumi, Apg1p, a novel protein kinase required for the autophagic process in Saccharomyces cerevisiae, *Gene*, **1997**, *192*, 245–250.

38. M. Straub, M. Bredschneider, and M. Thumm, AUT3, a serine/threonine kinase gene, is essential for autophagocytosis in Saccharomyces cerevisiae, *J. Bacteriol.*, **1997**, *179*, 3875–3883.

39. T. Funakoshi, A. Matsuura, T. Noda, and Y. Ohsumi, Analyses of APG13 gene involved in autophagy in yeast, Saccharomyces cerevisiae, *Gene*, **1997**, *192*, 207–213.

40. H. Abeliovich, W. A. Dunn, Jr., J. Kim and D. J. Klionsky, Dissection of autophagosome biogenesis into distinct nucleation and expansion steps, *J. Cell Biol.*, **2000**, *151*, 1025–1034.

41. A. Kihara, T. Noda, N. Ishihara, and Y. Ohsumi, Two distinct Vps34 phosphatidylinositol 3-kinase complexes function in autophagy and carboxypeptidase Y sorting in Saccharomyces cerevisiae, *J. Cell. Biol.*, **2001**, *152*, 519–530.

42. K. Suzuki, T. Kirisako, Y. Kamada, N. Mizushima, T. Noda, and Y. Ohsumi, The pre-autophagosomal structure organized by concerted functions of APG genes is essential for autophagosome formation, *EMBO J.*, **2001**, *20*, 5971–5981.

42a. U. D. Epple, I. Suriapranata, E. L. Eskelinen, and M. Thumm, 2001. Aut5/Cvt17p, a putative lipase essential for disintegration of autophagic bodies inside the vacuole. *J. Bacteriol.*, *183*, 5942–5955.

42b. I. Suriapranata, U. D. Epple, D. Bernreuther, M. Bredschneider, K. Sovarasteanu, and M. Thumm, 2000. The breakdown of autophagic vesicles inside the vacuole depends on Aut4p. *J. Cell. Sci.*, *113*, 4025–4033.

43. N. Nakamura, A. Matsuura, Y. Wada, and Y. Ohsumi, Acidification of vacuoles is repuired for autophagic degradation in the yeast, Saccharomyces cerevisiae, *J. Biochem.*, **1997**, *21*, 338–344.

44. N. Mizushima, H. Sugita, T. Yoshimori, and Y. Ohsumi, A new protein conjugation system in human. The counterpart of the yeast Apg12p conjugation system essential for autophagy, *J. Biol. Chem.*, **1998**, *273*, 33889–33892.

45. I. Tanida, E. Tanida-Miyake, T. Ueno, and E. Kominami, The human homolog of Saccharomyces cerevisiae Apg7p is a Protein-activating enzyme for multiple substrates including human Apg12p, GATE-16, GABARAP, and MAP-LC3, *J. Biol. Chem.*, **2001**, *276*, 1701–1706.

46. J. H. Doelling, J. M. Walker, E. M. Friedman, A. R. Thompson, and R. D. Vierstra, The Apg8/12-activating enzyme Apg7 is required for proper nutrient recycling and senescence in Arabidopsis thaliana, *J. Biol. Chem.*, **2002**, *277*, 33105–33114.

47. N. Mizushima, A. Yamamoto, M. Hatano, Y. Kobayashi, Y. Kabeya, K. Suzuki, T. Tokuhisa, Y. Ohsumi, and T. Yoshimori, Dissection of autophagosome formation using Apg5-deficient mouse embryonic stem cells, *J. Cell. Biol.*, **2001**, *152*, 657–668.

48. Y. Kabeya, N. Mizushima, T. Ueno, A. Yamamoto, T. Kirisako, T. Noda, E. Kominami, Y. Ohsumi, and T. Yoshimori, LC3, a mammalian homologue of yeast Apg8p, is localized in autophagosome membranes after processing, *EMBO J.*, **2000**, *19*, 5720–5728.

49. N. Mizushima, T. Yoshimori, and Y. Ohsumi, 2002. Mouse Apg10 as an Apg12 conjugating enzyme: Analysis by the conjugation-mediated yeast two-hybrid method. *FEBS Lett.*, **532**, 450–454.

Index

Handbook of ATPases. Edited by M. Futai, Y. Wada, J. H. Kaplan
Copyright © 2004 WILEY-VCH Verlag GmbH & Co. KGaA, Weinheim
ISBN 3-527-30689-7